T0329589

# A Course in Luminescence Measurements and Analyses for Radiation Dosimetry

# A Course in Luminescence Measurements and Analyses for Radiation Dosimetry

*by*

*Stephen W.S. McKeever*
Stillwater, US

WILEY

*Registered Offices*
John Wiley & Sons Ltd, The Atrium, Southern Gate, Chichester, West Sussex, PO19 8SQ, UK
John Wiley & Sons, Inc., 111 River Street, Hoboken, NJ 07030, USA

*Editorial Office*
John Wiley & Sons Ltd, The Atrium, Southern Gate, Chichester, West Sussex, PO19 8SQ, UK

For details of our global editorial offices, customer services, and more information about Wiley products visit us at www.wiley.com.

Wiley also publishes its books in a variety of electronic formats and by print-on-demand. Some content that appears in standard print versions of this book may not be available in other formats.

*Library of Congress Cataloging-in-Publication Data*
Names: McKeever, S. W. S., 1950 - author.
Title: A course in luminescence measurements and analyses for radiation dosimetry /
    by Stephen W. McKeever.
Description: Chichester, John Wiley & Sons, 2022. | Includes bibliographical references and index.
Identifiers: LCCN 2022000515 (print) | LCCN 2022000516 (ebook) | ISBN 9781119646891 (hardback) |
    ISBN 9781119646914 (pdf) | ISBN 9781119646921 (epub)
Subjects: LCSH: Luminescence. | Radiation dosimetry. | Radiation-Measurement.
Classification: LCC QC476.5. M35 2022 (print) | LCC QC476.5(ebook) |
    DDC 535/.35--dc2 3/eng20220218
LC record available at https://lccn.loc.gov/2022000515
LC ebook record available at https://lccn.loc.gov/2022000516

Cover image: Courtesy of Stephen W.S.McKeever
Cover design by Wiley

Set in 10/12pt and TimesLTStd by Integra software private Ltd, Puducherry, India.
Printed and bound by CPI Group (UK) Ltd, Croydon, CR0 4YY

C9781119646891_040522

To Claire, Declan, James and Lydia

# Contents

# Preface

The detection of ionizing radiation using luminescence methods has been at the forefront of radiation research ever since the earliest discoveries of radiation, radioactivity, and the structure of the atom. The list of pioneering names in the discovery of radiation is lengthy and impressive – Wilhelm Roentgen, Edmond and Henri Becquerel, Ernest Rutherford, Marie Curie, James Chadwick, and numerous others. Many of these pioneers used luminescence of one type or form in their studies, including (using the modern names) phosphorescence, thermoluminescence, and optically stimulated luminescence. Today's researchers use these techniques and others related to them in a wide range of radiation detection and measurement applications, including personal, environmental, medical, and space dosimetry, and embrace both very low as well as very high radiation dose regimes.

The genesis of this book is a course, given by the author, to graduate students at Oklahoma State University. The intended readers are graduate students (and undergraduate students who are performing research in these areas), and other beginning researchers in this field of study (e.g., postdoctoral fellows and new researchers entering the field). More experienced researchers may also find the book helpful to refresh their conceptual understanding of the topics. The benefit to the reader will be to see how the techniques relate in a holistic manner and to see how the fundamental processes that describe the phenomena can be used to explain many different experimental characteristics.

To new researchers entering the field, the array of measurement and data analysis techniques can be bewildering. Experimental collection of the data may seem simple on first introduction, but useful and reliable analysis requires attention to small details and knowledge of fundamental principles, along with very careful experimental technique. A phrase I often used with my students was that they will spend most of their time learning how to do the experiment properly before they have a reliable data set for analysis. And then what of that analysis? How should they interpret the results that they obtain? What is the best analytical approach, and why?

*What this book is*: This book presents a course of instruction for beginning students, or for more-experienced researchers new to the field, into three of the main and currently used luminescence phenomena in radiation dosimetry, namely, thermoluminescence (TL), optically stimulated luminescence (OSL), and radiophotoluminescence (RPL). The book outlines the theoretical background of each technique and stresses the connections between them. In doing so, the book treats luminescence techniques for radiation dosimetry holistically, beginning with the basic concepts and showing how the three techniques are related within the processes of energy storage, charge transfer, and defect interactions.

The book emphasizes pedagogy rather than the latest research. State-of-the-art research is included only if it demonstrates a principle or opens new insight into physical mechanisms. Thus, the primary purpose of the book is to teach beginning researchers in the field about the three techniques, their similarities and distinctions, and their applications. The intent is to provide the reader with the building blocks with which they can examine the latest research and have a sufficient understanding of it to enable them to raise questions and conceive future research programs to answer those questions.

*A note on the Exercises*: To emphasize the book as a teaching tool, several substantial Exercises are introduced at various points in the book for the reader to test their understanding of the subject matter being discussed. These include derivations of important relationships, or numerical simulations, or solutions to differential equations, etc., in order to demonstrate the various processes. Some are prescriptive, but others are open ended, requiring the reader to draw their own conclusions from the results of the Exercises. The Exercises also include analysis of real experimental data. For this purpose, the book has a companion website where the reader will find some original data sets (TL, OSL, RPL) that may be used by the reader to analyse using the material learned from the book. New data and new problems may be introduced via the web site in the future and in this way the book will be updated and refreshed for future students. In this sense, the author hopes that the book will serve as a "living" teaching tool.

Although it is recommended that the Exercises be worked one by one in the order given, the reader may wish to skip some of the Exercises if they do not align strongly with the reader's own research interests. Many of them are lengthy. While most are designed for the individual reader, supervisors and advisors may decide that some may be better solved as group exercises, perhaps as part of a tutorial. Some may even form the basis of internal research projects and reports.

*What this book is not*: Firstly, the book is not a course in radiation dosimetry. That topic is adequately covered in many excellent existing textbooks.

Secondly, the book is not an updated summary of the latest research on luminescence dosimetry. It is not a review. Several books exist on the topics of TL, OSL, and (to a much lesser extent) RPL. Each, by and large, is a summary of the latest research and newest developments and applications of the techniques (at the times that the books were written) and generally deal with the chosen topic (i.e. TL, OSL, or RPL) in isolation. For example, several texts exist on TL and related thermally stimulated processes, introducing the standard energy band diagrams and kinetic equations, analysis of TL glow curves, and subsequent applications. One can also identify similar texts on OSL. (Only one text, of which this author is aware, deals with RPL.) Rarely, however, do the published books relate one technique to another, except in passing. This is especially true of RPL and its relationship to TL and/or OSL.

Nor does the book present a list of references to all the latest research or pivotal developments. The original material is for the reader to find in independent study. Only when the author has judged that additional reading would be useful, or that the meaning requires more explanation, are references to the original literature included. Readers who wish to avail themselves of the latest research are advised to refer to the proceedings of relevant conferences in the field and to original, peer-reviewed literature. The author hopes that this book will assist in an understanding of the material readers find in those publications.

To achieve the goals outlined above, the book is constructed in two parts. Part I deals with theory, models, and kinetics. It can be considered as a "tool-box" into which the reader can delve to help form an understanding of the underlying principles which govern luminescence phenomena. In the same way that a designer of a new aircraft or of a sophisticated automobile will need an understanding of the basic principles of mechanics, materials, and aerodynamics, so too should the budding researcher into luminescence dosimetry have at hand a similar understanding of the basics of electronic processes in solids, particularly the kinetics of charge generation and storage, stimulation, and recombination. Armed with this "tool-box" the reader can more fully appreciate the experimental phenomena described in Part II.

Part II discusses several real examples of the fundamentals outlined in Part I. The intent is to illustrate how the principles developed in Part I have been used in experiments to measure, understand, and exploit the properties of luminescence materials, especially as they relate to radiation dosimetry. To learn from the wisdom of Albert Einstein, knowing the luminescence properties of a material is one thing, but real progress is made only when we understand them. The author's hope is that the reader can use the items in the "tool-box" and apply them to the properties of real materials in order to gain that understanding, and perhaps lead to greater creativity and innovation. As a caveat, however, the reader may be wise to recall the words of Prussian Field Marshal Helmuth von Moltke, which the author paraphrases as *few theories survive first contact with an experiment.*

# Acknowledgments

The author gratefully acknowledges the assistance of many colleagues in reading certain chapters and sections of the text, correcting mistakes, and making very positive suggestions. Also acknowledged are those who provided example data sets for use on the accompanying web site for reader analysis. Specific thanks go to Adrie Bos, Mayank Jain, Vasilis Pagonis, Nigel Poolton, Peter Townsend, Sergey Sholom, and Eduardo Yukihara, for their unfailing assistance in reading parts of the text and providing vital feedback to the author. Any remaining errors are the author's own. Special thanks, however, go to my friend and long-time colleague, Sergey Sholom, for not only reading multiple sections of the book, but also in answering my persistent calls for original data for inclusion in the book and for uploading to the web site for use in the Exercises and for analysis by the readers. I will remain eternally grateful.

Further thanks are due to the many contributors of data and figures for use in the book, including, again, Adrie Bos, Vasilis Pagonis, Nigel Poolton, and Sergey Sholom, plus Mark Akselrod, Ramona Gaza, Guerda Massilon, Kahli Remy, and Hannes Stadmann. I also thank my many collaborators throughout my career from whom I have learned so much and who have guided me on my own stumbling path through the topic. Thanks are also due to the editorial staff at Wiley for their professional guidance and assistance. Finally, I thank my many brilliant students who over the years have also taught me so much at the same time as I, hopefully, have taught them. Teaching is a two-way process, and I have loved every minute of it. I hope this book does them all justice.

# Disclaimer

Reference to commercial products does not imply or represent endorsement of those products on the part of the author. It is noted that the author's research has been funded at various points in his career by Landauer Inc. (USA) and Chiyoda Technol Corporation (Japan).

# About the Companion Website

This book is accompanied by a companion website.
www.wiley.com/go/mckeever/luminescence-measurements

This website includes:

- Exercises
- Figures
- Notes

# Part I

# Theory, Models, and Simulations

*When ... simulation and approximation yield similar results, the validity of the conclusions is strengthened.*

– R. Chen and V. Pagonis 2014

# 1

# Introduction

*I consider then, that generally speaking, to render a reason of an effect or Phaenomenon, is to deduce it from something else in Nature more known than it self, and that consequently there may be divers kinds of Degrees of Explication of the same thing.*

– R. Boyle 1669

## 1.1   How Did We Get Here?

Luminescence, the eerie glow of light emitted by many physical and biological substances, is familiar to us all. The bright speck of a firefly, the luminous glow from seawater in the evening, the glow of a watch dial in the dark – all are examples of luminescence phenomena that are familiar to most of us. Familiarity and understanding are not synonymous, however. Indeed, an understanding of the various luminescence phenomena has a very long genesis and over the centuries there have been several "divers kinds of Degrees of Explication". Luminescence has had, and continues to have, practical uses in both every-day and in more esoteric applications. Computer screens, electronic indicators, lighting, lasers, and many, many other examples are indications that the field of luminescence is very broad and potentially very useful.

One such field of use is in the detection and measurement of radiation – a field generally known as "dosimetry," or the act of measuring the dose of radiation absorbed by an object. The amount of radiation absorbed by an object and the subsequent amount of luminescence emitted from it is the basis of the use of luminescence in dosimetry. The connection between radiation and luminescence was made many years ago and, in fact, those of us active in the field of luminescence dosimetry can take pride in the fact that the study of luminescence can be traced to the beginning of the modern scientific method. Although it would be surprising if ancient Islamic or, perhaps, Chinese scholars had not already noted the phenomenon, in one of its many guises, it can nevertheless be argued that the first modern description of luminescence stems from the work of Robert Boyle in mid-seventeenth-century England, published in the Philosophical Transactions of the Royal Society. Boyle – considered to be the "father" of chemistry, as well as being a physicist, an inventor, a philosopher, and a theologian – gives an evocative description of (what we now term) luminescence emitted from a remarkable piece of diamond, loaned to him by a friend, John Clayton (Boyle 1664). The word "luminescence" was not used by Boyle who referred to it as the "glow" from the stone. In a later publication

*A Course in Luminescence Measurements and Analyses for Radiation Dosimetry,* First Edition.
Stephen W.S. McKeever.
© 2022 John Wiley & Sons Ltd. Published 2022 by John Wiley & Sons Ltd.
Companion Website: www.wiley.com/go/mckeever/luminescence-measurements

concerning luminescence from a liquid he uses the wonderfully suggestive term "self-shining" to describe the phenomenon (Boyle 1680).

Boyle's 1664 account of luminescence from diamond is generally accepted as the first scientific description of the phenomenon of **thermoluminescence** (TL). Boyle described various ways of heating the diamond to induce from it the emission of light. It is not clear, however, how Boyle energized the diamond in the first place. We now know that the TL phenomenon requires that the material must first absorb energy from an external energy source. The energy thus stored is then released by the application of a second source of energy (heat). As the initial energy is released, some of it is emitted in the form of visible light (thermoluminescence). Without that first energy storage step, no TL can be induced. Boyle may or may not have known that the process he was observing was, in fact, a two-step procedure, but he was vague on how he energized the diamond in his possession; readers are left to speculate how this may have been achieved. Possibilities include natural radioactivity or light, but perhaps the most likely source was physical stimulation (rubbing, scratching, etc.) producing what we now call tribo-thermoluminescence ("tribo-" from the Greek "tríbein," meaning "to rub"). In any case, once heated to release the TL, the material would have to be energized again and energy stored a second time before the TL phenomenon could be seen again during heating.

We may never know in sufficient detail how Boyle treated his diamond to be able to answer this question with certainty – and perhaps we should be satisfied with leaving it as an intriguing mystery. For our purposes here, we can be satisfied that the phenomenon that we discuss in this book was first reported in such vivid and expressive terms as long ago as the mid-seventeenth century, and by such a luminary as Robert Boyle.

McKeever (1985) traces several pre-twentieth century published descriptions of luminescence stimulated by heating and indicates that the term "thermoluminescence" can probably be attributed to Eilhardt Wiedemann (Wiedemann 1889) in his work on the luminescence properties of a wide variety of materials. Following Wiedemann's work, Wiedemann and Schmidt (1895) studied TL from an extensive series of materials following irradiation with electron beams, while Trowbridge and Burbank (1898), likewise, studied TL of fluorite following excitation by several different energy sources, including x-irradiation. These two early papers are examples where we can see the beginnings of the use of TL in radiation detection since, in each case, a source of radiation was used to provide the initial absorption of energy necessary for ultimate TL production. It is not surprising, therefore, to see the study of TL proceeding alongside the examination of radiation itself, with seminal works by Marie Curie and Ernest Rutherford, among others, including descriptions of thermoluminescence from minerals (Curie 1904; Rutherford 1913). Examinations of the color of the emitted light were also beginning around this time through studies of the spectra of the TL from various minerals (Morse 1905).

A point that should not go without mention is that Wiedemann (1889) and Wiedemann and Schmidt (1895) discussed the mechanism of luminescence in terms an "electric dissociation theory" wherein luminescence phenomena were explained on the basis of the separation and subsequent recombination of positive and negative charged species (specifically, positive and negative molecular ions). Others followed and adopted this initial and innovative suggestion to explain luminescence phenomena in a variety of materials (Nichols and Merritt 1912; Rutherford 1913). While the authors of the period attempted to apply this theory to all forms of luminescence, and while we now know that photoluminescence (i.e., fluorescence), for example, does not involve ionization and charge dissociation, the notion of charge dissociation and recombination nevertheless foreshadows our current understanding of the phenomena of TL, OSL, and phosphorescence. Bearing in mind that these early ideas initially suggested in the 1880s predate the birth of quantum mechanics, band theory, and the concepts of electron and hole generation, it is remarkable that the insight offered by these early pioneers aligns so well with our current understanding of the latter phenomena, which is given in terms of the creation of negative electrons and positive holes, followed by their ultimate recombination.

As described in McKeever (1985), the use of TL in the study of radiation accelerated in the beginning decades of the twentieth century. A key area of research was to examine the relationship between the point defects within the materials studied (e.g. color centers) and their role in localizing (trapping) the electrons and holes ionized from their host atoms during the absorption of radiation. A feature of TL is that the luminescence at first increases and then decreases, forming a series of characteristic TL peaks as the temperature increases. It was realized that the cause of the TL peaks was the thermal release of trapped charge from lattice defects – with the larger the trapping energy, the higher the temperature of the TL peak. In 1930, in Vienna, Austria, Urbach discussed the connection between the energy needed to release the trapped charge and the TL peak position in a series of papers on luminescence from the alkali halides (Urbach 1930). However, it was not until the work of the group at the University of Birmingham in the United Kingdom that the relationship was quantified through the development of mathematical descriptions of the process (Randall and Wilkins 1945a, 1945b; Garlick and Gibson 1948).

Not long afterwards, Farrington Daniels and the research group at the University of Wisconsin, United States of America, discussed several applications in which TL could be a useful research tool. Among them was radiation dosimetry. Daniels and colleagues wrote: "Since in many crystals the intensity of thermoluminescence is nearly proportional to the amount of $\gamma$-radiation received, a considerable effort has been devoted to developing a practical means of measuring the exposure to gamma radiation." (Daniels et al. 1953) – and so was born the field of thermoluminescence dosimetry. These authors specifically highlighted lithium fluoride as being the best crystal for this purpose and their work also initiated the parallel search for other TL dosimetry materials.

The growth of **optically stimulated luminescence** (OSL) as a method of radiation dosimetry had a similar genesis to that of TL and emerged as a potential dosimetry tool at about the same time. As described by Yukihara and McKeever (2011), the birth of OSL stems from the early work of the Becquerels, father and son, Edmond and Henri (E. Becquerel 1843; H. Becquerel 1883). These and similar studies through the late nineteenth and early twentieth centuries observed that phosphorescence could either be enhanced or quenched by the application of light to an irradiated material, the precise effect being dependent on the wavelength of the stimulating light. Observations of photoconductivity on some of these materials lead to the realization that free electrons were being produced during photostimulation and quenching of the phosphorescence (as discussed by Harvey 1957). Leverenz (1949) noted that when luminescence is enhanced by stimulating with an external light source, the eventual decay of the luminescence is unrelated to the characteristic fluorescence lifetime of the emitting species. As Leverenz (1949) states, for the luminescence to be produced, an "additional activation energy must be supplied to release the trapped electrons … This activation energy may be supplied by heat … or it may be supplied by additional photons." When the energy is supplied by heat, TL results; when it is supplied by photons, OSL results.

It seems that the name, optically stimulated luminescence (OSL), appeared in the literature only in 1963 with the work of Fowler (Fowler 1963). Earlier names for the phenomenon included photophosphorescence, radiophotostimulation, photostimulation phosphorescence, co-stimulation phosphorescence, and photostimulated emission (see Yukihara and McKeever 2011). Even today, one often sees the phrase photostimulated luminescence (PSL) instead of OSL, the two names being synonymous, referring to the same phenomenon. For clarity, the more popular and most frequently used name of OSL will be used throughout this book.

As with TL, the connection between these optically stimulated effects and the initial absorption of energy from radiation was also established in the mid-twentieth century. The first suggested use of OSL in radiation dosimetry appeared with the work of Antonov-Romanovsky and colleagues (Antonov-Romanovsky et al. 1955). These authors examined infra-red (IR) stimulated luminescence from irradiated sulfides and related the IR-induced luminescence to the initial dose of radiation absorbed. Other similar works followed, but OSL did not emerge as a popular radiation dosimetry tool at this time, primarily because the emphasis of these studies was on sulfide materials and infra-red stimulation. These materials contained defects from which the energy required to release the trapped electrons was quite small (and, hence, the electrons could be released through absorption of low-energy, infra-red light). As a result, room temperature thermal stimulation could also release the trapped charges, which were thus observed not to be stable. As a result, the OSL signal is said to have faded with time since irradiation.

With the advent of studies into wider-band-gap materials (e.g. oxides, alkali halides, and sulfates) it was found that OSL could be stimulated by shorter-wavelength, visible light from defects that required larger activation energies to release the trapped charge. Hence, the OSL signal was more stable and did not fade. Nevertheless, the breakthrough in OSL's application in radiation dosimetry came not in the radiation dosimetry field itself, but in the related field of geological dating. Huntley and colleagues (Huntley et al. 1985) demonstrated that OSL from quartz deposits in geological sedimentary layers could be used to determine the dose of natural radiation absorbed by the quartz grains since they were deposited in the layer. Analysis of the natural environmental dose rate then leads to a calculation of the age of the sediment (age = dose/dose rate). This paper, more than any other, opened the flood gates for the development of OSL in dosimetry. "Optical dating" is now an established technique (Aitken 1998; Bøtter-Jensen et al. 2003) and the method demonstrated that the OSL signal stimulated from defects with large activation energies could be stable for thousands of years in the right environmental circumstances.

Use of OSL in conventional radiation dosimetry started with the development of oxygen-deficient $Al_2O_3$, doped with carbon. This material was first suggested as a sensitive TL dosimeter at the Urals Polytechnical Institute in Russia (Akselrod and Kortov 1990), but the TL signal from this material was found to be very sensitive to visible light, such that exposure to daylight after irradiation reduced the subsequent TL signal. The group at Oklahoma State University in the United States then turned this apparent disadvantage into an advantage and showed that the material was a very sensitive OSL material (McKeever et al. 1996; Akselrod and McKeever 1999). The era of OSL dosimetry and the search for new OSL dosimetry materials had begun.

The term **radiophotoluminescence** (RPL) appears in the literature in the early 1920s with the work of Przibram and colleagues in Vienna (Przibram and Kara-Michailowa 1922; Przibram 1923). These researchers showed that photoluminescence (PL) can be induced in some materials only after exposure to ionizing radiation. Without irradiation, no PL is observed and, therefore, these authors gave their observation the name radiophotoluminescence. The distinction between RPL and OSL lies in the stability of the radiation-induced luminescence centers during stimulation with visible or infra-red light. In OSL, the luminescence signal decays under continued light stimulation, whereas in RPL, the signal remains constant. OSL is a destructive readout process (involving ionization) whereas RPL is a non-destructive process involving electron excitation, but not ionization. (These concepts will be discussed in more detail in later sections and chapters.)

The association between the RPL signal induced by the radiation and the initial dose of absorbed radiation was not exploited until the work of Schulman et al. (1951) who used RPL from a variety of materials as a means of determining the dose of radiation initially absorbed. (Interestingly, in addition to RPL, Schulman et al. (1951) also studied the other two major phenomena that comprise the subject of this book – namely, TL (termed radiothermoluminescence by Schulman and co-workers) and OSL (termed radiophotostimulation.) Schulman and colleagues examined the RPL properties of alkali halides and the relationship between RPL and the coloration of these materials after irradiation. These authors also examined Ag-doped phosphate glasses as RPL dosimeters, marking the introduction of what was to become the dominant RPL dosimeter material.

The development of RPL as a dosimetry tool expanded in Germany with the work of Becker (Becker 1968) and Piesch and colleagues (Piesch et al. 1986, 1990), and in Japan with Yokota and colleagues (Yokota et al. 1961; Yokota and Nakajima 1965). Emphasis was on the development of methods for reading the RPL signal as well as a search for improved materials. Although RPL dosimetry was slow to penetrate the dosimetry market because of the competition offered by TL dosimetry (in particular) and later OSL dosimetry, today RPL dosimetry retains an important place within the luminescence dosimetry community and the commercial marketplace.

---

**Exercise 1.1**

Landmark developments in the use of TL, OSL, and RPL in dosimetry have been scattered over many decades, beginning with the study of radioactivity. To understand where the field resides at present – and to ensure that one does not "re-invent the wheel" – familiarity with this background is very important. Some of the highlights and seminal papers have been referenced here; many other important publications are available.

Choose one of the three dosimetry methods – TL, OSL, or RPL – and perform a bibliography search to trace the development of the technique from its earliest beginnings to its present-day use.

---

## 1.2   Introductory Concepts for TL, OSL, and RPL

### 1.2.1   Equilibrium and Metastable States

Consider a luminescence dosimetry material in its equilibrium state (Figure 1.1) in which all electrons are in their equilibrium energy levels. Perturbation from the equilibrium state by absorption of energy from an external energy source (radiation) raises the system to a metastable state, in which some electrons now occupy higher, non-equilibrium energy levels. The metastable state is characterized by potential energy barrier(s), which need(s) to be overcome before the system can return to equilibrium. The system may be stable in the metastable state from fractions of a second to thousands of years, depending on the size of the potential energy barrier(s). Absorption by the system of energy from an external stimulus while in the metastable state can overcome the potential barrier(s) and cause the system to relax to its stable equilibrium state. The stored energy may then be released, usually in the form of heat, but a portion of it may be released in the form of visible light and, thus, luminescence may be emitted. The intensity of the luminescence is related to the amount of energy initially absorbed during the irradiation phase. If the stimulus provided is in the form of heat energy, the luminescence emission is TL. If the stimulus involves the absorption of optical energy, the luminescence emission is OSL. (An animated version of Figure 1.1 is given on the web site, under Exercises and Notes, Chapter 1.)

To examine the nature of the equilibrium and metastable states it is necessary to consider the energy band gap of the material and Fermi-Dirac statistics.

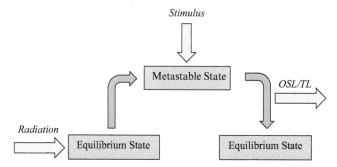

**Figure 1.1**  Conceptual notion of TL and OSL in which a system in an equilibrium state is perturbed from that state to a metastable state through the absorption of energy from radiation. Once a stimulus is applied, in the form of heat or light, the system is triggered to return to equilibrium, along with the release of a portion of the absorbed energy in the form of light. If the stimulus is heat, the light emission is TL; if the stimulus is light, the emission is OSL. An animated version of this figure is available on the web site, under Exercises and Notes, Chapter 1.

### 1.2.2    Fermi-Dirac Statistics

Insulators and semiconductors are characterized by an energy gap between the uppermost filled energy band (the valence band) and the next empty band (the conduction band). At absolute zero, the valence band is completely full and the conduction band is completely empty. For a "perfect" crystal, no energy states are allowed in the energy gap between the top of the valence band (at energy $E_v$) and the bottom of the conduction band (at energy $E_c$). That is, if $Z(E)$ is the density of available states at any energy $E$, then $Z(E) = 0$ for $E_v < E < E_c$, and the energy gap $(E_c - E_v)$ is known as the "forbidden" gap or zone. However, real crystals contain defects such that $Z(E) \neq 0$ in this forbidden zone. Energy states $E$ can exist for which $E_v < E < E_c$ and $Z(E) > 0$, and electrons can occupy energy states that are above the valence band but below the conduction band. Since such energy levels arise because of defects (e.g., impurities, vacancies, interstitials, and larger defect complexes), these states are localized at specific lattice sites within the crystal whereas the conduction and valence bands are delocalized. As a result, excitation of valence band electrons to one of these higher energy states, through the absorption of energy from a radiation field, requires not just a transition to a higher, excited energy level, but it also requires transport of the electron from one atomic or molecular site to another within the host crystal. That is, movement through the crystal is needed. This can only occur via a "transfer state" – in other words, via the conduction band (Figure 1.2a). Once the excited electrons have been transported to their new positions in the lattice, they relax into lower energy levels $E$, where $E_v < E < E_c$.

This description is only partially complete, however. Since electrons have been excited out of the valence band, delocalized electronic holes are created. These positive charge species can move via the valence band states until they too become localized at defects within the lattice. In effect, this can be considered as an electron from the defect transitioning to the valence band, or as a hole from the valence band transitioning to the defect. The net result is that a hole, that is, a lack of an electron, now exists at that localized state.

If the two localized states just described – i.e., the localized electron state and the localized hole state – are the same, that is to say at the same defect, then the electron and hole will recombine and the whole system will return directly to its equilibrium state. However, if the two localized states are at different defects (different defect types) then they will remain localized and the system will no longer be in equilibrium. This is the metastable state.

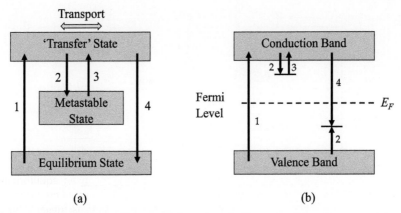

**Figure 1.2**   (a) Excitation from the equilibrium state (valence band) to the metastable state, via the conduction ("transport") band. Stimulation from the metastable state results in recombination and relaxation to the equilibrium state, again via the conduction band. (b) The metastable state can be thought of as two energy levels within the energy band gap, one above the Fermi Level and one below. At equilibrium, all energy levels above the Fermi Level are empty and all levels below the Fermi Level are full. Excitation of electrons to the conduction band results in "trapping" at localized states, above the Fermi Levels. Similarly, holes are localized ("trapped") at states below the Fermi Level. This is a non-equilibrium condition and represents the system in a metastable condition. Stimulation of the electron (say) from the localized state results in its recombination with the localized hole and the return of the system to equilibrium. Transitions: (1) Excitation (radiation); (2) Localization (trapping); (3) Stimulation (heat or light); (4) Relaxation (recombination). Animated versions of Figures 1.2a and 1.2b are available on the web site, under Exercises and Notes, Chapter 1.

The situation is illustrated in Figure 1.2b. (Animated versions of Figures 1.2a and 1.2b are available on the web site under Exercises and Notes, Chapter 1.) The two localized energy states, one above the Fermi Level and one below, localize excited electrons from the conduction band and free holes from the valence band, respectively. When localized in this way, the system is in a metastable condition. Absorption of energy from an external stimulus can free electrons (say) from the trap causing a transition to the conduction band, and these may subsequently recombine with the trapped holes, returning the system to equilibrium.

There may be multiple localized states available for electrons and holes. Consider an arbitrary distribution of available states $Z(E)$. According to Fermi-Dirac statistics, the occupancy of any energy level $E$, at temperature $T$, is given by the distribution function $f(E)$, where:

$$f(E) = \frac{1}{\exp\left\{\dfrac{E - E_F}{kT}\right\} + 1} \tag{1.1}$$

where $E_F$ is the Fermi Level and $k$ is Boltzmann's constant. At equilibrium (and at $T = 0$ K), $f(E < E_F) = 1$ (all states full), and $f(E > E_F) = 0$ (all states empty). The situation is illustrated in Figure 1.3a, for an arbitrary distribution $Z(E)$.

After irradiation (also at $T = 0$ K) the occupancy function $f(E)$ changes, as illustrated by the red line in Figure 1.3b. In this view, two new energy levels can be defined, known as quasi-Fermi levels, one for electrons $E_{Fe}$ and one for holes $E_{Fh}$. $E_{Fe}$ is defined such that all localized states at energy level $E$ are full when $E_F < E < E_{Fe}$, and are empty when $E_F > E > E_{Fh}$.

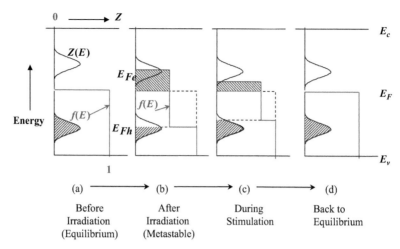

**Figure 1.3**    (a) Arbitrary distributions of available states $Z(E)$, at equilibrium and $T = 0$ K, with the Fermi-Dirac occupancy function $f(E)$. $f(E) = 1$ for $E < E_F$, and $f(E) = 0$ for $E > E_F$. (b) After irradiation some electrons occupy states above the Fermi level, and some states below this level are empty. Two quasi-fermi Levels can be defined, one for electrons $E_{Fe}$, where $E_F < E < E_{Fe}$ and one for holes $E_{Fh}$, where $E_F > E > E_{Fh}$, as shown. (c) During the return to equilibrium (i.e., during stimulation), the quasi-fermi Levels move toward $E_F$ as the occupancy of the localized states changes. (d) Eventually, the system returns to equilibrium.

During stimulation after irradiation, the states above $E_F$ empty while those below $E_F$ fill, and the two quasi-Fermi Levels move closer to the original Fermi Level, $E_F$ (Figure 1.3c). Eventually, when all localized states above $E_F$ are empty of electrons and all those below are full, $E_{Fe} = E_F = E_{Fh}$ and the system has returned to equilibrium (Figure 1.3d).

The above picture describes the broad, conceptual notions describing the perturbation of a system from equilibrium due to irradiation, and the return of the system to equilibrium during either thermal stimulation or optical stimulation. If the final relaxation processes are radiative, TL and OSL result. In the chapters to follow, the equations describing the changes in occupancy of the various energy levels during excitation and stimulation will be examined. First, however, related processes including radiophotoluminescence, RPL, are introduced.

### 1.2.3    Related Processes

Thermoluminescence is one of a family of thermally stimulated phenomena that include:

- Thermoluminescence (TL);
- Thermally Stimulated Conductivity (or Current) (TSC);
- Thermally Stimulated Capacitance (TSCap);
- Deep Level Transient Spectroscopy (DLTS);
- Thermally Stimulated Exo-Electron Emissions (TSEE).

Two additional and related phenomena should be added:

- Phosphorescence;
- Radioluminescence.

These have been described collectively in several text books (Bräunlich 1979; Chen and Kirsh 1981; Chen and McKeever 1997).

Similarly, Optically Stimulated Luminescence is one of a family of phenomena, including:

- Optically Stimulated Luminescence (OSL);
- Photoconductivity (PC);
- Optically Stimulated Exo-Electron Emission (OSEE).

As already noted, RPL does not involve ionization of the localized electrons from their trapping states and can be placed alongside other phenomena, including:

- Radiophotoluminescence (RPL);
- Electron Paramagnetic Resonance (EPR), also called Electron Spin Resonance (ESR);
- Photoluminescence (PL).

All of the above phenomena, except PL, require the initial absorption of energy from an external radiation field before the signal (TL, OSL, RPL, EPR, DLTS, etc.) can be observed. The radiation energy must be sufficiently high to cause ionization within the material – i.e. the creation of free electrons and holes. Photoluminescence is an intrinsic property of the material; that is, it may be observed before irradiation and is not created by it. Ionization is not required. (Photoluminescence can be an interference signal in both RPL and OSL.)

The differences between the phenomena are illustrated schematically in Figure 1.4. Here is represented the energy band diagram previously discussed for an insulator or semiconductor. Ionization by the radiation creates free electrons and holes that may be subsequently trapped at electron and hole localized states. Stimulation (heat or light) induces recombination of the electrons and holes. (Note: We can describe these phenomena in terms of releasing trapped electrons to recombine with holes, or vice-versa. In this description, for ease of explanation, the discussion is limited to releasing electrons to recombine with holes.) Production of TL or OSL only occurs when a portion of the energy released during the electron-hole recombination is radiative. Thus, the TL and OSL signals include information about both the trap and the recombination center. Both are required for TL and OSL to be observed.

Phosphorescence is in fact an unstable form of TL in which the electrons are trapped in the localized energy levels for short periods of time. If the traps are characterized by a small energy barrier then trapped charge can be thermally stimulated even at room temperature, leading to

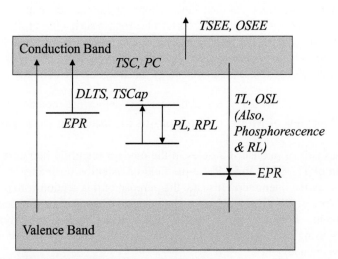

**Figure 1.4** Schematic diagram illustrating the differences between several thermally and optically stimulated phenomena. Only TL and OSL contain information about both the trap and the recombination center.

recombination and therefore luminescence emission. Similarly, radioluminescence (RL) results when electrons excited to the conduction band recombine with holes localized in hole traps, without the step of becoming trapped themselves. Emission of RL occurs during the irradiation and decays quickly after the cessation of the radiation, on a time frame governed by the recombination lifetime of the free electrons in the conduction band. In practice, both RL and phosphorescence can be observed during irradiation and the timescale for the decay of the luminescence after the radiation ends is then governed by several different lifetimes, including the recombination lifetime and the trapping lifetime of the electrons in the electron traps. The radioluminescence signal contains information about the recombination centers, whereas the phosphorescence signal includes information about both the traps and recombination centers.

In contrast to TL and OSL, TSC or PC require only the stimulated release of the trapped electrons to the conduction band. The free electrons now have the opportunity to participate in conductivity if an external electric field is applied. Thus, TSC and PC signals hold information about the traps only, not the recombination centers.

Similarly, DLTS and TSCap also monitor the release of electrons from traps, but in these cases the change in the electrical capacitance of the system is measured. Again, the signals contain no information about the recombination centers. Likewise, for TSEE and OSEE, which detect the thermally or optically stimulated emission of energetic exo-electrons from the material.

Electron paramagnetic resonance detects the trapped electrons (or holes) when the charges are localized at the defects, giving rise to an unpaired electron spin. Release of the charge from the traps is not required.

Photoluminescence occurs when electrons in a defect are raised to an excited state, but are not ionized. If relaxation to the ground state is radiative, luminescence (PL) results. If photoluminescence is observed before irradiation of the material, it is simply called PL. However, if irradiation is needed before PL is observed, the term used is RPL (radiophotoluminescence) – that is, PL from a defect that is created by the irradiation. An example helps to understand the distinction. In alkali halides, halide ion vacancies, or $F$-centers, are produced during irradiation. $F$-centers consist of trapped electrons localized by halide ion vacancies. At high enough doses, the concentration of $F$-centers is such that they cluster together to form $F_2$-, $F_3$-centers, etc. These higher-order clusters of $F$-centers produce photoluminescence when stimulated at the appropriate wavelength. Without radiation, however, these centers do not exist and so the PL signal from them is correctly termed RPL. In contrast, luminescent materials may be doped with tri-valent rare-earth (RE) ions, $RE^{3+}$. The 4f-electrons in such ions may be raised to excited states when stimulated with the right wavelength, and relaxation produces luminescence at wavelengths characteristic of the ion. Such signals are intrinsic to the phosphor, that is, they are not an effect of the absorption of radiation energy. Such signals are correctly called PL.

## 1.3   Brief Overview of Modern Applications in Radiation Dosimetry

There are thousands of published articles in the modern scientific literature demonstrating the application of TL, OSL, and RPL in the field of radiation dosimetry. A review of such applications is not the intent here. Instead, the purpose of this section is to give the reader a flavor of the types of use to which these techniques have been put in the general field of the radiation dosimetry. In this sense, we may broadly consider the following areas where radiation dosimetry using luminescence has been shown to be an essential and/or highly useful tool. The areas include:

- *Personal dosimetry* (detection and measurement of dose absorbed by people);
- *Medical dosimetry* (measurement of doses delivered to patients during medical treatments – diagnosis and therapy – to check and confirm the doses delivered);

- *Space dosimetry* (measurement of doses to astronauts and to space vehicles while in orbit or during interplanetary travel);
- *Retrospective dosimetry* (estimation of doses to people in the aftermath of radiation accidents, whether they are small accidents involving a handful of people, or large-scale events involving hundreds or thousands of people; also luminescence dating of geomorphological structures or archaeological artefacts);
- *Environmental dosimetry* (measurement of doses delivered to the environment – air, soil, built structures, and others).

There are other applications (e.g., detection of fake art objects) but those listed above are the main areas in which luminescence dosimetry has found important, not to say vital, application. One of the remarkable aspects of using luminescence in dosimetry is the fact that the phenomenon can be used to detect radiation levels as low as the naturally occurring environmental background levels on Earth, or as high as the radiation levels used in food irradiation or industrial processing, and everywhere in between. When expressed in units of Gray (Gy, where 1 Gy = 1 Joule of energy absorbed by one kilogram of a substance), doses ranging from $\mu$ Gy, for environmental radiation, to MGy, for industrial processes can be measured. This represents an amazing 12-orders of magnitude spread. While humans cannot survive in high radiation environments, homo sapiens evolved within a background of environmental radiation here on Earth. Humans can also survive for lengthy periods in more harsh radiation environments such as Space, where doses as high as several tens of mGy might be absorbed, depending on the mission. Much higher doses may be experienced by patients who may be treated with localized radiation doses of several kGy during medical radiotherapy. It is remarkable that luminescence dosimetry has found application in all of these areas.

In what follows, dosimetry applications are further discussed, albeit briefly, in order to give the reader a taste of some important examples.

### 1.3.1   Personal Dosimetry

Following the work by Farrington Daniels and colleagues, the application of TL to the world of dosimetry developed at a rapid pace. TL dosimetry (TLD), using a variety of TLD materials, became the foremost method to be used in the measurement of dose to people. The use of RPL in dosimetry developed slowly in parallel with the dominant application of TLD, while OSL did not become a popular dosimetry tool until the 1990s. Nevertheless, OSL dosimetry (OSLD) has since grown to become perhaps the dominant personal dosimetry method throughout the world, even though the availability of OSLD materials is rather limited. Today, in the world of luminescence personal dosimetry, TLD, OSLD, and radiophotoluminescence dosimetry (RPLD) each have their niches and one or other is the preferred dosimetry method in many institutions around the world.

TLD personal dosimeter badges have been designed by a wide range of companies and institutions, and a myriad of badge designs are available. Figure 1.5 displays several of them. (OSLD and RPLD badges are also included in this figure.) Each TLD badge contains a suitable TL material (the most popular of which is the first TLD material to be developed, namely LiF:Mg,Ti), along with an array of radiation filters (different metals or plastics of different thicknesses) to enable the analysis of the TLD badges to reveal not only dose, but also to provide some information about the radiation type and quality (i.e., energy). This is important to allow the dosimetry service provider to present information concerning how penetrative was the radiation to which the badge wearer was exposed.

TLD badges are used everywhere that requires the monitoring of radiation workers at radiation facilities (nuclear power stations, hospitals, industrial complexes, and many others), including guest visitors to those facilities. Although TLD badges are slowly being phased out

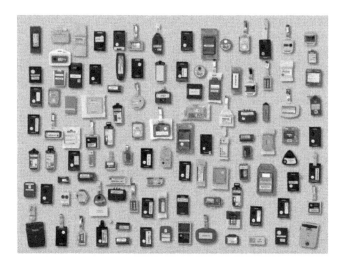

**Figure 1.5** Examples of personal dosimeters, including TLD, OSLD and RPLD badges. (Also included are some examples of electronic dosimeters.) Source: Dr. Hannes Stadtmann, © Hannes Stadtmann, European Radiation Dosimetry Group (EURADOS).

and replaced by other technologies (such as electronic dosimeters, some of which are also shown in Figure 1.5), they remain a powerful presence in the world of personal radiation dosimetry.

Also shown in Figure 1.5 are some of the OSLD and RPLD badges that are commercially available. While OSLD has become an extremely popular luminescence method for personal dosimetry, RPLD occupies a smaller portion of the luminescence dosimetry market share.

Commercial TLD materials include those based on LiF (e.g. LiF:Mg,Ti and LiF:Mg,Cu,P), $CaF_2$ (e.g. $CaF_2$:Mn, $CaF_2$:Dy, and $CaF_2$:Tm), $Li_2B_4O_7$ (e.g. $Li_2B_4O_7$:Cu), $MgB_4O_7$ (e.g. $MgB_4O_7$:Mn, $MgB_4O_7$:Dy, and $MgB_4O_7$:Cu), $Al_2O_3$ (e.g. $Al_2O_3$:C; $Al_2O_3$:Si,Ti, and $Al_2O_3$:Mg,Y) and $CaSO_4$ (e.g. $CaSO_4$:Tm and $CaSO_4$:Dy). Other, less well-known and utilized materials have often been described in the published literature, including natural minerals (e.g. calcite, fluorite, quartz). OSLD is dominated by $Al_2O_3$:C and BeO, while RPLD is dominated by Ag-doped alkali phosphate glass, and $Al_2O_3$:C,Mg.

The material properties that are important in personal dosimetry are a high sensitivity (large signal for a given dose), a linear dose-response relationship, angle independence, and energy independence. The latter is difficult to achieve at low radiation photon energies. However, the important property in this regard is that the material responds to low-energy photons in the same way as does human tissue – a property known as tissue-equivalence. The most important characteristic for tissue equivalence is the effective atomic number $Z_{eff}$ of the material. Here, materials such as LiF ($Z_{eff}$ = 8.14), BeO ($Z_{eff}$ = 7.4) and $Li_2B_4O_7$ ($Z_{eff}$ = 7.13) have advantages over other materials since these values are close to that of tissue ($Z_{eff}$ = 7.4). However, tissue equivalence is not essential as long as the response at low energies is well defined such that suitable corrections can be made (for example $Al_2O_3$:C with $Z_{eff}$ = 10.2).

### 1.3.2   Medical Dosimetry

Tissue equivalence is also a desirable characteristic of dosimeters used in medical dosimetry, but not an essential property. The high sensitivity of the TL, OSL, and RPL processes can be exploited in medical dosimetry applications since the dosimeters can be made small, which in turn gives them the property of high spatial resolution. This, in turn, means that they have the potential for dose measurement in regions of severe dose-gradients.

Modern advances in radiation medicine – in radiodiagnosis, radiotherapy, and interventional radiography – are now presenting dosimetry challenges that did not exist previously. For example, the movement toward the use of charged particles (protons and carbon ions) in radiotherapy has resulted in new requirements for luminescence dosimeters compared to the dosimetry of high-energy photons. Known responses to charged particles means careful calibration of the dosimeters in charged-particle fields. This consideration becomes especially significant beyond the Bragg peak for energy deposition by charged particles. Here, the linear energy transfer (LET) of the particles increases rapidly as the particles lose energy and slow down. Knowledge of the behavior of dosimeters in such regions, where the physical dose is rapidly decreasing but the relative efficiency of biological damage (compared to high-energy photons) is rapidly increasing, is a research area of growing importance. Secondary neutron production is also a feature of charged-particle irradiation and the response of the dosimeter in neutron fields also needs to be established.

Sophisticated intensity-modulation techniques in radiotherapy, whether using photons or particles, also create new challenges for dosimetry. This is due to the production of sharp dose gradients where the therapists try to protect healthy tissue surrounding the tumor as much as possible. It is also necessary to understand the response of the dosimeter in either steady or rapidly pulsed radiation fields.

The luminescence dosimeters may be used in vivo or ex vivo. Their small size enables their innovative use internally as the patient is treated (for example, in brachytherapy). The dosimeters may also be used externally to measure entrance and exit doses on the surface of the patient without disturbance of the radiation field.

Luminescence (particularly OSL) has also been used as an imaging technique in radiodiagnosis (where it has long been known by its alternate name of photostimulated luminescence, PSL). The sensitivity and speed of readout of the stimulated luminescence signal has given radiologists the ability to reduce radiation doses to patients and yet provide high-resolution images to aid diagnosis. However, the use of OSL in this way is not dosimetry. The actual dose to the patient is still determined by other methods.

In interventional radiography the real-time dose to patients (and surgeons) during surgical procedures can be significant. Adaptation of OSL techniques using fiber optics can provide dose measurements in real time, without disturbing the x-ray image.

In all of the above applications, the preferred properties of the needed dosimeters include tissue equivalence, high sensitivity, small size, rapid response, and known responses as functions of dose, angle, photon energy, particle LET, and neutrons.

### 1.3.3 Space Dosimetry

Dosimetry (for astronauts and equipment) in space environments is one of the most challenging applications for luminescence dosimetry. The space environment consists of an extremely wide range of highly energetic charged particles originating from cosmic rays (CR), solar particle events (SPE), and the radiation belts trapped by the Earth's magnetic field (Earth's radiation belts, ERB). Additionally, once these particles interact with the spacecraft structures, the astronauts' space suits, and the astronauts' bodies, secondary charged particles and neutrons are also produced. Personal and environmental dosimeters for use in space therefore need to respond to an extremely wide-ranging set of particles, each with its own efficiency of luminescence production. The dosimeter's response to each particle type needs to be well-known and calibrated.

TL dosimeters have been used since the beginnings of human space exploration, in both the American and former Soviet Union programs. More recently, OSL has been used in conjunction with TL in many international space programs. Neither TL nor OSL, however, is able to record dose over the wide range of charged-particle energies and types. The critical parameters are energy, charge, and the linear energy transfer (LET) of the particle. TLD and OSLD materials can record dose in fields of LET up to ~10 keV/µm with good efficiency. However, in space, particles of LET values of several hundred keV/µm can be expected. The efficiency of TLD and OSLD materials drops significantly as the LET increases beyond ~10 keV/µm and, therefore, TLD and OSLD devices are usually used in conjunction with dosimeters such as particle nuclear track detectors (PNTDs), which can measure doses and fluences for individual particles up of these high-LET values, but are insensitive below ~5 keV/µm. Thus, techniques to separate the doses in these different regions of the LET spectrum have been devised and used for several years by the various international space agencies. An example of a personal dosimeter used by the US National Aeronautics and Space Administration (NASA) is shown in Figure 1.6 and features LiF:Mg,Ti TL dosimeters, $Al_2O_3$:C OSL dosimeters and PNTD using a polycarbonate plastic known as CR-39. RPLD using $Al_2O_3$:C,Mg may have a future application in space as a PNTD, at least as a research tool.

### 1.3.4 Retrospective Dosimetry

In one sense, all passive methods of radiation dosimetry are retrospective in that they produce a measurement of dose only after the dose has been delivered. That is, they integrate the total dose received by the dosimeter since the dosimeter's last reading. (This is opposed to active dosimeters, which record the dose or dose rate in real time during the exposure period.) However, the term "retrospective" is here reserved for assessing the dose received by an individual who may have undergone an acute exposure during a radiation accident and who may not have been wearing a conventional dosimeter at the time. During radiation accidents (i.e., accidental exposure to a radiation source or contamination by radioactive pollutants), members of the public may be exposed, and since it is not likely that they would have been wearing personal dosimeters, such as those shown in Figure 1.5, methods have to be devised in which estimates may be made of the doses to which they may have been exposed. Regrettably, the potential for intentional exposure of members of the public has also be considered, for

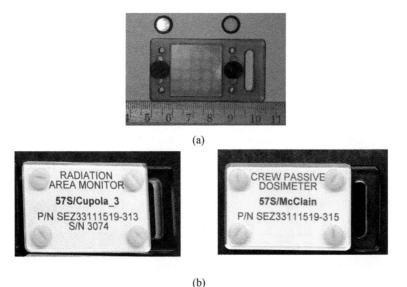

(a)

(b)

**Figure 1.6** (a) NASA's passive radiation area monitor (RAM), and (b) Radiation Area Monitor (RAM) and Crew Passive Dosimeter (CPD). Each consists of 20 TL dosimeters (LiF:Mg,Ti), two OSL dosimeters ($Al_2O_3$:C) and one PNTD (CR-39). In Figure 1.6a the TL dosimeters can be seen underneath the PNTD film, and the OSL dosimeters are inside the circular, black, Teflon holders. Source: NASA; reproduced with permission from NASA Export Control.

example, in terrorist events using so-called dirty bombs (radiation dispersal devices), or even improvised nuclear weapons. Some incidents, be they accidental or intentional, involve only a small number of people, as may the case in an over-exposure of a patient undergoing radio-therapy, or perhaps an accident with an industrial source. In other cases, the number of potentially exposed people could be very large, as in a nuclear power plant accident such as those at Chernobyl (Ukraine) and Fukushima (Japan), or in a terrorist attack.

One approach to these situations, especially to potential large-scale exposure, has been to examine "fortuitous" dosimeters. These are materials that may be found on a person and that may be used as conventional radiation dosimeters. Luminescence, especially TL and OSL, from such materials has been studied extensively for its potential in this application. Examples include TL or OSL from the components of electronic devices (in particular, smartphones), items of clothing and accessory apparel (especially synthetic fibers) and common household materials (such as table salt and ceramic objects). Figure 1.7 shows some examples. Sensitivity to radiation, dose-response characteristics, and fading of the radiation-induced TL or OSL signal are major properties of interest.

The materials shown in Figure 1.7 are examples of fortuitous luminescence dosimeters for personal dosimetry. That is, they may be used for dose assessment to the individual wearing or possessing the material. In other applications of retrospective dosimetry, such materials are no longer available. This may be the case in after-the-fact dose assessment following acute or chronic events that took place several months or years previously. In these cases, the dose to the built-environment may be determined by extracting suitable materials from that environment to be used as TL or OSL dosimetry materials. Examples include quartz grains extracted from bricks, or ceramic materials such as tiles or electrical insulators, or even washbasins and toilets. Such measurements, when combined with modeling, enable estimates of the dose to air in the vicinity of the building. Time-and-motion modeling of the movement of people within the environment then

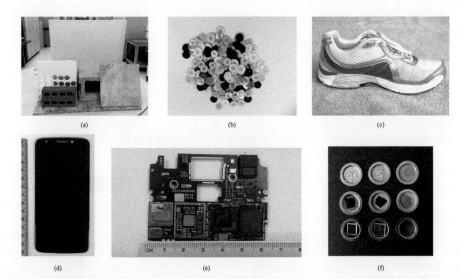

**Figure 1.7** Potential TL and/or OSL materials for personal retrospective dosimetry. (a) Building materials, including gypsum boards, bricks, ceramic tiles, and concrete blocks. (b) Buttons. (c) Shoes. (d) Smartphone (showing back glass; front protective glass and display glass can also be used). (e) Internal electronic components, including surface mount components and integrated circuits. (f) Materials prepared for TL and/or OSL analysis; top row – surface mount resistors from three different phone models; middle row – integrated circuit fragments; bottom row – protective front glass (left and middle) and display glass (right) from different phone models. Source: All photographs kindly provided by Sergey Sholom; except (c) – provided by the author.

allows estimation of the doses to which people may have been exposed. Example applications of this kind have included post-event dose assessment at Chernobyl, Hiroshima, and Nagasaki.

### 1.3.5    Environmental Dosimetry

Retrospective dose assessment to the built environment is but one example of environmental dosimetry – the assessment of dose to the environment and/or air. Another example, which is also a type of retrospective dosimetry, is measurement of the natural background dose as part of geological or archaeological dating. Here, an assessment of the dose rate in the soil of a sedimentary layer and the assessment of dose to the artefact found in that layer enables an estimate of the time the artefact has been buried (i.e., the age). Other examples might be the assessment of dose in the air or the soil surrounding a radioactive waste storage site where regular environmental dose assessments of the area surrounding the site are required for monitoring of waste leakage. A final example is the monitoring of doses in air surrounding a nuclear power plant. In each of these applications, both TL and OSL have found application and their use continues in this way.

---

**Exercise 1.2**

Choose an application from one of the many noted in Section 1.3 and write a paper, based on library research, to illustrate the development and usage of TL, OSL, or RPL in that application.

---

## 1.4    Bibliography of Luminescence Dosimetry Applications

A useful bibliography describing these applications, and more, is listed below (alphabetical order, by first author).

- Aitken, M.J. (1985). *Thermoluminescence Dating*. Academic Press, London.
- Aitken, M.J. (1998) *An Introduction to Optical Dating: The Dating of Quaternary Sediments by the Use of Photon-Stimulated Luminescence*. Oxford Science Publishers, Oxford.
- Bøtter-Jensen, L., McKeever, S.W.S., Wintle, A.G. (2003) *Optically Stimulated Luminescence Dosimetry*. Elsevier, Amsterdam.
- Chen, R., Pagonis, V. (eds.) (2019). *Advances in Physics and Applications of Optically and Thermally Stimulated Luminescence*. World Scientific, New Jersey.
- Furetta, C., Weng, P.-S. (1998) *Operational Thermoluminescence Dosimetry*. World Scientific, Singapore.
- Horowitz, Y.S. (ed.) (1984). *Thermoluminescence and Thermoluminescent Dosimetry, Vols I-III*. CRC Press, Boca Raton.
- McKeever, S.W.S., Moscovitch, M., Townsend, P.D. (1995). *Thermoluminescence Dosimetry Materials: Properties and Uses*. Nuclear Technology Publishing, Ashford.
- McKinlay, A.F. (1981). *Thermoluminescence Dosimetry*. Adam Hilger, Bristol.
- Oberhofer, M., Scharmann, A. (eds.) (1981). *Applied Thermoluminescence Dosimetry*. Adam Hilger, Bristol.
- Perry, J.A. (1987). *RPL Dosimetry: Radiophotoluminescence in Health Physics*. Adam Hilger, Bristol.
- Yukihara, E.G., McKeever, S.W.S. (2011). *Optically Stimulated Luminescence: Fundamentals and Applications*. Wiley, Chichester.

# 2

# Defects and Their Relation to Luminescence

*Crystals are like people, it is their imperfections which make them interesting.*

– P.D. Townsend 1992

## 2.1 Defects in Solids

### 2.1.1 Point Defects

Discussion of the electronic transitions that occur during TL, OSL, and RPL rely upon the simplified energy band model already discussed in Chapter 1 and illustrated again, in Figure 2.1. The depiction of a valence band completely full of electrons and a conduction band completely empty of electrons, with a constant band gap width throughout the crystal (Figure 2.1a) is a theoretical and ideal construct. It implies that all host atoms are exactly located in their equilibrium lattice positions dictated by their size and the nature of the bonds between them. No atoms other than the host atoms are assumed; that is, there are no impurity atoms. The conceptual picture also assumes an infinite crystal, with no surfaces. Of course, no such material exists in nature. Real materials contain defects, of which there are many possible types.

Point defects may be due to:

- vacancies, where a host atom is missing;
- interstitials, where a host atom occupies an off-lattice position between other host atoms;
- anti-site defects, where in a host of type AB, A atoms occupy B sites, and vice-versa;
- substitutional impurities, where a host atom is replaced by an impurity atom;
- interstitial impurities, where an impurity atom is located in an interstitial, off-lattice position;
- complex clusters of the above.

The archetypical materials for discussing defects in solids are the alkali halides, of the type $A^+B^-$. In such a material, vacancies can exist on the A or the B sub-lattice. Both A and B ions can occupy interstitial positions. Impurities can reside on either the A or the B lattice, depending on their valency, and they can also occupy interstitial sites. Depending on the valency of the

*A Course in Luminescence Measurements and Analyses for Radiation Dosimetry,* First Edition.
Stephen W.S. McKeever.
© 2022 John Wiley & Sons Ltd. Published 2022 by John Wiley & Sons Ltd.
Companion Website: www.wiley.com/go/mckeever/luminescence-measurements

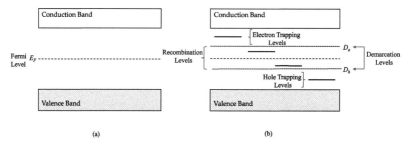

(a)                                              (b)

**Figure 2.1** (a) Idealized energy-band diagram for a perfect crystal at equilibrium, illustrating an empty conduction band and a filled valence band. For wide-band-gap insulators, the Fermi Level ($E_F$) is located mid gap. (b) A more-realistic energy-band model in which energy levels exist in the forbidden gap. Depending upon the location of the energy level (specifically, their position with respect to the band edges and the Fermi Level) the levels may be considered as "traps" or as "recombination centers," for either electrons or holes. The virtual demarcations between them are represented by Demarcation Levels, one for electrons ($D_e$) and one for holes ($D_h$).

impurity compared to that of the host atom, vacancies of type A or B may be required to co-exist with the impurities to maintain charge neutrality. Charge neutrality may also be maintained by the creation of Frenkel defects where vacancies of type A are charge compensated by interstitials, also of type A. Similarly, Frenkel defects involving the B sub-lattice may also occur. Alternatively, vacancies in the body of the crystal, on the A (or B) sub-lattice, may be charge compensated through the creation of equal numbers of vacancies on both the B (or A) sublattices and these are known as Schottky defects.

The local charge imbalance caused by vacancies, interstitials, and impurities may also be compensated by the localization of free charge carriers (electrons or holes) in cases where such delocalized free carriers exist. At equilibrium, there are negligible numbers of such carriers at normal temperatures (and none at zero Kelvin), but the defects can act as traps for whatever free charge carriers become available due to coulombic interactions between the free carriers and the traps. For example, trivalent rare-earth impurities ($RE^{3+}$) in an alkaline-earth halide (e.g. $CaF_2$) may substitute for host positive ions (anions, in this case $Ca^{2+}$). The charge imbalance results in a very strong coulombic attraction for free electrons forming divalent sites (i.e. $RE^{3+} + e^- = RE^{2+}$). In cases where the electrons are localized at defect sites they can attain energies which are higher than the valence band energies, but smaller than the conduction band energies. Thus, the energy band diagram of a real crystal, containing these simple defect types, would be characterized by allowed energy levels in the forbidden gap, as illustrated in Figure 2.1b. The Fermi-Dirac function (Equation 1.1 in Chapter 1) demonstrates that the occupancy of energy level $E$, be it localized or delocalized, depends upon the temperature $T$ and the value of $E$ relative to the Fermi Level $E_F$. If the band gap is such that $E_c - E_v \gg kT$ ($E_v$ = top of the valence band; $E_c$ = bottom of the conduction band), then all energy levels above $E_F$ are essentially empty of electrons at equilibrium, and all those below the Fermi level are essentially full.

Point defects allow a conceptual picture of what a defect might look like in a crystal. However, these simple descriptions are far from complete. They do not include ionic polarization effects and electronic interactions with electrons and nuclei in neighboring ions. Such effects mean that "point" defects in fact exert influences on the lattice out to several lattice spacings in all directions. Consider, for example, impurity $X^{2+}$ substituting for $A^+$ in ionic compound $A^+B^-$. In Figure 2.2a, a schematic ideal AB lattice is shown, typical of alkali halides. The alkali and halide sublattices are each face-centered-cubic and the picture shown in the figure is considered to stretch to infinity in all dimensions. Introduction of impurity $X^{2+}$ substituting for one of the $A^+$ host ions causes the surrounding $A^+$ ions to be repelled and the $B^-$ ions

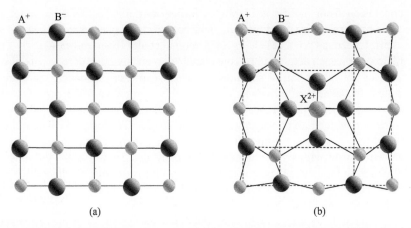

(a)                                                    (b)

**Figure 2.2**    (a) An idealized lattice for an ionic crystal of the type $A^+B^-$. (b) Stylized polarization effects caused by the substitution of an $A^+$ ion with a divalent impurity ion $X^{2+}$.

to be attracted to the $X^{2+}$ ion (Figure 2.2b). Such polarization effects cause a distortion to the lattice, spreading out over several atomic spacings. The polarization effects are dependent upon the dielectric constant of the medium and decrease rapidly ($1/r^2$) with inter-atomic spacing ($r$). The polarization energy affects the optical and electronic properties of the material and moving electronic charges also cause subsequent changes to the polarization and displacements (Hayes and Stoneham 1985).

The schematic illustration in Figure 2.2b is for illustrative purposes only. Modern density functional theory (DFT) calculations are able to calculate the positions of the ions when a divalent impurity is introduced into an $A^+B^-$ lattice. An example is shown in Figure 2.3 for LiF doped with $Mg^{2+}$, with the Mg ion in an interstitial position. For another example, Masillon et al. (2019) reveal that when $Mg^{2+}$ substitutes for $Li^+$ there is a resultant displacement of six nearest F atoms symmetrically in pairs, by 0.14 Å, 0.15 Å and 0.18 Å, while one F atom close to the Li-vacancy moves by 0.15 Å. At the same time, three neighboring Li atoms move away from their original position; two by 0.11 Å and one by 0.14 Å.

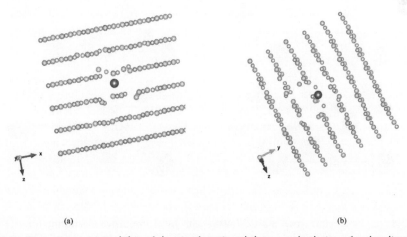

(a)                                                    (b)

**Figure 2.3**    Two views, (a) and (b), of density functional theory calculations for the distortions in the lattice caused by the addition of an interstitial $Mg^{2+}$ ion impurity into LiF. $Li^+$ ions are in grey; $F^-$ ions are in green, and the $Mg^{2+}$ impurity ion is in orange. (Original data kindly provided by Guerda Masillon, © Guerda Masillon, UNAM, Mexico.)

The conclusion from these considerations is that a "point" defect in a lattice can exert influence over several lattice spacings and, in the certain cases, over several thousand surrounding host ions. Indeed, a "point defect" is not a "point" at all (Townsend 1992).

In a dilute system each defect can be considered "unseen" by other defects. In this definition, no matter over how many lattice spacings the defect can exert influence, there are no other defects within this sphere of influence. However, this is not always the case. For example, a divalent impurity substituting for an alkali ion can be charge compensated locally through coulombic attraction with an alkali vacancy. Using the popular TL dosimetry material LiF:Mg as an example, $Mg^{2+}$ ions that substitute for host $Li^+$ ions are charge compensated by Li vacancies, which are effectively negatively charged and occupy nearest-neighbor positions along the <110> direction in the lattice. The clustering process forms dipolar complexes (Figure 2.4a), which are revealed by DFT calculations in LiF:Mg (Masillon et al. 2019). However, at room temperature this process of dipole formation does not lead to the final thermodynamic equilibrium state of the system, and further reduction in the system's free energy can be attained by further clustering of the dipoles to form trimers, consisting of three dipoles in one of several possible configurations, an example of which is shown in Figure 2.4b (Taylor and Lilley 1982a; McKeever 1985; Gavartin et al. 1991).

An additional consideration, not indicated in the conceptual Figures 2.2 and 2.4, is that the radius of the impurity ions generally do not match those of the host ions for which they substitute. For example, the radius of a $Mg^{2+}$ ion is $86 \times 10^{-12}$ m, whereas the radius of $Li^+$ is $90 \times 10^{-12}$ m. This small change in radius (<5%) causes a much larger change in ionic volume in that part of the lattice and substituting a $Li^+$ ion with a $Mg^{2+}$ ion results in a decrease of ionic volume by ~15%. Similarly, $Ti^{4+}$ results in a ~76% volume decrease, while $Y^{3+}$ in $CaF_2$ causes a ~69% decrease when substituting for $Ca^{2+}$. These effects immediately cause lattice distortions over and above distortions due to coulombic forces.[1]

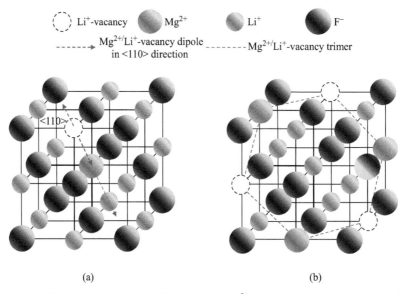

(a)                                                        (b)

**Figure 2.4**    (a) Schematic view of a LiF lattice with $Mg^{2+}$ impurity substituting for a $Li^+$ host ion and charge compensated by a Li-vacancy in a <110> direction, forming a dipolar complex; (b) example trimer cluster of three $Mg^{2+}$-$Li_{vac}$ dipoles.

---

[1] The author is grateful to P.D. Townsend for this comment.

Radiation-dosimetry-quality LiF does not contain only Mg impurities; the TL sensitivity of the material is strongly enhanced by the inclusion of $Ti^{4+}$ impurities. Charge compensation in this case appears through the incorporation of $OH^-$ impurities, forming Ti-OH complexes. The role of the Ti-OH complexes in the production of TL is still not precisely determined but they are believed to be active in the recombination processes occurring in this material. In particular, much evidence has accumulated in recent years to suggest that the main TL peak from this material (the so-called peak 5) is the result of a Mg-trimer/Ti-OH complex of an as-yet-undetermined crystalline structure. That is, the Ti-OH complexes and the Mg-$Li_{vac}$ trimers appear to be spatially associated in the LiF lattice and many of the characteristic TL properties of peak 5, and the TL glow curve in general, are uniquely affected by this association.

It is the breakdown in the lattice periodic potential caused by defects that gives rise to energy levels within the forbidden gap. The wavefunctions of electrons in a crystal with perfect periodicity are delocalized and extend throughout the material. States where the electron wavefunction is localized are not allowed. It is when the periodicity breaks down due to the presence of a defect that localized wavefunctions occur. These decay with distance away from the center of the defect over several lattice sites and the corresponding energies reside within the band gap.

Whether they are relatively simple defects or complex defects, interactions between the localized electrons and holes can take place either, or both, non-locally (i.e. via the delocalized bands) or locally (e.g. tunneling of charge between localized states) depending upon both the energy and the spatial association between the defects. The energy levels may be discrete (i.e. characterized by a single energy value) or, because of their complexity, can be distributed in energy with the exact energy value depending upon the nature of the surrounding environment and the presence of other defects. This is especially true of non-crystalline materials such as glasses, in which the surrounding lattice may display short- or long-range disorder, resulting in a range of energies for a particular defect type.

### 2.1.2    Extended Defects

In addition to point defects, one can also add even more complex defects consisting of line dislocations (boundaries between slipped and un-slipped lattice planes), grain boundaries, angular misfits between lattice planes, planar dislocations (internal surfaces), nanoparticle formations, inclusions, and precipitates. Another obvious cause of the breakdown in the periodicity of a lattice is the presence of a surface. Crystals are not infinitely large and at a surface the lattice periodicity ends abruptly giving rise to broken bonds and bonds passivated by the possible presence of foreign atoms, resulting in large concentrations of localized levels at the surface. Clearly, powdered materials with small grain sizes and large surface-to-volume ratios are more likely to exhibit effects due to surface states than larger, bulk materials.

### 2.1.3    Non-Crystalline Materials

An obvious example of a material in which the infinite periodicity of the lattice cannot be expected is a non-crystalline or glassy material. The amorphous nature of such materials means that, at best, only short-range order can be expected and once several lattice distances are considered the material cannot be defined by regular order. As a result, the presence of defects such as those described in the previous sections results in a range of potential energies to describe the defect, depending upon the local environment. Furthermore, the energy states available to free electrons and holes must be reconsidered in such materials, as shown in Figure 2.5 where the density of allowed states of a crystalline material (Figure 2.5a) is compared with that of an amorphous material (Figure 2.5b). As with crystalline materials, allowed

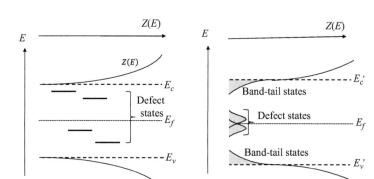

**Figure 2.5**   Density of states functions $Z(E)$ for (a) a crystalline solid, and (b) an amorphous (non-crystalline) solid. For non-crystalline materials the density of states extends into the band gap (band-tail states) and localized states (shaded) are distributed in energy.

energy zones do exist, but with non-crystalline materials the density of states $Z(E)$ partially extends into the gap, creating so-called band-tail states. Defect state energies are distributed rather than discrete and can extend above or below the Fermi Level, $E_F$, such that $Z(E_F) \neq 0$. Localized energy levels (shaded in Figure 2.5b) require absorption of external energy (thermal or optical) to ionize the defects, i.e. to excite the charge carriers (electrons or holes) into the delocalized bands (conduction and valence bands). Thus, a non-crystalline solid can be treated in much the same way as a crystalline solid in the sense that the delocalized bands are separated by an energy gap (more correctly termed a "mobility gap" since $Z(E)$ tail states extend into the gap) with localized states due to defects existing within the gap but with distributions of energy (band-tail states).

It should also be mentioned that even crystalline materials can exhibit band-tail states extending into the gap if they exhibit sufficient lattice disorder due to variations in bond lengths, bond angles, local density fluctuations, and/or contain high concentrations of impurities or other extrinsic defects. If the disorder is sufficiently high, it can give rise to a density of states $Z(E)$ that extends into the band gap. Such materials may include, for example, natural minerals, such as the feldspar family or the various polymorphs of silicon dioxide.

## 2.2   Trapping, Detrapping, and Recombination Processes

### 2.2.1   Excitation Probabilities

#### 2.2.1.1   *Thermal Excitation*

Consider an electron localized at a lattice defect, at energy $E_t$ below the conduction band and probability $p$ that the electron will absorb external energy and be excited from the trap into the conduction band. If the temperature of the system is $T$, the probability per second $p$ that the electron will be *thermally* excited into the conduction band is given by:

$$p = vK\exp\left\{-\frac{F}{kT}\right\}, \tag{2.1}$$

where $v$ is the lattice phonon vibration frequency, $K$ is the transition probability constant, $k$ is Boltzmann's constant, $F$ is the Helmholtz free energy $= E_t - \Delta ST$, and $\Delta S$ is the entropy change associated with the transition. Thus:

$$p = \nu K \exp\left\{\frac{\Delta S}{k}\right\} \exp\left\{-\frac{E_t}{kT}\right\} = s \exp\left\{-\frac{E_t}{kT}\right\}. \tag{2.2}$$

Here, $s$ is known as the "attempt-to-escape" frequency (also known as the "frequency factor," or the "pre-exponential factor"), with units of $s^{-1}$. It is the number of times per second that energy is absorbed from phonons in the lattice, and the term $\exp\left\{-\dfrac{E_t}{kT}\right\}$ is the probability that the energy absorbed is enough to cause a transition from the localized state to the conduction band. Typically, one can expect $s \sim 10^{12}$–$10^{14}$ $s^{-1}$; that is, of the order of the lattice vibration frequency, $v$, but differing from it by the factor $K \exp\left\{\dfrac{\Delta S}{k}\right\}$.

By applying equilibrium statistics, it is possible to show that:

$$s = N_c v_e \sigma, \tag{2.3}$$

where, $N_c$ is the concentration of available states in the conduction band (units, $m^{-3}$), $v_e$ is the thermal velocity of free electrons ($m.s^{-1}$) and $\sigma$ is the capture cross-section for the trap ($m^2$). The concentration of free electrons, $n_c$ at any given temperature may be written:

$$n_c = \int_{E_c}^{\infty} Z(E)f(E)dE \approx N_c \exp\left\{-\frac{E_c - E_F}{kT}\right\}. \tag{2.4}$$

Since the occupancy of the conduction band is essentially zero at the top of the conduction band, the integral can be taken to infinity. It is also assumed that $E_c - E_F >> kT$, which is a good approximation in insulators, even for $T = 1000s$ K. $N_c$ can, therefore, be considered to be the effective density of states of a fictional level lying at the conduction band edge and is defined by:

$$N_c = 2\left(\frac{2\pi m_e^* kT}{h^2}\right)^{\frac{3}{2}}. \tag{2.5}$$

Similarly, the number of free holes in the valence band is given by:

$$m_v \approx M_v \exp\left\{-\frac{E_F - E_v}{kT}\right\} \tag{2.6}$$

and

$$M_v = 2\left(\frac{2\pi m_h^* kT}{h^2}\right)^{\frac{3}{2}} \tag{2.7}$$

is the density of available states in the valence band. In Equations 2.5 and 2.7, $m_e^*$ and $m_h^*$ are the effective masses of the free electrons and free holes in the conduction and valence bands, respectively, and $h$ is Planck's constant.

The value of the capture cross-section $\sigma$ is critically dependent upon the potential distribution in the neighborhood of the trap and, in particular, upon whether the trap is coulombic attractive, neutral, or repulsive. Three representative cases are illustrated in Figure 2.6. The

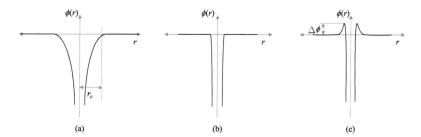

**Figure 2.6**    Potential $\phi$ as a function of distance $r$ in the vicinity of: (a) coulombic attractive, (b) neutral, and (c) repulsive localized states. The critical distance $r_c$, where $\sigma = \pi r_c^2$, is defined in the text. For coulombic repulsive centers the capture cross-section $\sigma$ is reduced exponentially as a function of the potential barrier $\Delta\phi$.

figures show the potential distribution around a coulombic attractive trap (a), a coulombic neutral trap (b), and a coulombic repulsive trap (c). The critical distance $r_c$ is that distance for which the energy of coulombic attraction equals the kinetic energy $KE$ of the free electron (Rose 1963). If the trap has a coulombic net charge of +1, then:

$$KE = \frac{q^2}{r_c \varepsilon},\tag{2.8}$$

where, $q$ is the charge on the electron and $\varepsilon$ is the dielectric constant of the material. From this:

$$\sigma = \pi r_c^2 = \pi \left(\frac{q^2}{KE\varepsilon}\right)^2.\tag{2.9}$$

For coulombic repulsive centers the capture cross-section $\sigma$ is reduced exponentially as a function of the potential barrier $\Delta\phi$ – i.e. by $\exp\left\{-\dfrac{q\Delta\phi}{kT}\right\}$. Thus, the capture cross-section of a repulsive trap is exponentially dependent on temperature.

Experimentally determined values for $\sigma$ range from $\sim 10^{-16}$ m$^2$ for attractive centers (so-called giant traps) to $\sim 10^{-26}$ m$^2$ for repulsive centers. Also, since $v_e \propto T^{1/2}$, then $KE \propto T$, and $\sigma \propto T^{-2}$ for attractive or neutral centers. It is clear that a coulombic attractive trap for a free electron is a repulsive trap for a free hole, and vice-versa. Thus, each localized state is represented by two cross-sections, one for electrons, $\sigma_e$ and one for holes, $\sigma_h$. If $\sigma_e \gg \sigma_h$ the state is defined as an electron trap, while for a hole trap $\sigma_h \gg \sigma_e$.

Apart from thermal excitation out of the localized state, there is also the possibility that the trapped electron might attract an oppositely charged hole and the two may recombine. If a recombination event is more likely that a detrapping event, the localized state is called a recombination center. If the opposite is true, it is a trap. Thus, one can imagine that at a given temperature $T$, there may exist a state for which the probabilities are equal, that is:

$$s_e \exp\left\{-\frac{D_e}{kT}\right\} = nB_h^e\tag{2.10}$$

for electrons, and

$$s_h \exp\left\{-\frac{D_h}{kT}\right\} = mB_e^h\tag{2.11}$$

for holes. In these expressions, the left-hand side of the equation represents the probability of thermal excitation from the trap and the right-hand side represents the probability of recombination with the carrier of opposite sign. The definitions of the terms are: $s_e$ and $s_h$ are the attempt-to-escape frequencies for the trapped electrons and the trapped holes, respectively; $n$ and $m$ are the concentrations of trapped electrons and holes; $B_h^e$ and $B_e^h$ are the probabilities of recombination for electrons with holes and vice-versa, expressed in units $m^3.s^{-1}$. In turn, these are defined as $B_h^e = \sigma_h^e v_e$ and $B_e^h = \sigma_e^h v_h$, where $\sigma_h^e$ and $\sigma_e^h$ are the cross-sections for the capture of electrons by holes, and holes by electrons, respectively and $v_e$ and $v_h$ are the free electron and free hole thermal velocities. The fictional energies $D_e$ and $D_h$ are those energies for which Equations 2.10 and 2.11 are true, and are known as the "Demarcation Levels," one for electrons and one for holes. They are illustrated in Figure 2.1b. For electron and hole localized levels at energy depths $E_{te}$ and $E_{th}$, such that $E_{te} > D_e$ and $E_{th} < D_h$, the localized states are traps; for energies $E_F < E_{te} < D_e$ and $E_F > E_{th} > D_h$, the localized states are recombination centers. In this way it is clear that recombination centers tend to have energies near the center of the band gap, whereas trap energies are toward the upper (for electrons) or lower (for holes) regions of the gap.

Also to be noted is that Equations 2.10 and 2.11 are highly dependent on temperature, and thus a localized level that is a trap at high temperatures, may be a recombination center at lower temperatures. The weaker temperature dependencies of the attempt-to-escape frequencies and the cross-sections are less significant than the exponential dependence on $T$.

The processes of trapping and recombination are represented schematically in Figure 2.7 based on the energy band model for a crystalline solid. In the simple model represented in Figure 2.7 there is only one species of electron localized state and one species of hole localized

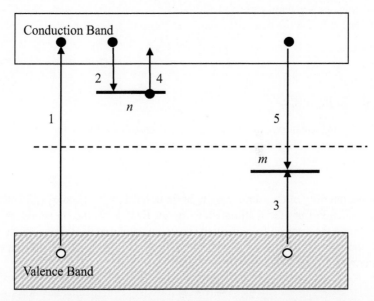

**Figure 2.7**   A simple model with one electron trap and one recombination center (the One-Trap/One-Recombination-Center, or OTOR, model). Transition (1) represents an ionization event in which one electron (black dot) is ionized from its host atom and becomes free to move throughout the crystal. At the same time, a free hole (open dot) is created in the valence band and it too becomes free to move. Transitions (2) and (3) represent trapping of the free electron and free hole at electron and hole traps respectively. In this model it is assumed that $s_e \exp\left\{-\dfrac{E_{te}}{kT}\right\} > nB_h^e$ and that $s_h \exp\left\{-\dfrac{E_{th}}{kT}\right\} < mB_e^h$

such that the trapped hole acts as a recombination center for free electrons. Upon thermal excitation from the trap (transition (4)), the freed electron may recombine with the trapped hole (transition (5)).

state. It is further assumed that $s_e \exp\left\{-\dfrac{E_{te}}{kT}\right\} > nB_h^e$ and that $s_h \exp\left\{-\dfrac{E_{th}}{kT}\right\} < mB_e^h$ such that the trapped hole acts as a recombination center for free electrons, whereas the electron center is defined as a trap. As a result, the model of Figure 2.7 is referred to as the One-Trap/One-Recombination-Center (OTOR) model. In this model, transition (1) represents an ionization event in which an electron is ionized from its host atom and becomes free to move throughout the crystal. At the same time, a free hole is created in the valence band, at which energies it also is free to move. Transitions (2) and (3) represent localization (i.e. trapping) of the free electron and free hole at the electron and hole traps respectively. Upon thermal excitation from the electron trap (transition (4)), the freed electron may recombine with the trapped hole (transition (5)).

---

**Exercise 2.1**

(a) Consider an electron trap at energy $E$ and energy depth $E_t = E_c - E$ (where $E_c$ is the bottom of the conduction band). The total concentration of traps is $N$, of which $n$ are filled with electrons. What will be the occupancy of this trap if $E = E_F$?

(b) If $N_c$ is the density of available states in the conduction band, $n_c$ is the concentration of free electrons, $v_e$ is the thermal velocity of free electrons, and $\sigma$ is the capture cross-section for the trap, show that the attempt-to-escape frequency, $s$ is given by Equation 2.3. (*Hint*: consider equilibrium between trap filling and trap emptying.)

(c) What is the expected $T$ dependence of $s$?

(d) If $m_e^* \approx m_h^*$, show that, at thermal equilibrium at $T > 0$ K, the Fermi Level lies mid-gap.

---

### 2.2.1.2   Optical Excitation

If, instead of heating a material, the trapped electrons are released from their traps via absorption of energy from photons, Equation 2.1 now becomes:

$$p = \Phi \sigma_p(E), \tag{2.12}$$

where $\Phi$ is the intensity of the stimulating light (in units of $m^{-2}s^{-1}$) and $\sigma_p(E)$ is the photoionization cross-section ($m^2$) for a stimulation energy $E$. If $E_o$ is the threshold photon energy required to excite the electron from the trap (i.e. the optical trap depth) one might expect $E_o = E_t$, that is, the thermal trap depth $E_t$ and the optical trap depth $E_o$ are the same. However, thermal energy is also absorbed by lattice phonons such that:

$$E_o = E_t + E_{ph}, \tag{2.13}$$

where $E_{ph}$ is the phonon energy given by:

$$E_{ph} = Shv_{ph}. \tag{2.14}$$

Here, $S$ is the Huang-Rhys factor, $h$ is Planck's constant and $v_{ph}$ is the phonon vibration frequency.

The usual way to consider the differences between the optical trap depth $E_o$ and the thermal trap depth $E_t$, is to use the configurational coordinate diagram, shown in Figure 2.8, where potential energy curves for the electron in the trap $E_g(Q)$ and in the conduction band $E_e(Q)$ are

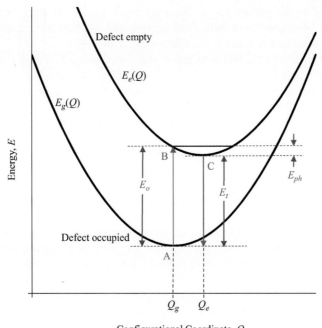

**Figure 2.8**   A configurational coordinate diagram showing the potential energy curves $E_g(Q)$ and $E_e(Q)$ in the region of the defect when the defect state is occupied by an electron, and when it is empty (ionized). When the level is occupied the energy is a minimum at configurational coordinate $Q_g$. Optical transitions take place vertically (transition AB) since the lattice does not have time to respond to the change in charge state of the defect and relax to its new configurational coordinate. The optical energy required to affect this transition is $E_o$. Once ionized, the lattice relaxes to new coordinate $Q_e$ and a new energy minimum at C, following emission of phonons of energy $E_{ph}$. Lattice relaxations are allowed during thermal excitations, however, and thermal stimulation can cause transitions directly from A to C. The required thermal energy is $E_t$, where $E_t=E_o - E_{ph}$.

expressed as a function of the configurational coordinate $Q$ of the system. Once the electron is excited into the conduction band, the charge state of the localized trap changes by +1 and significant lattice relaxation can occur. However, the Franck-Condon principle dictates that optical transitions occur without changes in the configuration and lattice relaxation only occurs once the electron is in its excited state. This is illustrated in Figure 2.8 by the vertical transition A→B, with energy $E_o$. After lattice relaxation, the excited electron loses an amount of energy $E_{ph}$, transition B→C. Thermal transitions, on the other hand allow for lattice relaxations and the net energy required is shown as $E_t$ in the figure, with the transition A→C. $E_t$ and $E_o$ are related as in Equation 2.13. Estimates reveal $E_o / E_t \approx \varepsilon_o / \varepsilon$, where $\varepsilon_o$ and $\varepsilon$ are the high frequency and static dielectric constants, respectively (Mott and Gurney 1948).

The form of the photoionization cross-section function $\sigma_p(E)$ depends upon several material-related factors and different models exist to describe it. The first point to remember is that it is a function, and its value depends on the wavelength of the stimulating light used to excite the electrons from their traps. Thus, if broad-band, polychromatic excitation light is used, the photoionization cross-section will be multivalued. At first glance, one might expect that the function might be a step function since as long as the photon energy $E = hv > E_o$, trap ionization will occur, whereas if $hv < E_o$, there will be no excitation of the electron from the trap. However, the shape of the $\sigma_p(E)$ function depends on the shape of the availability of states in the conduction band, i.e. the density of states function $Z(E)$ and a peak in the value of $\sigma_p$ is usually observed as the photon energy becomes too large for the corresponding states into which the electron could be excited. Furthermore, strong phonon coupling of the trapped electron to the lattice can give rise to excitation even if $hv < E_o$.

The various expressions for photoionization cross-section $\sigma_p(E)$ depend upon assumptions relating to the potential energy in the vicinity of the defect, the wavefunctions for the trapped and delocalized states, the density of states in the delocalized band, and the degree of phonon interaction. For a shallow (hydrogenic) electron trap:

$$\sigma_p(E) \propto \frac{(hv - E_o)^{\frac{3}{2}}}{(hv)^5}, \tag{2.15}$$

where $E = hv$ is the energy of the stimulating light (Blakemore and Rahimi 1984; Landsberg 2003). The coulombic attraction between the freed electron and the ionized defect is ignored when $hv$ is just larger than $E_o$. The cross-section reaches its maximum at $hv = 1.4 E_o$.

For deep traps, Lucovsky (1964) approximated the potential in the region of the defect to a delta function and assumed a plane wave excited-state wavefunction to derive:

$$\sigma_p(E) \propto \left[ \frac{4(hv - E_o)E_o}{(hv)^2} \right]^{\frac{3}{2}}. \tag{2.16}$$

The cross-section reaches a maximum at $hv = 2E_o$, and the coulombic field is taken into account.

A further assumption in the derivation of Equation (2.16) is that the effective mass $m_e^*$ of the electron in the conduction band can be used also for the electron in the localized state. By using the electron rest mass $m_o$ instead of $m_e^*$ while the electron is localized, Grimmeis and Ledebo (1975a, 1975b) derived:

$$\sigma_p(E) \propto \frac{(hv - E_o)^{3/2}}{hv\left[hv + E_o\left(m_o / m_e^* - 1\right)\right]^2} \tag{2.17}$$

also using a plane-wave final state and the assumption of parabolic bands.

By taking into account strong phonon coupling between the lattice and the trapped electron, Noras (1980) (see also Chruścińska 2010) derived:

$$\sigma_p(E) \propto \frac{\kappa}{\nu} \int_0^\infty \epsilon^{a-1} \left[ \exp\left\{ -\kappa^2 \left[ \epsilon - (hv - E_o) \right]^2 \right\} \right] d\epsilon. \tag{2.18}$$

The parameter $\epsilon$ is a dummy variable having the dimensions of energy, and $a = 5/2$ or $3/2$ for forbidden and allowed transitions, respectively. The parameter $\kappa$ is given by:

$$\kappa = \left[ 2S\left(hv_{ph}\right)^2 \coth\left(\frac{hv_{ph}}{2kT}\right) \right]^{-\frac{1}{2}}, \tag{2.19}$$

where again $S$ is the Huang-Rhys factor and $hv_{ph}$ is the energy of the phonon vibrational mode. For a purely electronic transition (no phonon coupling):

$$\sigma_p(E) = v^{-1}\left(hv - E_o\right)^{a-1} \tag{2.20}$$

for $hv >> E_o$, and $\sigma_p(E) = 0$ for $hv < E_o$. Compare Equation 2.20 with Equations 2.16 and 2.17.

Several other expressions for $\sigma_p(E)$ also exist (Jaros 1977; Blakemore and Rahimi 1984; Ridley 1988; Böer 1990; Landsberg 2003).

OSL signals from dosimetry materials originate from the release of electrons from deep trapping states and thus the expressions most frequently used to represent the photoionization cross-section of such centers are Equations 2.16, 2.17, and 2.18. A comparison of the shapes of some of these expressions is given in Figure 2.9.

**Exercise 2.2**

From the literature, look up as many expressions as you can find for the photoionization cross-section $\sigma_p(E)$. Plot each and compare shapes. Discuss and explain the differences, assumptions, limitations, etc.

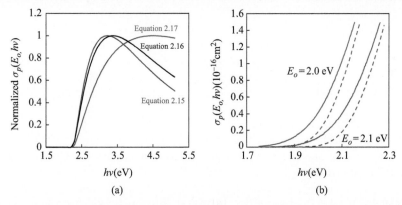

**Figure 2.9**   (a) Examples of postulated photoionization cross-sections as a function of stimulation energy. In this depiction, all curves are normalized to their maximum value and the optical trap depth is $E_o = 2.25$ eV. (b) Example photoionization cross-sections when phonon coupling is allowed. In this figure, the Huang-Rhys factor $S$ is 10 and the temperature is 300 K. Curves corresponding to two values for $E_o$ are illustrated, each with two curve shapes corresponding to values of $hv_{ph}$ of 20 meV (dashed lines) and 40 meV (full lines). (Adapted from Chruścińska 2010.)

### 2.2.2   Trapping and Recombination Processes

Once an electron-hole pair has been generated by radiation, trapping and recombination processes remove the free charge carriers (electrons in the conduction band and holes in the valence band) from their respective delocalized bands. Several mechanisms can lead to electron or hole capture and recombination. In principle, direct recombination of a free electron with a free hole (i.e. a transition directly across the gap) is possible, but is highly unlikely. Not only would a large amount of energy have to be dissipated, but conservation of momentum requires that the electron and hole recombine with oppositely directed momenta. If the direct recombination process is non-radiative, the energy would have to be dissipated via phonon vibrations. The materials used for radiation dosimetry are wide-band-gap insulators and the band gaps are several electron-volts wide. (For example, in LiF the band gap it is ~14 eV.) Considering that phonon energies in solids are about 0.03 eV at room temperature, thousands of phonons would be needed to couple with the electron in order to dissipate the energy. In contrast, if the energy was dissipated radiatively in the form of photons, the emitted photon wavelengths (using LiF as an example) would be around 80–90 nm, in contrast to the visible wavelengths observed in luminescence measurements. These considerations lead to the conclusion that recombination of charges occurs via localized states in a process known as Shockley-Read-Hall (SRH) recombination. The recombination process leaves the recombination site (defect) in an excited state, radiative relaxation of which produces the luminescence.

Considering only the recombination of free electrons with trapped holes, the rate of SRH recombination can be seen to be dependent on the free electron density, the density of trapped holes, and the temperature. The free carrier lifetime $\tau$ of an electron or a hole can be expressed as:

$$\frac{1}{\tau} = \frac{1}{\tau_{trap}} + \frac{1}{\tau_{recom}} \tag{2.21}$$

where $\tau_{trap}$ and $\tau_{recom}$ are trapping and recombination lifetimes respectively, given by:

$$\tau_{trap} = \frac{1}{v\sigma_t (N-n)} = \frac{1}{A(N-n)} \tag{2.22}$$

and

$$\tau_{recom} = \frac{1}{v\sigma_r n} = \frac{1}{Bn}. \tag{2.23}$$

In these expressions $v$ is the free carrier thermal velocity ($v_e$ for electrons in the conduction band or $v_h$ for holes in the valence band); $\sigma$ is the relative capture-cross section for either capture by $(N-n)$ empty traps, or for recombination with $n$ trapped charges of opposite sign (free electrons by trapped holes or free holes by trapped electrons) and $N$ and $n$ are the total concentrations of available traps and the concentration of filled traps, respectively. The products $A = v\sigma_t$ and $B = v\sigma_r$ are the respective trapping/recombination "probabilities" or "transition rates," in units of $m^3.s^{-1}$.

Capture of electrons or holes at localized states can be via multi-phonon processes, cascade capture, Auger emission or radiative recombination. (See Mott (1978) for an overview and Landsberg (2003) for more detailed treatments.) As argued above, radiative recombination can be neglected for SRH recombination at a level deep within the band gap of a wide-band-gap insulator. Auger emission, either direct or exciton-enhanced, occurs when the energy of the free carrier is transferred to another electron or hole. The process is not often considered in dosimetry materials, but is a possibility to be considered. The temperature dependence of the capture cross-section for the Auger process is usually a power law, $\sigma \propto T^{-a}$.

In cascade capture, the free charge loses its energy via the successive emission of phonons accompanied by loss of energy of the free charge by cascading down a series of discrete, excited states of the defect, eventually reaching the ground state. For example, a free electron may be captured by cascading down the trapped electron excited states, eventually reaching its ground state. The process requires that the electron will not be re-emitted from any of the excited states back to the conduction band and the probability of this happening will increase with temperature. This leads to a power-law dependence of the capture cross-section on temperature in which the capture cross-section decreases as $T$ increases, $\sigma \propto T^{-a}$.

The multi-phonon process requires a sufficient number of phonons to allow the defect energy level to cross into the delocalized band and capture the electron (and similarly for hole capture). Phonons are bosons and the distribution of phonons of energy $E_{ph}$ at temperature $T$ is given by Bose-Einstein statistics as $1/\left(1+ \exp\left\{-E_{ph}/kT\right\}\right)$. Thus, the availability of sufficiently energetic phonons increases with temperature, giving rise to an exponential dependence of the capture cross-section on temperature $\left(\exp\left\{-E_{ph}/kT\right\}\right)$. The cascade-capture mechanism is likely to be combined with the multi-phonon mechanism, especially for the last excited-state to ground-state transition. Thus, a combination of a power-law dependence on $T$ and an exponential dependence on $T$ might be expected. Note that if $E_{ph}$ is small, the dependence will approximate to the power law $T^{-a}$.

Shockley-Read-Hall recombination requires a combination of two of the above capture processes. First, capture of (say) a hole is required, followed by capture of an electron by the trapped hole and consequent electron-hole recombination.

---

**Exercise 2.3**

From the literature, look up discussions of radiative transitions, Auger processes, cascade capture, and multi-phonon relaxation. Under what conditions might one process be expected compared to another?

# 3

# TL and OSL

## *Models and Kinetics*

*...learning with mathematical models not only has practical applications, but also has philosophical and historical relevance in the construction of mathematical and scientific knowledge.*

– D.J. Carrejo and J. Marshall 2007

## 3.1 Rate Equations: OTOR Model

Considering the One-Trap/One-Recombination-Center (OTOR) model of Figure 2.7, let $G$ = rate of excitation across the gap (transition 1), $N$ = total concentration of available electron traps, $n$ = concentration of trapped electrons, $M$ = total concentration of available hole traps, $m$ = concentration of trapped holes, and $n_c$ and $m_v$ = the concentration of free electrons and holes in the conduction and valence bands, respectively. With $p$ = the probability per time of excitation out of the trap (thermal or optical), then the rate of excitation is:

$$\frac{dn^-}{dt} = np, \tag{3.1}$$

and the rate of electron trapping is:

$$\frac{dn^+}{dt} = n_c(N-n)\sigma_e v_e = n_c(N-n)A_n. \tag{3.2}$$

Here the cross-section for electron trapping into the trap is written $\sigma_e$ and $v_e$ is the electron thermal velocity. From the above, $A_n = \sigma_e v_e$ where $A_n$ is termed the electron trapping probability.

The net change in trapped electrons is then:

$$\frac{dn}{dt} = \frac{dn^+}{dt} - \frac{dn^-}{dt} = n_c(N-n)\sigma_e v_e - np. \tag{3.3}$$

*A Course in Luminescence Measurements and Analyses for Radiation Dosimetry,* First Edition.
Stephen W.S. McKeever.
© 2022 John Wiley & Sons Ltd. Published 2022 by John Wiley & Sons Ltd.
Companion Website: www.wiley.com/go/mckeever/luminescence-measurements

Similarly, the net change of holes in the hole trap is:

$$\frac{dm}{dt} = \frac{dm^+}{dt} - \frac{dm^-}{dt} = m_v (M-m)\sigma_h v_h - n_c m \sigma_r v_e = m_v (M-m) A_m - n_c m B. \quad (3.4)$$

Again, the cross sections are written as $\sigma_h$ for trapping of the hole into the hole trap, and $\sigma_r$ for recombination of the electron with the hole. In the above equations, $A_n$, $A_m$, and $B$ are the electron trapping, hole trapping, and electron-hole recombination transition probabilities, respectively. The time dependencies of $n(t)$ and $m(t)$ are understood, but not explicitly shown. Also, $v_e$ and $v_h$ are the thermal velocities of free electrons and free holes in the conduction and valence bands, respectively. The other terms have already been defined.

In Equation 3.3, the first term represents the rate of electron capture, at time $t$, into one of the $N-n$ empty electron traps. Similarly, the first term in Equation 3.4 represents the hole-capture rate into one of the $M-m$ empty hole traps. The second term in the latter equation represents the rate of recombination between free electrons $n_c$ and trapped holes $m$.

The rate of change of the concentrations of free electrons and holes in the conduction and valence bands is:

$$\frac{dn_c}{dt} = \frac{dn_c^+}{dt} - \frac{dn_c^-}{dt} = G + np - n_c (N-n)\sigma_e v_e - n_c m \sigma_r v_e \quad (3.5)$$

and

$$\frac{dm_v}{dt} = \frac{dm_v^+}{dt} - \frac{dm_v^-}{dt} = G - m_v (M-m)\sigma_h v_h, \quad (3.6)$$

respectively.

For charge neutrality, $n_c + n = m_v + m$. During irradiation, $G > 0$ and typically $p = 0$, whereas immediately after irradiation $G = 0$, $p = 0$, and $n_c$ and $m_v$ decay to zero as the remaining free holes become trapped and the free electrons become either trapped or recombine with trapped holes. The decays are on a time scale of the respective free carrier lifetimes. During subsequent stimulation (during either TL or OSL) $G = 0$, $p > 0$, and $m_v = 0$. The last equality is a consequence of the OTOR model in which only electrons are stimulated from traps; the number of trapped holes $m$ can only decrease if free electrons recombine with them. As a result, the luminescence intensity emitted (TL or OSL) is proportional to the rate of recombination, thus:

$$I(t) = \eta \left| \frac{dm}{dt} \right| = \eta n_c m \sigma_r v_e = \eta n_c m B. \quad (3.7)$$

The constant $\eta$ accounts for the fact that not all recombination events may result in an emitted photon, a phenomenon that will be discussed in a later section. For now, $\eta$ can be considered to equal unity.

Equations 3.3 to 3.7 comprise a set of coupled, non-linear, first-order differential equations and their solution requires the introduction of a critically important assumption, known as the Quasi-Equilibrium (QE) assumption. The QE assumption may be written:

$$\left| \frac{dn_c}{dt} \right|, \left| \frac{dm_v}{dt} \right| << \left| \frac{dn}{dt} \right|, \left| \frac{dm}{dt} \right| \quad (3.8)$$

and

$$n_c, m_v \approx 0. \quad (3.9)$$

Taken together, these approximations mean that the rate of change of free electrons and holes is much smaller than the rates of change of trapped electrons and trapped holes, and that free electrons and holes do not accumulate in the respective delocalized bands. Thus, as soon as an electron is released from the trap, it either recombines with a trapped hole or it is retrapped.

During thermal or optical stimulation ($G = 0$, $p > 0$ and $m_v = 0$) Equations 3.3, 3.4, and 3.5 may be combined to give:

$$\frac{dn_c}{dt} = -\frac{dn}{dt} + \frac{dm}{dt} = np - n_c(N-n)\sigma_e v_e - n_c m \sigma_r v_e \approx 0 \tag{3.10}$$

at quasi-equilibrium. Luminescence results when recombination occurs and thus the intensity is proportional to $|dm/dt|$. With the introduction of an additional assumption of slow retrapping, that is, $n_c(N-n)\sigma_e v_e \ll n_c m \sigma_r v_e$, or more simply, $(N-n)\sigma_e \ll m \sigma_r$, then:

$$I(t) \propto \left|\frac{dm}{dt}\right| = n_c m \sigma_r v_e = np. \tag{3.11}$$

Furthermore, since

$$\frac{dn}{dt} = -np \tag{3.12}$$

under conditions of slow retrapping, then

$$n = n_0 \exp\{-tp\}, \tag{3.13}$$

and so

$$I(t) = n_0 p \exp\{-tp\} = I_0 \exp\{-tp\}. \tag{3.14}$$

Here $n_0 = n$ at $t = 0$ (i.e. the initial concentration of trapped electrons at the start of the stimulation) while $I_0$ is the initial luminescence intensity at $t = 0$. For constant $p$, Equation 3.14 is a simple exponential decay (Figure 3.1). Note that the inequality $n_c(N-n)\sigma_e v_e \ll n_c m \sigma_r v_e$ is

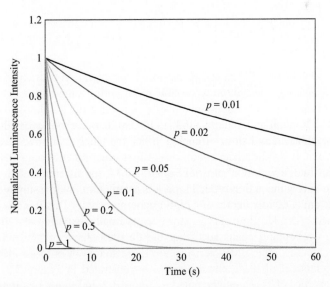

**Figure 3.1** Exponential decay of luminescence for fixed values of $p$, from 0.01 s$^{-1}$ to 1.0 s$^{-1}$. The luminescence intensity is normalized to a value of $I_0 = 1.0$ at $t = 0$. The stimulation may be thermal at a fixed temperature (isothermal luminescence decay) or optical at a fixed optical stimulation intensity (CW luminescence decay). In each case the decay constant $\tau = p^{-1}$.

a relationship between functions since $n$ and $m$ are both time (and temperature) dependent and the balance between the two rates will change as the trapped charge concentration changes during trap emptying.

In each case shown in Figure 3.1 the value of $p$ is given either by Equation 2.2 (thermal) or by Equation 2.12 (optical) from Chapter 2. For a given trap, a constant $p$ means a fixed temperature $T$ (isothermal excitation) for thermal stimulation, or a fixed stimulation light intensity $\Phi$ and wavelength $\lambda$ for optical stimulation. A constant light intensity is usually given the name "continuous wave" (CW) excitation.

## 3.2 Analytical Solutions: TL Equations

### 3.2.1 First-Order Kinetics

During a thermoluminescence experiment the sample is heated such that $T$ continuously increases. The usual heating scheme adopted is a linear increase in temperature with time in the following manner:

$$T(t) = T_0 + \beta_t t, \tag{3.15}$$

where $T_0$ is the temperature at the beginning of the heating (i.e. at $t = 0$) and $\beta_t$ is the constant heating rate. In principle, other heating schemes can also be adopted (e.g. exponential), but the linear scheme is the easiest to control experimentally in a reliable and reproducible fashion.

Specifying the luminescence intensity $I(t)$ to be the TL intensity $I_{TL}(t)$, Equation 3.11 becomes:

$$I_{TL}(t) = ns\exp\left\{-\frac{E_t}{kT}\right\}. \tag{3.16}$$

Changing the variable from $t$ to $T$ using Equation 3.15 and integrating from $T_0$ to $T$, the expression for the TL intensity as a function of temperature becomes:

$$I_{TL}(T) = n_0 s\exp\left\{-\frac{E_t}{kT}\right\}\exp\left\{-\frac{s}{\beta_t}\int_{T_0}^{T}\exp\left[-\frac{E_t}{k\theta}\right]d\theta\right\}, \tag{3.17}$$

where $n_0$ is the value of $n$ at $T = T_0$ and is the initial concentration of trapped electrons at the start of the linear heating; $\theta$ is a dummy variable representing temperature.

Equation 3.17 is an extremely important equation, known as the Randall-Wilkins First-Order TL equation, after Randall and Wilkins (1945a, 1945b). It is derived under the dual assumptions of quasi-equilibrium and slow retrapping. Since the latter assumption leads to Equation 3.12 and $dn/dt \propto n$, it is valid under conditions of first-order kinetics only.

Figure 3.2 illustrates a variety of plots of Equation 3.17, for different values of $E_t$, $s$, and $\beta_t$. The first thing to note is that a first-order TL peak is asymmetric, with a slowly increasing low-temperature side and a rapidly decreasing high-temperature side. This is a key observation for experimental, first-order TL glow curves. However, normally, experimental glow curves consist of multiple overlapping peaks such that the detailed shape of any individual peak is obscured. In such cases, other characteristics are to be sought in order to identify first-order peaks. An important set of these characteristics is illustrated in Figure 3.3, but first some fundamental properties of TL should be noted in Figure 3.2. In Figure 3.2a it may be observed that as the energy of the trap $E_t$ increases, so too does the temperature at which the TL maximum occurs $T_m$, for fixed values for $s$ and $\beta_t$. This is to be expected from Equation 2.2 (Chapter 2) where it may be seen that as $E_t$ increases, so the probability $p$ of release from the trap decreases.

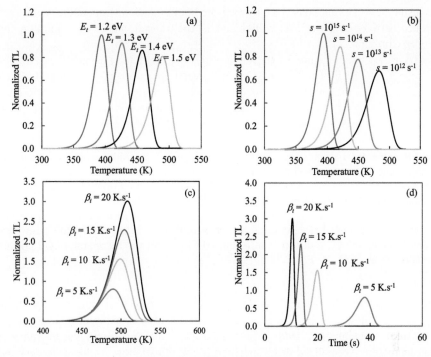

**Figure 3.2** First-order TL glow curves calculated using the Randall-Wilkins equation, Equation 3.17. (a) TL glow curves for fixed values of $s$ ($10^{15}$ s$^{-1}$) and heating rate $\beta_t$ (5 K.s$^{-1}$), but for different values of the trap depth $E_t$. (b) Fixed $E_t$ (1.3 eV) and $\beta_t$ (5 K.s$^{-1}$), but for different values of $s$. (c) Fixed $E_t$ (1.3 eV) and $s$ ($10^{15}$ s$^{-1}$), but for different heating rates $\beta_t$. (d) Same data as in figure (c), but as a function of heating time.

In Figure 3.2b it is observed that as $s$ increases, $T_m$ moves to lower temperatures. This too is predicted from Equation 2.2 since $p$ increases proportionally with $s$.

Less obvious is how the TL peak shape and position change with heating rate $\beta_t$. The relationship between the heating rate and the peak shape and position is shown in Figure 3.2c, for fixed $E_t$ and $s$. The peak shifts to higher temperature and gets larger as $\beta_t$ increases. A clearer understanding of the relationship between $\beta_t$ and the TL curve can be seen in Figure 3.2d in which the TL peak is represented as a function of time $t$, using the transformation $\beta_t = dT/dt$. At higher heating rates, the temperatures required for release of the trapped electrons are reached more quickly, and the temperature range over which the electron release is completed is passed through more quickly, compared to low heating rates. Hence, at high heating rates the peak occurs at lower times and is narrower than those observed at lower heating rates.

From Equation 3.11 it may be observed that the area under the TL peak, as a function of time, should be a constant and independent of heating rate, i.e.:

$$\int_0^t I(t)\,dt = m_0,\tag{3.18}$$

where $m_0 = n_0$ is the concentration of trapped charges at the beginning of heating (for the OTOR case). Thus, the areas under the different TL peaks in Figure 3.2d are equal. This is not the case in Figure 3.2c where the areas scale with the heating rate used, i.e.:

$$\int_0^T I(T)\,dT = \beta_t m_0.\tag{3.19}$$

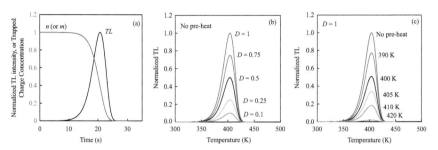

**Figure 3.3** First-order TL glow curves: (a) Comparison of the position and shape of the TL peak with the change in the trapped charge concentration (electrons $n$, or holes $m$) for the OTOR model. (b) Changes in the TL peak as a function of dose, in arbitrary dose units. The peak maintains the same shape (but of different intensities) and the same position as the dose increases. (c) Changes in the TL peak as a function of pre-heat temperature after irradiation. If the sample is pre-heated to the temperatures shown, before cooling and then re-heating to record the full glow curve, the resultant TL peak will be reduced in size, but will maintain the same shape and position. The parameters values used are: $E_t = 1.23$ eV, $s = 10^{15}$ s$^{-1}$, and $\beta_t = 5$ K.s$^{-1}$.

Figure 3.3a illustrates the relationship of the first-order peak position and shape and the trapped charge concentration $n$ (or $m$). The peak position occurs at the maximum slope of the decrease in $m$ (i.e. at the maximum of d$m$/d$t$). Figure 3.3b shows the changes seen as the initial applied dose is increased. Normalized doses are used in this illustration. As the dose decreases, the peak changes intensity accordingly, but maintains the same shape and position. The same attribute can be observed by pre-heating the sample after irradiation. The pre-heat partially depletes the trapped charge concentration. If the sample is cooled after the pre-heat, and then re-heated to record the full, remaining, TL glow curve, the observed peak is reduced in size, but again maintains the same shape and position. In principle, the characteristics shown here can be experimentally verified and used as a test for first-order kinetics.

The calculations required to produce Figures 3.2 and 3.3 require a computation of the integral $\int_{T_0}^{T} \exp\left[-\dfrac{E_t}{k\theta}\right] d\theta$ in Equation 3.17. For these particular example calculations the method described by Kitis et al. (1998) was used, where:

$$\int_{T_0}^{T} \exp\left[-\frac{E_t}{k\theta}\right] d\theta \approx \frac{kT^2}{E_t} \exp\left\{-\frac{E_t}{kT}\right\} \left(1 - \frac{2kT}{E_t}\right). \tag{3.20}$$

However, other approximations to the integral exist in the literature and several are summarized in Bos et al. (1993), Chen and McKeever (1997), and elsewhere.

---

**Exercise 3.1   The Exponential Integral**

Figures 3.2 and 3.3 use the method described by Kitis et al. (1998), for the evaluation of the exponential integral term in Equation 3.17, namely

$$F\left(E_t, T\right) = \int_{T_0}^{T} \exp\left[-\frac{E_t}{k\theta}\right] d\theta \approx \frac{kT^2}{E_t} \exp\left\{-\frac{E_t}{kT}\right\} \left(1 - \frac{2kT}{E_t}\right).$$

However, some other approximations to the integral exist in the literature and are summarized by Chen and McKeever (1997), Chapter 6, and Bos et al. (1993). Alternatively, (for example, Kitis and Pagonis 2019) the integral may be approximated using the exponential integral function Ei[..], which is available in many modern software packages, thus:

$$F(E_t,T) = \int_{T_0}^{T} \exp\left[-\frac{E_t}{k\theta}\right] d\theta = T\exp\left\{-\frac{E_t}{kT}\right\} + \frac{E_t}{k} \text{Ei}\left[-\frac{E_t}{kT}\right]$$

Search the literature (find the original papers) and select 4 or 5 of the integral approximations given in the publications. Include the approximation of Kitis et al. (1998) as one of the 4 or 5 approximations; you are free to choose the remainder.

(a) Plot each approximation of $F(E_t,T)$ as a function of $T$ for an activation energy $E_t = 1.5$ eV. Comment on your observations.
(b) For each of these approximations, evaluate and plot the shapes of a 1st-order TL peak using $E_t = 1.5$ eV, $s = 10^{12}$ s$^{-1}$ and $\beta_t = 1$ K.s$^{-1}$. Are they identical?
(c) Numerically integrate the area under the curve (using Excel or a similar program) to show how $n(T)$ $(= m(T))$ varies compared to $I_{TL}(T)$.
(d) Using a suitable curve-fitting program (e.g. Solver in Excel; Afouxenidis et al. (2012)) and fit each of the curves obtained in part (b) to the first-order expression obtained when using the Kitis et al. (1998) approximation. What errors, if any, are induced in the evaluated $E_t$ and $s$ values? (This exercise will give you a feel for the effect of the chosen approximation for the integral on the accuracy of the results.)
(e) Repeat part (b), but vary $n_o$ (3 or 4 different values) and re-plot the TL peaks to verify that the position and shape of the peak are independent of the initial trapped charge concentration, as shown in Figure 3.3b.

### 3.2.2 Second-Order and General-Order Kinetics

A critical assumption in the foregoing analysis was that of slow retrapping, namely $(N-n)\sigma_e << m\sigma_r$. If the opposite is true, namely $(N-n)\sigma_e >> m\sigma_r$, retrapping will dominate over recombination. This case, known as fast retrapping, is perhaps a strange condition since, when taken alongside the QE approximation, it might be expected that the electrons would never escape the traps since, once they have done so, they are immediately retrapped. However, the assumption does not require that the recombination term $m\sigma_r = 0$, and always some charge will recombine as the temperature increases, especially as $np = ns\exp\{-E_t / kT\}$ rapidly increases as the temperature rises.

Garlick and Gibson (1948) examined this case and demonstrated that the dual conditions of fast retrapping and QE lead to an expression for TL, thus:

$$I_{TL}(T) = \frac{n_0^2 s' \exp\{-E_t / kT\}}{\left[1 + \left(\frac{n_0 s'}{\beta_t}\right) \int_{T_0}^{T} \exp\{-E_t / k\theta\} d\theta\right]^2}. \tag{3.21}$$

This is known as the Garlick-Gibson Second-Order TL equation, in which:

$$s' = \frac{s\sigma_r}{N\sigma_e} = \frac{s}{RN}, \tag{3.22}$$

with $R = \sigma_e/\sigma_r$. Note that the term $s'n_0$ in Equation 3.21 replaces $s$ in Equation 3.17, and is known as an "effective" pre-exponential factor. The main consequence of this is that the

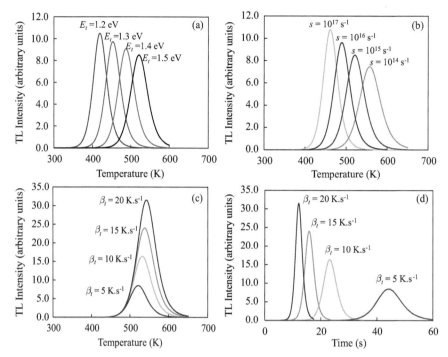

**Figure 3.4**  Second-order TL glow curves calculated using the Garlick-Gibson equation, Equation 3.21. (a) TL glow curves for fixed values of $s$ ($10^{15}$ s$^{-1}$), heating rate $\beta_t$ (5 K.s$^{-1}$) and $n_0/N$ (0.1), but for different values of the trap depth $E_t$. (b) Fixed $E_t$ (1.5 eV), $\beta_t$ (5 K.s$^{-1}$) and $n_0/N$ (0.1), but for different values of $s$. (c) Fixed $E_t$ (1.5 eV), $s$ ($10^{15}$ s$^{-1}$) and $n_0/N$ (0.1), but for different heating rates $\beta_t$. (d) Same data as in figure (c), but as a function of heating time.

position of the peak will depend on $n_0$. The effective pre-exponential factor may be written $s'' = s'n_0$, which has units of s$^{-1}$.

The shape of a second-order TL peak, following Equation 3.21, is shown in Figure 3.4, where the integral is again approximated using the expression of Kitis et al. (1998). To be noted is that the second-order TL peak is much more symmetrical compared to the first-order case. This is to be expected since retrapping "delays" the recombination event and so the TL peak appears spread out over a wider temperature range. Figure 3.4a illustrates a variety of second-order TL peaks for a range of $E_t$ values, with $s = 10^{15}$ s$^{-1}$, $\beta_t = 5$ K.s$^{-1}$ and $n_0/N = 0.1$. Figure 3.4b illustrates the change in peak shape for fixed $E_t = 1.3$ eV (and $\beta_t = 5$ K.s$^{-1}$ and $n_0/N = 0.1$) but variable $s$; Figure 3.4c shows how the peak shape and position will change as a function of heating rate; and Figure 3.4d shows the same data as Figure 3.4c, but as a function of heating time. Again, the area under the different peaks in Figure 3.4d is the same, and is proportional to $n_0 = m_0$.

Following the discussion of the first-order case, Figure 4.5 shows the relationship between $n(t)$ (or $m(t)$) and the TL peak for second-order kinetics. In Figure 3.5b it can be observed that the second-order peak shifts to lower temperature as the dose $D$ increases, whereas Figure 3.5c similarly shows that if the sample was to be pre-heated to the temperatures shown, the resulting TL peak position shifts to higher temperature as the size of the peak decreases. This is to be contrasted to the behavior of a first-order peak in Figure 3.3 and is a result of the term $s'n_0$ replacing $s$ in the expression for TL. Taken together, the figures show the characteristic behaviors of 1st- and 2nd-order TL peaks that can be used to distinguish between them experimentally.

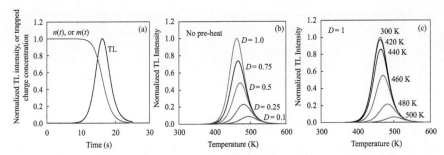

**Figure 3.5** Second-order TL glow curves: (a) Comparison of the position and shape of the TL peak with the change in the trapped charge concentration (electrons *n*, or holes *m*) for the OTOR model. (b) Changes in the TL peak as a function of dose, in arbitrary dose units. The peak shifts to lower temperatures and increases in size as the dose increases. (c) Changes in the TL peak as a function of pre-heat temperature after irradiation. If the sample is pre-heated to the temperatures shown, before cooling and then re-heating to record the full glow curve, the resultant TL peak is reduced in size and shifts to higher temperature.

---

**Exercise 3.2    The General One-Trap (GOT) Equation**

(a) Under Quasi-Equilibrium Equation 3.10 means that $np = n_c (N - n)\sigma_e v_e + n_c m\sigma_r v_e$. Since $I_{TL}(t) \propto |dm/dt| = n_c m\sigma_r v_e$, show that:

$$I_{TL}(t) = \frac{ns\exp\left\{-\dfrac{E_t}{kT}\right\} m\sigma_r}{\left[(N-n)\sigma_e + m\sigma_r\right]} \qquad (A)$$

and that:

$$I_{TL}(t) = ns\exp\left\{-\frac{E_t}{kT}\right\}\left[1 - \frac{(N-n)\sigma_e}{(N-n)\sigma_e + m\sigma_r}\right] \qquad (B)$$

(b) Equations (A) and (B) are two different forms of what is known as the GOT equation. Apply the slow-retrapping assumption to either equation and show that:

$$I_{TL}(T) = n_0 s\exp\left\{-\frac{E_t}{kT}\right\}\exp\left\{-\frac{s}{\beta_t}\int_{T_0}^{T}\exp\left[-\frac{E_t}{k\theta}\right]d\theta\right\},$$

where $\beta_t = dT/dt$, and $n_0 = n$ at $t = t_0$.

(c) Similarly, apply the fast-retrapping assumption and demonstrate that

$$I_{TL}(T) = \frac{n_0^2 s'\exp\left\{-E_t/kT\right\}}{\left[1 + \left(\dfrac{n_0 s'}{\beta_t}\right)\displaystyle\int_{T_0}^{T}\exp\left\{-E_t/k\theta\right\}d\theta\right]^2},$$

where $s' = \dfrac{s\sigma_r}{N\sigma_e}$.

The Randall-Wilkins TL equation (Equation 3.17) is derived from the QE and slow retrapping $(N-n)\sigma_e << m\sigma_r$ assumptions, leading to the expression given in Equation 3.16, namely:

$$I_{TL}(t) = \left|\frac{dn}{dt}\right| = ns\exp\left\{-\frac{E_t}{kT}\right\}. \tag{3.16}$$

That is, the rate of trap emptying $dn/dt$ is proportional to the concentration of trapped charge $n$, showing that the kinetics are first-order.

Similarly, the Garlick-Gibson TL expression (Equation 3.21) is derived from the QE and fast retrapping $n_c(N-n)\sigma_e v_e >> n_c m\sigma_r v_e$ assumptions. If it is further assumed that $N >> n$, the TL intensity is given by:

$$I_{TL}(t) = \left|\frac{dn}{dt}\right| = n^2 s'\exp\left\{-\frac{E_t}{kT}\right\} \tag{3.23}$$

where $s' = \dfrac{s\sigma_r}{N\sigma_e}$. Here, the rate of trap emptying is proportional to the square of the trapped

charge concentration and the kinetics are second-order. However, it is sometimes observed that TL peak shapes can be described by neither the Randall-Wilkins first-order kinetics equation (Equation 3.17) nor the Garlick-Gibson second-order kinetics equation (Equation 3.21). This prompted May and Partridge (1964) to devise a purely empirical formula, termed the general-order kinetics TL equation, thus:

$$I_{TL}(t) = \left|\frac{dn}{dt}\right| = n^b s'\exp\left\{-\frac{E_t}{kT}\right\}, \tag{3.24}$$

where $b$ is termed the order-of-kinetics and $s'$ has the dimensions of $m^{3(b-1)}s^{-1}$. The "general order" parameter $b$ can have a value of neither 1 nor 2. Integration of Equation 3.24 for $b \neq 1$ yields:

$$I_{TL}(T) = \frac{s''n_0\exp\{-E_t / kT\}}{\left[1+(b-1)\dfrac{s''}{\beta_t}\displaystyle\int_{T_0}^{T}\exp\{-E_t / k\theta\}d\theta\right]^{\frac{b}{b-1}}}, \tag{3.25}$$

where now $s'' = s'n_0^{(b-1)} = \dfrac{s}{N}n_0^{(b-1)}$.

Use of Equation 3.25, known as the General-Order TL equation, is popular when experimental TL peaks appear to be described by neither the first-order nor the second-order expressions. It is easy to imagine that there may be combinations of values for $N$, $n$, $m$, $\sigma_r$ and $\sigma_e$, for which neither the slow- nor the fast-retrapping assumptions are valid. Nevertheless, it is to be emphasized that there is no theoretical basis for Equations 3.24 and 3.25 and the general-order expression is entirely empirical. It is also burdened with strange dimensions for $s'$, which change with kinetic order.

It has been proposed (Rasheedy 1993) that Equation 3.24 be recast as:

$$I_{TL}(t) = \left|\frac{dn}{dt}\right| = \left(\frac{n^b}{N^{b-1}}\right)s\exp\left\{-\frac{E_t}{kT}\right\}, \tag{3.26}$$

which reduces to the first- and second-order cases when $b = 1$ or $b = 2$, respectively. Integration now gives:

$$I_{TL}(t) = \frac{n_0^b s \exp\left\{-\dfrac{E_t}{kT}\right\} N^{(1-b)}}{\left[1 + \dfrac{s(b-1)(n_0/N)^{(b-1)}}{\beta_t} \displaystyle\int_{T_0}^T \exp\left\{-\dfrac{E_t}{k\theta}\right\} d\theta\right]^{\frac{b}{b-1}}}. \tag{3.27}$$

This too is an empirical expression, but it avoids the difficulty of understanding the meaning of $s'$.

Finally, it should be noted that Equations 3.25 and 3.27 both give the second-order TL equation when $b = 2$, and, in the limit, reduce to the first-order equation as $b \to 1$.

Figure 3.6 illustrates several TL peak shapes as predicted from Equation 3.25 for different values of the general-order parameter $b$, ranging from $b = 1$ to $b = 2$. (Note: the first-order peak was obtained using Equation 3.17.) The parameters were chosen such that the peak position is the same for all peaks and the areas under the curves are equal. It was discussed with reference to Figure 3.5 that the position of the peak for second-order kinetics shifts if the initial concentration of trapped charge changes due to the corresponding change in the effective frequency factor $s'$. In a similar way, the position of the peak for values of $b>1$ will also shift because of the change in the effective frequency factor. The first-order kinetics TL peak, however, will not change with variations in $n_0$ (as shown in Figure 3.3).

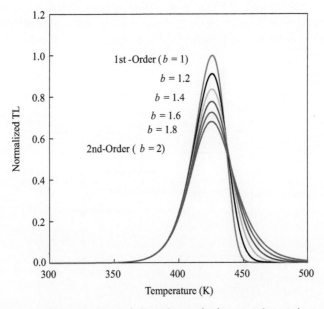

**Figure 3.6**   Changes in shape of a TL peak from first-order kinetics ($b = 1$) through general-order kinetics ($b = 1.2$, 1.4, 1.6, and 1.8) to second-order kinetics ($b = 2$). $E_t = 1.3$ eV, $s = 10^{15}$ s$^{-1}$, $\beta_t = 5$ K.s$^{-1}$, $\sigma_e / \sigma_t = 1$, and $n_0/N = 1$.

---

**Exercise 3.3    Shift in Peak Position with Dose**

Bos and Dielhof (1991) derived the following approximate[1] expression for the shift in peak position $\Delta T$ with dose:

$$\Delta T = T_1 - T_2 \approx T_1 T_2 \frac{k(b-1)}{E_t} \ln(f) \tag{1}$$

where $T_1 = T_m$ is the peak maximum position at dose $D_1$ and $T_2$ is the position at dose $D_2$, where $D_2 > D_1$ and $T_2 < T_1$. Parameter $f$ is the fractional change in dose, $f = D_2/D_1$. Clearly, if $b = 1$ (first-order kinetics), $\Delta T = 0$ and there is no shift in peak position with dose. The equation shows that the position shift is significant, depending on the value of $b$.

(a)  Using Equation 3.25, evaluate several TL curves for dose values ranging from $D = 1$ to $D = 100$ for $b = 2$.
(b)  Apply Equation 1 above and determine how accurately it describes the peak shift.
(c)  Repeat the calculation using different values for $b$, namely $b = 1.25, 1.5$, and $1.75$.

---

[1]  Bos and Dielhof (1991) derive Equation 1 for a hyperbolic heating rate. It is accurate for a linear heating rate for large values of $f$.

---

### 3.2.3    Mixed-Order Kinetics

The primary difficulty with using Equation 3.25 or Equation 3.27 is that both equations are empirical, with no theoretical definition of the value of $b$, other than to simply state that the combination of parameter values is such that neither the slow-retrapping nor the fast-retrapping assumption is valid. This has prompted the development of mixed-order kinetics, as follows.

The OTOR model of Figure 2.7 may be modified by the addition of a second trap, which is stable over the temperatures at which the first trap empties. This is illustrated in Figure 3.7, and the second trap is described as "thermally disconnected" from the first. In this figure, the

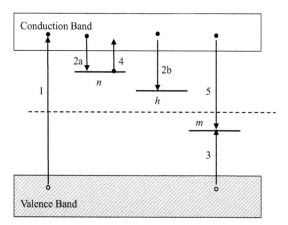

**Figure 3.7**  A model including thermally disconnected traps. Electrons may be trapped in the thermally active trap (transition 2a) or the deeper, thermally disconnected trap (transition 2b). "Thermally disconnected" means that electrons $h$ trapped in the deeper trap are stable over the same temperature range that electrons $n$ in the shallower trap are released. After irradiation, charge neutrality means that $n + h = m$.

second "deeper" trap is able to trap free electrons (transition 2b) in the same way as the first trap (transition 2a), but a thermally stimulated detrapping transition is not allowed (i.e., there is no transition equivalent to transition 4).

In the OTOR model after irradiation and before heating, charge neutrality means that the number of trapped electrons equals the number of trapped holes, i.e. $m = n$. In Exercise 3.2, the GOT Equation was developed, thus:

$$I_{TL}(t) = \frac{ns\exp\left\{-\dfrac{E_t}{kT}\right\}m\sigma_r}{\left[(N-n)\sigma_e + m\sigma_r\right]}.$$

(3.28)

With $m = n$, and $\sigma_e / \sigma_r = R$, this becomes:

$$I_{TL}(t) = \frac{n^2 s\exp\left\{-\dfrac{E_t}{kT}\right\}}{\left[R(N-n)+n\right]}.$$

(3.29)

With the addition of a thermally disconnected trap, charge neutrality now means the $m = n + h$ and Equation 3.29 becomes:

$$I_{TL}(t) = \frac{n(n+h)s\exp\left\{-\dfrac{E_t}{kT}\right\}}{\left[n+h+R(N-n)\right]}.$$

(3.30)

Slow retrapping in this case means that:

$$R << \frac{n+h}{N-n},$$

(3.31)

which, if applied to Equation 3.30, leads to the first-order expression and the Randall-Wilkins TL Equation 3.17.

However, if fast retrapping applies:

$$R >> \frac{n+h}{N-n}$$

(3.32)

and

$$I_{TL}(t) = \frac{n(n+h)s\exp\left\{-\dfrac{E_t}{kT}\right\}}{R(N-n)}.$$

(3.33)

With $N >> n$, this becomes:

$$I_{TL}(t) = \frac{n(n+h)s\exp\left\{-\dfrac{E_t}{kT}\right\}}{RN}.$$

(3.34)

Alternatively, with $R = 1$, Equation 3.30 becomes:

$$I_{TL}(t) = \frac{n(n+h)s\exp\left\{-\dfrac{E_t}{kT}\right\}}{N+h}.$$

(3.35)

Equations 3.34 and 3.35 can both be represented in the form:

$$I_{TL}(t) = n(n+h)s'\exp\left\{-\frac{E_t}{kT}\right\}, \tag{3.36}$$

where $s'$ is given by $s' = s / RN$, or $s' = s / (N+h)$, respectively.

Expansion of Equation 3.36 yields:

$$I_{TL}(t) = n^2 s'\exp\left\{-\frac{E_t}{kT}\right\} + nhs'\exp\left\{-\frac{E_t}{kT}\right\}. \tag{3.37}$$

The first term on the left-hand side looks like second-order kinetics while the second term looks like first-order kinetics. Thus, this expression has been termed "mixed-order" kinetics. If $h \ll n$, the first term dominates and the expression reduces to purely second order, whereas if $h \gg n$, it becomes purely first order.

The solution of Equation 3.37 is:

$$I_{TL}(T) = \frac{s'h^2\alpha\exp\left\{(hs'/\beta_t)\int_{T_0}^{T}\exp\{-E_t/k\theta\}d\theta\right\}\exp\{-E_t/kT\}}{\left[\exp\left\{(hs'/\beta_t)\int_{T_0}^{T}\exp\{-E_t/k\theta\}d\theta\right\}-\alpha\right]^2}, \tag{3.38}$$

where $\alpha = n_0 / (n_0 + h)$. Equation 3.38 becomes identical to the Randall-Wilkins equation (Equation 3.17) if $h \gg n_0$ (and $\alpha \to 0$), and identical to the Garlick-Gibson equation (Equation 3.21) if $h \ll n_0$ (i.e. when $\alpha = 1$).

From the foregoing analyses it may be observed that, under conditions of QE, a TL peak shape can be described by Equations 3.17 (first order), 3.21 (second order), 3.25 (general order) or 3.38 (mixed order). The shape, size, position (temperature of peak maximum), and behavior (changes as a function of initial trapped charge concentration and/or heating rate) can be described by the parameters $E_t$, $s$ (or $s'$ or $s''$), $\beta_t$, $n_0$ and $b$ (or $\alpha$). Thus, for a given $n_0$ and $\beta_t$, an experimental TL peak can be described by one or other of the above expressions with suitable values for $E_t$, $s$ (or $s'$ or $s''$), and $b$ (or $\alpha$).

Some further insight into the meaning of $b$ can be obtained by equating the general-order expression, Equation 3.24 with Equation 3.30, to give:

$$n^b s' = ns\gamma, \tag{3.39}$$

where

$$\gamma = \frac{(n+h)}{[n+h+R(N-n)]}. \tag{3.40}$$

With $s' = n_0^{1-b}s$, and $n_0 = N$, then $b$ can be shown as:

$$b = \frac{\ln(\gamma n / N)}{\ln(n / N)}. \tag{3.41}$$

As noted earlier, the value of $b$ depends upon the degree of trap filling $n/N$, which changes as the trap empties. Attempts to fit experimental TL peaks to the general-order expression Equation 3.25 assuming a constant value for $b$ are therefore problematic and the meaning of the obtained value for $b$ is unclear.

Similarly, from the expressions for $\gamma$ (Equation 3.40) and for $\alpha$, the assumption that $R = 1$, leads to:

$$\alpha = \frac{n_0}{\gamma(N+h)}.$$ (3.42)

As trap emptying proceeds and the value of $\gamma$ changes, so too does $\alpha$. Furthermore, if $h \gg n_0$, then $\alpha = n_0/h \to 0$, and $\gamma = h/(N + h)$. Alternatively, if $n_0 \gg h$, then $\alpha = 1$ and $\gamma = n_0/(N + h)$.

The analyses presented so far have started with assumptions concerning the retrapping rates (e.g. fast or slow retrapping) and derived the shape and character of the resulting TL peak. It is valid to ask the reverse question – i.e. if a TL peak is obtained with a certain shape, can one infer the relative retrapping rates? Specifically, if a TL peak has a first-order shape, does this always mean that retrapping is much slower than recombination? Are the two synonymous? The answer to this question will be deferred until a later section when models more complex than the OTOR model are analyzed (see Section 3.4).

---

### Exercise 3.4    General-Order and Mixed-Order Kinetics

(a) Use the general-order expression for TL (Equation 3.25) and plot TL curves for different values of $n_o$ and $b$. How does the peak shape depend on $n_0$ for fixed $b$?
(b) How does the peak shape depend on $b$ for fixed $n_0$?
(c) Do you understand the differences between $s$, $s'$ and $s''$ and their relationship to $n_0$ and $b$?
(d) Using the curves you obtained in parts (a) and (b), and considering these to be experimental TL peaks, fit them with the mixed-order expression (Equation 3.38). What values of the parameter $\alpha$ do you obtain for the different values of $b$? Does $\alpha \to 0$ as $b \to 1$? Does $\alpha = 1$ when $b = 2$?

---

## 3.3    Analytical Solutions: OSL Equations

The OTOR model can also be used to predict the shape of OSL curves. Examination of OSL curve shapes starts with Equation 2.12 (Chapter 2) describing the probability of optical release of trapped charge from a trap, namely:

$$p = \Phi\sigma_p(E).$$ (2.12)

As noted in Section 2.2.1.2 (Chapter 2), the photoionization cross-section $\sigma_p$ is dependent on the energy $E$ of the stimulation light, i.e. it is a function of the wavelength of the stimulating light, $\lambda$. Thus, $p$ can be varied in time by either changing the intensity of the stimulating light:

$$\Phi = \Phi_0 + \beta_\Phi t,$$ (3.43a)

or by changing the wavelength:

$$\lambda = \lambda_0 - \beta_\lambda t,$$ (3.43b)

where $\lambda_0$ and $\Phi_0$ are the initial values at $t = t_0$. The wavelength is, of course, energy dependent $(\lambda(E))$

If $\beta_\Phi = 0$ and $\beta_\lambda = 0$ (i.e. if $p$ is a constant) then the OSL measured is termed CW-OSL (where CW = Continuous Wave) and light of constant wavelength and constant intensity is used to stimulate the luminescence emission (OSL).

However, if $\beta_\Phi$ is a constant and $> 0$, then $p$ is time variant according to Equation 2.12 and Equation 3.43a. In this circumstance the OSL is recorded as LM-OSL (where LM = Linear Modulation). Alternatively, if $\Phi$ is pulsed (e.g. in a square-wave pattern), then the OSL is termed Pulsed OSL (POSL).

If $\beta_\Phi = 0$ but the incident wavelength is scanned ($\beta_\lambda > 0$) the measurement is called OSL stimulation (or excitation) spectroscopy. A variation of this is to scan the incident energy (while keeping the intensity constant). This is termed VE-OSL (where VE = Variable Energy) and the energy is increased linearly according to:

$$E = E_0 + \beta_E t, \tag{3.43c}$$

where $E_0$ is the initial energy of the stimulating light and $\beta_E$ is the rate at which the energy in increased.

Each of these stimulation schemes is shown schematically in Figure 3.8.

Thus, OSL offers a range of experimental options, each of which has its own advantages and degrees of experimental difficulty. VE-OSL is perhaps the most difficult, since the number of photons per unit area per unit time falling on the sample has to be the same at all energies. Also, increasing the energy $E$ is implemented in practice by decreasing the wavelength $\lambda$. In order for the energy to change linearly, $\lambda$ has to change nonlinearly according to $E = hc / \lambda$. Furthermore, the wavelength bandwidth $\Delta\lambda$ has to be continually changed in order for the energy bandwidth $\Delta E$ to remain constant at each wavelength. For these reasons, VE-OSL is not performed often.

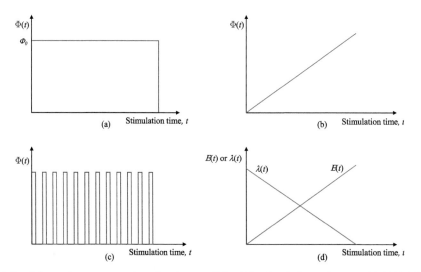

**Figure 3.8**   Schematic representations for the possible OSL stimulation schemes. (a) CW-OSL, with constant $\Phi(t) = \Phi_0$ throughout the stimulation. (b) LM-OSL, with a linearly increasing stimulation intensity, $\Phi = \Phi_0 + \beta_\Phi t$ at a fixed wavelength and constant $\beta_\Phi$. (c) POSL, with stimulation pulses of pulse width $t_p << \tau_d$, where $\tau_d$ is the luminescence decay time between stimulation and emission. (d) VE-OSL, where the energy of the stimulation light is increase linearly, according to $E(t) = E_0 + \beta_E t$. Also shown is the normal scheme adopted during OSL stimulation spectroscopy where the wavelength of the stimulating light is decreased according to $\lambda = \lambda_0 - \beta_\lambda t$.

### 3.3.1 First-Order Kinetics

Under conditions of QE, and slow retrapping, it was demonstrated above that:

$$\frac{dn}{dt} = -np \tag{3.12}$$

and therefore

$$I_{OSL}(t) = n_0 p \exp\{-tp\} = n_0 \sigma_p(E)\Phi \exp\{-t\sigma_p(E)\Phi\}. \tag{3.14}$$

The initial OSL intensity at the start of the decay is $I_0 = n_0\sigma_p(E)\Phi$.

#### 3.3.1.1 Expressions for CW-OSL

For CW-OSL, $p$ is a constant and so a CW-OSL curve $I_{CW}(t)$ for the OTOR model is a simple exponential decay (Equation 3.14) during the stimulation period (illustrated in Figure 3.9a).

#### 3.3.1.2 Expressions for LM-OSL

For LM-OSL, $p$ is non-constant according to:

$$p = \sigma_p(E)(\Phi_0 + \beta_\Phi t) \tag{3.44}$$

(from Equations 2.12 and 3.43a). If $\Phi$ increases linearly from $\Phi_0 = 0$ at $t = 0$ to $\Phi_m$ at $t = t_m$, then $\beta_\Phi = \Phi_m / t_m$ and

$$\frac{dn}{dt} = -n\sigma_p(E)\frac{\Phi_m}{t_m}t \tag{3.45}$$

the solution of which is:

$$I_{LM-OSL}(t) = n_0\sigma_p(E)\frac{\Phi_m}{t_m}t\exp\left\{\sigma_p(E)\frac{\Phi_m}{2t_m}t^2\right\}. \tag{3.46}$$

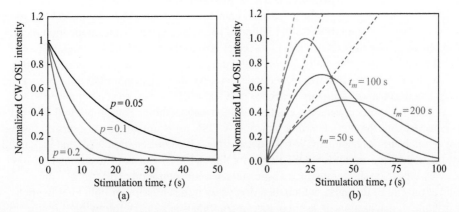

**Figure 3.9** (a) Normalized CW-OSL curves of three values of the stimulation probability $p$ (arbitrary units). (b) LM-OSL curves for three values of the rate of change in stimulation intensity, defined as $\Phi_m / t_m$, with $\Phi_m = 0.1$ arbitrary units and $t_m$ values as shown on the figure. Also shown in each case is the initial linear increase in the LM-OSL intensity, given by $n_0\sigma(E_o)\frac{\Phi_m}{t_m}t$ (dashed lines). The data have been normalized to a maximum of 1.0 for the case of $t_m = 50$ s.

Equation 3.46 has the form of a peak, with the LM-OSL intensity increasing linearly initially from $t = 0$ towards a maximum, before decreasing as the depletion of the trap concentration becomes dominant (Bulur 1996). The shape of $I_{LM}(t)$ is illustrated in Figure 3.9b.

---

**Exercise 3.5    LM-OSL**

(a) Using Equation 3.46 for LM-OSL, show that the time $t_{max}$ at which the LM-OSL is a maximum is given by:

$$t_{max} = \sqrt{\frac{t_m}{\sigma_p(E)\Phi_m}} \tag{A}$$

and that the LM-OSL intensity $I_{max}$ at this point is:

$$I_{max} = n_o\sqrt{\frac{\sigma_p(E)\Phi_m}{t_m}}\exp\left\{-\frac{1}{2}\right\} = \frac{n_o}{t_{max}}\exp\left\{-\frac{1}{2}\right\} \tag{B}$$

(b) Use the transformation: $u = \sqrt{2tt_m}$, or time, $t = \dfrac{u^2}{2t_m}$ and show that

$$I(u) = \left|\frac{dn}{du}\right| = \left|\frac{dn}{dt}\right|\left|\frac{dt}{du}\right| = n_0\sigma_p(E)\frac{\Phi}{t_m}u\exp\left\{-\sigma_p(E)\frac{\Phi}{t_m}u^2/2t_m\right\}. \tag{C}$$

This is identical in form to Equation 3.46, but with $u$ instead of $t$ and $\Phi$ instead of $\Phi_m$.

(c) Use a spreadsheet to plot a first-order CW-OSL curve using Equation 3.14. Then use the transformation (C) to convert the CW-OSL curve into the form of an LM-OSL curve. Such transformations are termed "pseudo-LM-OSL" curves (Bulur 2000).

---

### 3.3.1.3    *Expressions for POSL*

In CW-OSL mode, the emission appears to start almost immediately when the stimulation starts (Figure 3.9a). Actually, there is a delay between stimulation and emission due to several factors. The first is the time for electrons to be stimulated into the conduction band, given by $p^{-1} = 1/\Phi\sigma_p(E)$. Then there is the lifetime of the free electrons in the conduction band before recombination; for the OTOR model this is the same as the recombination lifetime, i.e. $\tau = \tau_{rec}$. (For the model with a thermally disconnected deep trap (TDDT) it is the net lifetime due to the probability of trapping into the TDDT and the probability of recombination.) Another delay is caused by the characteristic luminescence lifetime $\tau_e$ of the excited state of the emitting center after recombination between the electron and the trapped hole. Thus, there will be a net delay $\tau_d$ caused by all these effects. In a typical CW-OSL measurement, this occurs on a timescale that is much shorter than the CW-OSL decay time, given in Equation 3.14, and is ignored, but the method of Pulsed OSL (POSL) exploits this delay by pulsing the stimulation light with pulse widths much shorter than $\tau_d$. This results in a concentration of electrons in the conduction band at the end of the pulse, which then decays with the emission of light after the pulse, and/or a finite concentration of excited state recombination centers following the electron-hole recombination, depending on which is larger, $\tau_{rec}$ or $\tau_e$.

Depending on the material, one of the above parameters may dominate over the others. For example, in $Al_2O_3$:C, recombination occurs at $F^+$-centers, creating excited $F$-centers ($F^*$, i.e. $e^- + F^+ = F^*$). The slowest mechanism in this process is the relaxation from the excited state of

$F^*$ to its ground state accompanied by characteristic luminescence (i.e. OSL emission). The excited state lifetime $\tau_e$ in this case is 35 ms, due to it being a spin-forbidden transition. Thus, $\tau_d \approx \tau_e$ and POSL from this material uses pulse widths $t_p <<$ 35 ms (Akselrod and McKeever 1999). Under these circumstances the OSL emitted during each pulse is a rising curve following:

$$I_{POSL}(t) = I_0\left(1 - \exp\{-t/\tau_d\}\right), \tag{3.47}$$

where $I_0$ is the POSL intensity that would be obtained for long-enough stimulation pulse widths and is proportional to the peak stimulation power $\Phi$, i.e. $I_0 = n_0\Phi\sigma_p(E)$. It is that value that is attained in a CW-OSL measurement at the start of the CW-OSL decay (assuming the same recombination process).

The OSL emitted between the pulses follows:

$$I_{POSL}(t) = I_{t_p}\exp\{-t/\tau_d\}. \tag{3.48}$$

Here, $I_{t_p}$ is the value of $I_{POSL}$ obtained at the end of the stimulation pulse. It is given by Equation 3.47 at $t = t_p$. By ensuring that $t_p << \tau_d$, it can be arranged that much more luminescence is emitted after the pulse than is emitted during it. Experimentally, this feature means that the OSL can be efficiently detected between pulses when the stimulation beam is off, in contrast to both CW-OSL and LM-OSL when, in both cases, the OSL is detected while the stimulation beam is on.

Figure 3.10 illustrates this schematically where three stimulation pulse widths are shown such that the total energy delivered during each pulse $\Phi t_p$ is the same, but the pulse width $t_p$ is different. It can be seen that the ratio of the net emission after the pulse (area A, shown here for the case where $\Phi = 200$ units for example) to the emission during the pulse (area B) increases as $t_p$ decreases, for the same energy input.

During the stimulation pulse a certain concentration of trapped electrons $\Delta n$ is removed from the trap. For first-order kinetics (Equation 3.12) this is given by:

$$\Delta n = \int_0^{t_p} n\sigma_p(E)\Phi dt. \tag{3.49}$$

The rate of trap emptying is directly proportional to the stimulation power $\Phi$ and, for weak stimulation ($\Delta n << n_0$; $n$ approximately constant), Equation 3.49 yields $\Delta n = n\sigma_p(E)\Phi t_p$ and thus, in this situation, the total charge released is approximately the same per pulse.

Figure 3.11a shows schematic POSL data for a series of stimulation pulses where $t_p << \tau_d$. Figures 3.11b–3.11c illustrate the case where $\Delta n << n_0$. A series of pulses (a) impinge on the sample which has an initial trapped charge concentration $n_0$. The net POSL signal is the sum of each of the individual emissions following each pulse (b). The recorded POSL signal is the integrated net signal between pulses, following the end of each pulse and before the start of the next one (c). Since $n \approx n_0$ and is constant throughout (the trap approximately does not deplete), the integrated signal (c) increases with stimulation time approaching an asymptotic value.

In Figure 3.11d–3.11e, however, it is assumed that the trapped electron concentration decreases with each pulse, i.e. $\Delta n < n_0$ and is measurable such that the remaining $n$ decreases with each subsequent pulse. As a result, the POSL signal intensity per pulse decreases as the pulses continue (d) and, after an initial increase as the POSL signals pile up, the depletion of $n$

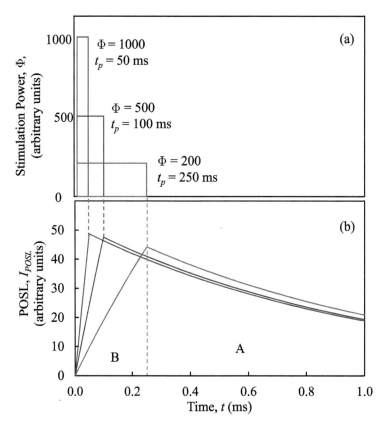

**Figure 3.10** Schematic representation of POSL showing (a) three different stimulation powers $\Phi$ and pulse widths $t_p$ such that $\Phi t_p$ remains the same. In (b) are shown the POSL signals corresponding to each of these pulses. As the value of $t_p$ decreases more POSL light is emitted after the pulse than during it. For efficient POSL detection, the stimulation pulse width $t_p$ must be such that $t_p \ll \tau_d$.

becomes dominant and the integrated signal also decreases (e). Under first-order kinetics the depletion rate is $\mathrm{d}n / \mathrm{d}t = -np = -n\sigma_p (E)\Phi$.

### 3.3.1.4　Expressions for VE-OSL

Scanning the wavelength enables the stimulation spectrum to be obtained where the OSL intensity is plotted as a function of the stimulation wavelength. Clearly, during the period that the sample undergoes stimulation, the trapped charge concentration is depleted and thus to obtain a true stimulation spectrum only a weak stimulation intensity can be tolerated, such that again the change in the trap concentration is minimal compared to the original concentration, i.e. $\Delta n \ll n_0$.

The related VE-OSL monitors the change in OSL intensity as a function of the stimulation energy, using the experimental constraints already noted. Using the QE and first-order assumptions, and $p = \Phi\sigma_p (E)$, and scanning the stimulation energy at a rate $\beta_E = \mathrm{d}E / \mathrm{d}t$, the OSL intensity as a function of $E$ is given by:

$$I_{VE-OSL} (E) = np = n_0 \Phi\sigma_p (E)\exp\left\{-\frac{\Phi}{\beta_E}\int_{h\nu_0}^{h\nu}\sigma_p (\epsilon)\mathrm{d}\epsilon\right\} \qquad (3.50)$$

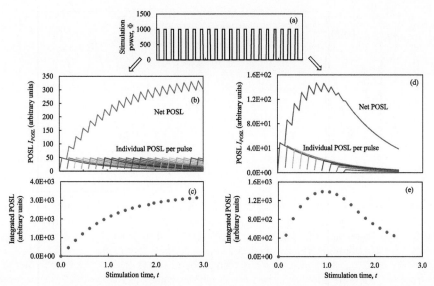

**Figure 3.11** (a) A train of stimulation pulses used in a typical POSL experiment. (b) Assuming the change in trap occupancy per pulse is such that $\Delta n << n_0$, the net POSL is an increasing signal caused by the sum of the POSL emissions for each pulse. By integrating the POSL signal only between the pulses and not during them, the net integrated signal varies as shown in part (c). (d) On the other hand, if measurable depletion occurs with each pulse, the net POSL signal at first increases as the individual pulses are summed (between pulses) until the trap becomes depleted, at which point the net POSL shows a simple decay, since additional stimulation pulses do not add anything to the net POSL signal once $n = 0$. The integrated OSL is now of the form of a peak, as shown in part (e).

where $\epsilon$ is a dummy variable representing the optical stimulation energy, and $\Phi$ is a constant at all values of $E$. The stimulation energy is scanned from optical frequency $\nu_0$ to frequency $\nu$, according to $E = h\nu_0 + \beta_E t$.

As emphasized by Chruścińska (2010), a major difference between this mode of OSL measurement and that of scanning the wavelength $\lambda$ to obtain a stimulation spectrum is that the restriction of $\Delta n << n_0$ is not necessary and, as with TL, the traps become fully depleted as the energy is scanned.

The analogy with TL is very strong when Equation 3.50 is compared to the Randall-Wilkins expression, Equation 3.17, when it can be seen that the forms of the expressions are very similar. However, unlike TL where the temperature at any time $t$ is a discrete, single value (and thus, $p$ at any time $t$ is also a fixed single value) in optical measurements the stimulation light is usually a narrow spectrum, the bandwidth of which is defined by the equipment parameters used in the experiment. Thus, for an energy bandwidth of $2\Delta h\nu$, the stimulation probability $p(E)$ is:

$$p\left(E\right) = \frac{\Phi}{2\Delta h\nu} \int\limits_{h\nu - \Delta h\nu}^{h\nu + \Delta h\nu} \sigma_p\left(\epsilon\right) \mathrm{d}\epsilon \tag{3.51}$$

This leads to slightly broadened $I_{VE-OSL}\left(E\right)$ peaks compared to the use of truly monochromatic light (i.e. a discrete value for $E = h\nu$).

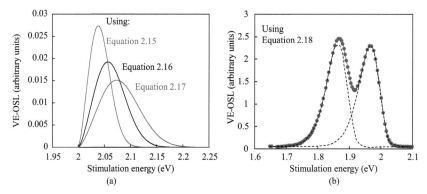

**Figure 3.12**   Example VE-OSL peak shapes from solutions to Equation 3.50 and using Equations 2.15–2.18 to describe the function $\sigma_p(E)$. (a) Equations 2.15–2.17 are used, with a scan rate $\beta_E$ = 0.3 eV.s$^{-1}$ and an optical trap depth $E_o$ = 2.0 eV. (b) Equation 2.18 is used with a Huang-Rhys parameter $S$ = 10, a phonon vibrational energy $h\nu_{ph}$ = 20 meV, temperature $T$ = 300 K, a scan rate $\beta_E$ = 3 × 10$^{-5}$ eV.s$^{-1}$, and a stimulation wavelength band width of $\Delta h\nu$ = 0.05 eV. Two VE-OSL signals from two traps are shown (dashed lines), with $E_o$ values of 2.0 eV and 2.05 eV. Note that under these conditions the VE-OSL peaks appear at $h\nu < E_o$. (Adapted from Chruścińska (2010).)

The solution to Equations 3.50 and 3.51 depends upon the chosen form of the function $\sigma_p(E)$. Chruścińska (2010) adopts the expression given in Chapter 2, Equation 2.18, to derive the VE-OSL curve shapes. Illustrated in Figure 3.12a are the VE-OSL curves for the three photoionization cross-section curves shown in Figure 2.9a, using Equations 2.15, 2.16, and 2.17. Figure 3.12b shows the VE-OSL for the cross-section given in Equation 2.18 (Figure 2.9b). Note that in Figure 3.12a there is no signal below a stimulation energy equal to the optical trap depth (i.e. below $h\nu = E_o = 2.0$ eV). In Figure 3.12b, however, a VE-OSL signal is observed below $h\nu = E_o$ since the photoionization cross-section given in Equation 2.18 includes phonon-assistance. Whereas in Figure 3.12a the start of the VE-OSL peak is abrupt, in Figure 3.12b it is much more gradual. Furthermore, the stimulation light is not perfectly monochromatic in the simulation of Figure 3.12b.

---

**Exercise 3.6   Non-monochromatic stimulation light for CW-OSL and LM-OSL**

In the discussion of VE-OSL, Chruścińska (2010) assumed that the stimulation light had an energy bandwidth $2\Delta h\nu$, meaning that the stimulation probability is given by:

$$p(E) = \frac{\Phi}{2\Delta h\nu} \int_{h\nu - \Delta h\nu}^{h\nu + \Delta h\nu} \sigma_p(\epsilon)\, d\epsilon \tag{3.51}$$

Develop the expressions for CW-OSL and LM-OSL using Equation 3.51 for $p(E)$, for different values of $\Delta h\nu$. To estimate $\Delta h\nu$, look up the bandwidths of the common filters used in OSL experiments, as described in the literature, e.g. Bøtter-Jensen et al. (2003). Also include a laser in your calculations. For this exercise use any one of the expressions given in Chapter 2, Equations 2.15–2.20, for the function $\sigma_p(E)$. (The choice is yours.) From your calculations what differences might you expect to see in CW-OSL and LM-OSL curve shapes when using a polychromatic light source and a filter, versus using a laser?

### 3.3.2 Non-First-Order Kinetics

In Exercise 3.2 for the OTOR model, Equation A gives the GOT equation, namely:

$$I_{OSL}(t) = \frac{npm\sigma_r}{\left[(N-n)\sigma_e + m\sigma_r\right]} = n\sigma_p(E)\Phi\left[\frac{m\sigma_r}{(N-n)\sigma_e + m\sigma_r}\right]. \tag{3.52}$$

where the thermal stimulation term has been replaced by $p = \sigma_p(E)\Phi$ for optical stimulation. If the conditions $\sigma_e = \sigma_r$ and $n = m$ are introduced Equation 3.52 becomes:

$$I_{OSL}(t) = \frac{n^2\sigma_p(E)\Phi}{N}, \tag{3.53}$$

the solution of which is:

$$I_{OSL}(t) = I_{CW\text{-}OSL}(t) = \frac{n_0^2\sigma_p(E)\Phi}{N}\left(1 + \frac{n_0\sigma_p(E)\Phi}{N}\right)^2. \tag{3.54}$$

With $\sigma_p(E)\Phi$ constant, this is the second-order expression for a CW-OSL curve.

Following May and Partridge for TL, a general-order expression can also be proposed for OSL (Yukihara and McKeever 2011), namely:

$$I_{OSL}(t) = \frac{n^b\sigma_p(E)\Phi}{N^{b-1}} \tag{3.55}$$

yielding:

$$I_{CW\text{-}OSL}(t) = \frac{n_0^b\sigma_p(E)\Phi}{N^{b-1}}\left[1 + (b-1)\left(\frac{n_0}{N}\right)^{b-1}\sigma_p(E)\Phi t\right]^{-\frac{b}{b-1}}, \tag{3.56}$$

for constant $\sigma_p(E)\Phi$. This gives the first-order CW-OSL expression when $b \to 1$ and the second-order expression when $b = 2$. Corresponding CW-OSL curve shapes are shown in Figure 3.13. Note how the increase in the degree of re trapping slows down the decay rate for $b > 1$, on the time scale shown. However, as with the equivalent TL expression, Equation 3.56 should be used with great caution, if at all, since the kinetic order $b$ changes as the trap empties, as discussed above for TL. This is not accounted for in Equation 3.56, nor in the calculation of the curves in Figure 3.13. Nevertheless, the general sense of retrapping on the decay rate is clear. Also note that in the OTOR model, since $I_{CW\text{-}OSL} = dm/dt$, then $\int_0^\infty I_{CW\text{-}OSL}dt = m_0 = n_0$ always, independent of the kinetic order, and so the absolute area under each curve in Figure 3.13 is the same when integrated over long enough time.

## 3.4 More Complex Models: Interactive Kinetics

### 3.4.1 Thermoluminescence

The analysis so far has generally assumed the simplest of all models, namely the OTOR model with just one trap and one recombination center (with the exception of the discussion relating to mixed-order kinetics). It is highly unlikely, however, that such a simple model

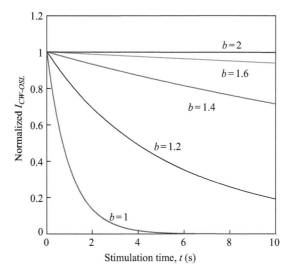

**Figure 3.13**   Normalized CW-OSL curves for different kinetic orders, with $p = 1$ s$^{-1}$ in all cases. Equation 3.14 was used for $b = 1$, and Equation 3.55 was used for $b > 1$. For these calculations, $n_0 = 10^{15}$ m$^{-3}$ and $N = 10^{16}$ m$^{-3}$. (Note: Since the decay rates decrease as the kinetic order increases, the absolute initial starting values are lower, but the area under each curve is the same when integrated to long stimulation times.)

could be representative of any real system. A realistic model, for real materials, might also contain:

- Multiple electron traps (both shallow and deep).
- Multiple hole traps (both shallow and deep).
- Multiple recombination centers (both radiative and non-radiative).
- Interactive kinetics (in which electrons or holes released from one trapping energy level can be re-trapped by another).
- Localized transitions (in which charge is released from a trap and recombines with charges of opposite sign, but does not do so via a transition to the delocalized band).
- Non-equilibrium kinetics (in which the QE assumption is not valid).
- Distributions of trapping states (not discrete).

A generalized model is illustrated in Figure 3.14.

If $N(E)$ is the density of states function in the energy gap, then the occupancy of each state is $N(E)f(E)$, where $f(E)$ is the Fermi-Dirac distribution function. The net rate of electron trap emptying is then:

$$R_{ex} = \int_{D_e}^{E_c} p_n(E) N(E) f(E) dE, \tag{3.57}$$

with the excitation rate for the $n$th trap ($p_n$) either due to thermal or optical stimulation.

The rate of trapping is:

$$R_{trap} = n_c v_e \int_{D_e}^{E_c} \sigma_e(E) N(E) (1 - f(E)) dE, \tag{3.58}$$

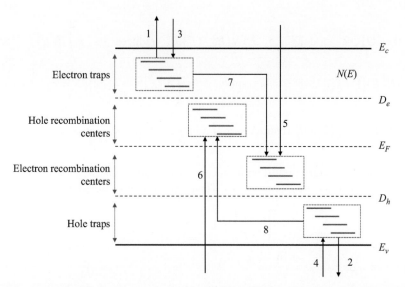

**Figure 3.14** Generalized phenomenological model for TL/OSL showing multiple trapping and recombination centers between the conduction band and valence band edges. Arrows indicate electron and hole excitation (1 and 2), electron and hole trapping or retrapping (3 and 4), and electron and hole recombination (5 and 6), and potential localized transitions (7 and 8).

and the rate of recombination of electrons with trapped holes is:

$$R_{recom} = n_c v_e \int_{D_h}^{E_F} \sigma_r^e(E) N(E) (1 - f(E)) dE. \tag{3.59}$$

Note the integration limits in Equation 3.59 compared to Equations 3.57 and 3.58. The term $N(E)(1 - f(E))$ is the number of empty states. In Equation 3.59, between the limits $D_h$ and $E_F$, this is the number of trapped holes available for recombination with electrons. Between the limits $D_e$ and $E_c$ (Equation 3.58) it is the number of empty electron traps. The net rate of change of free electrons is thus:

$$\frac{dn_c}{dt} = R_{ex} - R_{trap} - R_{recom}. \tag{3.60}$$

Similarly, the rate of change for free holes may be written:

$$\frac{dn_v}{dt} = \int_{E_v}^{D_h} p_m(E) N(E) (1 - f(E)) dE - n_v v_h \int_{E_v}^{D_h} \sigma_h(E) N(E) f(E) dE$$

$$- n_v v_h \int_{E_F}^{D_e} \sigma_r^h(E) N(E) f(E) dE. \tag{3.61}$$

Again, pay attention to the integration limits.

The nomenclature used for the cross-sections is:

$\sigma_e$ – capture cross-section for electrons;

$\sigma_h$ – capture cross-section for holes;

$\sigma_r^e$ – cross-section for recombination of electrons (with trapped holes);

$\sigma_r^h$ – cross-section for recombination of holes (with trapped electrons).

For the OTOR model, $\displaystyle\int_{D_e}^{E_c} p_n(E)N(E)f(E)dE$ reduces to $\displaystyle p\int_{D_e}^{E_c} N(E)f(E)dE = pn$,

$\displaystyle\int_{D_e}^{E_c}\sigma_e(E)N(E)\bigl(1-f(E)\bigr)dE$    becomes    $\displaystyle\sigma_e\int_{D_e}^{E_c}N(E)\bigl(1-f(E)\bigr)dE = \sigma_e(N-n)$,    and

$\displaystyle\int_{D_h}^{E_F}\sigma_r^e(E)N(E)\bigl(1-f(E)\bigr)dE$ reduces to $\displaystyle\sigma_r\int_{D_h}^{E_F}N(E)\bigl(1-f(E)\bigr)dE = \sigma_r m$.

Assume a set of $i$ electron traps, with $i = 1 \rightarrow u$ and $j$ hole traps, with $j = 1 \rightarrow v$, which act as potential recombination sites for electrons. Then, for thermal stimulation:

$$\frac{dn_c}{dt} = \sum_{i=1}^{u} n_i s_{ei} \exp\left\{-\frac{E_{ti}}{kT}\right\} - n_c v_e \left[\sum_{j=1}^{v} m_j \sigma_{rj} - \sum_{i=1}^{u}(N_i - n_i)\sigma_{ei}\right], \qquad (3.62)$$

along with $dn_v/dt = 0$.

Applying QE ($dn_c/dt \approx 0$), ($n_c \ll n_i, m_j$) and $\displaystyle I_{TL} = \sum_{j=1}^{v} dm_j/dt$ gives:

$$I_{TL}(t) = A\sum_{j=1}^{v}\frac{m_j \sigma_{rj}}{(B+C)}, \qquad (3.63)$$

with:

$$A = \sum_{i=1}^{u} n_i s_{ei}\exp\left\{-\frac{E_{ti}}{kT}\right\}, \qquad (3.64a)$$

$$B = \sum_{j=1}^{v} m_j \sigma_{rj} \qquad (3.64b)$$

and

$$C = \sum_{i=1}^{u}(N_i - n_i)\sigma_{ei}. \qquad (3.64c)$$

This is termed an Interactive Multiple Trap System (IMTS) and Equation 3.63 is the multi-trap equivalent of the GOT equation, introduced in Exercise 3.2.

It is clear that such a complex model will yield several TL peaks as each trap empties in turn. If all the released electrons are stimulated into the delocalized bands they may recombine with any of the trapped holes at any of the recombination sites, with priority given to the recombination sites with the largest cross-section and/or largest concentration. If each of the recombination events is radiative then each TL peak will emit luminescence at multiple wavelengths, corresponding to the different recombination centers. The released electrons may also be trapped in those traps that are not yet emptying, i.e. deeper traps. With this situation in mind, several authors have examined the TL glow curves predicted from the IMTS model, and compared the results with the Randall-Wilkins (RW) and Garlick-Gibson (GG) equations, each of which was derived for the simplest OTOR case. The essential question is: can a complex TL glow curve resulting from an IMTS model be described by the simple superposition of RW or GG equations? Figure 3.15 examines this issue.

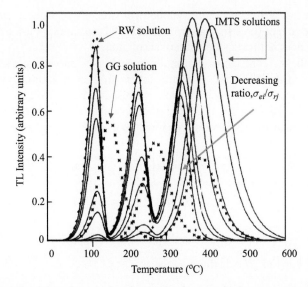

**Figure 3.15**   First-order (+) and second-order (x) TL glow peaks for a non-interactive trap system consisting of three trapping levels and one recombination center, calculated using Equations 3.17 and 3.21, respectively. The curves are compared with the calculations for an interactive model using Equations 3.62–3.64, with $u = 3$ and $v = 1$ (three traps and one recombination center). It is also assumed that the values of $\sigma_{ei}$ (with $i = 1, 2,$ or 3) are equal and that the ratio $\sigma_{ei} / \sigma_{rj}$ (with $j = 1$) is varied, thus: 10, 5.0, 1.0. 0.5, 0.1, 0.05, and 0.01. For these calculations, $E_{t1} = 1.0$ eV, $E_{t2} = 1.25$ eV, $E_{t3} = 1.5$ eV, $s_i$ $(i = 1, 2$ or 3$) = 10^{13}$ s$^{-1}$, $N_i = 10^{16}$ cm$^{-3}$, $n_{0i} = 5 \times 10^{15}$ cm$^{-3}$. (Reproduced from Levy (1984) with permission from Elsevier.)

Figure 3.15 represents the results of solutions to Equations 3.60–3.64 describing interactive kinetics for a model with three traps and one recombination center.[2] QE is assumed. Solutions are obtained for several values of the ratio of the trapping and recombination cross-sections, as indicated in the caption. The conclusions drawn from such data are: (1) When the ratio of the cross-sections is high (corresponding to fast retrapping) the net TL curve in an interactive system is not described by the sum of multiple GG solutions. (2) When the cross-section ratio is small, approaching the case of slow retrapping, the net glow curve can more accurately be described by the sum of three RW solutions. (3) For intermediate ratios, the lower temperature peaks are similar in shape to first-order RW peaks, but the last peak is similar to a second-order peak.

The data of Figure 3.15 were obtained using Equations 3.62–3.64, which in turn used the QE assumption. By solving the relevant differential equations numerically, however, the imposition of the QE assumption can be avoided. An example solution of this type is shown in Figure 3.16. In this case there are three electron-traps and one hole-trap ($u = 3$, $v = 1$). In Figure 3.15a, the parameters chosen are such that retrapping is weak compared to recombination. In such a case, the solution to the rate equations yields a glow curve that can be well described by the summation of three independent RW equations. However, in Figure 3.15b, the parameters chosen are such that the recombination and retrapping coefficients are now equal. Even in this case, the lower temperature peaks have an asymmetric shape more characteristic of first-order

---

[2] The use of one recombination center instead of multiple recombination centers does not undermine the IMTS model since, once recombined, the electrons are removed from the system and are unable to undergo further transitions. This is true whether there are multiple recombination centers or just one. The end result is the same.

peaks whereas the last peak has a more symmetrical structure more characteristic of second-order kinetics. The sum of three second-order peaks (GG solution) is shown for comparison; it can be seen that it does not match the numerical solution.

These data highlight a characteristic of multiple peak glow curves – namely that the lower-temperature peaks in the TL curve generally follow first-order kinetics independent of the probabilities of retrapping and recombination. As discussed by several authors (Sunta et al. 2001, 2005; Chen and Pagonis 2013) this observation can be easily understood from an examination of the rate equations. Expanding Equation 3.63:

$$I_{TL}(t) = \frac{\sum_{i=1}^{u} n_i s_{ei} \exp\left\{-\dfrac{E_{ti}}{kT}\right\} \sum_{j=1}^{v} m_j \sigma_{rj}}{\sum_{j=1}^{v} m_j \sigma_{rj} + \sum_{i=1}^{u}(N_i - n_i)\sigma_{ei}}, \tag{3.65}$$

with $\sum_{i=1}^{u} n_i = \sum_{j=1}^{v} m_i$ for charge neutrality. (Compare with Equation 3.28 for the OTOR case.) For the first trap only ($i = 1$) this becomes:

$$I_{TL1}(T) = n_1 s_{e1} \exp\left\{-\frac{E_{t1}}{kT}\right\}\left[\frac{\sum_{j=1}^{v} m_j \sigma_{rj}}{\sum_{j=1}^{v} m_j \sigma_{rj} + \sum_{i=1}^{u}(N_i - n_i)\sigma_{ei}}\right]. \tag{3.66}$$

Note that the denominator in the term in square brackets contains $\sum_{i=1}^{u}(N_i - n_i)\sigma_{ei}$ – i.e. a sum over all $i$ – since any electron released from trap 1 may be retrapped or trapped in any of the other traps, as well as recombining. All possible transitions must be included in the denominator.

If $n_1 << \sum_{j=1}^{v} m_j$ that is, if the proportion of electrons released from the first trap and that subsequently undergo recombination is much less than the total number of available recombination centers, then the term in square brackets in Equation 3.66 can be considered to be approximately constant. In this case, Equation 3.66 is first order and the TL peak for the first trap will be described by the expression:

$$I_{TL1}(T) = n_{01} s_1 \exp\left\{-\frac{E_{ti}}{kT}\right\} \exp\left\{-\left(\frac{s_1}{\beta_t}\right)\int_{T_0}^{T}\exp\left\{-\frac{E_{t1}}{k\theta}\right\}d\theta\right\}\left[\frac{\sum_{j=1}^{v} m_j \sigma_{rj}}{\sum_{j=1}^{v} m_j \sigma_{rj} + \sum_{i=1}^{u}(N_i - n_i)\sigma_{ei}}\right]. \tag{3.67}$$

This is the RW expression modified by $\sum_{j=1}^{v} m_j \sigma_{rj} / \left(\sum_{j=1}^{v} m_j \sigma_{rj} + \sum_{i=1}^{u}(N_i - n_i)\sigma_{ei}\right)$. The latter term represents the probability of the freed electron undergoing recombination compared to any other transition.

It should be emphasized that this will be the case even if retrapping is strong compared to recombination. That is, a first-order peak shape does not necessarily mean that recombination dominates. It is important to realize that the terms "first-order kinetics," "slow retrapping" and "fast recombination" are not synonymous. Slow retrapping means fast recombination (thereby giving first-order kinetics) only in the case of the OTOR model, or in the case of non-interactive thermally disconnected traps (a non-interactive, multiple trap system, NMTS). When interactive kinetics apply, IMTS, these three terms become disconnected and first-order kinetics can be obtained even when retrapping is strong.

Similar arguments can be applied to all of the TL peaks – except the last one. For the last peak, the term in square brackets in Equation 3.67 is non-constant and will vary strongly as the

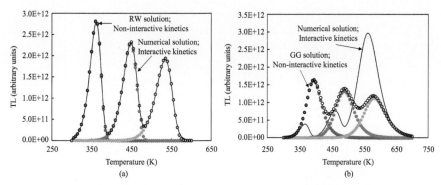

**Figure 3.16**   Comparison of TL curves obtained using (a) Randall-Wilkins (RW) and (b) Garlick-Gibson (GG) solutions, with numerical solutions of the relevant rate equations and using a Three-Trap/One-Recombination-Center (3TOR) model. In each figure, three individual peaks are generated using either the RW (Equation 3.17) or the GG (Equation 3.21) expressions (colored circles). The sum of the three peaks obtained is also shown (black open circles). They are compared with numerical solutions to the rate equations describing the flow of charge between the different levels (black line). The sum of the RW or GG solutions represents entirely non-interactive kinetics (i.e. the OTOR model is applied three times, and the results are added). The numerical solutions include trap interaction (IMTS model). One recombination center is assumed. The parameters used in these calculations are; (a) $E_{t1} = 0.8$ eV, $E_{t2} = 1.0$ eV, $E_{t3} = 1.2$ eV, $s_1 = s_2 = s_3 = 10^{10}$ s$^{-1}$, $n_{10} = n_{20} = n_{30} = 10^{14}$ m$^{-3}$, $N_1 = N_2 = N_3 = 10^{15}$ m$^{-3}$, $m_0 = 3 \times 10^{14}$ m$^{-3}$, $A_{e1} = A_{e2} = A_{e3} = 10^{-16}$ m$^3$s$^{-1}$, (where $A_{ei} = \sigma_{ei}v_e$), and $A_r = 10^{-13}$ m$^3$s$^{-1}$. (b) The same, except $A_{e1} = A_{e2} = A_{e3} = A_r = 10^{-13}$ m$^3$s$^{-1}$.

last trap empties; the TL peak will approach the second-order shape. For the next-to-last peak, a situation intermediate between first- and second-order shapes may apply. Furthermore, especially if conditions of strong retrapping apply, the concentration of charge in the conduction band $n_c$ cannot be ignored during emptying of the last trap and may be similar to the remaining value of $m_j$. In other words, the conditions of QE may not apply to the last peak in the glow curve. Thus, although second-order kinetics may apply, the GG equation may still not be the appropriate expression to describe the peak (as illustrated in Figure 3.16b).

Also to be noted in Figure 3.16 is that when retrapping is slow compared to recombination, the glow curve from interactive kinetics can be described as the sum of individual RW components (Figure 3.16a). However, when retrapping becomes significant the glow curve cannot be described as the sum of GG components. It is observed in Figure 3.16b that the individual GG components do not correspond, in either position or shape, to the actual glow curves produced, in this case, by numerical solution of Equations 3.60–3.64. This is easy to understand since electrons released from the first trap become retrapped in the deeper traps, such that the largest TL peak corresponds to the deepest trap. As a result, the position of each peak as calculated by GG analysis is smaller and at a higher temperature than that produced using interactive kinetics, as is expected from an understanding of the behavior illustrated in Figure 3.5b.

The data of Figure 3.17 were calculated using equal capture-cross sections and each trap filled to 10% of its maximum. Other similar examples can be found in Bull et al. (1986) demonstrating the same conclusions but for different situations.

A similar situation arises when there is a high population of recombination centers pre-existing in the material even before irradiation. An example of this type of situation is the popular luminescence dosimetry material, $Al_2O_3$:C. The recombination centers in this material are $F^+$-centers created during crystal growth in a reducing atmosphere. Charge compensation is provided by an excess of electrons at atomic species elsewhere in the crystal. Experiments have

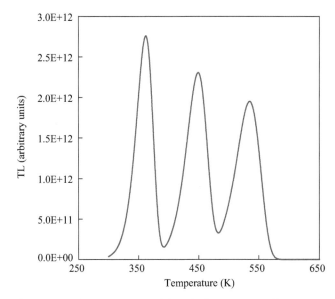

**Figure 3.17**   TL glow curves for an IMTS model with three traps and one recombination center, assuming the concentration of recombination centers is much bigger than the net concentration of trapped electrons (i.e. $m_0 >> \sum_{i=1}^{3} n_{0i}$). The parameters are the same as in Figure 3.16b, but with $m_0 = 10^{17}$ m$^{-3}$. The glow curve is now more like that of Figure 3.16a, with the peaks asymmetric and first-order like.

shown that, during TL, electrons recombine with the $F^+$-centers yielding excited state $F$-centers, ($F^*$-centers) which decay with the emission of luminescence (Akselrod and Kortov 1990). During the irradiation phase, additional $F^+$-centers may be created (via the localization of holes at $F$-centers) but the total number of $F^+$-centers available at the start of the stimulation (heating) phase $m_0$ is given by $m = m_{unirr} + m_{irr}$, where $m_{unirr}$ is the initial number of $F^+$-centers before irradiation, and $m_{irr}$ is the number created by irradiation. Often, $m_{irr} << m_{unirr}$. In such circumstances, $m >> \sum_{i=1}^{u} n_i$ – that is, the total number of available recombination sites is much larger than the total number of trapped electrons due to the irradiation – and the term in square brackets in Equation 3.67 can be considered to be approximately unity. In these circumstances, the TL processes are again first order, independent of the probability of retrapping. A simulated glow curve representing this situation is illustrated in Figure 3.17.

---

**Exercise 3.7   Interactive Kinetics**

Construct an interactive kinetics model using multiple electron traps (at least 3) and multiple hole traps (also at least 3). Examine the literature to establish reasonable values for the various parameters ($E_t$, $s$, $N$, $\sigma$) for each trap. (See examples by Levy (1984), Bull et al. (1986), Sunta et al. (2001, 2005), Chen and Pagonis (2013), and others.)

(a) Assume each hole trap acts as a recombination center. Write the rate equations describing the changes in the trap concentrations during: (i) irradiation and (ii) heating. Show the equation for the net TL intensity.

(b) Often it can be found in the published literature that these equations are numerically solved during the heating phase only. To do so, guesses of the initial trapped charge concentrations are usually made without considering if these concentrations would actually be obtained using the parameter values ($N$, $\sigma$, etc.) that have been assumed. (For example, it might be assumed that the initial trapped charge concentration in two traps is the same at the start of the heating, but the assumed $N$ and $\sigma$ values for one trap might be much larger than for the other, making it unlikely that the initial concentrations would be equal at the end of the irradiation.) To overcome this problem, once the parameter values have been chosen, it is necessary to solve the equations during irradiation and during relaxation after irradiation, before solving them for the heating stage.

Use a suitable software program (e.g. Mathlab or Mathematica) to numerically solve the rate equations during irradiation, relaxation, and heating. (The time used for the relaxation stage will depend on your chosen values for the parameters. How might you estimate this time?)

(c) Vary the rates of retrapping and recombination such that (i) first-order, and (ii) second-order, might be the prevailing kinetics. Compare the different glow curves obtained from your solutions. Can the obtained solutions be described by the sum of RW equations, or GG equations, or neither?

### 3.4.2 Optically Stimulated Luminescence

Similar considerations to those above for TL can also be made for OSL, but using optical stimulation ($p = \sigma_p(E)\Phi$) instead of thermal stimulation ($p = s\exp\{-E_t / kT\}$). There are several caveats, however. Firstly, the order in which traps empty thermally does not necessarily match the order in which they empty optically. Trap 1 (say) may thermally empty before trap 2, but trap 2 may optically empty before trap 1. This can be understood by considering Figure 3.18 in which are plotted two photoionization cross-sections, using Equation 2.16, with $E_o$ values of 1.25 eV and 1.4 eV, respectively. It may be seen that below 1.25 eV (region A), $p_1 = p_2 = 0$. In region B, $p_1 > 0$, $p_2 = 0$; in region C, $p_1 > p_2 > 0$; and in region D, $p_2 > p_1 > 0$. Therefore, which traps empty during stimulation and the relative rates at which they empty depend critically upon the chosen stimulation energy (wavelength). Although the relative values of the photoionization cross-section $\sigma_p(E)$, and the value of $p$, will depend on the shape of the $\sigma_p(E)$ function (Equation 2.16 was used in this illustration), it is critical to note that, feasibly, one can get the deeper trap optically emptying more rapidly than the shallower trap. Also note, that around the transition between regions C and D it may appear from OSL analysis that only one trap produces the OSL signal at the chosen stimulation wavelength (energy) when in fact two traps are active.

Secondly, in TL the traps thermally empty as governed by the relative detrapping rates $p = s\exp\{-E_t / kT\}$ for each trap. The detrapping rates change exponentially as the temperature changes and, importantly, the ratio of the rates also varies exponentially. For two traps with $E_{t1}$ and $s_1$ and $E_{t2}$ and $s_2$ the ratio of the detrapping rates $p_1 / p_2 = (s_1 / s_2)\exp\{-(E_{t1} - E_{t2}) / kT\}$ can vary over several orders of magnitude as the temperature increases. In OSL, however, for a fixed stimulation wavelength, the traps empty simultaneously at relative rates dictated by the value of $p = \sigma_p(E)\Phi$ for each trap type. For multiple traps, the relative rates remain fixed. For example, for two traps, the ratio $p_1 / p_2 = \sigma_{p1}(E)\Phi_1 / \sigma_{p2}(E)\Phi_2$ is a constant throughout the emptying process. As shown in Figure 3.18 and for a given value of $\Phi$, the value of $\sigma_p(E)$, and therefore $p = \sigma_p(E)\Phi$, can be non-zero for several traps at the same time (regions C and D),

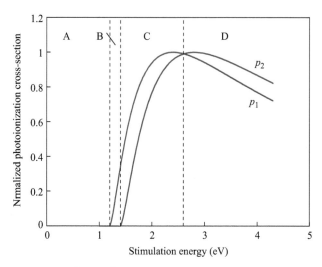

**Figure 3.18**    A comparison of two photoionization cross-sections as a function of stimulation energy, for optical trap energies $E_o$ of 1.25 eV and 1.4eV. In region A, $p_1 = p_2 = 0$; in region B, $p_1 > 0$, $p_2 = 0$; in region C, $p_1 > p_2 > 0$; in region D, $p_2 > p_1 > 0$.

even though the optical trap depths $E_o$ may be different, but the ratio of the $p$ values will not change during trap emptying when $\lambda$ is fixed.

Thus, in a TL experiment, the TL intensity at any given point is the sum of a few distinct trap emptying processes and, in many cases, may be dominated by just one trap at a given temperature. In contrast, in a CW-OSL experiment, the value of the OSL emission intensity is always the sum of several processes occurring simultaneously. Even with LM-OSL, where $\Phi$ is ramped linearly, the ratios of the various values of $p$ for the different traps will remain the same. With TL, however, the ratios vary exponentially as the temperature is increased.

With these caveats in mind the characteristics of OSL under interactive kinetics can be examined. It was discussed above how the lower-temperature TL peaks in a multi-trap system are generally described by first-order kinetics, with the last trap to empty being governed by non-first-order kinetics. Given that the notion of a "last trap to empty" is less clear in a CW-OSL experiment, since all traps empty simultaneously but at different rates, it is pertinent to analyze the kinetics of optical emptying of traps in an interactive multi-trap system (IMTS). Examples of these considerations are shown in Figure 3.19 for a variety of cases involving three traps and one recombination center (a model designated here as 3TOR). In this figure, the time-variant values for the different electron and hole concentrations were obtained by numerically solving the relevant differential equations. See the accompanying web site for a list of the equations, the details of the parameters used to calculate the data in Figure 3.19, and a more detailed description of the results. (Exercise and Notes, Chapter 3, Explanation of Figure 3.19.)

For CW-OSL under conditions of QE, the equivalent of Equation 3.65 for $u = 3$ traps and $v = 1$ recombination center (3TOR), and $m = n_1 + n_2 + n_3$ is:

$$I_{CW-OSL}\left(t\right) = \frac{\sum_{i=1}^{3} n_i p_i m\sigma_r}{m\sigma_r + \sum_{i=1}^{3}\left(N_i - n_i\right)\sigma_{ei}} = Rm.$$    (3.68)

In Figures 3.19a, 3.19b, and 3.19c, $\sigma_{ei} \ll \sigma_r$ for all traps and each trap empties exponentially. For the three traps, the numerical calculations used values of $p_1 = 1.0$, $p_2 = 0.33$ and $p_3 = 0.1$, respectively. Best fits to the decay curves for $n_1$, $n_2$, and $n_3$ (Figure 3.19a) yielded the

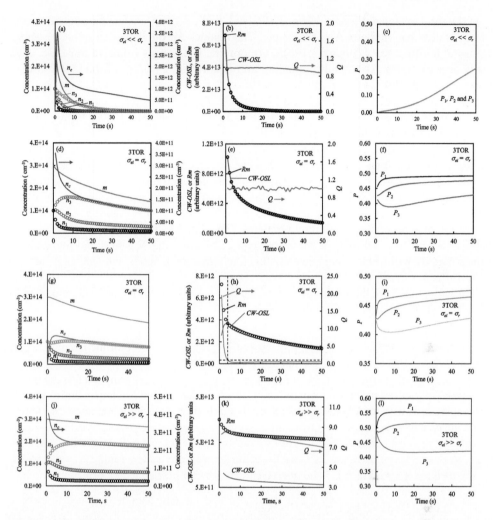

**Figure 3.19** CW-OSL: Example numerical solutions of the rate equations for an IMTS, 3TOR model. For all cases shown, $n_{i0} = 0.1 N_i$. Also, in all examples, $p_1 = 1.0$, $p_2 = 0.33$, and $p_3 = 0.1$. Figures a-c represent examples of QE and a sum of first-order signals. Figures d-f represent QE and non-first order kinetics. Figures g-i represent non-QE and non-first order kinetics. Figures j-l represent highly non-QE and non-first order kinetics. The definitions of the terms $P$ and $Q$ are described in the text. Detailed parameter values and further explanation are given in the web site.

inputted values for $p_1$, $p_2$, and $p_3$, as expected in this situation. Furthermore, in Figure 3.19b, it is demonstrated that the QE holds and thus Equation 3.68 is an accurate reflection of the CW-OSL intensity $(I_{CW-OSL}(t) = Rm)$. The functions $Q$ and $P$ shown in Figure 3.19 are explained in a later section.

For Figures 3.19d, 3.19e, and 3.19f, however, the cross-sections were made equal $(\sigma_{ei} = \sigma_r)$. In this situation significant retrapping of charge from trap 1 occurs into traps 2 and 3, and from trap 2 into trap 3. The result is that none of the curves for $n_1$, $n_2$, and $n_3$ can be described by simple exponential decays (Figure 3.19e). However, QE still holds for this case, and again Equation 3.68 is valid for the CW-OSL intensity (Figure 3.19e).

In the third example, Figures 3.19g, 3.19h, and 3.19i, $\sigma_{ei}$ and $\sigma_r$ are also equal, but $10^3$ times smaller than in the previous example. The kinetics for the depletion of $n_1$, $n_2$, and $n_3$ are again not first order (Figure 3.19g). Furthermore, QE does not hold in this case, especially at early times. As a result, Equation 3.68 is not valid and $I_{CW-OSL}(t) \neq Rm$ at early times (Figure 3.19h).

In the final example, Figures 3.19j, 3.19k, and 3.19l, $\sigma_{ei}$ is $10^3$ times larger than $\sigma_r$. The depletion kinetics are not first order (Figure 3.19j), the system is far from QE, and $I_{CW-OSL}(t) \neq Rm$ at all times (Figure 3.19k).

Clearly, many other IMTS examples can be conceived; for example, there may be much deeper traps with high concentrations that are not taking part in the OSL process (i.e. optically disconnected traps) or there may be a large concentration of pre-existing trapped-hole centers before the irradiation. In either case, $m \gg n_1 + n_2 + n_3$. A wide array of values for the various parameters can be imagined. In the calculations for Figure 3.19, $n_0/N = 0.1$ and is the same for all three different trap types. (This latter equality is true in this case only because it was assumed that the three trap types had the same $N$ values and the same $\sigma_e$ values. Reality will be more complex and it is likely that the kinetics will change if the $n_0/N$ ratios differ.)

From Equation 3.68, the relationship between $I_{CW-OSL}(t)(= dm/dt)$ and $m$ is modified by the term $R$. This term can be generalized as:

$$R = \frac{\sigma_r \sum_{i=1}^{u} n_i p_i}{\sum_{j=1}^{v} m\sigma_{rj} + \sum_{i=1}^{u}(N_i - n_i)\sigma_{ei}}, \tag{3.69}$$

for $u$ types of trap and $v$ types of recombination center. When $m$ is multiplied by this modifying term, agreement with the CW-OSL is obtained, but *only* if QE (i.e. $dn_c/dt \ll dm_j/dt$, $dn_i/dt$ and $n_c \ll m_j$, $n_i$) applies.

---

**Exercise 3.8    CW-OSL interactive kinetics with large $m$**

(a)  With the model used in Figures 3.19a–3.19c, and the same parameter values, numerically solve the relevant rate equations in order to determine the time-dependencies of the trapped electron concentrations ($n_1$, $n_2$, and $n_3$), the free electron concentrations ($n_c$), and the recombination center concentration ($m$), but assuming that the initial concentration of trapped holes $m_0 = 10^{17}$ $m^{-3}$ (i.e. $m_0 \gg n_{10}$, $n_{20}$, or $n_{30}$). All other parameter values stay the same as for Figures 3.19a, 3.19b, 3.19c.

(b)  Plot the CW-OSL shape as a function of stimulation time over the period 1–50 s and compare with a calculation of $Rm$.

(c)  Repeat the calculations of parts (a) and (b) but using the parameter values of Figures 3.19d–3.19f, and again with $m_0 = 10^{17}$ $m^{-3}$.

---

## 3.5   Trap Distributions

An obvious extension of a model that has a multiplicity of trap and recombination sites is to consider continuous distributions of traps and centers. Following the discussion of defects in solids in Section 2.1 in Chapter 2, it is clear that amorphous, non-crystalline materials are more likely than crystalline materials to exhibit localized states that are characterized by a distribution of thermal and optical trap depths, $\Delta E_t$ and $\Delta E_o$, rather than discrete values, $E_t$ and $E_o$. The frequency factors $s$ and photoionization cross-sections $\sigma_p$ will likewise be distributed functions. It is also possible that in crystalline or semi-crystalline materials with high defect densities, distributions of these parameters may also be expected due to electronic and ionic interactions between the defects when in concentrated numbers. It becomes important, therefore, to consider the cases of localized state distributions and to consider their impact on the TL and OSL processes.

Distribution functions that might be considered, include the following forms:

*Linear Distribution*: $N(E_t) = N_m \left[ \dfrac{E_t - E_A}{E_B - E_A} \right];$

*Uniform Distribution*: $N(E_t) = N;$

*Exponential Distribution*: $N(E_t) = N_m \exp\left\{ -\dfrac{E_t}{kT_c} \right\}$, where $T_c$ is a "characteristic temperature" for the distribution;

*Gaussian Distribution*: $N(E_t) = N_m \exp\left\{ -a(E_t - E_{tm})^2 \right\}$, where $a$ is a constant.

Other distribution functions can be imagined. The terms used in these expressions are defined in Figure 3.20.

What will a TL glow curve look like if such distributions are prevalent? As a first approach to the answer, consider the case of TL being produced under conditions of first-order kinetics and quasi-equilibrium, and that the net TL is composed of a weighted sum of signals from a narrow range of trap energies. Further assume, for simplicity, and that each component is characterized by the same value for the frequency factor, $s$. The net TL is then:

$$I_{TL}(T) = \int_0^\infty g(E_t) I_{RW}(T) dE_t, \tag{3.70}$$

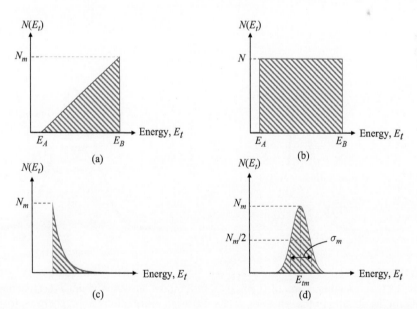

**Figure 3.20** Potential trap energy distributions: (a) Linear, (b) uniform, (c) exponential, and (d) Gaussian.

where $g(E_t)$ is a weighting (or distribution) function, and $I_{RW}(T)$ is the RW equation (Equation 3.17). The weighting function is given by $g(E_t) = N(E_t) f(E_t)$, with $N(E_t)$ being the available trap distribution function and $f(E_t)$ being the Fermi-Dirac occupation function.

For a discrete distribution (single valued $E_t$), $g(E_t) = \delta(E_t - E_0)$ for which Equation 3.70 reduces to Equation 3.17 when $E_t = E_0$. However, as described by Hornyak and Chen (1989), and Coleman and Yukihara (2018), for a uniform distribution of energy width $\Delta E = E_B - E_A$, Equation 3.70 becomes:

$$I_{TL}(T) = \frac{n_0 s}{\Delta E} \int_{E_A}^{E_B} \exp\left[-\frac{E_t}{kT} - \frac{s}{\beta_t} \int_{T_0}^{T} \exp\left\{-\frac{E_t}{k\theta}\right\} d\theta\right] dE_t, \tag{3.71}$$

whereas for a Gaussian distribution, $g(E_t) = \dfrac{1}{\sqrt{2\pi}\sigma_m} \exp\left\{-\dfrac{(E_t - E_{tm})^2}{2\sigma_m^2}\right\}$, where $\sigma_m$ is the

half-width of the distribution, centered around trap energy $E_t = E_{tm}$, the following TL equation is obtained:

$$I_{TL}(T) = \frac{n_0}{\sqrt{2\pi}\sigma_m} \int_{0}^{\infty} \exp\left\{-\frac{(E_t - E_{tm})^2}{2\sigma_m^2}\right\} \exp\left\{-\frac{E_t}{kT} - \frac{s}{\beta_t} \int_{T_0}^{T} \exp\left\{-\frac{E_t}{k\theta}\right\} d\theta\right\} dE_t. \tag{3.72}$$

A comparison of first-order TL peaks for discrete, uniform, and Gaussian distributions is shown in Figure 3.21a. To be noted is that the TL peaks for the distributions in $E_t$ are broader than that for a discrete $E_t$, despite the fact that the kinetics are first-order. Thus, if the kinetic order is being evaluated based on the shape (symmetry) of the peak, an incorrect conclusion regarding kinetics may be reached. However, the change in the position of the peak will still be invariant with dose if the kinetics are indeed first order. A broad peak that does not change shape or position with dose is a clue that there may in fact be a distribution of trapping levels in the system being studied.

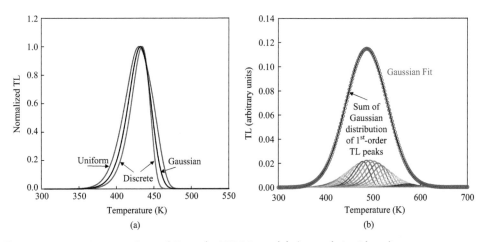

**Figure 3.21**   (a) A comparison of TL peaks (OTOR model, first order) with a discrete trap energy, a uniform distribution of trap energies, and a Gaussian distribution of trap energies. (b) Sum of a Gaussian distribution of first-order TL peaks (black line, with $E_{tm} = 1.2$ eV and $s_m = 10^{11}$ s$^{-1}$), and a fit to the net TL using a Gaussian-shaped TL peak (red circles, with $T_m = 486$ K).

One useful point about a Gaussian distribution of trap depths is that a sum of first-order TL peaks (Equation 3.72) will produce a near-Gaussian-shaped TL peak. This is illustrated in Figure 3.21b where several (41 in this case) glow "curvelets," each with its own value of $E_t$ (and the same value of $s$) where $E_t$ is weighted by a Gaussian function, are added to produce a net TL peak. Curve fitting shows that the net TL peak can be fitted well by a Gaussian-shaped TL curve, centered at the peak maximum temperature $T_m$. Thus, if a TL peak is broad (wider symmetry than first-order) but its peak position and shape is invariant with dose (indicating first order) and if it can be fitted to a Gaussian shape, then it is probable that the trap distribution is itself Gaussian, lending itself to straightforward curve fitting. However, it should be noted that for a narrow distribution of trap depths, the TL peak shape may still not be quite a Gaussian. This is explained and further explored on the accompanying web site (under Exercises and Notes, Chapter 3, Gaussian TL peaks).

For OSL, a distribution of optical trap depths $\Delta E_o$ leads to a similar distribution in the values of $p = \sigma_p(E)\Phi$, for a fixed stimulation intensity $\Phi$ at a single stimulation wavelength. This in turn leads to a change in shape of CW-OSL decay curve compared to that obtained with a discrete value for $E_o$. At longer stimulation times the slower components, corresponding to smaller values of $p$, become evident, whereas at short stimulation times, those components for which $p$ is large are stronger. These effects are shown in Figure 3.22 for the OTOR model assuming first-order kinetics. The particular mix of slower and faster $p$ values (compared to the discrete case) will depend on the shape of the $\sigma_p$ curve (Figure 2.9) and the chosen stimulation wavelength/energy.

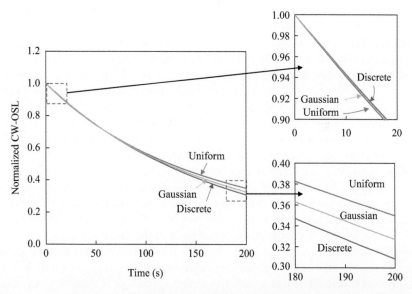

**Figure 3.22** A comparison of CW-OSL curves (OTOR model, first-order) with a discrete optical trap energy, a uniform distribution of energies, and a Gaussian distribution of energies. In these calculations the simulation energy is fixed at $h\nu = 1.4$ eV. $E_o = 1.3$ eV is the optical trap depth used in the discrete case and is also the center of the uniform and Gaussian distributions. In the distribution cases, the early part of the CW-OSL curves differ slightly from that for the discrete case due to those faster decaying components that correspond to larger values of the photoionization cross-section $\sigma_p$, and, therefore, to higher values of $p$. At longer stimulation times, the slower components dominate, due to smaller values of $\sigma_p$ and $p$.

**Exercise 3.9    LM-OSL with a distribution in *E***

Repeat the calculation shown in Figure 3.22, but this time for LM-OSL curves. Assume: (a) a uniform distribution in $E_o$ from 1.24 eV to 1.36 eV, with the center at $E_o = 1.3$eV, and (b) a Gaussian distribution, centered at $E_o = 1.3$eV and with a standard deviation of 0.02 eV. Choose a suitable formula for the function $\sigma_p(E)$, and assume a rate of increase in $\Phi$ of $\beta_\Phi = 3.5$ intensity units/s. Finally, let $n_0 = 1$ concentration unit. Compare the LM-OSL curve shapes obtained with the LM-OSL curve for a discrete value for $E_o$ of 1.3 eV.

The difficulty with using Equation 3.70 to represent the net TL glow curve is that it assumes only a distribution in $E_t$ and a common, fixed value for $s$. Distributed values of $s$ are also to be expected in real systems and thus, a better expression would be:

$$I_{TL}(T, \beta_t) = \int_{s_1}^{s_2} \int_{E_{t1}}^{E_{t2}} g(E_t, s) I_{RW}(T, \beta_t) \mathrm{d}E_t \mathrm{d}s, \tag{3.73}$$

where $E_t$ varies over the range $E_1$ to $E_2$, and $s$ varies over the range $s_1$ to $s_2$. The heating rate $\beta_t$ is included here as a parameter since the position of a TL peak, for a given pair of values for $E_t$ and $s$, depends on $\beta_t$. That is, for a TL peak with a maximum at a particular temperature and a particular value of $\beta_t$, the parameters $E_t$ and $s$ are not independent variables. If $E_t$ changes, so must $s$ in order for the peak maximum to correspond to the one observed. There are multiple combinations of $E_t$ and $s$ than can give a TL peak at the same temperature, for a given heating rate. Therefore, fitting a TL peak to an equation of the form given in Equation 3.73 is problematic. The TL data is a one-dimensional data set, whereas Equation 3.73, known as the Fredholm equation, is a 2-dimensional expression.

The solution is to obtain several TL curves, each at a different heating rate $\beta_t$, and to find that combination of $E_t$ and $s$ that gives the correct peak position at all values of $\beta_t$. This combination of $E_t$ and $s$ is then unique. The principle was seen in Figure 3.2c in which it was observed that the position of the peak changes as $\beta_t$ changes, but $E_t$ and $s$ remain constant. Only one pairing of $E_t$ and $s$ will give the peaks in the positions shown for each value of $\beta_t$.

An additional consideration regarding the use of Equation 3.73 is that this expression assumes the superposition principle – that is, the overall glow curve can be described by a simple sum of individual, first-order, Randall-Wilkins equations. It has already been discussed how a glow curve from an IMTS cannot be described by the sum of second-order Garlick-Gibson peaks (e.g. Figure 3.16) and so superposition is not valid with non-first-order kinetics. However, it has also been shown that when there is a multiplicity of traps the first traps to empty generally follow first-order kinetics whereas the last traps to empty may not necessarily do so. Thus, for a broad distribution, the initial parts of the distribution may empty via first-order kinetics whereas the latter parts may not. This is a limitation on the use of Equation 3.73. One the other hand, if the number of recombination sites is very large, for example, if there are several different distributions due to different defect species, or if there is a large number of pre-existing recombination sites, first-order kinetics may apply throughout the distribution.

Unfortunately, there are few examples in the published literature where there has been an attempt to fit several glow curves simultaneously, each glow curve having been obtained at a different $\beta_t$. One example is shown in Figure 3.23, which illustrates the principle of the technique of fitting multiple glow curves, each one recorded at a different heating rate. In this example, thermally stimulated conductivity (TSC), rather than thermoluminescence, was used in order to avoid complications due to thermal quenching of the TL. Nevertheless, since TSC

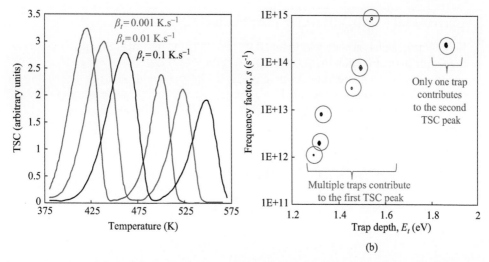

**Figure 3.23** An example of use of the Fredholm equation (Equation 3.73) to fit multiple TSC curves obtained different heating rates $\beta_t$. The TSC curves are shown in (a), while (b) illustrates the result of the analysis showing the distribution in $E_t$ and $s$ and the various $E_t/s$ combinations that are required to give the measured TSC curves. (Adapted from Whitley et al. (2002).)

and TL curves are each described by equations of the same form (Chen and McKeever 1997; Chen and Pagonis 2013) the same analytical methods apply. The figures show the shift in the TSC peak positions as a function of heating rate. By fitting several such curves simultaneously using the Fredholm equation, Equation 3.73, Whitley et al. (2002) determined those sets of $E_t/s$ pairs that were necessary to fit each of the various TSC curves simultaneously. In the example given, the $E_t/s$ pairs revealed that the traps are discrete but that the first TSC peak was the result of multiple, overlapping signals, whereas the second major TSC peak could be described by just one component.

A simpler, alternative strategy to using a 2-D Fredholm equation is to assume a fixed value to $s$ and to use the one-dimensional form (Equation 3.70) with each trap in the distribution assumed to have the same $s$. This is unlikely to be true in practice. Nevertheless, several publications have proceeded in this manner to analyze TL glow curves (e.g. Rudlof et al. 1978; Hornyak and Chen 1989; Coleman and Yukihara 2018).

A similar analysis was performed by Whitley and McKeever (2001) for LM-OSL. Here, however, the distribution is in $E_o$ only and the one-dimensional version of the Fredholm equation suffices. For LM-OSL this is:

$$I_{LM-OSL}(t) = \int_{0}^{\infty} g(E_o) I'_{LM-OSL}(t) dE_o, \tag{3.74}$$

where $I'_{LM-OSL}(t)$ is the LM-OSL expression for a fixed value of $E_o$ and is of the form given in Equation 3.46. Using Equation 3.74 with LM-OSL, Whitley and McKeever (2001) produced the distribution of photoionization cross-sections $\sigma_p$, shown in Figure 3.24 for $Al_2O_3$, where the distribution is compared to the LM-OSL curve showing that three traps contribute to the LM-OSL signal, with the photoionization cross-sections indicated. The stimulation wavelength was 526 nm ($E = 2.36$ eV) and Equation 2.16 was used to determine $\sigma_p(E)$. The determined $\sigma_p$ values are $5.8 \times 10^{-19}$ cm$^2$, $1.33 \times 10^{-19}$ cm$^2$ and $3.3 \times 10^{-20}$ cm$^2$, and the corresponding $E_o$ values are 1.3 eV, 1.8 eV and 2.1 eV, respectively.

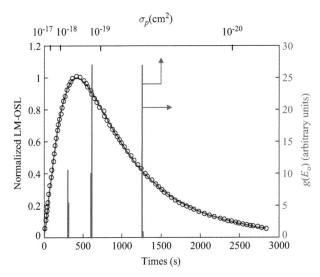

**Figure 3.24**  Deconvolution of an LM-OSL curve to reveal the photoionization cross-section distribution for OSL from $Al_2O_3$. (Adapted from Whitley and McKeever (2001).)

## Exercise 3.10   One-Dimensional Fredholm Equation

1)  Consider the 3TOR model with the following parameters:

|  | Peak 1 | Peak 2 | Peak 3 |
|---|---|---|---|
| $E_t$ (eV) | 0.8 | 1.0 | 1.2 |
| $s$ (s$^{-1}$) | $10^{10}$ | $10^{10}$ | $10^{10}$ |
| $A_e = v_e\,\sigma_e$ (m$^3$s$^{-1}$) | $10^{-16}$ | $10^{-16}$ | $10^{-16}$ |
| $n_0$ (m$^{-3}$) | $10^{14}$ | $10^{14}$ | $10^{14}$ |
| $N$ (m$^{-3}$) | $10^{15}$ | $10^{15}$ | $10^{15}$ |

with

| | |
|---|---|
| $m_0$ (m$^{-3}$) | $10^{17}$ |
| $A_r = v_e\,\sigma_r$ (m$^3$s$^{-1}$) | $10^{-13}$ |

(a)  Write the rate equations governing the flow of charge between the trap, recombination center and conduction and valence bands during trap emptying.

(b)  Using a suitable mathematical software package, numerically solve the rate equations and calculate the TL intensity, given by:

$$I_{TL} = \left|\frac{dm}{dt}\right| = n_c m A_r$$

as a function of temperature $T$, using $\beta_t = dT\,/\,dt = 1.0$ K.s$^{-1}$.

(c)  Repeat the exercise twice more using $\beta_t = 3.0$ K.s$^{-1}$ and $\beta_t = 10.0$ K.s$^{-1}$.

2)  A glow curve can be represented as a one-dimensional Fredholm equation, namely:

$$I_{TL}(T) = \int_0^\infty g(E_t)\, I_{RW}(T)\, dE_t = \sum_{i=1}^3 g_i I_{RW}^i(T)$$

i.e. the sum of three RW expressions, one for each trap.

3) Use the numerical glow curves obtained in Part (1), at the three heating rates, to simultaneously fit the three glow curves in order to evaluate the unique set of $E_t/s$ pairs required to produce the TL glow peaks at the temperatures determined from the numerical solutions. (For the fitting, choose starting values for $E_t$ and $s$ to be different from those used in the numerical calculations.)

## 3.6 Quasi-Equilibrium (QE)

### 3.6.1 Numerical Solutions: No QE Assumption

One of the most contentious assumptions in analyzing solutions to the rate equations in order to determine analytical forms of TL and OSL, is the assumption of quasi-equilibrium, namely: $dn_c/dt \ll dn_i/dt$, $dm_j/dt$ and $n_c \ll n_i$, $m_j$. There have been several examinations of this assumption under a wide variety of parameter values and models, including OTOR, NMTS, and IMTS. For TL, such investigations have usually been carried out by numerically solving the rate equations for each of the models and searching for combinations of parameters that provide TL peaks but for which QE does not hold. Early examples have been provided by Shenker and Chen (1972), Kelly et al. (1971), and later by Sunta et al. (2001, 2005), along with several others. The general consensus is that TL peaks can be obtained without QE, but their shape does not generally conform to that predicted by the simple analytical expressions. Chen and McKeever (1997) have summarized some of these studies.

Examples of the same type of analysis, but for CW-OSL, have already been described in Figure 3.19. Here, some CW-OSL curves were included for cases where the QE assumption was not valid. Assuming QE in these cases could easily lead to erroneous analyses of an experimental CW-OSL curve obtained under these conditions.

### 3.6.2 *P* and *Q* Analysis

These observations lead to the question of how the equations may be analyzed without the QE assumption. One approach to this issue (Lewandowski and McKeever 1991; Lewandowski et al. 1994) is to replace the QE assumption with the identity:

$$-\frac{dn_c}{dt} = q\frac{dm}{dt} \tag{3.75a}$$

or

$$Q\frac{dm}{dt} = \frac{dn}{dt} \tag{3.75b}$$

for the OTOR or NMTS models. It may be determined that $Q = q + 1$ and it should be noted that, for TL, $Q$ and $q$ are both functionals, $Q\{T(t)\}$ and $q\{T(t)\}$, in that they each depend on $T$, and $T$ depends on time $t$. For CW-OSL and POSL, we may write them as functions $Q(t)$ and $q(t)$, whereas for LM-OSL and VE-OSL we may write $Q\{p(t)\}$ and $q\{p(t)\}$, and $Q\{E(t)\}$ and $q\{E(t)\}$, respectively. From the above definitions, it may be concluded that the value of $Q$ indicates how close the system is to QE, with $Q \approx 1$ ($q \approx 0$) meaning QE. If $R_{ex}$ is the rate of excitation from the trap, $R_{retrap}$ is the rate of retrapping into the same trap, and $R_{recom}$ is the rate of recombination, then:

$$Q = \frac{R_{ex} - R_{retrap}}{R_{recom}}. \tag{3.76}$$

Additionally, instead of assuming or estimating a specific kinetic order, a parameter $P$ can be introduced, such that:

$$P = \frac{R_{retrap}}{R_{recom}}. \tag{3.77}$$

It is straightforward to show that:

$$Q + P = \frac{R_{ex}}{R_{recom}}, \tag{3.78}$$

and

$$\frac{Q}{P} = \frac{R_{ex}}{R_{retrap}} - 1. \tag{3.79}$$

$P$ is also a functional, $P\{T(t)\}$ for TL, $P\{p(t)\}$ for LM-OSL and $P\{E(t)\}$ for VE-OSL, but $P(t)$ for CW-OSL.

The above expressions are for the OTOR or the NMTS models only. In these models after an electron has been released from its trap the only options for it are to be retrapped or to recombine. However, the expressions require modification for the various IMTS model types. For example, consider a model consisting of one active trap (thermally or optically active, concentration of trapped electrons $n$), one thermally or optically disconnected deep trap (TDDT or ODDT, concentration of trapped electrons $h$) and one recombination center (concentration of trapped holes $m$). Whereas the charge neutrality condition in the OTOR model is simply $m = n_c + n$, in the IMTS model it is $m = n_c + n + h$, from which the rate of change of trapped holes is:

$$\frac{dm}{dt} = \frac{dn_c}{dt} + \frac{dn}{dt} + \frac{dh}{dt}. \tag{3.80}$$

Using the definitions of $q$ and $Q$:

$$\frac{dn_c}{dt} = \frac{dm}{dt} - \frac{dn}{dt} - \frac{dh}{dt} = -q\frac{dm}{dt} \tag{3.81}$$

from which, using $Q = q+1$:

$$Q = \frac{\frac{dn}{dt} + \frac{dh}{dt}}{\frac{dm}{dt}}. \tag{3.82}$$

Using $dn/dt = -np + n_c(N-n)A_e$, $dh/dt = n_c(H-h)A_{eh}$, and $dm/dt = -n_c m A_r$, leads to:

$$Q = \frac{np - n_c(N-n)A_e - n_c(H-h)A_{eh}}{n_c m A_r} = \frac{R_{ex} - R_{retrap} - R_{trap}}{R_{recom}}. \tag{3.83}$$

For clarity, the terms are defined:

$R_{ex}$ = rate of excitation from the active trap = $np$;
$R_{retrap}$ = rate of retrapping into the active trap = $n_c(N-n)A_e$;

$R_{trap}$ = rate of trapping by the *TDDT / ODDT* $= n_c (H - h) A_{eh}$;
$R_{recom}$ = rate of recombination of an electron with a trapped hole $= n_c m A_r$;
$n_c$ = concentration of free electrons in the conduction band;
$n$ = concentration of trapped electrons in the active traps;
$N$ = concentration of active traps;
$h$ = concentration of trapped electrons in the disconnected traps (TDDT/ODDT);
$H$ = concentration of disconnected traps;
$m$ = concentration of trapped holes in the recombination centers;
$A_e$ = probability of capture of an electron in the active traps $= v_e \sigma_e$;
$A_{eh}$ = probability of capture of an electron in the disconnected traps $= v_e \sigma_{eh}$;
$A_r$ = probability of capture of recombination of a free electron with a trapped hole $= v_e \sigma_r$.

Similarly, *P* for this IMTS model may be defined as the ratio of the rate of retrapping to the net rates of all other possible transitions. For the OTOR model the only other available transition is recombination, and therefore Equation 3.77 results. For the IMTS model, however, trapping into the TDDT/ODDT is also possible and therefore *P* is given by:

$$P = \frac{R_{retrap}}{R_{recom} + R_{trap}} = \frac{n_c (N - n) A_e}{n_c m A_r + n_c (H - h) A_{eh}}. \tag{3.84}$$

The simple relationships shown in Equations 3.78 and 3.79 now no longer apply.

Table 3.1 summarizes and defines the relevant expressions for *P* and *Q* for several models. Note that when there are multiple active traps (for example, models IMTS(3) and IMTS(4)) individual values of the kinetic parameter *P* must be defined for each trap. (As has been shown in previous illustrations, one peak can empty under first-order conditions and another under non-first-order conditions.) *P* is always defined as the ratio of the rate of retrapping to the rate of transitions into all other possible energy levels. *Q* is always defined as the ratio of the net rate of trapping and retrapping to the rate of recombination. Only if $P \ll 1$ and $Q \approx 1$ do first-order and QE conditions hold, in which case the RW equations can be used to describe the peak. (The reader is prompted to be aware that several publications on the topic of kinetic order of TL have used expressions for *P* and *Q* derived from the NMTS model and incorrectly applied them to the IMTS model, and as a result have arrived at erroneous and misleading conclusions.)

Some further, logical considerations of the various potential values of *P* and *Q* are given on the accompanying web site (in the document "*P* and *Q* Analysis") indicating the self-consistencies of various paired combinations of *P* and *Q* values (Exercises and Notes, Chapter 3, Discussion of Section 3.6.2 *P* and *Q* analysis).

---

**Exercise 3.11   P & Q**

(1) Consider the 3TOR model with the following parameters:

|  | Peak 1 | Peak 2 | Peak 3 |
|---|---|---|---|
| $E_t$ (eV) | 0.8 | 1.0 | 1.2 |
| $s$ (s$^{-1}$) | $10^{10}$ | $10^{10}$ | $10^{10}$ |
| $A_e = v_e \sigma_e$ (m$^3$s$^{-1}$) | $10^{-16}$ | $10^{-16}$ | $10^{-16}$ |
| $n_0$ (m$^{-3}$) | $10^{14}$ | $10^{14}$ | $10^{14}$ |
| $N$ (m$^{-3}$) | $10^{15}$ | $10^{15}$ | $10^{15}$ |

with

| | |
|---|---|
| $m_0$ (m$^{-3}$) | $3 \times 10^{14}$ |
| $A_r = v_e \, \sigma_r$ (m$^3$s$^{-1}$) | $10^{-13}$ |

The TL glow curve for this case has already been given in Figure 3.16a.

Numerically solve the relevant rate equations and calculate $P$ for each trap and $Q$ for the system. Is the system in QE? Are the peaks first-order? Are these observations consistent with the observations in Figure 3.16a?

(2) Consider the same model and the same parameters, except that the $A_e$ values now equal to $A_r = 10^{-13}$ m$^3$s$^{-1}$. The TL glow curve for this set of parameters has already been given in Figure 3.16b.

Recalculate $P$ for each trap and $Q$ for the system. What do the values tell you about the kinetic order, and whether or not the system is in QE? Are these conclusions consistent with the observations in Figure 3.16b?

With these definitions it is possible to examine several cases for TL and for CW-OSL. Using model IMTS(4) from Table 3.1 as an example, Figure 3.25 shows the expected TL peaks obtained by numerical solution of the relevant rate equations, and using the parameter values given in the figure caption. Three examples are illustrated. In the first example $n_{10} = n_{20} = h_0$ and $m_0 = n_{10} + n_{20} + h_0$. Also, $A_{e1} = A_{e2} = A_{eh} \ll A_r$. In this case, the values of $P$ for the two traps are always $\ll 1$, and $Q \approx 1$ throughout the process. Therefore the RW expression may be used to describe the two TL peaks. In the second example, $n_{10} = n_{20} = h_0$ and $m_0 = n_{10} + n_{20} + h_0$, as before, but now $A_{e1} = A_{e2} = A_{eh} = A_r$. QE also applies, but values of $P$ for each peak are $< 1$, but not $\ll 1$. In these conditions, the RW equation is a poor fit to the data. (Note that the TL peaks in the RW solution appear at different temperatures than in the numerical solution, and that the intensities of the numerical solution peaks are smaller than those for the RW peaks, due to trapping of charge in the TDDT centers. In the final example, $A_{e1} = A_{e2} = A_{eh} = A_r$, but here $m_0 \gg n_{10} + n_{20} + h_0$ since a large pre-existing concentration of holes is assumed. Again $P \ll 1$ even though the $A$ values are all the same as in the previous example due to $m$ remaining approximately constant throughout the process. Again, $Q \approx 1$ and the system is in QE. The RW equation is still a reasonable fit to the data.

### 3.6.3   Analytical Solutions: No QE Assumption

Lewandowski et al. (1994) examined the NMTS model when $P$ and $Q$ are unknown. They showed analytically that:

$$I_{TL}(T) = n_0 \left( \frac{s}{Q(T) + P(T)} \right) \exp\left\{ -\frac{E_t}{kT} \right\} \exp\left\{ -\frac{s}{\beta_t} \int_{T_0}^{T} \left( \frac{Q(\theta)}{Q(\theta) + P(\theta)} \right) \exp\left[ -\frac{E_t}{k\theta} \right] d\theta \right\}. \quad (3.85)$$

This is a perfectly general solution describing the TL peak shape for either of the OTOR or NMTS models since it makes no assumptions as to the value of $Q$ or $P$ – i.e. no assumptions of QE or kinetic order. The difficulty with applying Equation 3.85 to analyze TL peaks is that the functions $Q(T)$ and $P(T)$ are unknown a priori.

For $Q \approx 1$ and $P \ll 1$ it is easy to see that Equation 3.85 reduces to the RW solution, Equation 3.17, but for other values of $Q$ and $P$ the solution is unknown since $Q(T)$ and $P(T)$ are unknown. For the specific first-order case ($P \ll 1$) and non-quasi-equilibrium ($Q \neq 1$), it may be shown that:

**Table 3.1** Example expressions for $P$ and $Q$ for several simple models, and the corresponding charge neutrality conditions. All terms are defined in the text. Models include the OTOR, the NMTS, and four variants of the IMTS (IMTS(1) to IMTS(4)). In the OTOR model there is just one main trap type, with initial trapped electron concentration $n_0$, and one recombination center, with initial trapped hole concentration $m_0$. For NMTS there is a thermally (or optically) disconnected deep trap (TDDT/ODDT) which is considered to be full after irradiation ($h_0 = H$). Thus, there are no transitions into it during emptying of the main trap. For this model, $m_0 = n_0 + h_0$, and $h_0 = H$. For IMTS(1), transitions into empty TDDTs/ODDTs are allowed during trap emptying but it is assumed that a large concentration of recombination centers already exists in the material before irradiation, $m_{unirr}$, making a total of $m_0 = m_{unirr} + m_{irr}$ at the start of trap emptying. Here, $n_0 + h_0 = m_{irr} < m_0$. In the IMTS(2) model, transitions into empty TDDTs/ODDTs are allowed during trap emptying and $m_0 = n_0 + h_0$ and $m_0 < m_0$. In IMTS(3) there are no disconnected traps. Instead there are two thermally or optically active traps (OATs), of concentrations $N_1$ and $N_2$. In IMTS(4), there are disconnected traps in addition to two active traps. The $P$ and $Q$ expressions are based on the rates of thermal excitation from the trap ($R_{ex}$), the rates of retrapping by the active trap or traps ($R_{retrap}$), the rates of recombination at the trapped hole sites ($R_{recom}$), and the rate of trapping by the thermally (or optically) disconnected traps ($R_{trap}$)

| Model | $P$ and $Q$ | Neutrality condition |
|---|---|---|
| OTOR | $P = \dfrac{R_{retrap}}{R_{recom}} = \dfrac{(N-n)\,A_e}{m A_r}$ <br><br> $Q = \dfrac{R_{ex} - R_{retrap}}{R_{recom}} = \dfrac{np - n_c(N-n)\,A_e}{n_c m A_r}$ | $n_0 = m_0$ |
| NMTS | $P = \dfrac{R_{retrap}}{R_{recom}} = \dfrac{(N-n)\,A_e}{m A_r}$ <br><br> $Q = \dfrac{R_{ex} - R_{retrap}}{R_{recom}} = \dfrac{np - n_c(N-n)\,A_e}{n_c m A_r}$ | $n_0 + h_0 = m_0$ <br><br> $h_0 = H$ |

*(Continued)*

**Table 3.1** (Continued)

IMTS(1)

$$P = \frac{R_{retrap}}{R_{recom} + R_{trap}} = \frac{(N-n)A_e}{mA_r + (H-h)A_{eh}}$$

$$Q = \frac{R_{ex} - R_{retrap} - R_{trap}}{R_{recom}} = \frac{np - n_c(N-n)A_e - n_c(H-h)A_{eh}}{n_c.mA_r}$$

$n_0 + h_0 = m_{irr}$

$m_0 = m_{irr} + m_{unirr}$

$h_0 < H$

$m_{unirr} >> 0$

IMTS(2)

$$P = \frac{R_{retrap}}{R_{recom} + R_{trap}} = \frac{(N-n)A_e}{mA_r + (H-h)A_{eh}}$$

$$Q = \frac{R_{ex} - R_{retrap} - R_{trap}}{R_{recom}} = \frac{np - n_c(N-n)A_e - n_c(H-h)A_{eh}}{n_c.mA_r}$$

$n_0 + h_0 = m_0$

$h_0 < H$

IMTS(3)

For Trap 1

$$P_1 = \frac{R_{retrap1}}{R_{recom} + R_{retrap2}} = \frac{(N_1-n_1)A_{e1}}{mA_{mr} + (N_2-n_2)A_{e2}}$$

$$Q = \frac{R_{ex} - R_{retrap}}{R_{recom}} = \frac{n_1p_1 + n_2p_2 - n_c(N_1-n_1)A_{e1} - n_c(N_2-n_2)A_{e2}}{n_c.mA_r}$$

For Trap 2

$$P_2 = \frac{R_{retrap2}}{R_{recom} + R_{retrap1}} = \frac{(N_2-n_2)A_{e2}}{mA_r + (N_1-n_1)A_{e1}}$$

$n_{10} + n_{20} = m_0$

IMTS(4)

For Trap 1

$$P_1 = \frac{R_{retrap1}}{R_{recom} + R_{retrap2} + R_{trap}} = \frac{(N_1-n_1)A_{e1}}{mA_r + (N_2-n_2)A_{e2} + (H-h)A_{eh}}$$

$$Q = \frac{R_{ex1} + R_{ex2} - R_{retrap1} - R_{retrap2} - R_{trap}}{R_{recom}} = \frac{n_1p_1 + n_2p_2 - n_c(N_1-n_1)A_{e1} - n_c(N_2-n_2)A_{e2} - n_c(H-h)A_{eh}}{n_c.mA_r}$$

For Trap 2

$$P_2 = \frac{R_{retrap2}}{R_{recom} + R_{retrap1} + R_{trap}} = \frac{(N_2-n_2)A_{e2}}{mA_r + (N_1-n_1)A_{e1} + (H-h)A_{eh}}$$

$n_{10} + n_{20} + h_0 = m_0$

$h_0 < H$

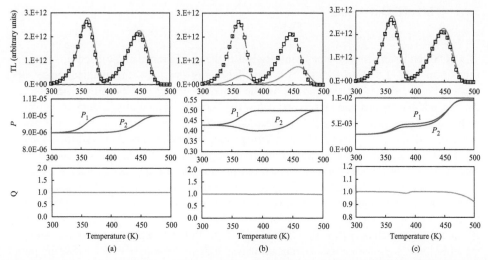

**Figure 3.25**  TL curves and $P$ and $Q$ factors for a variety of conditions using the IMTS(4) model given in Table 3.1. The parameters for (a) are: $E_{t1} = 0.8$ eV, $E_{t2} = 1.0$ eV, $s_1 = s_2 = 10^{10}$ s$^{-1}$, $n_{10} = n_{20}$ $= h_0 = 10^{14}$ m$^{-3}$, $m_0 = 3 \times 10^{14}$ m$^{-3}$, $N_1 = N_2 = H = 10^{15}$ m$^{-3}$, $A_{e1} = A_{e2} = A_{eh} = 10^{-16}$ m$^3$s$^{-1}$, (where $A_{ei} = \sigma_{ei} v_e$), and $A_r = 10^{-13}$ m$^3$s$^{-1}$. For (b), the parameters are the same, except $A_{e1} = A_{e2} = A_{eh} = A_r$ $= 10^{-16}$ m$^3$s$^{-1}$. For (c), the parameters are the same as (b), except $m_0 = 10^{17}$ m$^{-3}$. In each column, the top figure shows the numerical solution to the relevant rate equations (in green), the individual Randall-Wilkins (RW) TL peaks using the parameters given above (dashed blue and red curves) and the sum of the RW peaks (open squares). The middle figure shows the $P$ values for each trap, using the formulae given in Table 3.1 for IMTS(4). The bottom figure gives the value of $Q$ for the system in each case, also using the formula given in Table 3.1 for IMTS(4).

$$I_{TL}(T) = \frac{I_{RW}(T)}{Q(T)}. \tag{3.86}$$

Once again, however, $Q(T)$ is unknown and therefore this simple relationship has little practical utility.

The definitions for $P$ and $Q$ as listed in Table 3.1 can also be applied to CW-OSL. An example analysis of this type has already been given in Figure 3.19 and explained in detail on the accompanying web site.

## 3.7  Thermal and Optical Effects

The occupancy of any given trap is related to temperature through the term $\exp\{-E_t / kT\}$, and to optical stimulation through the term $\sigma_p(E)\Phi$. As a result, it can be expected that TL glow curves can be affected by light exposure after irradiation, but before TL readout. Similarly, it can be expected that OSL curves may be affected by heating after the irradiation but before OSL measurement. Furthermore, the temperature at which the OSL stimulation takes place can be expected to affect the resultant OSL data. Finally, both TL and OSL may also be affected by the phenomenon of thermal quenching in which the efficiency of luminescence decreases with increasing temperature. These thermo-opto-phenomena are now discussed, beginning with thermal quenching.

### 3.7.1   Thermal Quenching

#### 3.7.1.1   Mott-Seitz Model

TL or OSL emission from dosimetry materials is in the visible range of the electromagnetic spectrum, from, say, ~1.5 eV to ~3.5 eV. However, the energy levels of the recombination centers are clustered between the mid-gap and the middle of the lower half of the energy band – i.e. above the trapped hole Demarcation levels and below the Fermi level (assuming electrons recombining with trapped holes; see Figure 3.14). Since dosimetry materials are generally wide band-gap insulators, the energy lost by the electron during a recombination transition (transition 5 in Figure 3.14) is much greater than the energy emitted during TL or OSL. The electron energy is dissipated via phonons (lattice vibrations) in a multi-phonon process, cascade capture, or Auger process (as discussed in Chapter 2) and results in the excitation of the recombination site into an excited state. An example (mentioned earlier in Section 3.4.1) is the TL and/or OSL emission from $Al_2O_3$:C in which the recombination of an electron with an $F^+$-center results in an excited $F$-center ($F^*$), which decays with a lifetime of 35 ms to the $F$-center ground state, along with the emission of light with a spectrum peaking near 420 nm ($e^- + F^+ = F^* = F + 420$ nm photon). Similar considerations apply to, for example, rare earth impurities, and radiative recombination sites in general. (The word "radiative" here refers to the emission of luminescence; not all recombination events are radiative.)

Once in the excited state, the electron has more than one pathway to return to the ground state. In the Mott-Seitz model of thermal quenching (Mott and Gurney 1948), this can be understood from the configurational coordinate diagram, displayed here in Figure 3.26 for the general case. The excited ($E_e$) and ground ($E_g$) state energies of the center are shown as a function of configuration coordinate. For a radiative transition, the electron decays from its relaxed excited state (at energy $E_{e0}$) to the equilibrium ground state (at energy $E_{g0}$) with the emission of a photon of energy $E_{em}$ and phonons. However, if the temperature is high enough the probability exists for the electron to absorb an amount of energy $W$ from phonons while in its excited state and be raised to energy level $E_x$, from where it may relax via the emission of phonons of energy $E_x$-$E_{g0}$. In this way, relaxation may occur via the emission of phonons only, without the emission of photons. Thus, at a given temperature, there exists the probability that an electron in an excited state can relax to the ground state either radiatively or non-radiatively.

In general, an electron in an excited state can decay radiatively via spontaneous emission or phonon-assisted (vibronic) emission, or it can decay non-radiatively via the emission phonons. The net probability of decay (e.g. Di Bartolo 1968) is:

$$p_d(T) = \frac{1}{\tau} = \frac{1}{\tau_0} + a\coth\left(\frac{h\nu_{ph}}{kT}\right) + s_q\exp\left\{-\frac{W}{kT}\right\} \qquad (3.87)$$

where $\tau$ is the excited state lifetime, $\tau_0$ is the radiative lifetime, $a$ is a temperature-independent constant, $\nu_{ph}$ is the phonon vibration frequency, and $s_q$ and $W$ are the frequency factor and activation energy for the non-radiative transition – i.e. for thermal quenching.

Assuming the vibronic component is negligible, the overall decay time becomes:

$$\tau = \frac{\tau_0}{1 + \tau_0 s_q \exp\left\{-W / kT\right\}} \qquad (3.88)$$

meaning that the decay time is an exponentially decreasing function of temperature, with activation energy $W$. The luminescence efficiency $\eta(T)$ is the ratio of the radiation probability to all other probabilities and is a function of temperature thus:

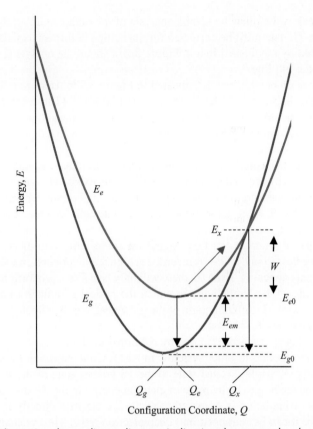

**Figure 3.26** Configurational coordinate diagram indicating the energy levels of the excited ($E_e$) and ground states ($E_g$) of a radiative recombination center. Emission of a photon (of energy $E_{em}$) occurs when the electron transitions from its minimum excited state energy $E_{e0}$ to the ground state, at the same configuration coordinate (vertical downward arrow in the figure). Final relaxation to the ground state minimum $E_{g0}$ occurs with the further emission of phonons and a change in configuration coordinate (from $Q_e$ to $Q_g$). However, if, while in the excited state, the electron absorbs enough thermal energy $W$ such that it attains the energy $E_x$, a transition to the ground state can occur at a configuration coordinate $Q_x$, and the energy minimum $E_{g0}$ is then attained via the emission of phonons of total energy $E_x - E_{g0}$, and a change in the coordinate from $Q_x$ to $Q_g$.

$$\eta(T) = \frac{\tau}{\tau_0} = \frac{1}{1 + C\exp\{-W/kT\}}, \tag{3.89}$$

where $C = \tau_0 s_q$.

If thermal quenching is occurring during the production of TL, the measured TL is then $I_{TL}^Q(T) = I_{TL}^{UQ}(T)\eta(T)$, where $I_{TL}^Q(T)$ is the quenched TL and $I_{TL}^{UQ}(T)$ is the unquenched TL.

Figure 3.27 illustrates the effect of thermal quenching on the shape of a single TL glow peak, using first-order kinetics as an example. In Figure 3.27a, the expected TL peak without thermal quenching has a maximum at ~480 K, whereas with quenching the peak appears at a lower temperature (at ~476 K). Furthermore, the shape of the peak is distorted from the RW shape, and is reduced in height. The extent to which the peak is reduced in height, by how much the peak shifts to lower temperature, and the degree of peak distortion depend on the values of $C$ and $W$ for the quenching term, and $E_t$, $s$ and $\beta_t$ for the TL peak. Since the measured TL peak is the quenched peak, analysis of the obtained TL glow curve will be of no value and the TL data must be corrected for quenching before analysis can begin. That is, the peak indicated by the dashed line in Figure 3.27a should be analyzed, not the measured peak shown by the full line.

The dependence on heating rate is indicated in Figure 3.27b. Here four TL peaks are indicated for $\beta_t$ values of 5 K.s$^{-1}$, 10 K.s$^{-1}$, 15 K.s$^{-1}$ and 20 K.s$^{-1}$. (These are the same data as in Figure 3.2.) Since the efficiency curve $\eta(T)$ is fixed, the TL peak "moves" through the $\eta(T)$ curve as $\beta_t$ changes, from temperatures where quenching is low to temperatures where quenching is high as $\beta_t$ increases.

Since TL requires the sample to be heated in order to stimulate the emission, it is inevitable that it will be affected by thermal quenching, if quenching exists. The same is not true for OSL. Normally, most OSL measurements take place at room temperature, whereas many quenching phenomena occur above room temperature. As a result, OSL is often recorded with greater efficiency than TL. An example of this is Al$_2$O$_3$:C where thermal quenching of the TL signal reduces the emission whereas the OSL signal (at room temperature) is unaffected by the phenomenon. Needless to say, if OSL is recorded at higher temperatures, it too may be adversely affected by thermal quenching. Even so, there will only be a fixed, proportional reduction in the OSL signal as long as the temperature at which the optical stimulation takes place remains constant, whereas with TL different parts of the glow peak will be affected to different extents depending on parameters $E_t$, $s$ and $\beta_t$.

A variation on this model was proposed by Nikiforov et al. (2001) for thermal quenching in Al$_2$O$_3$:C, and since applied to other systems. In Al$_2$O$_3$:C the emission (at 420 nm) is due to relaxation from the 3P triplet excited state to the 1S ground state of the $F$-center. Instead of phonon absorption leading to an increase of electron energy in the 3P state to level $E_x$ (Figure 3.26), the thermal stimulation is proposed to excite the electron directly into the conduction band, from where it can be retrapped into deep electron traps. This mechanism also yields a luminescence efficiency function as in Equation 3.89, with $W$ being the activation energy

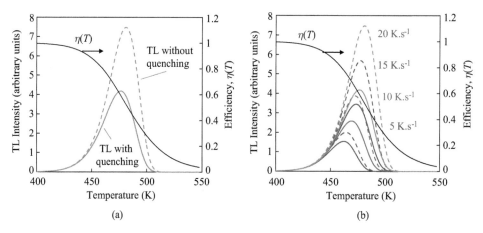

**Figure 3.27**    (a) The luminescence efficiency function $\eta(T)$ (for $W = 1.15$ eV, and $C = 1\times10^{12}$) and a TL peak for $E_t = 1.5$ eV, $s = 1\times10^{15}$ s$^{-1}$ and $\beta_t = 1$ K.s$^{-1}$. The TL peak without quenching $\left(I_{TL}^{UQ}(T)\right)$ is given by the dashed curve, whereas the peak with quenching $\left(I_{TL}^{Q}(T) = \eta(T)I_{TL}^{UQ}(T)\right)$ is shown by the full line. (b) Changes in the TL peak shape as a function of heating rate ($\beta_t = 5$ K.s$^{-1}$, 10 K.s$^{-1}$, 15 K.s$^{-1}$ or 20 K.s$^{-1}$) both with (full lines) and without (dashed lines) thermal quenching.

necessary for ionization of the *F*-center. Unlike the Mott-Seitz mechanism, the Nikiforov et al. model leads to a change in charge state from *F* to $F^+$ and predicts that the occupancy of the deep traps will affect the degree of thermal quenching observed.

### 3.7.1.2  Schön-Klasens Model

An alternative mechanism for thermal quenching is the so-called Schön-Klasens mechanism (Klasens 1946; Schön 1951). So far, the discussion of the various models for TL and OSL has assumed that electrons are released from the "active" trap and recombine with trapped holes, which are considered thermally and optically stable under the conditions of the TL or OSL measurement. The Schön-Klasens model considers the situation where the trapped holes are in fact thermally unstable in the temperature range over which the TL signal appears (or at the temperature of stimulation in OSL). The principle is illustrated in Figure 3.28a using the OTOR model. Here electrons can be thermally emptied (with $p_e = s_e \exp\{-E_{te}/kT\}$) or optically emptied (with $p_e = \sigma_p(E)\Phi$) emptied at the same time that the trapped holes, which act as the recombination sites, are also released thermally (with $p_h = s_h \exp\{-E_{th}/kT\}$). The subscripts "*e*" and "*h*" refer to the electrons and holes, respectively, and otherwise the terms have their usual meanings. Holes that are freed by this mechanism are presumed to recombine non-radiatively with trapped electrons. This can occur elsewhere in the sample at other electron centers, or at the electron trap, as indicated in Figure 3.28a. Note that the electron- and hole-centers act as both traps and recombination centers in this model and thus the energy levels must be located near the respective Demarcation levels, at which energies $n_c m A_{er} \approx s_h \exp\{-E_{th}/kT\}$ and $n_v n A_{hr} \approx s_e \exp\{-E_{te}/kT\}$. Here, $A_{er}$ and $A_{hr}$ are the free electron and free hole recombination probabilities, respectively.

Although the TL or OSL signal is still given by $n_c m A_{er}$, this term is no longer equal to $|dm/dt|$, since now:

$$\frac{dm}{dt} = -m s_h \exp\left\{\frac{-E_{th}}{kT}\right\} - n_c m A_{er}. \tag{3.90}$$

The efficiency of producing luminescence is then:

$$\eta(T) = \frac{n_c m A_{er}}{n_c m A_{er} + m s_h \exp\left\{\dfrac{-E_{th}}{kT}\right\}} = \frac{1}{1 + C^* \exp\left\{\dfrac{-E_{th}}{kT}\right\}}, \tag{3.91}$$

where $C^* = s_h / n_c A_{er}$ is a dimensionless, but temperature-dependent quantity.

Although Equation 3.91 has the same form as Equation 3.89 (with $E_{th}$ replacing *W*, and $C^*$ replacing *C*) it is emphasized that $C^*$ is a function of temperature since $n_c$ is a function of temperature. Under QE conditions, $dn_c/dt \ll dm/dt$, $dn/dt$, and therefore it might be expected that $n_c$, and therefore $C^*$, is approximately constant. However, this is certainly not true throughout the temperature range. Furthermore, since the Schön-Klasens mechanism is a "dynamic" one in which the rates at which the relative processes occur depend on the heating rate, then one can expect that Equation 3.91 will also depend on the heating rate – that is, $C^*$ must also be dependent on $\beta_t$.

Note that whereas in the Mott-Seitz model of quenching, the obtained TL peaks can be corrected by dividing by the Mott-Seitz efficiency curve, Equation 3.89 (that is, $I_{TL}^{UQ}(T) = I_{TL}^{Q}(T)/\eta(T)$) the obtained TL peak in Schön-Klasens model the cannot be corrected by dividing by Equation 3.91. Doing so will simply produce $dm/dT$.

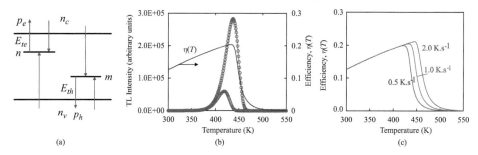

**Figure 3.28** (a) Schön-Klasens mechanism for thermal quenching in the OTOR model. The trapped holes are thermally unstable during electron-hole recombination giving rise to a luminescence efficiency, $\eta(T) = 1/\left(1 + C^* \exp\left\{-E_{th}/kT\right\}\right)$ (terms defined in the text). (b) Numerical solutions to the model of figure (a), showing the obtained TL curve (blue). The input parameters were $E_{te} = 1.2$ eV, $E_{th} = 1.15$ eV, $s_e = s_h = 5 \times 10^{12}$ s$^{-1}$, $n_0 = m_0 = 1 \times 10^7$ cm$^{-3}$, and $\beta_t = 1$ K.s$^{-1}$. It is compared with the TL curve that is obtained if the holes are thermally stable (orange) (same parameters, except for a large value of $E_{th}$). The efficiency curve $\eta(T)$ (Equation 3.91) is also shown. Fits to the TL curves are also shown (open circles) using the RW expression. The obtained parameter values for the Schön-Klasens model are $E_t = 1.18$ eV and $s = 1.63 \times 10^{12}$ s$^{-1}$ and $E_t = 1.20$ eV and $s = 0.6 \times 10^{12}$ s$^{-1}$ for the TL peak without thermal release of holes. (c) Changes in the efficiency curves $\eta(T) = 1/\left(1 + C^* \exp\left\{-E_{th}/kT\right\}\right)$ with heating rate. $C^*$ is calculated from $C^* = s_h/n_c A_r$.

The effect of this model on the TL curve shape is illustrated in Figure 3.28b in which is illustrated example numerical solutions to the Schön-Klasens model, along with the efficiency curve $\eta(T)$ calculated using Equation 3.91. The quenched TL peak is shown for a heating rate of 1 K.s$^{-1}$. For comparison, the TL peak that would be obtained with the same trapping values of $E_{te}$, $s_e$, and $\beta_t$ is also shown. This is obtained by choosing a large value for $E_{th}$ such that $n_c m A_{er} \gg s_h \exp\left\{-E_{th}/kT\right\}$ and thus the hole center plays the role of a recombination center only, not a trap.

Fits to the TL peaks using the RW equation (Equation 3.17) are shown in Figure 3.28. The value of $E_{te}$ obtained does not equal the input value. It has been determined that, assuming electron-hole recombination is radiative, that when $E_{te} < E_{th}$, accurate values for $E_{te}$ may be obtained from fitting, whereas when $E_{te} > E_{th}$ (as in the data for Figure 3.28) inaccurate values are obtained. The opposite applies if it is the hole-electron recombination transition that is radiative.

As the heating rate changes (Figure 3.28c) so too does the shape of the function $\eta(T)$ (Equation 3.91). This is not true of the Mott-Seitz model. Comparison of Figure 3.27 ($W = 1.15$ eV) with Figure 3.28 ($E_{th} = 1.15$ eV) indicates that the Mott-Seitz efficiency curve is different in shape and decreases more slowly than the Schön-Klasens efficiency curve of Figure 3.28. The Schön-Klasens efficiency curves cannot be fitted to an equation of the form of Equation 3.89 using a constant value of $C$.

It can be difficult to determine which mechanism is responsible for the thermal quenching phenomenon. However, the Schön-Klasens mechanism should be suspected if it is found that either: (i) the efficiency curve (measured using one of the methods described in the next section) is unable to account consistently for the observed quenching of the TL signal when the TL is measured at different heating rates, and/or (ii) it is difficult to fit the obtained efficiency curve with Equation 3.89 with a constant value of $C$. An examination of the charge state of the recombination site may separate the Mott-Seitz mechanism from the Nikiforov et al. mechanism.

### 3.7.1.3 Tests for Thermal Quenching

Regardless of the mechanism of thermal quenching, it is important that tests are made for the phenomenon before analysis of TL is attempted. (For OSL, at a constant temperature, the shape of the OSL curve (whether it is CW-OSL, LM-OSL, VE-OSL, or POSL) is unaffected by quenching; the OSL signal is simply reduced in intensity.) Several potential methods are described below.

#### 3.7.1.3.1 τ-versus-T

Equation 3.89 indicates that a direct method for determining if thermal quenching is occurring, and over what temperature range, is to measure the excited state lifetime $\tau$ at different temperatures. The preferred method is to probe the luminescence center directly using photoluminescence (PL). Using fast stimulation pulses (of pulse width $< \tau$) the decay of the PL can be followed after the pulse and the lifetime $\tau$ determined. By repeating the measurement at different temperatures, a plot of $\tau$-versus-$T$ can be obtained. If thermal quenching is occurring the function $\tau(T)$ will follow Equation 3.89, and $C$ and $W$ can be determined.

If the induced center is not photoluminescent, however, an alternative approach is to irradiate the sample and to use POSL. By measuring the intensity of the POSL signal after the stimulating pulse (again, using a pulse width $< \tau$) as a function of time, at different $T$, $\tau$ can be determined and a plot of $\tau$-versus-$T$ can again be obtained. There are several caveats with this approach, however.

As mentioned previously, the delay between the stimulating pulse and the POSL emission is governed by several lifetimes. Only if the excited state lifetime $\tau$ is the dominating lifetime can this method directly yield $\tau$-versus-$T$. For example, consider a sample with shallow traps in addition to the active trap. Since the stimulation pulse causes electrons to be released into the conduction band, freed electrons can be retrapped into shallow traps. At low temperatures, when the residence time in the shallow trap $\tau_{trap} = s^{-1}\exp\{E/kT\}$ is long, electrons trapped by the shallow traps are lost to the POSL signal and do not interfere with the measurement of $\tau$. As $T$ increases, and electrons are released from the shallow trap, the measured lifetime is a convolution of the shallow trap residence time $\tau_{trap}$ and the excited state lifetime $\tau$. Finally, as $T$ is increased further, the POSL signal is free of the effects of the shallow traps and $\tau$ alone is measured.

An example of this type of measurement, for POSL from samples of $Al_2O_3$:C, one with and two without shallow traps, is shown in Figure 3.29.

#### 3.7.1.3.2 Luminescence Intensity versus T

A simple method to examine thermal quenching is to measure the luminescence intensity, not lifetime, as a function of $T$. As examples, the intensity of PL, RL, or OSL could each be measured at different temperatures, taking care to ensure that all other conditions are the same for each measurement. The simplest is PL, since this does not require irradiation of the sample and only the emitting center is probed. Charge transport via the delocalized bands is not involved. RL and OSL are more complex since each requires irradiation of the sample. The intensity is then a function of the occupancy of all the traps present in the system. RL, for example, normally increases with irradiation time as the traps fill, only reaching a constant value after a long enough irradiation time (and subject to the concentrations of available traps). Once steady state has been achieved, the intensity can be measured; the temperature can then be changed, steady-state re-established at the new temperature, and the RL intensity measured again. In this way the RL intensity as a function of temperature can be monitored. Note, however, that as traps become thermally unstable and empty as the temperature increases, the intensity of the

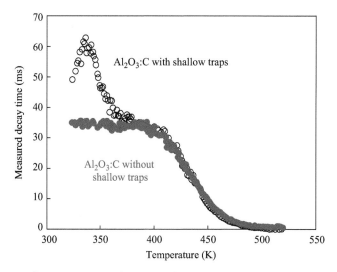

**Figure 3.29** POSL decay time as a function of temperature for *F*-centers in $Al_2O_3$:C, after a stimulating light pulse. When shallow traps are present an increasing decay time is observed as the traps empty with increasing temperature and the measured decay time is a convolution of the residence time in the shallow trap $\tau_{trap}$ and the *F*-center excited state lifetime $\tau$. With further increase in *T*, the value of $\tau_{trap}$ decreases until the measured lifetime corresponds to the *F*-center lifetime of 35 ms. In the samples without shallow traps, only $\tau = 35$ ms is measured at low *T*. With an increase in temperature, $\tau$ decreases due to thermal quenching according to Equation 3.89. (Adapted from Akselrod et al. (1998).)

measured luminescence can transiently increase with temperature, distorting the expected decrease if thermal quenching is present. This additional signal is technically TL superimposed on top of the RL signal.

For OSL, the situation is even more complex. Here the measurement mode itself empties the traps and a noticeable change in the trap occupancies may occur during the measurement. Since this also affects the OSL intensity, it may be necessary to take one OSL measurement at one stimulation temperature, then completely empty all traps by heating to high temperatures, re-irradiate and measure the OSL again at a different stimulation temperature. The assumption is that all initial conditions are reliably reproduced after each irradiation, which may not be true. Even if true, an increase in OSL with increase in *T* can often be seen as shallow traps become less effective and more electrons released from the active trap produce luminescence.

For the above reasons, the observation of thermal quenching can be confused when measuring RL or OSL as a function of temperature such that analysis of the quenching curve to find *W* (or $E_{th}$) may be unreliable. Notwithstanding, the observation of a decrease in RL or OSL intensity as a function of temperature is a clear indication of thermal quenching, even if the analysis of the activation energy may be compromised.

### Exercise 3.12    Thermal Quenching

(a) Consider an IMTS model with at least two electron traps and one hole trap (which acts as the recombination center). One of the electron traps should be a shallow trap. You choose the concentrations and parameter values to be used. Numerically solve the rate equations and simulate irradiation of the system, at a rate *G* (electron-holepairs.cm$^{-3}$.s$^{-1}$) in order to calculate the luminescence emitted during irradiation

> (i.e. radioluminescence, RL). Change the temperature, and calculate how the RL intensity changes with temperature. Make sure you understand the results.
> (b) Now add thermal quenching, as in the Mott-Seitz model, and repeat the calculations. How does the RL-versus-*T* curve change?
> (c) Repeat, altering the concentrations and parameters of the various traps, and the *C* and *W* parameters of thermal quenching (see Equation 3.89) to gain an understanding of how the various parameters change the results observed.

### 3.7.2   Thermal Effects on OSL

In addition to thermal quenching, the factors that can affect the temperature dependence of CW-OSL and LM-OSL curves and are most easily described by the IMTS model include: (1) retrapping by shallow traps; (2) retrapping by deep traps; (3) thermally assisted optical stimulation; (4) simultaneous hole and electron release; and (5) localized donor–acceptor type recombination. The last of these is dealt with in the next section. Possibilities (1)–(4) are discussed here.

#### 3.7.2.1   *Effects of Shallow Traps*

An example model with a shallow trap in addition to the main, optically active trap (OAT) is shown in Figure 3.30a. Also included in this example is a deep, thermally and optically disconnected trap (TDDT/ODDT, trap 3, concentration $N_3$). In this model, the shallow traps (trap 1, concentration $N_1$) are presumed to thermally empty according to $p_1 = \exp\{-E_{t1}/kT\}$, whereas the OAT (trap 2, concentration $N_2$) empties optically, according to $p_2 = \sigma_{p2}\Phi$ (where the notation indicating the stimulation energy dependence of $\sigma_{p2}$ has been dropped for simplicity). Optical emptying from the shallow trap (level 1) is ignored in this example, as is thermal emptying from level 2. Since $m = n_1 + n_2 + n_3 + n_c$ and assuming QE, the OSL intensity as a function of stimulation time can be written:

$$I_{OSL}(t) = \left|\frac{dm}{dt}\right| \approx \frac{dn_1}{dt} + \frac{dn_2}{dt} + \frac{dn_3}{dt} = -n_1 s_1 \exp\left\{-\frac{E_{t1}}{kT}\right\} + n_c (N_1 - n_1) A_{e1}$$
$$- n_2 \sigma_{p2}\Phi + n_c (N_2 - n_2) A_{e2} + n_c (N_3 - n_3) A_{e3}. \tag{3.92}$$

Equation 3.92 indicates that electrons released from the OAT can be retrapped into the shallow trap, whereupon, if the temperature *T* is high enough, they can then be thermally re-released before they ultimately recombine. This causes a delay in the emission of OSL. Retrapping into the deep trap is also allowed and this leads to a "loss" of electrons from the OSL process. An example solution to this model is shown in Figure 3.30(b) showing an initial slow rise of the OSL signal due to the delay caused by retrapping and subsequent thermal release of electrons from the shallow trap.

It is clear that many variations on this model can be contemplated – for example, optical as well as thermal emptying from the shallow trap; thermal as well as optical emptying from the OAT; multiple recombination centers, etc. In this sense, the model and solutions shown in Figure 3.30 are examples only. The last possibility mentioned here (i.e. multiple recombination centers) does not lead directly to a temperature dependence but can create a non-luminescent, competing recombination pathway. If the concentration of holes trapped in such a center is $m_2$, while the concentration of holes in the radiative center is $m_1$, then the charge neutrality condition must be modified to $m_1 + m_2 = n_1 + n_2 + n_3 + n_c$. However the OSL emission is

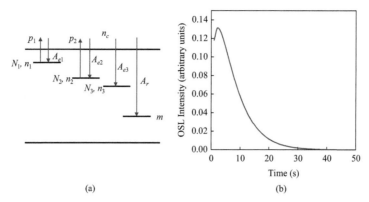

(a)                                              (b)

**Figure 3.30** (a) IMST model with a shallow trap (trap 1), and OAT (trap 2), and an optically disconnected deep trap (trap3; ODDT). The indicated parameters are: $N_1$, $N_2$, $N_3$ – available concentrations of traps; $n_1$, $n_2$, $n_3$ – the concentrations of trapped electrons; $n_c$ – concentration of electrons in the conduction band; $m$ – the concentration of trapped holes; $A_{e1}$, $A_{e2}$, $A_{e3}$ – retrapping coefficients ( $v_e\sigma_{e1}$, $v_e\sigma_{e2}$ and $v_e\sigma_{e3}$, respectively); $A_r$ – recombination coefficient of trapped electrons with trapped holes. (b) Solution to the rate equations for this model describing the CW-OSL emission at 300 K, with parameter values: $N_1$, $N_2 = 10^{15}$ m$^{-3}$; $N_3 = 10^{14}$ m$^{-3}$; $s_1 = 5\times10^{11}$ s$^{-1}$; $E_{t1} = 0.7$ eV; $s_2 = 10^{12}$ s$^{-1}$; $E_{t2} = 1.5$ eV; $A_{e1} = A_{e2} = 10^{-15}$ m$^3$.s$^{-1}$; $A_{e3} = 10^{-16}$ m$^3$.s$^{-1}$; $A_r = 10^{-13}$ m$^3$.s$^{-1}$; $R_{ex} = 10^{10}$ m$^3$.s$^{-1}$; $p_2 = 0.1$ s$^{-1}$. (Note: At 300 K, thermal release of electrons from trap 2 ($s_2\exp\{-E_{t2}/kT\}$) is negligible and is not included in Equation 3.92. Also, the system was first simulated from all traps empty and irradiated at a rate $G$, followed by a relaxation period to allow all trapping levels to reach their equilibrium concentrations, and then the OSL stimulation was simulated).

only $|dm_1/dt| = n_c m_1 A_{r1}$, and an extra "loss" term appears on the right-hand side of the equation for OSL, namely $n_c m_2 A_{r2}$. Two terms on the right-hand side of Equation 3.92 would then represent permanent losses for OSL, namely $n_c (N_3 - n_3) A_{e3}$ and $n_c m_2 A_{r2}$, whereas two others, $n_c (N_1 - n_1) A_{e1}$ and $n_c (N_2 - n_2) A_{e2}$, lead to a re-cycling of electrons still with the potential to contribute to OSL but with a delay in OSL emission.

---

**Exercise 3.13   Modelling OSL emission with the IMTS model, including shallow traps**

Consider the 3TOR model of Figure 3.30a. Assume values for the various parameters (perhaps using the parameter values in the figure caption as a starting point) and simulate the shape of the CW-OSL curve. Vary the parameter values (e.g. increasing/decreasing the concentrations of the various centers ($N_i$, with $i = 1$, 2, or 3), the trap depths ($E_{ti}$), and the transition probabilities ($A_{ei}$ and $A_r$), and examine how the CW-OSL curve shape changes as these parameters change. Make sure you can interpret the changes observed in terms of the changes to the rates of the various transitions. Note: the simulation has to start with all traps empty (why?) and the irradiation phase must also be simulated, followed by a relaxation phase before simulating the OSL measurement phase. For the irradiation phase, the rate of generation of electron-hole pairs across the band gap, $G$, has to be included. During the relaxation and OSL phases, $G = 0$.

Since the delay in the CW-OSL emission is a result of trapping and thermal detrapping of electrons from the shallow trap, it can be expected that this delay will increase if the temperature of stimulation is decreased, and decrease if the temperature is increased. At low temperatures, the shallow traps may be quite stable, therefore any charges transferred to them from the active traps will remain in the shallow traps for a longer period than they will at higher temperatures. As a result, the shallow traps will act as competitors to luminescence and the sensitivity will be low. As the stimulation temperature increases the sensitivity also increases, but there will be a delay in reaching the maximum OSL (Figure 3.30b). At still higher temperatures, the shallow traps will no longer hold charge for any significant period and there will be little evidence of a delay in reaching the OSL maximum; the sensitivity will be increased since the shallow traps will no longer act as competitors. The delays observed and the changes in the sensitivity as the temperature increases depend on the various concentrations and parameters for the sample. This discussion assumes thermal quenching is not a factor.

Another, sometimes-observed phenomenon with OSL is that after complete readout of the OSL signal (full bleaching), the signal can re-appear without additional irradiation. The phenomenon is called "recuperation." If the sample is held at a fixed temperature, during and after stimulation, it is possible that during stimulation, some of the released electrons become trapped by the shallow traps. After the end of the stimulation, the shallow traps then slowly release their charge and some of these electrons become re-trapped back into the OATs. A second stimulation now reveals a second ("recuperated") OSL signal, albeit much weaker than the first OSL signal.

### 3.7.2.2   Effects of Deep Traps: Thermally Transferred OSL (TT-OSL)

It is possible for a second OSL signal to be observed even if there are no shallow traps. In this case, the cause of the phenomenon is the thermal transfer of trapped electrons from hard-to-bleach, more thermally stable traps into the OATs that were emptied during OSL measurement. (Here the term "thermally stable" means stable at the temperature of the OSL stimulation. The charge trapped in these traps becomes unstable as the sample is heated post-stimulation.) The recovered OSL in this case termed thermally transferred OSL, or TT-OSL. The critical parameters in this are the relative thermal stabilities of the OAT and the hard-to-bleach trap (HBT). Consider heating the sample after OSL stimulation to a temperature $T$, at which temperature the lifetime in a trap is given by $\tau = s^{-1}\exp\{E_t / kT\}$. If the lifetimes for the traps are such that $\tau_{OAT} \approx \tau_{HBT}$, or $\tau_{OAT} > \tau_{HBT}$, then thermal transfer from HBTs to OATs will be possible. If $\tau_{OAT} < \tau_{HBT}$, it is likely that the electrons are insufficiently stable in the OATs such that no subsequent OSL signal would be observed.

Beryllium oxide provides an example of the effect (Yukihara 2020). In this material, OSL is produced by stimulating irradiated BeO at 470 nm. The trapped electrons in OATs that give rise to this signal are stable up to ~250 °C, above which temperature the trapped electrons are thermally emptied. However, if the OSL is stimulated at, say, room temperature, and then heated to above ~225 °C followed by cooling to room temperature, an OSL signal is observed again, without the need to re-irradiate. In BeO, this observation is explained by the proposed existence of two traps, OATs and the HBTs, which are of similar thermal stabilities (i.e. $\tau_{OAT} \approx \tau_{HBT}$), but only the OATs are sensitive to the 470 nm stimulation light; the HBTs are emptied only by a few percent during optical stimulation.

A model for the effect is illustrated in Figure 3.31. Vertical blue arrows indicate sensitivity to light; vertical red arrows indicate sensitivity to heat; trapping and recombination transitions are indicated in black. The key element is that the trapping lifetime of the electrons in the OATs and HBTs are similar at the temperature to which the sample is heated following OSL stimulation. Thus, during optical stimulation, the OATs are emptied but the HBTs are not. Upon

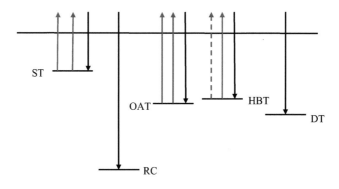

**Figure 3.31**  Model for TT-OSL. The model consists of shallow traps (ST), optically active traps (OAT), hard-to-bleach traps (HBT), deep, thermally and optically disconnected traps (DT) and recombination centers (RC). Blue, vertical, upward arrows indicate optical stimulation transitions. (The dashed blue arrow from the HBT level indicates a weak transition.) Red, vertical, upward arrows indicate thermal stimulation transitions. Black, downward, vertical arrows indicate either trapping or recombination transitions. The OAT and HBT levels have very similar trapping lifetimes at the temperature to which the system is heated following initial OSL stimulation (i.e. $\tau_{OAT} \approx \tau_{HBT}$). (The model is based on that by Yukihara (2020). Yukihara's model is slightly more complex, but the main elements are the same.)

heating after stimulation, the HBTs partly empty and a proportion of them are transferred to the OATs. Upon subsequent stimulation, OSL from the OATs is seen again. Modelling of the system reveals that even when the OATs can empty thermally during post-stimulation heating, electrons can be transferred into them from the HBTs. An optimal time exists for the transfer to take place; if the time at the elevated temperature is too long, the concentration of electrons in the OATs will eventually decay to zero.

### 3.7.3    More Temperature Effects for TL and OSL

Thermal quenching results in a decrease in the OSL intensity with increasing temperature, but the opposite effect is often seen, namely an increase in OSL efficiency with an increase in stimulation temperature. There are multiple possible mechanisms that may explain this effect, generally termed "thermal assistance." The role of shallow traps has already been discussed, but an additional important observation is that not only is the efficiency of OSL production seen to be temperature dependent, but the rate of decay, in CW-OSL, is also temperature dependent. Since the rate of decay is related to the optical stimulation probability, $p = \sigma \Phi$, then $p$ must also be temperature dependent. In some cases, the stimulation probability is observed to be thermally activated according to the expression:

$$p = p_0 \exp\left\{-\frac{\Delta E}{kT}\right\}, \qquad (3.93)$$

where $p_0$ is the stimulation probability at 0 K and $\Delta E$ is the observed thermal activation energy. By monitoring the CW-OSL decay rate as a function of stimulation temperature, for a given stimulation wavelength, a range of values of $\Delta E$ has been observed for different materials. Several phenomena are described in the following sections that may contribute to a temperature dependence for the production of OSL.

### 3.7.3.1 Phonon-coupling

For purely electronic transitions, OSL will be stimulated once the stimulating photon energy $h\nu$ is greater than the optical trap depth, $E_o$. However, with systems with strong phonon coupling, transitions can take place when $h\nu < E_o$, with the extra energy being supplied by phonons of energy $h\nu_{ph}$ (see Section 2.2.1.2 and Exercise 2.2). The photoionization cross-section is then given by an expression such as Equation 2.18, reproduced here:

$$\sigma_p(E) \propto \frac{\kappa}{\nu} \int_0^\infty \epsilon^{a-1} \left[ \exp\left\{ -\kappa^2 \left[ \epsilon - (h\nu - E_o) \right]^2 \right\} \right] d\epsilon, \tag{2.18}$$

where the phonon broadening factor $\kappa$ is related to temperature via:

$$\kappa = \left[ 2S \left( h\nu_{ph} \right)^2 \coth\left( \frac{h\nu_{ph}}{2kT} \right) \right]^{-\frac{1}{2}}, \tag{2.19}$$

with all terms defined in Section 2.2.1.2.

Some authors suggest changing the temperature (at a fixed heating rate) during the continuous stimulation of OSL. This is termed thermally modulated-OSL (TM-OSL) and the effect of a changing value for $\sigma_p$ due to phonon-coupling is convolved with the thermal release of the trapped electrons. By tuning the rate of heating, the stimulation power and the stimulation wavelength the impact of particular traps on the obtained luminescence data can be emphasized as a method of obtaining greater detrapping resolution (Chruścińska 2019).

### 3.7.3.2 Shallow Traps

Some of the other models explaining temperature-dependent effects are illustrated schematically in Figure 3.32. Figure 3.32a illustrates the effect of shallow traps, as already described in Section 3.7.2.1. With just one type of shallow trap, the temperature dependence has an activation energy $\Delta E = E_a$, where $E_a$ is the thermal trap depth of the shallow trap. In real materials, there may by multiple traps depths and a single value for $\Delta E$ may not be observed experimentally.

### 3.7.3.3 Sub-Conduction Band Excitation

A well-known observation in feldspar is that OSL can be stimulated by infra-red (IR) light (so called IRSL, first observed by Hütt et al. 1988). By scanning the IR wavelengths, a resonance is observed around 1.4 eV, with the value varying with feldspar type. The interpretation initially given is shown in Figure 3.32b, where the IR photon raises the trapped electron to an excited state from where phonon coupling allows thermal excitation to the conduction band, with an energy $\Delta E = E_b$.

An alternative model was provided by Poolton et al. (2002a) who estimated the wavefunction overlap between the excited states of the OSL center and interpreted the observed activation energy to be the necessary energy to hop over the potential barrier between overlapping states, i.e. $\Delta E = E_c$ from Figure 3.32c. The model is described as a donor–acceptor recombination process, with the final recombination transition occurring, possibly via tunneling, from the excited state to the recombination center.

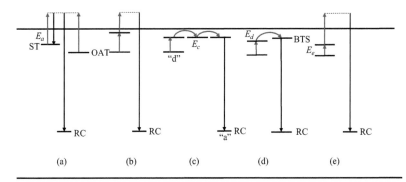

**Figure 3.32**   Schematic representation using the energy-band model for possible sources of the temperature dependence of OSL, all of which have been proposed by different research groups to account for the observed temperature dependence of the probability of optical stimulation to produce OSL. Optical transitions are in blue; thermal transitions are in red; trapping and recombination transitions are in black. ST = shallow trap; OAT = optically active trap; BTS = band-tail states; and RC = recombination center. Figure (a), shows the role of shallow traps, as discussed in Section 3.7.2.1. Figure (b) illustrates the optical excitation to a resonant excited state, at thermal energy $E_b$ below the conduction band. In (c) is illustrated the notion of donor (d)-to-acceptor (a) recombination where optical excitation raises the electron to an excited state, and overlap of the excited state wavefunctions of similar sites allows hopping of the electron, with activation energy $E_c$, from one site to another until recombination with at the recombination center (acceptor) occurs. Figure (d) shows excitation of the electron into the band-tail states. Hopping between band-tail states, with a mean energy $E_d$, occurs before recombination at the recombination center. The notion of band-tail states is discussed in further detail in Chapter 10. In (e), thermal excitation to a higher energy level occurs from which optical excitation to the conduction band occurs, followed by recombination. The thermal activation energy is $E_e$.

An additional model from Poolton et al. (2002b) described the process of excitation by the IR photon into the band-tail states, discussed for disordered materials (see Section 2.1.3 and Figure 2.5). Natural feldspars are crystalline but show considerable local disorder; random fluctuations in bond lengths and angles give rise to the band-tail states. The activation energy in this model ($\Delta E = E_d$ from Figure 3.32d) is caused by hopping within the band-tail states and in this way significant transport throughout the material can occur, resulting finally in a recombination transition.

A model proposed by Spooner (1994) for thermally assisted OSL in quartz is illustrated in Figure 3.32e. Here, an array of ground state energy levels may be populated to different extents, dependent on temperature, from which optical excitation to the conduction band can occur. The activation energy is associated with the thermal activation energies necessary to populate the different ground states (($\Delta E = E_e$) and scanning the stimulation wavelengths indicates that $\Delta E$ varies with stimulation wavelength and does not indicate a resonance at any particular wavelength. In a sense, this is the opposite of the Hütt et al. model.

Experiments to determine the details of thermally assisted OSL are usually performed by holding the temperature constant and monitoring rate of optical stimulation of luminescence. Alternatively, the temperature can be ramped during the stimulation. The procedure is the same as with TM-OSL mentioned above, but the physical effects being monitored are different. Such experiments are normally called thermally assisted-OSL (TA-OSL; Polymeris and Kitis 2019). The effects of phonon-coupling and thermal activation can be occurring simultaneously and interpretation of the data may not be straightforward.

**Thermally assisted OSL (TA-OSL)**

Consider the OTOR model shown in Figure 3.32b, in which a trapped electron can be raised from its ground state to an excited state by the absorption of a photon of energy $h\nu$. The electron can then be thermally excited from its excited state to the conduction band with a thermal activation energy $E_b$ and frequency factor $\nu$, and subsequently undergo recombination to produce OSL. However, the efficiency of OSL emission ($\eta(T)$) is also thermally quenched according to the Mott-Seitz model,

$$\text{i.e., } \eta(T) = \frac{1}{1 + C\exp\{-W/kT\}}.$$

(a) Set up the rate equations governing the optical stimulation from the ground state to the excited state, the thermal stimulation from the excited state to the conduction band, and the recombination of the free electron with the trapped hole at the recombination site. Ignore relaxation from the excited state to the ground state.

(b) Devise values for the various rate parameters (you choose the numbers) and solve the rate equations in order to evaluate the OSL intensity as a function of stimulation temperature, from −200 °C to + 200 °C in 50 °C intervals, using values of $E_b = 0.2$ eV with $\nu = 10^{12}$ s$^{-1}$, and $W = 0.6$ eV with $C = 10^{12}$ s$^{-1}$. Explain the shape of the OSL($T$) curve so obtained.

### 3.7.3.4   Random Local Potential Fluctuations (RLPF)

Solutions to the Schrödinger equation with periodically varying potentials lead to well-defined allowed energy bands, normally drawn as flat lines, as in any of the energy band models described so far in this text. However, breakdown in the periodicity of the lattice not only gives rise to allowed energies in the forbidden zones (traps and recombination centers), but it can also produce random and local changes in the forbidden gap width. RLPF may be expected to occur in any highly disordered material in which there are variations in bond angles, bond lengths, lattice disorder, local microdensity fluctuations, and high defect densities. One result is the already-mentioned band-tail states extending into the band gap below the mobility edge, where the mobility edge defines the separation between localized and delocalized band states (Mott and Davis 2012; see also Figure 2.5). Materials exhibiting these properties that are commonly used for dosimetry include some natural materials, but also synthetic materials including glasses, and materials with nano- and micro-structures, grain boundaries, and non-uniform impurity distributions.

The notion is illustrated schematically in Figure 3.33, illustrating how the effective band gap can change due to the RLPF. The minima in the conduction band edge and the maxima in the valence band edge do not necessarily coincide in the same region of the crystal. As a result of such fluctuations, electrons (say) raised to the conduction band can become localized in the conduction band minima. Similarly, holes in the valence band can be localized in the valence band maxima. One result of this is persistent photoconductivity (PPC) in which electrons and holes in the conduction and valence band can percolate or hop through the various band minima/maxima but not recombine such that the photostimulated conductivity is long-lived even after the stimulation light has been removed (Jiang and Lin 1990; Firszt et al. 2004; Grossberg et al. 2007).

From the point of view of OSL, it can be imagined that this may result in long recombination lifetimes and slow decays of, say, POSL after the stimulation light pulse has ended due to extended electron lifetimes in the conduction band before recombination occurs. PPC can be subsequently quenched by stimulation with IR wavelengths by exciting the electrons and/or

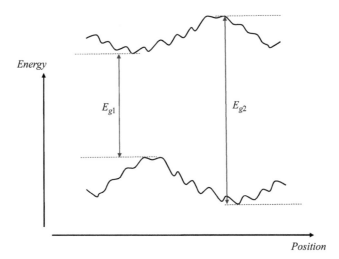

**Figure 3.33**   Schematic representation of the fluctuations in the band gap (not to scale) that might be expected due to RLPF. The resulting band gap is variable throughout the crystal and can vary from a minimum ($E_{g1}$) to a maximum ($E_{g2}$).

holes from the band minima/maxima and thereby stimulating recombination. It may be speculated that similar effects might be seen with POSL. It may also be imagined that the effect could even be seen in CW-OSL if the stimulating light corresponds to excitation of electrons (say) to the minimum of the conduction band, into the band-tail states, but is not energetic enough to overcome the random potential wells at the band edge; that is, the stimulating light does not raise the energy of the electrons high into the conduction band above the mobility edge.

Most observations of RLPF and PPC in the literature have been in highly disordered semiconductor structures. However, Ching and colleagues (Li and Ching 1985; Xu and Ching 1991) calculated band gap fluctuations by as much as several tenths of an electron volt in various polymorphs of $SiO_2$ as a function of Si-O bond length, mass density, and, weakly, on Si-O-Si bond angle, implying that highly disordered insulators or composite materials with multiple grain boundaries and internal nano- or micro-structures, may exhibit fluctuations in the band gap and expected effects due to band-tail states. Such band edge fluctuations may be especially significant when compared to the trap depths and a temperature dependence of OSL due to these effects could be expected. These possibilities and their impact on OSL are discussed further in Chapter 10.

### 3.7.4   Optical Effects on TL

In the same way that temperature can affect the observed OSL properties, so too can light affect the observed TL properties. Two possible effects are discussed next.

#### 3.7.4.1   Bleaching

In the event that a trapped electron can be stimulated from its trap by both heat (for TL) or light (for OSL), the question arises concerning what happens to the TL curve if a material is optically stimulated after irradiation, but before TL readout – known as "bleaching" of the TL.

Examples of this type of behavior abound in the literature and multiple effects have been observed with different materials. In the simplest case of a material described by the OTOR model, the TL signal from the trap will simply be reduced with optical stimulation before the

TL measurement is taken. For the NMTS model, a similar effect will be seen. Here, the additional deeper trap (ODDT/TDDT) does not change, neither increasing nor decreasing, if it is already full. For IMTS, however, the deep interacting trap can capture electrons released during optical stimulation and whether the number of electrons it holds decreases or increases depends upon whether it too is optically emptied during stimulation. Several more complex scenarios can be imagined, with trapped charge both increasing and decreasing as multiple traps are stimulated.

So much for the traps. What about the recombination centers? Even if electrons from a particular trap are not emptied during the optical stimulation phase the number of available recombination centers may be decreased because of stimulation of electrons from other traps. How will this affect the resulting TL for the trap in question? What about competing, non-radiative recombination centers? How will altering the concentration of these affect the TL? The overall picture is one of complex, interacting systems in which the net result (the TL intensity) may be difficult to predict in a given situation.

Numerical simulations can provide some insight. Consider the IMTS model of Figure 3.34 as an example. It consists of three traps (3T) and two recombination centers (2R) in a 3T2R model. One trap is shallow (ST), one is the main trap used for TL dosimetry (MT), and one is a deep trap (DT). One of the recombination centers is radiative, yielding the TL signal (RRC), and one is non-radiative (NRRC) and acts as a competitor to the RRC. Using this model, several scenarios can be imagined. For example, the bleaching light may stimulate electrons from the ST, but not from the MT or DT; or, both ST and MT may empty at the same time (during bleaching); or all three may empty at once, at rates dictated by the individual photoionization cross-sections and the stimulating wavelength. Furthermore, it was shown in Figure 3.18 that the deeper trap can have a larger photoionization cross-section than a shallower trap and so the rate of bleaching of DT may be faster than those of ST and MT.

Some example solutions are represented in Figures 3.35 and 3.36. To produce these results, the following equations were solved:

$$\frac{dn_1}{dt} = n_c A_1 \left( N_1 - n_1 \right) - n_1 s_1 \exp\left\{ -\frac{E_{t1}}{kT} \right\} - n_1 p_1, \tag{3.94a}$$

$$\frac{dn_2}{dt} = n_c A_2 \left( N_2 - n_2 \right) - n_2 s_2 \exp\left\{ -\frac{E_{t2}}{kT} \right\} - n_2 p_2, \tag{3.94b}$$

$$\frac{dn_3}{dt} = n_c A_3 \left( N_3 - n_3 \right) - n_3 s_3 \exp\left\{ -\frac{E_{t3}}{kT} \right\} - n_3 p_3, \tag{3.94c}$$

$$\frac{dm_1}{dt} = n_v A_{m1} \left( M_1 - m_1 \right) - n_c B_1 m_1, \tag{3.94d}$$

$$\frac{dm_2}{dt} = n_v A_{m2} \left( M_2 - m_2 \right) - n_c B_2 m_2, \tag{3.94e}$$

$$\frac{dn_c}{dt} = G + n_1 s_1 \exp\left\{ -\frac{E_{t1}}{kT} \right\} + n_1 p_1 + n_2 s_2 \exp\left\{ -\frac{E_{t2}}{kT} \right\}$$
$$+ n_2 p_2 + n_3 s_3 \exp\left\{ -\frac{E_{t3}}{kT} \right\} + n_3 p_3 - n_c A_1 \left( N_1 - n_1 \right) - n_c A_2 \left( N_2 - n_2 \right) \tag{3.94f}$$
$$- n_c A_3 \left( N_3 - n_3 \right) - n_c B_1 m_1 - n_c B_2 m_2,$$

and

$$\frac{dn_v}{dt} = G - n_v A_{m1}\left(M_1 - m_1\right) - n_v A_{m2}\left(M_2 - m_2\right). \tag{3.94g}$$

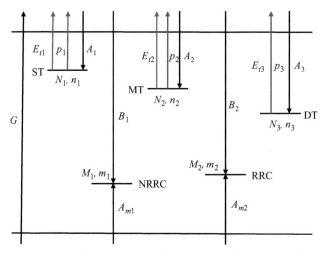

**Figure 3.34**   A 3T2R, IMTS model consisting of three traps and two recombination centers, only one of which is radiative. ST = shallow traps; MT = main dosimetry traps for TL; DT = deep, thermally disconnected traps. RRC represents the radiative recombination centers and NRRC are the non-radiative centers. For the electron traps, $N_i$ and $n_i$ ($i$ = 1,2,3) represent the concentration of available traps and the concentration of trapped electrons in that trap, respectively. For the recombination centers, $m_j$ ($j$ = 1,2) represents the concentration of trapped holes at each center and $M_j$ are the available concentrations of hole traps. $A_i$ and $A_{mj}$ are the trapping probabilities for the respective electron and hole traps, and $B_j$ are the recombination probabilities for free electrons recombining with trapped holes. $G$ is the electron-hole pair generation rate during irradiation; $p_i$ are the optical stimulation probabilities.

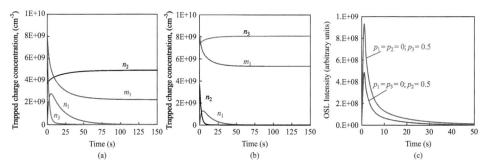

**Figure 3.35**   Bleaching of the various trap concentrations during optical stimulation. In (a) $p_3$ = 0.5 s$^{-1}$, while $p_1 = p_2 = 0$; in (b) $p_2 = 0.5$ s$^{-1}$, while $p_1 = p_3 = 0$. In (c) are shown the CW-OSL curves during optical stimulation. The remaining parameters used in these calculations are: $N_1 = N_2 = N_3$ = $M_1 = M_2 = 10^{11}$ cm$^{-3}$; $E_{t1} = 0.8$ eV, $E_{t2} = 1.2$ eV and $E_{t3} = 5.0$ eV; $s_1 = s_2 = s_3 = 5 \times 10^{12}$ s$^{-1}$; $A_1 = A_{m1} = 10^{-9}$ cm$^3$.s$^{-1}$; $A_2 = A_{m2} = 10^{-10}$ cm$^3$.s$^{-1}$; $A_3 = B_2 = 2 \times 10^{-10}$ cm$^3$.s$^{-1}$; $B_1 = 10^{-8}$ cm$^3$.s$^{-1}$; $G = 10^8$ cm$^{-3}$.s$^{-1}$; and irradiation time = 300 s. Note that in (a) and (b), the ST are initial empty but during bleaching at first increase due to transfer from either the DT (a), or from the MT (b). Then $n_1$ decays due to thermal emptying at the temperature of the simulation (300 K).

In the above set of equations:

$E_{ti}$ = electron traps depths (eV);
$s_i$ = electron trap frequency factors (s$^{-1}$);
$N_i$ = available electron trap concentrations (cm$^{-3}$);
$n_i$ = trapped electron concentrations (cm$^{-3}$);
$A_i$ = electron trapping probabilities (cm$^3$s$^{-1}$);
$M_j$ =available concentrations of hole traps (cm$^{-3}$);
$m_j$ = trapped hole concentrations = concentrations of recombination centers (cm$^{-3}$);
$A_{mj}$ = hole trapping probabilities (cm$^3$s$^{-1}$);
$B_j$ = recombination probabilities for free electrons with trapped holes (cm$^3$s$^{-1}$);
$G$ = electron-hole pair generation rate (cm$^{-3}$s$^{-1}$);

with $i = 1, 2$ or $3$ and $j = 1$ or $2$. Level $i = 1$ corresponds to the shallow trap (ST) in Figure 3.34, $i = 2$ is the main trap (MT) while $i = 3$ is the deep trap (DT). Level $j = 1$ is the radiative recombination center (RRC), while $j = 2$ is the non-radiative center (NRRC).

In solving Equations 3.94a–3.94g, five separate stages need to be considered: (i) Irradiation phase ($G > 0$, $p_i = 0$, $\beta_t = 0$); (ii) first relaxation stage ($G = 0$, $p_i = 0$, $\beta_t = 0$); (iii) bleaching stage ($G = 0$, $p_i > 0$, $\beta_t = 0$); (iv) second relaxation stage ($G = 0$, $p_i = 0$, $\beta_t = 0$); and (v) heating stage ($G = 0$, $p_i = 0$, $\beta_t > 0$). Figure 3.35a shows the changes in the trapped charge concentration when emptying of the deep trap only is allowed ($p_3 = 0.5$; $p_1 = p_2 = 0$). The deep trap concentration ($n_3$) decays rapidly to zero, but the concentration of holes in the radiative-recombination center ($m_1$) also decays commensurately, but not to zero. (The concentration $m_2$, not shown, also decays, but not to zero.) During this process, the concentration of electrons in the main trap, $n_2$, actually increases due to trapping of electrons released from the DT. The trapped electron concentration in the shallow trap, $n_1$, starts at zero due to the fact that this trap is so shallow that the trapped charge is unstable during the irradiation and relaxation periods, but before OSL measurement. During optical stimulation $n_1$ first grows due to the transfer of electrons from the DT, and then decays due to the thermal instability of the trap. Note that $m_1$ and $n_2$ remain stable and constant at long stimulation times. (All processes are calculated at 300 K; the parameters used in the calculation are given in the figure caption.)

In contrast, if optical stimulation from the MT only is allowed (Figure 3.35b), $n_2$ decays rapidly to zero ($p_2 = 0.5$; $p_1 = p_3 = 0$). Again $m_1$ is reduced (as is $m_2$, not shown). Figure 3.35c illustrates the resultant CW-OSL decay curves in each case.

Figure 3.36a shows the changes in the concentrations of $n_2$, $n_3$, and $m_1$ during heating, while Figure 3.36b shows the subsequent TL. The concentrations shown are for the case of stimulation from the DT only ($p_3 = 0.5$; $p_1 = p_2 = 0$). These are compared with the case when there is no bleaching at all (i.e. when the material is heated after irradiation without an optical stimulation period.) The arrows show the increases or decreases seen as a result of the bleaching and during subsequent heating. To be noted is that the change in the value of $m_1$ during heating (i.e. $\Delta m_1$) is larger for the unbleached case than it is for the bleached case. Thus, the TL from the main dosimetry trap (MT) with bleaching is less than it would have been without bleaching – even though the MT was not bleached and, in fact, $n_2$ actually increased during the bleaching process. Also to be noted is that no further reduction in the TL signal is observed with further optical bleaching. In Figure 3.35a, the values for all the concentrations have reached steady-state such that no further changes are observed. This means that the reduction in TL seen in Figure 3.36b is all that is going to be observed.

For the case when the electrons are directly stimulated from the MT ($p_2 = 0.5$; $p_1 = p_3 = 0$), the subsequent TL can be bleached completely, as shown in Figure 3.36b.

If it is observed experimentally that a TL peak can be reduced to zero during bleaching, direct stimulation from the trap should be suspected. On the other hand, if stimulation

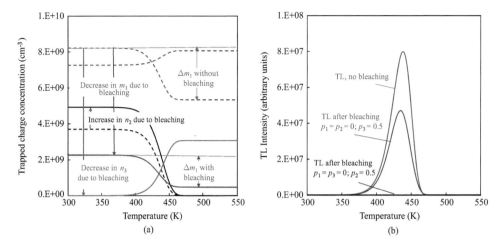

**Figure 3.36** (a) Changes in the various concentrations during TL, at $\beta_t = 1$ K.s$^{-1}$. Green lines indicate $n_3$ (concentration of electrons trapped at DT); blue lines indicate $m_1$ (holes in the radiative recombination center); black lines indicate $n_2$ (electrons trapped at MT). The data for no bleaching are shown by the dashed lines, and after bleaching ($p_3 = 0.5$ s$^{-1}$, $p_1 = p_2 = 0$) by full lines. It is seen that $m_1$ decreases during bleaching (as does $m_2$, not shown), while $n_3$ decreases to zero, and $n_2$ increases. During heating, $n_2$ decreases to zero due to thermal emptying; $n_3$ increases due to transfer from electrons from MT, and $m_1$ decreases due to recombination. The changes in $m_1$ (the concentration of RRC) are shown by $\Delta m_1$. (b) Subsequent TL peaks with no bleaching (orange) and after bleaching the DT only (blue curve). There is no TL if the MT is bleached instead of the DT ($p_2 = 0.5$ s$^{-1}$, $p_1 = p_3 = 0$).

produces a reduced TL peak, but a residual, or "impossible-to-bleach" signal remains, even after long bleaching times, then a more complex process, such as that described in Figures 3.35 and 3.36, should be suspected. It is important to note that these examples are but a small fraction of the examples that could be imagined, even with this one model. For example, changes in the various concentrations and cross-sections could produce different results, as could allowance for optical stimulation from several of the traps at the same time. In reality, the latter is likely.

---

**Exercise 3.15   Modelling optical bleaching of TL**

Using the 3T2R model of Figure 3.34, solve the relevant rate equations, Equations 3.94a–3.94g, and show the effects of optical bleaching on the subsequent TL glow curves. Start the calculations from all traps empty and use the same parameters as used for Figures 3.35 and 3.36, with the following exceptions:

(a) Change the trap depth $E_t$ for the deep trap (DT) to 1.6 eV.
(b) Change the values of $p_1$, $p_2$, and $p_3$ such that two of the three traps empty during optical stimulation, but the third does not (e.g. $p_1$, $p_2 > 0$ and $p_3 = 0$; and $p_2$, $p_3 > 0$ and $p_1 = 0$).
(c) Also calculate what happens if $p_1$, $p_2$, and $p_3 > 0$. In this case choose values for $p_1$, $p_2$, and $p_3$ commensurate with reasonable photoionization cross-sections for the traps and assume $E_{o1} < E_{o2} < E_{o3}$.

### 3.7.4.2    Phototransferred TL (PTTL)

During the bleaching of the DT or the MT in the above examples, charge was transferred to the empty trap, ST. The trap was empty at the start of the optical stimulation because the values of $E_{t1}$ and $s_1$ were such that the lifetime (residence time) of the charge trapped in the ST was short and thus ST emptied in the period between the end of the irradiation and the start of the bleaching. During optical stimulation and bleaching of the electrons from either the DTs or the MTs electrons were transferred to the ST. This can be seen in Figures 3.35a and 3.35b where an initial increase in the concentration $n_1$ is observed due to optical transfer from either DT (Figure 3.35a) or MT (Figure 3.35b), followed by a slower decrease due to the thermal instability of the trapped electrons in the ST. This phenomenon is generally called phototransfer and it is of particular interest because it enables access to the electrons in deep, thermally disconnected traps, which cannot be accessed by heating, by optically transferring the trapped charge to shallower traps, which had been previously thermally emptied. If the electrons in the shallower trap were sufficiently stable, a TL signal could be recorded due to the transferred electrons. The TL in this case is termed "phototransferred TL," or PTTL.

In general, the sequence for recording PTTL is to first irradiate, then pre-heat to a given temperature to empty the traps and record a first TL signal; then stimulate with light of a particular wavelength – OSL may or may not be recorded during the stimulation; and then finally to record a second TL signal from the same traps. This is the PTTL signal. It is the conceptual opposite of Thermally Transferred OSL (TT-OSL) described in Section 3.7.2.2.

The simplest model required to explain PTTL is a two-trap, one recombination center model (2T1R), consisting of trap 1 (with trapped electron concentration $n_1$), trap 2 (with trapped electron concentration $n_2$) and one recombination center (with trapped hole concentration $m$). Assume trap 1 to be initially empty ($n_1 = 0$) and $n_2 > 0$ at the start of the optical stimulation. For charge neutrality, $n_1 + n_2 = m$, from which it can be seen that always $n_1$ or $n_2 \leq m$. During optical stimulation $n_2$ decreases, $m$ decreases due to recombination, and $n_1$ increases due to phototransfer. During subsequent heating, $n_1$ empties, and recombination produces PTTL. The PTTL signal is given by:

$$I_{PTTL} = \left| \frac{dm}{dt} \right|, \tag{3.95}$$

and the area under the PTTL peak is:

$$S_{PTTL} = \int_0^\infty I_{PTTL} dt = \int_0^\infty \frac{dm}{dt} dt = m_0 - m_\infty. \tag{3.96}$$

If $n_{10}$ and $n_{1\infty} = 0$ are the concentrations of $n_1$ at the start and end of the heating, then the area under the PTTL curve is proportional to $\Delta n_1 = n_{10} - n_{1\infty} = n_{10} = m_0 - m_\infty$. (Note that $m_\infty \neq 0$ since there are still electrons trapped in trap 2.) Thus, the PTTL peak area is proportional to $n_{10}$, the concentration of electrons trapped in trap 1 at the end of the stimulation and the start of the heating.

During optical stimulation, the rate equations for the 2T1R model may be written:

$$\frac{dn_1}{dt} = n_c \left( N_1 - n_1 \right) A_1, \tag{3.97a}$$

$$\frac{dn_2}{dt} = -p_2 n_2 + n_c \left(N_2 - n_2\right) A_2, \tag{3.97b}$$

and

$$\frac{dm}{dt} = -n_c m B, \tag{3.97c}$$

where the terms have their usual meaning and optical excitation from trap 2 only is allowed ($p_2 > 0$; $p_1 = 0$).

The solution to Equation 3.97a is:

$$n_1 = N_1 \left[1 - \exp\left\{-n_c A_1 t\right\}\right], \tag{3.98}$$

where $t$ is the optical stimulation time. Under conditions of quasi-equilibrium, $n_c$ is approximately constant compared to $n_1$. Therefore, the PTTL intensity will vary with stimulation time according to Equation (3.98) and is a saturating exponential up to a fixed level.

In many PTTL experiments it is observed that the PTTL varies as a function of stimulation time in a more complex manner than this. Often, a peak response is found after a certain stimulation time, after which the PTTL decreases with stimulation. Sometimes, the PTTL decreases to zero after a long enough stimulation time; sometimes it decreases to a fixed, constant level. Both of these observations are now explained.

The simplest and most obvious explanation of a decrease to zero at long stimulation times is that the trap is thermally unstable. This was the situation noted for the STs in Figure 3.35a and b. However, in the 2T1R model currently being discussed, thermal excitation from trap 1 has not been included. If thermal excitation is ruled out, a second obvious explanation is that there is concurrent optical stimulation from trap 1, as well as from trap 2, during the stimulation period – i.e. $p_2$ and $p_1$ are both > 0. Since the area under the PTTL peak is proportional to $n_1$ at the end of the stimulation period, i.e. to $n_{10}$, then, if the stimulation period is long enough, $n_1$ will decay to zero, $n_{10} = 0$, and therefore the PTTL will equal zero at the end of the stimulation.

An example is shown in Figure 3.37 where optical stimulation from both trap 1 and trap 2 is allowed, for different values of $p_1$ and $p_2$. Thermal excitation is ignored.

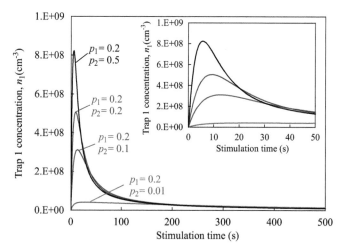

**Figure 3.37**   Changes in the concentration of trapped electrons in trap 1 during optical stimulation of electrons from trap 1 and trap 2, for the 2T1R model. Initial concentrations of trapped electrons at the start of the stimulation were $n_1 = 0$ cm$^{-3}$ and $n_2 = m = 10^{10}$ cm$^{-3}$. The values shown on the figure for $p_1$ and $p_2$ are in s$^{-1}$. Other parameters were: $N_1 = N_2 = M = 10^{11}$ cm$^{-3}$; $A_1 = A_2 = 10^{-9}$ cm$^3$.s$^{-1}$; $B_1 = 10^{-8}$ cm$^3$.s$^{-1}$. The inset shows the same data over the first 50 s of stimulation.

To explain how a PTTL peak can grow with stimulation time, then decreases but only to a fixed, non-zero level, it is necessary to add a non-radiative recombination center to the model, i.e. a 2T2R model. In the simpler 2T1R model, it is always true that $n_1 < m$, in which case the PTTL follows $n_1$ as a function of stimulation time. For the 2T2R case, however, it is no longer guaranteed that $n_1 < m_1$, where $m_1$ is the concentration of holes at the radiative recombination center (and $m_2$ is the hole concentration at the non-radiative center). The suffix "0" is used to indicate the concentrations at the end of the stimulation period and the start of the TL heating, and "$\infty$" is used to indicate the final concentration at the end of the TL peak. If $n_{10} < m_{10}$, then the TL emission ends because the electrons in trap 1 are exhausted first. From Equation 3.96, the area under the PTTL curve is given by $S_{PTTL} = m_{10} - m_{1\infty} = n_{10} - n_{1\infty} = n_{10}$. On the other hand, if $m_{10} < n_{10}$, then the TL peak ends because $m_1$ is exhausted first, and thus $S_{PTTL} = m_{10} - m_{1\infty} = m_{10}$. This discussion can be summarized by recognizing that the size of the PTTL peak is equal to the minimum of either $n_{10}$ or $m_{10}$, or:

$$S_{PTTL} = \min(n_{10}, m_{10}) \qquad (3.99)$$

The principles can be illustrated by solving the necessary rate equations for the 2T2R model. An example is shown in Figures 3.38 and 3.39. In this example, the irradiation phase was first simulated (using $G = 5 \times 10^8$ cm$^{-3}$.s$^{-1}$ for 300 s), followed by a relaxation phase and a heating phase (to 500 K) to empty trap 1. Then the optical stimulation phase was simulated using $p_1 = 0$ and $p_2 = 0.01$ s$^{-1}$ (black lines) or 0.05 s$^{-1}$ (red lines), as shown in the figure. In each case, the full lines represent $m_1$ and the dashed lines represent $n_1$ during the stimulation period. As electrons are optically emptied from trap 2, some are trapped into trap 1 while others recombine at both the radiative and the non-radiative recombination centers. In each simulation it is to be noted that $m_1 > n_1$ until time $t^*$, beyond which $m_1 < n_1$. Bearing in mind Equation 3.99, it can be expected that the PTTL will follow $n_1$ for stimulation times less than $t^*$ and then follow $m_1$ for stimulation times greater than $t^*$. The resultant PTTL will first increase with stimulation time, and then decrease.

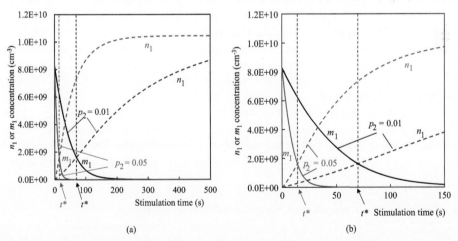

**Figure 3.38** (a) Simulation of the changes in the concentration of trapped electrons in trap 1 ($n_1$) and holes in the radiative recombination centers ($m_1$) for the 2T2R model, as a function of stimulation times. Part (b) shows the same data but over the first 150 s for greater clarity. The parameters used in these calculations were: $G = 5 \times 10^8$ cm$^{-3}$.s$^{-1}$; $N_1 = N_2 = M_1 = M_2 = 10^{11}$ cm$^{-3}$; $A_1 = A_{m1} = 10^{-9}$ cm$^3$.s$^{-1}$; $A_2 = A_{m2} = 10^{-10}$ cm$^3$.s$^{-1}$; $E_1 = 1.2$ eV; $s_1 = 5 \times 10^{12}$ s$^{-1}$; $B_1 = 10^{-8}$ cm$^3$.s$^{-1}$; $B_2 = 2 \times 10^{-10}$ cm$^3$.s$^{-1}$; $p_1 = 0$; and $p_2 = 0.01$ s$^{-1}$ or 0.05 s$^{-1}$, as indicated.

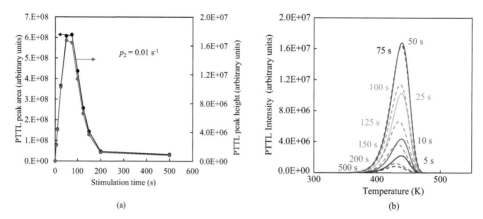

**Figure 3.39**   (a) Changes in the PTTL peak area (black dots) or peak height (red dots) at different stimulation times for the simulations shown in Figure 3.38 and for the case where $p_2 = 0.01$ s$^{-1}$. During PTTL, $\beta_t = 1.0$ K.s$^{-1}$. (b) The PTTL peaks, observed at different stimulation times. (Increasing peaks, full lines; decreasing peaks, dashed lines.)

This behavior is shown in Figure 3.39a in which the area under the resultant PTTL peak area (back dots) and the PTTL peak intensity (red dots) are plotted against stimulation time, for the case where $p_2 = 0.01$ s$^{-1}$. It can be observed that the PTTL peak increases initially as electrons are transferred into trap 1 and $n_1 < m_1$. The PTTL passes through a maximum (defined when $n_1 = m_1$ at stimulation time $t^*$) and then decreases as $m_1 < n_1$, and $m_1$ depletes due to recombination. However, note that the PTTL peak never gets fully depleted, since $m_1$ is never fully depleted due to the requirement of charge neutrality ($n_1 + n_2 = m_1 + m_2$, and $n_c \approx 0$). This is unlike the case in Figure 3.37 where $n_1 < m_1$ throughout the stimulation period and the PTTL peak must go to zero as $n_1$ goes to zero due to the fact that $p_1 > 0$.

The PTTL peak shapes are shown in Figure 3.39b. As the intensity decreases a shift in the peak position to higher temperatures is seen and the PTTL peak shape gets a little wider. This is a result of a changing order of kinetics, from first order to non-first order. This effect is not general, but is observed in this case as the recombination sites deplete. For other models, or other values of the various parameters and concentrations, in which there is a large reservoir of non-radiative recombination centers, the change in kinetic order and shift in the PTTL peak position may not be observed.

In summary, PTTL peak position, shape, and size, and their behavior as a function of stimulation time, depends upon the details of the kinetic model and the values of the various parameters. A variety of behaviors can be expected, as indicated in the examples shown here. The PTTL results can be used as tests for various models employed to explain real experimental data.

## 3.8   Tunneling, Localized and Semi-Localized Transitions

In all the models for TL and OSL discussed so far, the transition of (say) an electron from its trap to recombine with a trapped hole at a recombination site takes place via the conduction band. On the other hand, if either a direct transition from the trap to the recombination center was allowed, or if an indirect recombination route was allowed via an intermediate energy level, but not via the conduction band, then "leakage" of the electrons from the trap resulting in recombination may be possible.

This type of localized transition manifests itself in TL and OSL in various ways. One is an unexpected loss of electrons from the electron traps. Considering only thermal excitation from the trap to the conduction band, the expected lifetime in the trap is given by:

$$\tau_{lifetime} = p^{-1} = s^{-1}\exp\left\{\frac{E_t}{kT}\right\}. \tag{3.100}$$

A "half-life" can be defined, at temperature $T$, equal to the time taken for the trapped-electron concentration ($n$) to decay to half of its original concentration ($n_0$):

$$\tau_{1/2} = \ln(2)\tau_{lifetime} = \ln(2)s^{-1}\exp\left\{\frac{E_t}{kT}\right\}. \tag{3.101}$$

For known values for $E_t$ and $s$, the half-life can be calculated at any value of $T$. For some TL/OSL materials, the half-life of the trap as calculated using Equation 3.101 is much longer than that actually observed, and the signal loss is observed to follow a relationship with time $t$ and temperature $T$ that is not predicted by Equation 3.101. The loss of signal ("fading") in such cases is said to be "anomalous," or "athermal." Anomalous fading of TL signals was one of the first indications to appear in the literature that recombination was occurring via a mechanism not involving transitions to the conduction band.

During the leakage of electrons, a weak "after-glow" following irradiation is often seen. This weak luminescence occurs due to the escape of the electrons from the trap to the recombination site, either directly or indirectly, immediately following (and indeed during) the irradiation period.

Such transitions, whether direct or indirect, but not involving the conduction band, can generally be termed "localized transitions" in that they occur between traps and recombination sites that are present within a local volume. Different models have been devised to account for such localized transitions, including direct recombination via quantum mechanical tunneling, and indirect recombination via a local excited state. (There may be some confusion over nomenclature, in that although both types of recombination occur within a local volume, the term "localized transition" is usually reserved for the indirect route involving a local excited state, whereas the term "tunneling" applies specifically to the direct route of localized recombination via the tunneling.)

In addition to anomalous fading, this kind of process has also been suggested to be responsible for anomalous values of $E_t$ and $s$ obtained from conventional analysis of the TL glow curves, and unexpected dependencies of the TL peak positions on heating rate. Furthermore, in some of these models, both localized and delocalized transitions from the same trap are included. Known as "semi-localized" transition models, these have been successful in describing many TL/OSL phenomena in different materials, including the dependence of the TL or OSL response on increasing radiation dose (the "dose-response" relationships) and the changes observed in these relationships as a function of ionization density of the absorbed radiation.

Transitions not involving the conduction band can occur in disordered materials such as natural minerals (of which feldspar is the archetypical material), glasses, or in nano-structured materials where the distances between the electron and hole centers are confined to the nanoscale. They can also occur in high-quality, doped, macroscopic crystals when the electron and hole traps are each part of a defect cluster, which may form during doping and crystal growth, or in post-growth thermal annealing. They may also occur when the ionization density of the radiation is high, such that electrons and holes traps are created close to each other. An example might be energetic heavy-ion irradiation in which the ion deposits its energy in the material along densely ionized tracks giving rise to close spatial association between defects.

While some of these phenomena will be discussed in more detail in later chapters, the following sections describe tunneling, localized transition, and semi-localized transition models in general and discuss some of the predicted behavior that might be expected.

### 3.8.1   Tunneling

#### 3.8.1.1   *General Considerations*

Once an electron has been excited into the conduction band its wavefunction is delocalized. The electron can move throughout the material and prediction and identity of the trapped hole center at which the recombination event will occur is not possible. In contrast, when the electron is trapped, its wavefunction is localized in the volume occupied by the defect and confined by the potential well of that defect. The position of the trapped electron is thus well defined. As noted above, some materials display TL and OSL properties that cannot be explained by invoking delocalized transitions of the charge carriers (electrons and holes) and an explanation of these properties requires consideration of recombination only at specific hole centers for any given trapped electron. Since the trapped electron wavefunction is localized, this notion requires that the hole center is located sufficiently close to the electron trap that a transition of the electron from its trap to the nearby trapped hole center is possible via the process of quantum mechanical tunneling. The notion is illustrated conceptually in Figures 3.40 and 3.41. In Figure 3.40a, the electron wavefunction $\Psi(r)$ is entirely delocalized such that the position of the electron cannot be defined. In Figure 3.40b, however, the electron wavefunction is confined by the potential well associated with the electron trap.

When the location of the hole trap is spatially close enough to that of the electron trap that the potential barriers overlap, a situation such as that shown in Figure 3.41 can occur. Here, although the wavefunction $\Psi(r)$ for the trapped electron is still localized within the electron trap, the overlap of potential barriers between the electron trap and the hole trap can lead to the situation where part of the electron wavefunction extends into the potential well of the recombination site. When this occurs there is a finite probability of finding the electron in the hole center where it can recombine with the hole, without ever having passed through the conduction (delocalized) band. The electron is said to have "tunneled" through the potential barrier. While there is a finite probability for this to happen when the electron is in the ground state of the trap, at energy $E_g$ (where $E_c - E_g = E_t$, the trap depth), tunneling is more likely to occur if the electron is in an excited state, $E_{el}$ in Figure 3.41. Not only is the potential barrier width narrower at this energy (compare $r_2$ to $r_1$ in Figure 3.41), but the wavefunction of the excited state electron is broader (less localized), making tunneling more probable.

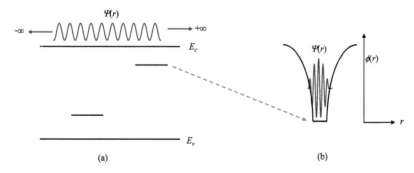

(a)                                                         (b)

**Figure 3.40**   (a) Delocalized wavefunction $\Psi(r)$ for an electron in the delocalized band (in blue). The functions are everywhere the same and as a result definition of the location of the electron is not possible. Likewise, prediction of where the electron may recombine is also not possible. (b) While confined by the potential well of the trap, the electron wavefunction is localized and drops off exponentially with distance from the trap.

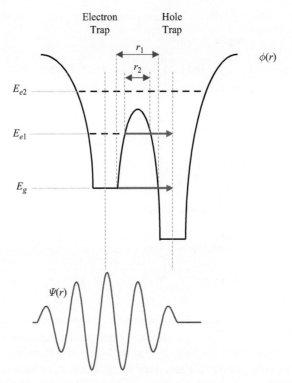

**Figure 3.41** When the electron and hole traps are spatially close to each other, overlap of their potential wells occurs. The localized wavefunction ($\Psi(r)$) now may extend into the potential well of the recombination site resulting in a finite probability that the electron may be found in the hole center, where recombination occurs. The transition is said to occur via tunneling (red arrow). The probability of tunneling is increased if the electron resides in an excited state of the trap (energy $E_{e1}$) compared to that when the electron is in the ground state ($E_g$).

For more detail and further discussion of tunneling phenomena and their effects on TL and OSL, the reader is referred to Visocekas (2002), Jain et al. (2012), Pagonis et al. (2013), and Pagonis (2019).

### 3.8.1.2 Ground-State Tunneling

The probability per second of tunneling between the trap and the recombination center, separated by distance $r$, can be written:

$$P(r) = P_0 \exp\{-\alpha r\}, \tag{3.102}$$

where $P_0$ is a frequency factor ($s^{-1}$). By approximating the potential well to a square well, the constant $\alpha$ (in units of $m^{-1}$) is approximately $4\pi\sqrt{2m^*E}/h$; $m^*$ is the electron effective mass (kg), $E$ is the barrier height between the electron and hole traps (in J), and $h$ is Planck's constant (in J.s).

The probability of emptying the trap is $np$ where $n$ is the concentration of trapped electrons and in the case of tunneling $p$ is replaced by $P(r)$ from Equation 3.102. Thus, since all recombination events are radiative in this model, and since there is no "back-reaction" (retrapping), then the luminescence emitted during the tunneling process, known as "after-glow," is given by:

$$I_{after-glow}(r,t) = \left|\frac{\delta n(r,t)}{\delta t}\right| = n(r,t)P(r), \tag{3.103}$$

from which

$$n(r,t) = n_0(r)\exp\{-P(r)t\}. \tag{3.104}$$

Combining Equations 3.103 and 3.104 gives

$$I_{after-glow}(r,t) = n_0(r)P(r)\exp\{-P(r)t\}. \tag{3.105}$$

The total luminescence from the sample is then obtained by summing over all values of $r$:

$$I_{after-glow}(t) = \int_0^\infty n_0(r)P(r)\exp\{-P(r)t\}dr. \tag{3.106}$$

Figure 3.42a illustrates the exponential dependence of the tunneling rate $P(r)$ on distance, while Figure 3.42b shows the result of numerically integrating Equation (3.106), for different times, over a wide range of $r$ values (up to $1 \times 10^{-8}$ m, at which point $P(r) \sim 0$ and beyond which no further tunneling occurs). The values of the parameters used in the calculation are given in the figure caption.

An important feature of Figure 3.42b is the approximately $t^{-1}$ dependence of the after-glow, especially at long times. More generally, it is found experimentally that the decay is of the form $t^{-k}$ with $k$ varying from ~0.95 to ~1.5.

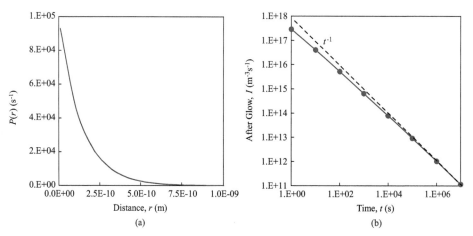

(a)                                    (b)

**Figure 3.42**    (a) Probability of tunneling, given by Equation 3.102, with parameter values $P_0 = 10^5$ s$^{-1}$, and for a barrier height of $E = 0.5$ eV. (b) Intensity of the after-glow during tunneling, calculated from a numerical summation of the integral in Equation 3.106 and with the parameters as in part (a), with $n_0 = 10^{24}$ m$^{-3}$ and a numerical integration step of $\Delta r = 10^{-10}$ m. The dashed line shows a $t^{-1}$ relationship.

For randomly distributed acceptors (recombination centers) the probability distribution of the donor–acceptor nearest-neighbor (NN) separation, $r$, is given by:

$$g_{NN}(r) = 4\pi n_0 r^2 \exp\left\{-\frac{4}{3}\pi n_0 r^3\right\}.$$ 

(3.107)

This is shown in Figure 3.43a. Here, the trapped electron and hole densities are assumed to be the same ($n_0$).

The number of electron traps remaining in the distance interval $dr$, at a separation $r$, after time $t$, is:

$$dn(r,t) = n_0^2 4\pi r^2 \exp\left\{-\frac{4}{3}\pi n_0 r^3\right\} \exp\left\{-P(r)t\right\} dr$$ 

(3.108)

Integrating Equation 3.108 over all possible $r$ values gives the time dependence of the surviving electron traps. Figure 4.43b shows the change in the distribution of surviving traps as a function of time. As time proceeds, the "tunneling front" penetrates further into the distribution. The shape of the tunneling front is a result of the convolution of the original distribution $g_{NN}(r)$ (illustrated by the dotted line) and the sharply increasing double exponential function, $\exp\left\{-P(r)t\right\} = \exp\left\{-P_0\exp\left\{-\alpha r\right\}t\right\}$. The position of the front can be defined at $r = r_c$, where the critical radius $r_c$ is defined as that distance for which $P(r_c)t = 1$. As argued by Huntley (2006), this is equivalent to assuming that the tunneling front in Figure 3.43b can be replaced by a vertical line and the critical radius in this case is then written $r_c = \alpha^{-1}\ln\left[P_0 t\right]$, at times $t$ such that $P_0 t > 1$.

Equation 3.106 represents the luminescence "lost" in the specimen (i.e. faded) and therefore the subsequent TL or OSL signal originates from the surviving concentration of trapped electrons and holes (donor–acceptor pairs). Since the after-glow decays with a $t^{-k}$ law, so too will the surviving TL or OSL. Experimental observation of the near-$t^{-k}$ behavior (of either the after-glow, or the surviving TL or OSL signal) is an indication of tunneling. Note that tunneling from the ground state, as described in this section, is temperature independent, but is sensitive to the concentrations $n_0$ (and $m_0$). Thus, there is a dose dependence, with faster tunneling occurring at higher doses.

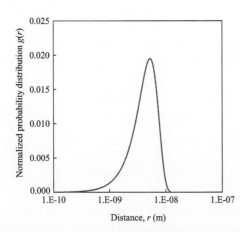
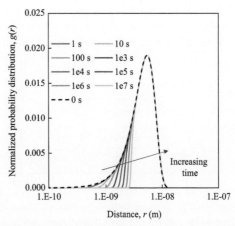

**Figure 3.43**    (a) Normalized distribution $g_{NN}(r)$ (Equation 3.107) using the same parameters as for Figure 3.42. (b) The "tunneling front" calculated from Equation 3.108, using the same parameters, for the different tunneling times shown in the figure.

---

### Exercise 3.15  Tunneling (1)

(a) The degree of tunneling from the ground state depends critically on the concentration of donor–acceptor pairs. Assuming $n_0$ is the concentration of both trapped electrons and trapped holes, and using the same parameters as used in Figures 3.42 and 3.43, calculate the nearest-neighbor distribution and the tunneling front as a function of tunneling time for different values of $n_0$.

(b) The tunneling probability also depends on the value of potential-well depth $E$. Show how the tunneling front changes for a fixed $n_0$ but for variable $E$.

(c) Show analytically that, for long times, the after-glow can be written:

$$I_{after-glow}(t) \approx \frac{K}{t},$$

where $K$ is a constant.

---

#### 3.8.1.3  Excited-State Tunneling

As noted in Section 3.8.1.1, the probability of tunneling increases if the electron is excited from the ground state to the excited state (energy level $E_{e1}$ in Figure 3.41). The transition from level $E_g$ to $E_{e1}$ can be thermally activated according to the probability $s\exp\{-(E_{e1} - E_g)/kT\}$ at temperature $T$; $s$ is the usual frequency factor. This leads to the concept of thermally activated tunneling and explains why some materials exhibit a tunneling behavior (as illustrated by the $\sim t^{-k}$ law) but also a temperature dependence. Likewise, optically assisted tunneling can be expected with external stimulation light of suitable wavelength.

Figure 3.44(a) is based on Figure 3.41 and shows the energy level $E_g$, $E_{e1}$, and $E_{e2}$, with the transitions considered for thermally assisted tunneling, with transition probabilities:

$$A = s\exp\left\{-\frac{E_{e1} - E_g}{kT}\right\},\ B = \text{relaxation rate from the excited state back to the ground state,}$$

$$P_e(r) = P_0\exp\{-\alpha_e r\} = \text{the tunneling probability from the excited state,}$$

and $P_g(r) = P_0\exp\{-\alpha_g r\}$ = the tunneling probability from the ground state (assuming $P_0$ is the same for both). All terms have their usual meanings. Since the primary concern of this section is tunneling from the excited state, tunneling from the ground state will be ignored.

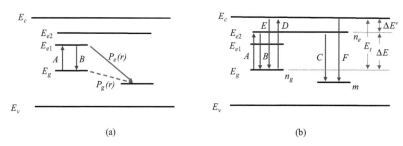

(a)  (b)

**Figure 3.44** (a) Model for excited state tunneling, showing an excitation transition at rate $A$, a relaxation transition at rate $B$, and a tunneling transition (red arrow) at a rate $P_e(r)$ from energy level $E_{e1}$. Ground state tunneling (dashed red arrow) is also shown, from level $E_g$, at a rate $P_g(r)$. (b) The localized transition model, where excitation (thermal or optical) to the common energy level $E_{e2}$ at rate $A$ enables relaxation to the hole center at rate B.

With the explicit dependencies on distance $r$ and time $t$ indicated, the rate equations describing the transitions in Figure 3.44a are:

$$\frac{\delta n_g(r,t)}{\delta t} = -An_g(r,t) + Bn_e(r,t),$$ (3.109a)

$$\frac{\delta n_e(r,t)}{\delta t} = An_g(r,t) - Bn_e(r,t) - P_e(r)n_e(r,t),$$ (3.109b)

$$\frac{\delta m(r,t)}{\delta t} = -P_e(r)n_e(r,t),$$ (3.109c)

and the luminescence emitted during tunneling is:

$$I(r,t) = \left|\frac{\delta m}{\delta t}\right| = P_e(r)n_e(r,t).$$ (3.109d)

The total concentration of trapped electrons is $n(r,t) = n_g(r,t) + n_e(r,t)$.

Note that the relaxation rate is a constant $B$ (in s$^{-1}$), and not the more familiar term of the form $A[n(r,t) - n_e(r,t)]$, where $A$ is in units of m$^{-3}$.s$^{-1}$. This is because the latter term implies that the excited electron can be de-excited into any trapped electron empty ground state, whereas it remains in the trap and can only de-excite into the specific trapped electron ground state from which it came. Thus, the de-excitation rate is a constant, $B$.

The rate $A$ is equal to $s\exp\{-\Delta E / kT\}$, where $\Delta E = E_{e1} - E_g$, for thermal excitation, or $\sigma_p \Phi$ for optical excitation. The tunneling probability $P_e(r)$ is equal to $P_0\exp\{-\alpha_e r\}$.

If the luminescence $I(r,t)$ is emitted between irradiation and heating (or optical stimulation) at constant $T$, it is the after-glow. If it is emitted during heating (or optical stimulation) after irradiation it is TL (or OSL).

With the equilibrium condition $\delta n_e(r,t)/\delta t \approx 0$,

$$n_e(r,t) = \frac{An_g(r,t)}{(B + P_e(r))}$$ (3.110)

so that

$$\frac{\delta n_g(r,t)}{\delta t} = -An_g(r,t) + \frac{BAn_g(r,t)}{(B + P_e(r))} = -An_g(r,t)\left(\frac{P_e(r)}{B + P_e(r)}\right).$$ (3.111)

Assuming most excited electrons decay to the ground state before tunneling ($B \gg P_e(r)$):

$$\frac{\delta n_g(r,t)}{\delta t} = -\frac{AP_e(r)n_g(r,t)}{B}.$$ (3.112)

Equation 3.112 is a partial differential equation with respect to $t$, for a fixed value of $r$. The solution to Equation (3.112) is:

$$n_g(r,t) = n_g(r,0)\exp\left\{-\frac{AP_e(r)}{B}t\right\}.$$ (3.113)

The term $n_g(r,0)$ is the initial distribution of the electron traps at $t = 0$, described by Equation 3.107. The rates $A$, $B$, and $P_e(r)$ are constants with time.

For thermal stimulation, and expanding the expression for $P_e(r)$, Equation 3.113 is re-written as:

$$n_g(r,t) = n_g(r,0)\exp\left\{-\left(\frac{s}{B}\right)\exp\{-\Delta E/kT\}P_0\exp\{-\alpha_e r\}t\right\}. \qquad (3.114)$$

Using equilibrium statistics and the Law of Detailed Balance it may be written that $s = B$, although this is not essential to the analysis that follows. (Note, that the value of $s$ is much smaller for a transition to the excited state than the corresponding value that would be expected for a transition to the delocalized band since the entropy change is much smaller.)

Equation 3.114 now becomes:

$$n_g(r,t) = 4\pi n_0^2 r^2 \exp\left\{-\frac{4}{3}\pi n_0 r^3\right\}\exp\left\{-\left(\frac{s}{B}\right)\exp\{-\Delta E/kT\}P_0\exp\{-\alpha_e r\}t\right\}, \quad (3.115)$$

where $n_0(r)$ is the original total number of electron traps.

Integrating 3.115 over all $r$, the surviving concentration of electron traps in the ground state is:

$$n_g(t) = \int_0^\infty 4\pi n_0^2 r^2 \exp\left\{-\frac{4}{3}\pi n_0 r^3\right\}\exp\left\{-\left(\frac{s}{B}\right)\exp\{-\Delta E/kT\}P_0\exp\{-\alpha_e r\}t\right\}dr. \quad (3.116)$$

Since

$$I(r,t) = \left|\frac{\delta m(r,t)}{\delta t}\right| = \frac{\delta n_g(r,t)}{\delta t} + \frac{\delta n_e(r,t)}{\delta t} \approx \frac{\delta n_g(r,t)}{\delta t}, \qquad (3.117)$$

then the time-dependent luminescence intensity $I(t)$ is:

$$I(t) = \int_0^\infty I(r,t)dr = \int_0^\infty \frac{\delta n_g(r,t)}{\delta t}dr, \qquad (3.118)$$

which, from Equation 3.116, is:

$$I(t) = \int_0^\infty \frac{\delta}{\delta t}\left[4\pi n_0^2 r^2 \exp\left\{-\frac{4}{3}\pi n_0 r^3\right\}\exp\left\{-\left(\frac{s}{B}\right)\exp\{-\Delta E/kT\}P_0\exp\{-\alpha_e r\}t\right\}\right]dr. \quad (3.119)$$

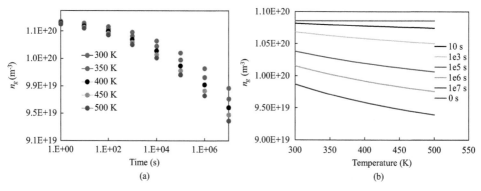

**Figure 3.45** Excited state tunneling showing (a) the change in the concentration of electrons in the ground state as a function of time, for different temperatures. (b) The same data as in (a), but illustrated as a function of temperature for different decay times.

Figure 3.45(a) shows the isothermal ($T$ = constant) decay of $n_g(t)$ with time after irradiation, for different temperatures. The integration over distance $r$ was performed numerically. The parameters used in the calculation are listed in the figure caption. Figure 3.45b illustrates the same data, but expressed as a function of temperature for different decay times. As with ground-state tunneling, the after-glow luminescence intensity for tunneling from the excited state also decays with time following a $t^{-k}$ law.

---

### Exercise 3.16   Tunneling (2)

Excited state tunneling requires excitation from the ground state $E_g$ to energy level $E_{e1}$ (Figure 3.44a). Assuming that excitation from the ground state to the excited state can take place both thermally and optically, consider exciting an irradiated sample with light of intensity $\Phi$ and a corresponding optical cross-section $\sigma$ at temperature $T$. (Note the use of "optical cross-section" not "photoionization cross-section," since the electron is not ionized from the trap, but only excited into the excited state.) The net luminescence emitted due to tunneling from the excited state will be due both the optically and thermally excited tunneling.
(a)  Assuming a discrete energy level (i.e. not a distribution of states) derive an expression for the luminescence emitted for a fixed wavelength and temperature.
(b)  Assume values for the various parameters. Using numerical integration over NN distances $r$, plot the luminescence intensity as a function of time during the stimulation. Change the parameters to give results under different conditions. What do you observe?

---

### 3.8.1.4   Decay during Irradiation

In the foregoing sections, the decay of trapped electrons due to tunneling, whether it be from the ground state or the excited state, has been after the irradiation has ended. However, for long enough irradiation times decay of the trapped electrons must also take place during irradiation. If the irradiation time $t_i$ is also taken into consideration, then the approximation $I_{after-glow}(t) = K/t$ now becomes:

$$I_{after-glow}(t) = \frac{K}{t_i}\int_{-t_i}^{0}\frac{dt'}{(t-t')} = \frac{K}{t_i}\ln\left(1+\frac{t_i}{t}\right).$$

(3.120)

Figure 3.46a shows a comparison between the $t^{-1}$ law (see Exercise 3.15) and Equation 3.120 assuming $t_i = 10$ s. and shows that for times > ~$10t_i$, the irradiation time can be ignored.

An alternative approximation is to use a total decay time of $t + t_i/2$ and to use $I_{after-glow}(t) \approx K/(t + t_i/2)$. This is also shown in Figure 3.46a and is seen to approximate Equation 3.120 better than a simple $t^{-1}$ law and is indistinguishable from the solution to Equation 3.120 for $t > t_i$. An alternative representation is to plot the after-glow as a function of $t + t_i/2$, instead of $t$ (Figure 3.46b). In this plot Equation 3.120 is seen to produce faster decay at small times, but $(t + t_i/2)^{-1}$ is a good approximation to Equation 3.120 for $(t + t_i/2)$ values > ~$t_i$.

### 3.8.1.5   Effect of Tunneling on TL and OSL

As already noted, since the after-glow can be considered to be the luminescence "lost" during the time before TL or OSL measurement, then the TL or OSL will also follow a similar time dependence to that of the after-glow. In all instances, however, tunneling will not continue to

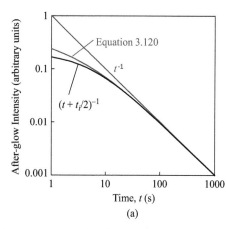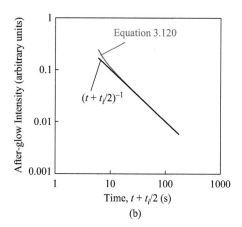

**Figure 3.46**  (a) The $t^{-1}$ approximation compared to Equation 3.120 and a $(t + t_i/2)^{-1}$ approximation for the after-glow decay due to tunneling. In this calculation the constant $K$ was assumed to be unity and $t_i = 10$ s. (b) Alternatively, the same data can be plotted against $(t + t_i/2)$. The data according to Equation 3.120 shows a faster decay at short times (while, of course, $(t + t_i/2)^{-1}$ shows a slope of –1).

infinite time and will stop once all pairs separated by NN distances have been exhausted and the remaining pairs are too far apart for tunneling to occur. Defining a time $t = t_m$ at which this occurs (which corresponds to the distances at which the electron wavefunction ceases to overlap with the neighboring recombination center) the ratio $R$ of the residual TL (or OSL) remaining at time $t$ after irradiation compared to that at a short time $t_0$ after irradiation is given by:

$$R = \frac{I_{TL/OSL}(t)}{I_{TL/OSL}(t_0)} = \frac{\ln(t_m/t)}{\ln(t_m/t_0)},\tag{3.121}$$

where the $t^{-1}$ law is assumed.

Alternatively, an arbitrary time $t_c$ can be defined such that the TL or OSL intensity at $t_c$ is related to the intensity at any other time $t$ by:

$$I_{TL/OSL}(t) = I_c \left[1 - g \log_{10}\left(\frac{t}{t_c}\right)\right],\tag{3.122a}$$

where $I_c$ is the TL or OSL intensity at $t = t_c$, and $g$ is the fractional loss of intensity per decade in time (i.e. from $t = 10t_c$). Equation 3.122a can also be written:

$$I_{TL/OSL}(t) = I_c \left[1 - \kappa \ln\left(\frac{t}{t_c}\right)\right].\tag{3.122b}$$

where $\kappa = g/\ln(10)$ is the fractional loss per time interval from $t_c$ to $2.3t_c$.

Also note that for a defect-rich medium in which there is a distribution of NN distances (Figure 3.43a) it should be expected that the fluctuations in the donor–acceptor distances may also give rise to perturbations in the trap depths, resulting in a distribution of trap depths. As a schematic example, Figure 3.47 shows two possible situations in which the potential barrier height between the donor and the acceptor $E$ is the same in each case. The trap depth $E_t$ is smaller in (a) than in (b), but the NN distance $r$ is smaller in (b) than in (a), so that the potential barrier is thinner with a commensurate higher probability of tunneling. Such a situation would give rise to a TL signal from (a) that would appear at a lower temperature than (b), but there

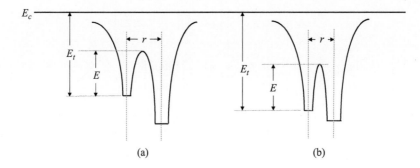

**Figure 3.47** Two possible configurations of donor (electron trap) and acceptor (recombination center), showing (a) a shallower trap (small $E_t$) but a wider potential barrier $r$, and (b) a deeper trap (larger $E_t$) but with a narrower potential barrier. The potential barrier height $E$ is the same in both cases.

would be greater tunneling, and therefore greater fading, for the higher temperature TL peak than the low temperature TL peak. Similar arguments can be made for OSL; in this case electron traps with greater optical trap depths could fade more rapidly than those with smaller optical trap depths. The exact degrees of perturbation caused by the closeness of the donor and acceptor are difficult to predict in a generalized sense, but greater fading from deeper traps than from shallower traps is certainly possible.

### 3.8.2 Localized and Semi-Localized Transition Models

Another possibility for a recombination transition that does not involve the delocalized band is a localized transition via an energy level common to both the trap and the recombination site. Such a situation is possible from Figure 3.41 where it can be seen that excitation to energy level $E_{e2}$ may allow direct relaxation into the hole center. This is indicated schematically in Figure 3.44b in transition A, followed by either relaxation (transition $B$) or recombination (transition C). Transition A can be optical or thermal. This situation is termed a "localized transition model" and results in luminescence emission for excitation energies less than those that would be expected considering the trap depth $E_t$. However, for suitable temperatures, a delocalized transition D can also occur, followed by either a retrapping transition E or a recombination transition F. A situation such as this, where a recombination event can be triggered by either localized or delocalized transitions is termed "semi-localized transition model." Both situations are now examined. The energy level scheme shown in Figure 3.44b was discussed generally by Bräunlich and Scharmann (1964) and further information on such models can be found in several books on TL or OSL (Chen and McKeever 1997; Chen and Pagonis 2011), and also in the thesis by Templer (1986) and in papers by Mandowski and colleagues (Mandowski 2004, 2006, 2008), and the references therein.

#### 3.8.2.1 Localized Transition Model

Using Figure 3.44b, the rate equations describing transitions A, B, and C, in the case of a thermal stimulation for transition A, are:

$$\frac{dn_g}{dt} = Bn_e - An_g = sn_e - n_g s \exp\left\{-\frac{\Delta E}{kT}\right\}, \tag{3.123a}$$

$$\frac{dn_e}{dt} = An_g - Bn_e - Cn_e = n_g s \exp\left\{-\frac{\Delta E}{kT}\right\} - sn_e - \alpha n_e, \tag{3.123b}$$

and

$$\frac{dm}{dt} = -Cn_e = -\alpha n_e. \tag{3.123c}$$

The concentrations $n_g$, $n_e$, and $m$ are defined in Figure 3.44b, and $\Delta E = E_{e2} - E_g$. A, B, and C are the transitions rates (in s$^{-1}$) of transitions A, B, and C, respectively; the relaxation rate B is again taken to be equal to $s$, and $C = \alpha$ is the recombination rate.

As usual, the TL intensity is:

$$I_{TL} = \left|\frac{dm}{dt}\right| = Cn_e = \alpha n_e. \tag{3.123d}$$

At quasi-equilibrium, $n_e << n_g, m$ and $|dn_e / dt| << |dn_g / dt|, |dm / dt|$, which gives:

$$I_{TL} = \left|\frac{dm}{dt}\right| = \left(\frac{\alpha s}{\alpha + s}\right) m \exp\left\{-\frac{\Delta E}{kT}\right\}. \tag{3.124}$$

The equivalent expression for OSL is:

$$I_{OSL} = \left|\frac{dm}{dt}\right| = \left(\frac{\alpha}{\alpha + s}\right) m \sigma_p \Phi. \tag{3.125}$$

Both expressions, for TL and OSL, are first order since $\alpha$ and $s$ are both constants. The TL peak will be a first-order shape with a frequency factor $\bar{s} = \alpha s / (\alpha + s)$. The frequency factor becomes $\bar{s} = s$ if $\alpha >> s$, but if $s >> \alpha$ it means that the relaxation transition B is much faster than the recombination transition C. This may result in the QE approximation not holding and therefore Equations 3.124 and 3.125 may not be accurate in this case. This is equivalent to "fast retrapping" in the case of a delocalized transition model.

Note that if $s >> \alpha$, $\bar{s} = \alpha$ and can be small (perhaps as small as $10^5$ s$^{-1}$). This results in a broad TL peak that rapidly changes its peak position as the heating rate $\beta_t$ is changed. This may be a way in which TL due to localized transitions can be distinguished from TL due to delocalized transitions.

If the temperature is a constant, Equation 3.124 becomes the expression for after-glow (or phosphorescence, $I$). Solving for $m = n_{g0}$ at constant $T$ gives,

$$I = n_{g0}\bar{s}\exp\left\{-\frac{\Delta E}{kT}\right\}\exp\left\{-\bar{s}t\exp\left\{-\frac{\Delta E}{kT}\right\}\right\}, \tag{3.126}$$

For a discrete trap, this yields a first-order luminescence decay, which is to be distinguished from the $t^{-k}$ power law decay of luminescence due to tunneling.

### 3.8.2.2  Semi-Localized Transition Model

#### 3.8.2.2.1  Templer Model

Templar's semi-localized transition model includes transitions D, E, and F in addition to transitions A, B, and C, as illustrated in Figure 3.44b. The rate equations now become:

$$\frac{dn_g}{dt} = -n_g A + n_e B - n_g D + n_c E, \tag{3.127a}$$

$$\frac{dn_e}{dt} = n_g A - n_e B - n_e C, \tag{3.127b}$$

$$\frac{dn_c}{dt} = n_g D - n_c E - n_c F, \tag{3.127c}$$

$$\frac{dm}{dt} = -n_c F - n_e C. \tag{3.127d}$$

As before, $D$, $E$, and $F$ (in $s^{-1}$) are the transition rates for transitions D, E, and F, respectively. The transition rates for thermal excitation are defined as:

$$A = s\exp\left\{-\frac{\Delta E}{kT}\right\}, \tag{3.128a}$$

$$B = s, \tag{3.128b}$$

$$C = \alpha, \tag{3.128c}$$

$$D = s_t\exp\left\{-\frac{E_t}{kT}\right\}, \tag{3.128d}$$

$$E = A_n\left(N - n_g - n_e\right), \tag{3.129e}$$

and

$$F = A_m m. \tag{3.129f}$$

In these expressions, all terms have their usual meanings with the following additions: $s_t$ is the frequency factor for transition D, with energy gap $E_t$; $A_n$ is the transition E probability (in $m^{-3}.s^{-1}$); and $A_m$ is the probability for the recombination transition F (also in $m^{-3}.s^{-1}$). $N$ is the total concentration (in $m^{-3}$) of electron centers, thus $(N - n_g)$ is the concentration of available, empty ground states. Note that $s_t$ is not equal to $s$ since one is a localized excitation only to the excited state, while the other is a delocalized excitation to the conduction band. Although both transitions gain their energy from phonon absorption, the entropy change associated with each transition is entirely different and is larger for delocalization (ionization) than for simple excitation.

Two possible transitions, C and F, can produce luminescence:

$$I_{TL}^C = n_e C = n_e \alpha \tag{3.130a}$$

and

$$I_{TL}^F = n_c F = n_c A_m m. \tag{3.130b}$$

The first requires stimulation across the gap $\Delta E$, while the second requires stimulation across the gap $E_t$. Each will produce its own TL peak and thus two peaks are expected from this model with their relative sizes depending upon the balance between the various relaxation, retrapping, and recombination transitions. As shown above, transition C is first order and should produce a first-order TL peak or OSL decay curve. It is also clear that transition F gives a first-order TL or OSL curve if QE holds.

---

**Exercise 3.16    Semi-Localized Transitions**

(a) Using the Templer model of Figure 3.44, solve the rate equations for the various transitions and generate TL glow curves, using Equations 3.127–3.130. Assign your own values to the various parameters.

(b) By changing the values of the various parameters, study the changes in the TL glow-curve shapes.

(c) Fit the resulting glow curves to a normal RW equation, and examine how the activation energy or energies obtained change as the various parameters are changed.

---

*3.8.2.2.2 Mandowski Model*

No discussion of the spatial association between the trap and the center is included in the Templer model, other than to assume a random distribution of electron (donor) sites and hole (acceptor) sites. In the Mandowski model (Mandowski 2008), the electron and hole sites are considered to be part of the same defect complex, an example of which for a dosimetry material is LiF doped with Mg and Ti. The case of LiF:Mg,Ti will be discussed in more detail in a later chapter, but for now suffice it to say that the defect complex that gives rise to the TL peak structure known as "peaks 5 and 5a" are understood to be due to a large complex of Mg-vacancy trimers (see Chapter 2) and Ti-OH centers (e.g. Horowitz et al. 2019). The exact structure is unknown.

In the Mandowski model, each defect complex can trap one electron, or one hole, or one electron and one hole, or neither. Furthermore, the trapped electron can be either in the ground state or the excited state. To describe this more complex situation, Mandowski introduced the following notation:

$$\begin{bmatrix} \overline{n_e} \\ \overline{n_g} \\ \overline{m} \end{bmatrix}$$

where the $\overline{n_e}$, $\overline{n_g}$ and $\overline{m}$ notation represents the integer number (0 or 1) of electrons in the excited state, electrons in the ground state, and holes, respectively. A value of 1 for each number indicates the electron or hole is present, while a value of 0 indicates it is not. Thus:

$$\begin{bmatrix} 0 \\ 1 \\ 0 \end{bmatrix}$$

means that the defect complex has trapped one electron, which is in the ground state. Regarding all possibilities, terms of the form $H^{\overline{n_e}}_{\overline{n_g}}$ and $E^{\overline{n_e}}_{\overline{n_g}}$, are then defined, where:

$$H^{\overline{n_e}}_{\overline{n_g}} = \begin{bmatrix} \overline{n_e} \\ \overline{n_g} \\ 1 \end{bmatrix} \text{ and } E^{\overline{n_e}}_{\overline{n_g}} = \begin{bmatrix} \overline{n_e} \\ \overline{n_g} \\ 0 \end{bmatrix}.$$

$H$ indicates a hole center since $\bar{m} = 1$, always, and $E$ indicates an electron center since $\bar{m} = 0$, always. Furthermore, since the electron must be in either the ground state or the excited state, $n_e = n_g = 1$ is not allowed.

The possible states for the defect are thus:

$$H_0^0 = \begin{Bmatrix} 0 \\ 0 \\ 1 \end{Bmatrix};\cdot H_0^1 = \begin{Bmatrix} 1 \\ 0 \\ 1 \end{Bmatrix};\cdot H_1^0 = \begin{Bmatrix} 0 \\ 1 \\ 1 \end{Bmatrix}; \tag{3.131a}$$

and

$$E_0^0 = \begin{Bmatrix} 0 \\ 0 \\ 0 \end{Bmatrix};\cdot E_0^1 = \begin{Bmatrix} 1 \\ 0 \\ 0 \end{Bmatrix};\cdot E_1^0 = \begin{Bmatrix} 0 \\ 1 \\ 0 \end{Bmatrix}. \tag{3.131b}$$

The Mandowski model also assumes a different set of transitions between levels than those considered by Templer. The transitions allowed in the model are shown in Figure 3.48 in which the level $E_{e1}$ has been removed for clarity and $E_{e2}$ renamed simply $E_e$. Each dashed box indicates a separate defect complex consisting of both the electron trap and the hole trap. Many such complexes exist within the crystal host. The dashed lines connecting the conduction and valence band edges are meant to indicate that the energy gap within the defect complex is not necessarily the same as that in the rest of the host. The complex overlap of potentials in the region of the defect may result in a local band gap that is different from that in the host matrix.

Although in the figure $\Delta E' < \Delta E$, in general $\Delta E'$ could be greater than $\Delta E$, or they could be equal. The actual situation will depend upon the nature of the shared energy level, $E_e$. Several possibilities exist. Level $E_e$ could be an excited state in which the electron wavefunction overlaps the region of the hole center, but with no potential barrier through which the electron needs to tunnel. In this case, it may be considered to be a shared excited state and recombination directly into the hole center is possible. Alternatively, the level may represent band-tail states, as discussed previously in relation to Figure 3.32. Finally, the energy band within the confines of the complex may be lower than that in the rest of the host crystal. Although the electron may

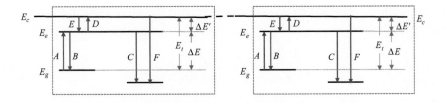

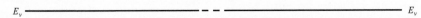

**Figure 3.48** The semi-localized transition model of Mandowski, showing energy levels $E_g$ and $E_e$ for the ground and excited state of the electron trap, and the nearby hole trap. The possible transitions A-to-F are indicated. The dashed boxes indicate that each set of levels belongs to one defect complex consisting of the trap and the recombination center. Many such complexes exist throughout the material.

be raised to the "delocalized" band within the complex, the wavefunction will only be "quasi-delocalized" in that it is confined by the potential well at the bottom of the conduction band, similar to the situation depicted in Figure 3.33. In this situation $E_e$ represents the bottom of the local potential well at the bottom of the conduction band and $\Delta E'$ represents the energy needed to raise the electron energy to values higher in the conduction band whence it becomes entirely delocalized. A similar situation may arise in any dosimeter material after irradiation with a heavy charged particle (HCP) in which the energy absorbed is confined to a narrow track, the width of which depends upon the particle energy and mass. The high ionization density within the particle track may give rise to close spatial association between the electron and hole traps, and to local potentials that change the local band gap such that the "freed" electron is only quasi-delocalized and recombination is similarly confined to the region inside the track. Escape from the track would occur only if the energy absorbed raises the electron energy to higher levels in the conduction band, whereupon it becomes delocalized and is free to recombine anywhere.

A further major difference between the models of Templer and Mandowski concerns transitions E and D. These are now between the excited state and the conduction band, not the ground state and the conduction band.

With the above notation the following equivalencies can be written:

$$n_g = \sum H_1^0 + \sum E_1^0, \tag{3.132a}$$

$$n_e = \sum H_0^1 + \sum E_0^1, \tag{3.132b}$$

$$m = \sum H_0^0 + \sum H_0^1 + \sum H_1^0, \tag{3.132c}$$

where $n_g$, $n_e$, and $m$ are the total numbers of electrons and holes trapped within all the defect complexes in the crystal, and the summations are over the crystal volume. Charge neutrality means that:

$$n_c = m - n_e - n_g = \sum H_0^0 - \sum E_0^1 - \sum E_1^0. \tag{3.133}$$

There are several ways each state can change. For example, $H_0^1$ can change to $H_1^0$ via transition B, or to $H_0^0$ via transition D, or to $E_0^1$ via transition F, or to $E_0^0$ via transition C. Using Figure 3.48, the equations expressing all possible changes during stimulation (thermal or optical) after irradiation are:

$$\frac{dH_0^0}{dt} = DH_0^1 - FH_0^0 - EH_0^0, \tag{3.134a}$$

$$\frac{dH_0^1}{dt} = AH_1^0 + EH_0^0 - DH_0^1 - CH_0^1 - BH_0^1 - FH_0^1, \tag{3.134b}$$

$$\frac{dH_1^0}{dt} = BH_0^1 - AH_1^0 - FH_1^0, \tag{3.134c}$$

$$\frac{dE_0^0}{dt} = CH_0^1 + FH_0^0 + DE_0^1 - EE_0^0, \tag{3.134d}$$

$$\frac{dE_0^1}{dt} = AE_1^0 + FH_0^1 + EE_0^0 - DE_0^1 - BE_0^1, \tag{3.134e}$$

$$\frac{\mathrm{d}E_1^0}{\mathrm{d}t} = BE_0^1 + FH_1^0 - AE_1^0. \tag{3.134f}$$

The above reactions are presented diagrammatically below.

If the stimulation is thermal, then:

$$A = s\exp\left\{-\frac{\Delta E}{kT}\right\} \tag{3.135a}$$

and

$$D = s'\exp\left\{-\frac{\Delta E'}{kT}\right\}. \tag{3.135b}$$

Note that $s'$ will be larger $s$ than since the entropy increase is larger in transition D compared to transition A.

If the stimulation is optical, then:

$$A = \sigma_p \Phi \tag{3.136a}$$

and

$$D = \sigma_p' \, \Phi. \tag{3.136b}$$

As with the Templer model, two pathways (C and F) are possible for luminescence. They give TL or OSL intensities:

$$I_{TL/OSL}^C = CH_0^1 \tag{3.137a}$$

and

$$I_{TL/OSL}^F = F\left(H_0^0 + H_1^0 + H_0^1\right). \tag{3.137b}$$

For TL and under conditions of quasi-equilibrium, each pathway will produce its own TL first-order peak, and CW-OSL will be a superposition of two OSL exponential decay curves. Likewise, all other OSL measurement types will be a superposition of two first-order OSL signals.

A luminescence efficiency $\eta$ can be defined:

$$\eta = \frac{CH_0^1}{DH_0^1 + BH_0^1} + \frac{F\left(H_0^0 + H_1^0 + H_0^1\right)}{E\left(E_0^0 + H_0^0\right)} = \left(\frac{C}{D+B}\right) + \frac{F\left(H_0^0 + H_1^0 + H_0^1\right)}{E\left(E_0^0 + H_0^0\right)}. \tag{3.138}$$

The first term on the left-hand side of this equation represents the efficiency of recombination of the excited state electron with a hole within the same defect complex (path C), compared to the alternative transitions of relaxation to the ground state (path B) or delocalization (path D).

Only $H_0^1$ appears in this part of the equation since a hole has to be present and an excited state electron has to be present. The second term represents the possibilities after delocalization. In this, the free electron can recombine with any trapped hole in any defect cluster ($H_0^0$, $H_1^0$ or $H_0^1$) through pathway F, or it can be recaptured by any empty excited state through path E, as long as a ground state electron is not present. (i.e. it can be captured by $E_0^0$ or $H_0^0$). The ratios of these probabilities gives the luminescence efficiency for the system.

A so-called retrapping ratio $r$ can also be defined such that $r = B/C$. (This is something of a misnomer in that the transition B is a relaxation transition, not a retrapping transition. The electron has not necessarily left the trap when it is at energy level $E_e$. However, since it has a probability of doing so, through transition C or D, the term "retrapping ratio" is used.) Finally, a delocalized recombination ratio $\delta$ is defined as $\delta = E / F$. This is identical to the parameter $P = R_{retrap}/R_{recom}$ introduced in Equation 3.77.

### 3.8.2.3   Semi-Localized Transitions and the TL Glow Curve

Example solutions to Equations 3.134 and 3.137 using Equations 3.135 were generated by Mandowski (2004) and examples are shown here in Figure 3.49. (In this calculation Mandowski considered that the $\Delta E$ values (from Figure 3.47) are dependent on whether or not the hole is present in the defect complex. Mandowski also allowed $\Delta E'$ to remain unchanged and $s = s'$. This is somewhat arbitrary, however, and it is perhaps more likely that $\Delta E'$ will be dependent on the presence of the hole since this energy represents transition D, which is an ionization event, whereas transition A is not and, as has been mentioned, $s$ is likely to be less than $s'$.)

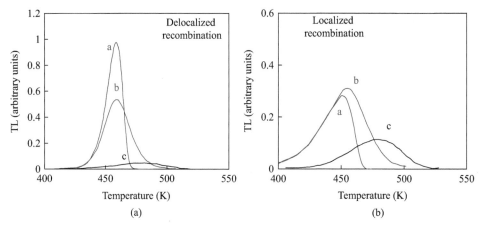

**Figure 3.49**   Solutions to Equations 3.134 and 3.137, using thermal stimulation obtained by Mandowski (2008), who, for purposes of this calculation, assumed that the value of $\Delta E$ in the model of Figure 3.48 depends upon whether or not a hole is present in the nearby hole trap. Thus, Mandowski used $\Delta E = 0.9$ eV when the hole is present (for $H_{n_g}^{n_e}$-type centers) and $\Delta E = 0.6$ eV (for $E_{n_g}^{n_e}$-type centers). $\Delta E'$ was set at 0.7 eV. Values for $s$ and $s'$ were $10^{10}$ s$^{-1}$. The retrapping ratio $r$ was $10^2$, and the heating rate $\beta_t$ was 1 K.s$^{-1}$. Figure (a) represents the TL from delocalized recombination ($I_{TL}^F = F\left(H_0^0 + H_1^0 + H_0^1\right)$, Equation 1.137b), while (b) shows the TL from the localized transition ($I_{TL}^C = CH_0^1$, Equation 3.137a). The notations a, b, and c correspond to three different values of $\delta$, namely 0, 1, and $10^2$. (Adapted from Mandowski 2008.)

Also note in the figure that a, b, and c correspond to three different values of $\delta$, namely 0, 1, and $10^2$. Since $\delta$ and $P$ are identical, only $\delta = P = 0$ represents a first-order reaction. This can be confirmed by the asymmetrical nature of the TL peaks for case a, and the almost symmetrical shape of the TL peaks for cases b and c, indicating non-first-order kinetics. Fitting curve a in Figure 3.49a to the RW, first-order peak shape yielded a single activation energy of 2.90 eV and an extremely large frequency factor of $1.1 \times 10^{31}$ s$^{-1}$ (Mandowski 2008). Such large values are reminiscent of the $E_t$ and $s$ values obtained from fitting the main TL peak in LiF:Mg,Ti ("peak 5") to the RW equation (discussed further in Chapter 7).

## 3.9   Master Equations

So-called master equations for TL or OSL are those that include the function $p$ for the stimulation rate, without specifying the form of this function. As has been discussed, $p$ can be of the form $s\exp\{-E_t / kT\}$ for thermal stimulation where $T = T_0 + \beta_t t$; or $p = \sigma_p \Phi$ for optical stimulation, with $\Phi = \Phi_0 + \beta_\Phi t$, or $\lambda = \lambda_0 - \beta_\lambda t$ (Equations 3.43a,b). An example "master equation" for the OTOR model is:

$$I_L\left(t\right) = \frac{npm\sigma_r}{\left[(N-n)\sigma_e + m\sigma_r\right]} = p\frac{n^2}{(N-n)R+n} \tag{3.139}$$

with $n = m$ and $R = \sigma_e / \sigma_r$. The subscript $L$ denotes that this expression can be used for $L = TL$, or $L = CW\text{-}OSL$, or $LM\text{-}OSL$, etc., with the appropriate function for $p$. Equation 3.139 is also known as the GOT expression for the OTOR model (see Exercise 3.2).

With the appropriate term for $p$, this general expression can yield the RW equation, Garlick-Gibson equation, CW-OSL expression, LM-OSL expression, etc. Similar equations can be written for other models, such as IMTS, NMTS, localized transitions, tunneling, etc.

# 4

# RPL

## *Models and Kinetics*

*Mit der Vermehrung der gelben Zentren geht die Steigerung der Thermolumineszenz und Radiophotolumineszenz parallel.*
[*With the multiplication of the yellow centers, the increase in thermoluminescence and radiophotoluminescence goes in parallel.*]

– K. Przibram 1927

## 4.1 Radiophotoluminescence and Its Differences with TL and OSL

Early pioneers in the development of what is now known as radiophotoluminescence (RPL) recognized the relationship that existed between RPL, TL, and OSL phenomena following irradiation of a substance with gamma or x-irradiation. The word "radiophotoluminescence" was coined by Przibram (Przibram and Kara-Michailowa 1922; Przibram 1923) and is meant to convey the property of photoluminescence that only appears after absorption of ionizing radiation. Before irradiation, no photoluminescence properties are present; the radiation induces photoluminescent defects that were not present in the material before irradiation. Unlike TL or OSL, however, the registering of an RPL signal after irradiation does not require ionizing the radiation-induced defect, but only elevating the energy of the trapped electron to an excited state, followed by relaxation and the emission of light. Importantly, unlike TL or OSL, it is a non-destructive measurement technique and the RPL signal can be re-measured over and over again without loss of sensitivity. Schulman and colleagues (Schulman et al. 1951) recognized that RPL, TL, and OSL can each be exploited as dosimetry methods.

## 4.2 Background Considerations

The distinction between the energy level transitions that are necessary for RPL and those that are needed for TL or OSL was described in Figure 1.4, where it was mentioned that transfer of charge between defects does not take place. The transitions can be best described using a

*A Course in Luminescence Measurements and Analyses for Radiation Dosimetry,* First Edition.
Stephen W.S. McKeever.
© 2022 John Wiley & Sons Ltd. Published 2022 by John Wiley & Sons Ltd.
Companion Website: www.wiley.com/go/mckeever/luminescence-measurements

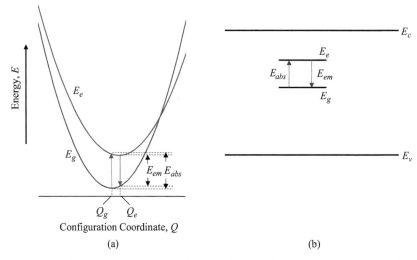

Configuration Coordinate, $Q$

(a)

(b)

**Figure 4.1** (a) Configurational coordinate diagram for photoluminescence emission showing the absorption transition at coordinate $Q_g$ and the emission transition at $Q_e$, with $E_{abs} > E_{em}$. (b) The same transitions represented on an energy band diagram.

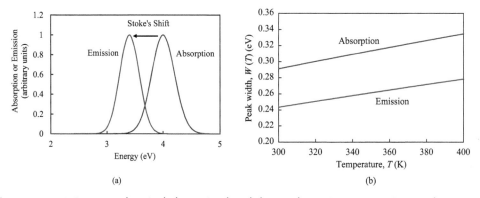

(a)

(b)

**Figure 4.2** (a) Conceptual optical absorption band due to absorption at a maximum of $E_{abs} = 4$ eV and $W_{abs}(300K) = 0.291$ eV, and emission at lower energy at a maximum $E_{em} = 3.4$ eV, with $W_{em}(300K) = 0.244$ eV. (a) Typical temperature dependence of the full-width at half-maximum for the absorption and emission bands, using Equation 4.2 with arbitrary values of $\omega_g = 2\times10^{13}$ s$^{-1}$, $\omega_e = 3\times10^{13}$ s$^{-1}$, $S_g = 22$, and $S_e = 10$.

configurational coordinate diagram, re-drawn here in Figure 4.1a, specifically for photoluminescence. Figure 4.1b shows the same transitions represented on a conventional flat-band diagram. Since all optical transitions are vertical in the configurational coordinate diagram, the absorption of a photon of energy $E_{abs}$ raises the energy of the ground state electron to the excited state at coordinate $Q_g$. Relaxation from the excited state to the ground state emits phonons such that relaxation occurs to coordinate $Q_e$, at which a transition of the electron from the excited state to the ground state occurs, with the emission of a photon of energy $E_{em}$. Final phonon emission allows the system to relax to its original coordinate, at $Q_g$. It can be seen that $E_{em} < E_{abs}$, a relationship known as Stoke's shift. Absorption and emission bands for such a model are shown conceptually in Figure 4.2a.

The width of the absorption and emission bands is a function of the strength of the phonon coupling between the electron and the lattice, as given by the dimensionless Huang-Rhys factor, $S$. Treating the ground and excited states as classical quantum-mechanical oscillators with quantum energy $\hbar\omega$, the full-width at half maximum at temperature $T$ is:

$$W(T) = W(0)\left[\coth\left(\frac{\hbar\omega}{2kT}\right)\right]^{1/2},$$

(4.1)

which, for a Gaussian-shaped band, becomes:

$$W(T) = \hbar\omega\left[8S\ln(2)\coth\left(\frac{\hbar\omega}{2kT}\right)\right]^{1/2}.$$

(4.2)

Figure 4.2b shows typical temperature dependencies for the full-width at half-maximum for the absorption and emission bands. For the purposes of this illustration it is assumed that the lattice coupling to the ground states and excited states are not the same, as illustrated with the parameters listed in the figure caption (with subscripts $g$ and $e$).

The oscillator strength $f$ of the absorption transition is related to the peak of the absorption band by Smakula's equation (Townsend and Kelly 1973) thus:

$$Nf = K\frac{n}{\left(n^2 + 2\right)^2}\alpha_{max}W,$$

(4.3)

where $K$ is a constant, $\alpha_{max}$ is the maximum absorption coefficient (in cm$^{-1}$), $n$ = the refractive index and $N$ = the concentration of absorbing centers (in cm$^{-3}$).

Several dosimetry-related materials exhibit RPL. The most well-known are the alkali halides (especially LiF for dosimetry), $Al_2O_3$, and phosphate glasses. Example absorption and emission bands for radiation-induced photoluminescence (RPL) centers are shown in Figure 4.3. In LiF doped with Mg, (Figure 4.3a) the creation of $F$-centers (a F$^-$ ion vacancy with a trapped electron) leads to clustering of these centers to form $F_2$-, $F_3^+$-, $F_3$-centers, etc., with differing numbers of $F$-centers in a variety of charge states. Excitation with light corresponding to the peak of the $F_3^+/F_2$ band, around 450 nm, induces emission at 520 nm (from $F_3^+$-centers) and 670 nm (from $F_2$-centers).

In $Al_2O_3$ doped with C and Mg (Figure 4.3b), $F_2^{2+}$(2Mg)-centers are formed during crystal growth, consisting of two oxygen vacancies charge compensated by two Mg$^{2+}$ ions. During irradiation these defects efficiently capture free electrons to become $F_2^+$(2Mg)-centers with absorption bands at 335 nm and 620 nm. Subsequent stimulation, usually at 335 nm, produces RPL emission centered at 750 nm.

In Ag-doped alkali phosphate glasses (Figure 4.3c), Ag$^+$ ions are converted to Ag$^0$ atoms and Ag$^{2+}$ ions by trapping of electrons and holes respectively during irradiation. Subsequent clustering of the Ag-related species form Ag$_m^{n+}$ complexes (where m and n are integers and m = n + 1). RPL results from absorption of stimulation light around 310 nm-to-340 nm by Ag$_2^+$ centers and emission at 630 nm.

In each of the RPL materials mentioned, a frequent observation is that the RPL intensity increases after the irradiation ceases in a phenomenon known as "buildup," on a time scale of minutes to hours depending on the material. This is caused by specific, post-irradiation defect reactions leading to the growth of the RPL emitting centers. The defect reactions may consist of clustering and/or charge trapping/detrapping processes and the timescales over which these reactions occur for any given material is generally dependent upon both temperature and defect concentration. The kinetics of these post-irradiation reactions are discussed in the following section.

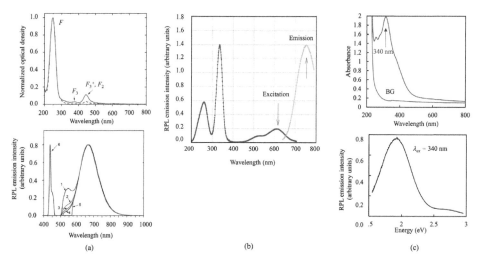

**Figure 4.3** (a) Absorption (upper graph) and RPL emission bands (lower graph) in LiF:Mg following alpha particle irradiation (red line) and beta particle irradiation (blue dashed line). The lower graph shows the RPL emission band following alpha particle irradiation. The blue line is the excitation spectrum and corresponds to the $F_3^+/F_2$ absorption bands shown in the upper graph. The other numbered lines show the measured spectra recorded using different filter combinations and excitation powers. (Reproduced from Bilski et al. (2019) with permission from Elsevier.) (b) RPL excitation and emission spectra in beta-particle irradiated $Al_2O_3$:C,Mg. (Reproduced from Akselrod and Kouwenberg (2018) with permission from Elsevier.) (c) Absorption (top figure) and emission bands in beta-irradiated Ag-doped alkali phosphate glass. (Data kindly provided by S. Sholom.)

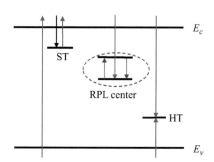

**Figure 4.4** Conceptual diagram illustrating the buildup phenomenon using electronic processes only. The dashed line indicates the RPL center with its ground and excited states, and the blue arrows indicate the stimulation and relaxation transitions within the RPL center, producing RPL emission. Red arrows indicate electron and hole generation, trapping, detrapping, and recombination pathways.

## 4.3  Buildup Kinetics

### 4.3.1  Electronic Processes

For buildup models involving purely electronic processes, a two-electron-trap/one-hole-trap (2T1R) example serves as a suitable model to demonstrate the principles involved. The example is shown in Figure 4.4. Irradiation creates free electrons and holes. The latter become trapped at hole traps (HT), which act as charge compensators to the trapped electrons. The electrons may become trapped in the ground states of the RPL centers, or in shallow traps (ST). After irradiation, electrons in the shallow traps are unstable and are thermally stimulated from the traps. The electrons that leak from these traps may recombine with the trapped holes or become trapped in the RPL sites. Thus, the RPL centers increase in number (i.e. buildup) after the end of the irradiation on a time scale of the lifetime of the electrons in the shallow traps; the shallower the trap, the faster the process. The extent to which the RPL centers grow depends upon the concentrations of the shallow traps and the RPL precursors, and the relative trapping and recombination cross-sections of the RPL precursors and the hole traps.

The rate equations describing the irradiation and buildup periods for such a system are:

$$\frac{dn_{ST}}{dt} = n_c A_{ST} \left( N_{ST} - n_{ST} \right) - n_{ST} s_{ST} \exp\left\{ -\frac{E_{ST}}{kT} \right\}, \tag{4.4a}$$

$$\frac{dn_{RPL}}{dt} = n_c A_{RPL} \left( N_{RPL} - n_{RPL} \right), \tag{4.4b}$$

$$\frac{dm_{HT}}{dt} = n_v A_{HT} \left( M_{HT} - m_{HT} \right) - n_c m_{HT} B, \tag{4.4c}$$

$$\frac{dn_c}{dt} = G - n_c A_{ST} \left( N_{ST} - n_{ST} \right) - n_c A_{RPL} \left( N_{RPL} - n_{RPL} \right) + n_{ST} s_{ST} \exp\left\{ -\frac{E_{ST}}{kT} \right\} - n_c m_{HT} B, \tag{4.4d}$$

and

$$\frac{dn_v}{dt} = G - n_v A_{HT} \left( M_{HT} - m_{HT} \right). \tag{4.4e}$$

In these equations the subscripts *ST*, *RPL*, and *HT* refer to the shallow traps, RPL centers and hole traps, respectively; otherwise all terms have their usual meaning. During irradiation, the generation rate $G > 0$, while during post-irradiation buildup, $G = 0$.

Typical solutions to these equations for the irradiation and corresponding buildup periods after irradiation are shown in Figure 4.5. In these simulations, the concentration of trapped electrons in the shallow traps (dashed lines) and the RPL centers (full lines) are shown. The

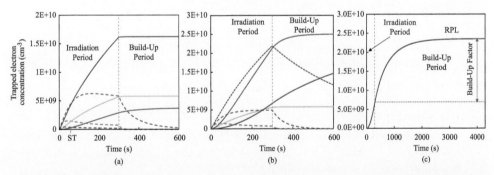

**Figure 4.5** Simulations of the growth of the shallow traps (ST) and the RPL centers (RPL) during irradiation and in the buildup period following irradiation. The full lines indicate the RPL centers (proportional to the RPL signal) and the dashed lines indicate the ST centers. The parameters used in the calculations are $G = 10^8$ cm$^{-3}$.s$^{-1}$; $E_{ST} = 0.8$ eV; and $s_{ST} = 5\times10^{12}$ s$^{-1}$. In (a), yellow lines: $A_{ST} = A_{RPL} = A_{HT} = 10^{-10}$ cm$^3$.s$^{-1}$, $B = 10^{-8}$ cm$^3$.s$^{-1}$, and $N_{ST} = N_{RPL} = M_{HT} = 10^{11}$ cm$^{-3}$; red lines: same, except for $N_{ST} = 10^{12}$ cm$^{-3}$; blue lines: as for yellow lines, except, $N_{RPL} = 10^{12}$ cm$^{-3}$. (b) Yellow lines – same as in (a); red lines: $B = 10^{-10}$ cm$^3$.s$^{-1}$; blue lines: $N_{ST} = 10^{12}$ cm$^{-3}$. (c) Same as blue line in (b), but extended to long buildup times.

RPL intensity is proportional to the final RPL center concentration. As speculated, the buildup factor is dependent upon the relative concentration of defects (Figure 4.5a), and the relative capture cross-sections (Figure 4.5b). The largest of the buildup factors is highlighted in Figure 4.5c where it can be seen that the buildup is slow, but large, reaching a maximum after approximately 1 hour after the irradiation has ended. The buildup rate will also depend on temperature due to the exponential dependence on $T$ of the detrapping term for the ST.

Also to be noted (for example, the full blue lines in Figures 4.5b and 4.5c) is the upward curvature of the growth of the RPL centers during the irradiation period. This is a result of the creation of the RPL centers by the radiation plus the buildup occurring simultaneously during irradiation.

An example of a material that demonstrates post-irradiation buildup of the RPL signal is $Al_2O_3$:C,Mg for which a 10% increase is observed over a period of days after irradiation. The buildup effect was also observed to be larger and faster if the sample is exposed to light after the irradiation period (Eller et al. 2013). The model used by Eller et al. to explain the observations in $Al_2O_3$:C,Mg is shown in Figure 4.6 where a set of electron traps, denoted shallow (ST), main dosimetric (MDT) and deep (DET) capture electrons during irradiation. The STs lose their trapped electrons at room temperature due to thermal stimulation, while the main traps (MDT) also lose their trapped electrons due to optical stimulation if the material is exposed to light. Either or both processes contribute to an increase in the number of electrons trapped in the RPL centers via the process $F_2^{2+}(2Mg) + e^- = F_2^+(2Mg)$ leading to the phenomenon of post-irradiation buildup.

### 4.3.2   Ionic Processes

The growth of RPL after irradiation (buildup) involves the continued increase in the number of RPL centers even after the irradiation ends. In the previous section this was illustrated using

---

### Exercise 4.1    RPL BuildUp (1)

Consider a model consisting of:
- One shallow trap (ST) unstable at room temperature;
- one deeper trap (denoted the main trap, MT) and stable at room temperature;
- one RPL center (RPL);
- one deep hole trap.

Enable the electrons trapped in the MT to be optically freed upon stimulation with light.

(a) Write the rate equations describing the buildup of trapped electrons and holes during irradiation.
(b) Use estimated values for the various parameters (concentrations, cross-sections, trap depths, frequency factors, electron-hole generation rate) and solve the equations numerically for (i) the irradiation period, and (ii) the post-irradiation, buildup period. Assume buildup takes place in the dark and at room temperature.
(c) Repeat the buildup calculation (part (b)ii), but now assume exposure to light and allow electrons to be optically stimulated from MT to the conduction band at a rate $p$. (You choose the value of $p$.)
(d) Repeat these calculations varying the values of the various concentrations, cross-sections, trap depth, and light stimulation rate and show how the observed post-irradiation RPL buildup depends upon the parameters chosen.

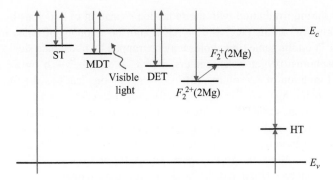

**Figure 4.6**   The Eller et al. (2013) model for RPL and buildup in Al$_2$O$_3$:C,Mg.

transfer of electrons from shallow traps to the RPL precursors to form the RPL centers. It is also possible that the continued growth of the RPL centers may be due to ionic diffusion processes in which ionic species A diffuses to species B to form RPL species AB. In LiF, RPL centers are identified as $F_2$- and $F_3^+$-centers. These may be formed during irradiation by the creation of a second *F*-center in the vicinity of an existing *F*-center. However, a more likely mechanism is the diffusion of empty F⁻-ion vacancies (anion vacancies, $V_a$-centers) to existing *F*-centers creating $F_2^+$-centers ($V_a + F = F_2^+$). Subsequent trapping of an electron by the $F_2^+$-center yields an $F_2$-center ($F_2^+ + e^- = F_2$). Further diffusion of anion vacancies leads to the creation of $F_3^+$-centers ($V_a + F_2 = F_3^+$), etc. These processes occur during irradiation, but may also continue after irradiation, leading to the post-radiation buildup of RPL and changes to other color center concentrations (McLaughlin et al. 1979).

A material exhibiting strong post-irradiation buildup is Ag-doped alkali phosphate glass (Yamamoto 2011). The process of RPL center production in this material is a multi-step process involving both electronic and ionic processes. In the model suggested by Dmitryuk et al. (1986, 1989), Ag$^+$ ionic diffusion is critical for the production of RPL in this material. The mechanism can be envisioned conceptually in Figure 4.7 in which it can be observed that the

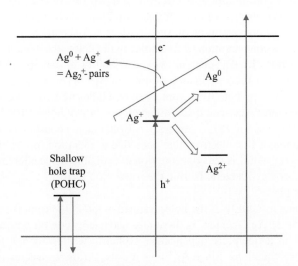

**Figure 4.7**   A conceptual model for the trapping of electrons and holes, and the production of Ag$_2^+$ RPL centers, in Ag-doped alkali phosphate glass.

process begins during irradiation with the reduction/oxidation of $Ag^+$ ions by the trapping of electrons to form $Ag^0$ atoms ($Ag^+ + e^- = Ag^0$), and the trapping of holes to produce $Ag^{2+}$ ($Ag^+ + h^+ = Ag^{2+}$). At the same time, holes may be trapped by phosphate sites to form what are known as phosphorus-oxygen-holes centers, POHC (i.e. $PO_4^{3-} + h^+ = PO_4^{2-}$, or POHC). Following the formation of $Ag^0$ atoms, diffusion of $Ag^+$ ions leads to the formation of $Ag_2^+$-centers (or $Ag_2^+$-pairs; $Ag^+ + Ag^0 = Ag_2^+$). It is the $Ag_2^+$-centers that act as the main RPL emitters. During irradiation, some $Ag_2^+$-centers are formed, but more are formed after irradiation due to continued $Ag^+$-ion diffusion to the $Ag^0$-centers. (Further reactions may occur to form $Ag_3^{2+}$-trimers and, generally, $Ag_m^{n+}$ particles, with m = n + 1. Condensation occurs via this diffusion process and ultimately Ag-nanoparticles are formed.)

Rate equations can be written to describe these processes, namely:

$$\frac{dn_c}{dt} = G - n_c A_1 \left( N_1 - n_1 - m_1 - n_2 \right) - n_c A_4 m_1; \tag{4.5a}$$

$$\frac{dm_v}{dt} = G - m_v A_2 \left( N_1 - n_1 - m_1 - n_2 \right) - m_v A_3 n_1 - m_v A_5 \left( M_2 - m_2 \right) + m_2 s_h \exp\left\{ -\frac{E_h}{kT} \right\}; \tag{4.5b}$$

$$\frac{dn_1}{dt} = n_c A_1 \left( N_1 - n_1 - m_1 - n_2 \right) - m_v A_3 n_1 - \frac{dn_2}{dt}; \tag{4.5c}$$

$$\frac{dn_2}{dt} = n_1 A_6 \left( N_1 - n_1 - m_1 - n_2 \right); \tag{4.5d}$$

$$\frac{dm_1}{dt} = m_v A_2 \left( N_1 - n_1 - m_1 - n_2 \right) - n_c A_4 m_1; \tag{4.5e}$$

$$\frac{dm_2}{dt} = m_v A_5 \left( M_2 - m_2 \right) - m_2 s_h \exp\left\{ -\frac{E_h}{kT} \right\}; \tag{4.5f}$$

where: $N_1$ = original concentration of $Ag^+$ centers; $M_2$ = concentration of POHC center precursors (*i.e.* $PO_4^{3-}$ -centers); $m_1$ = concentration of holes trapped by $Ag^+$-centers = concentration of $Ag^{2+}$-centers; $n_1$ = concentration of electrons trapped by $Ag^+$-centers = concentration of $Ag^0$-centers; $m_2$ = concentration of holes trapped by $PO_4^{3-}$ = concentration of POHC-centers (*i.e.* $PO_4^{2-}$-centers); $n_2$ = concentration of $Ag_2^+$-pairs; $n_c$ = concentration of free electrons in the delocalized band; $m_v$ = concentration of free holes in the delocalized band; $A_1$ = probability of capture of free electrons by $Ag^+$-centers; $A_2$ = probability of capture of free holes by $Ag^+$-centers; $A_3$ = probability of recombination of holes at $Ag^0$; $A_4$ = probability of recombination of electrons at $Ag^{2+}$; $A_5$ = probability of capture of free holes by $PO_4^{3-}$; $A_6$ = probability of $Ag^0$ reacting with $Ag^+$ to form $Ag_2^+$-pairs; $s_h$ = escape frequency of holes from $PO_4^{2-}$-centers; $E_h$ = trap depth for holes trapped at $PO_4^{2-}$-centers; $k$ = Boltzmann's constant; $T$ = temperature; and $G$ = rate of generation of free electron-hole pairs due to radiation (proportional to dose rate). The term $\left( N_1 - n_1 - m_1 - n_2 \right)$ is the concentration of available $Ag^+$ centers not yet converted to either $Ag^0$-, $Ag^{2+}$-, or $Ag_2^+$-centers. During the irradiation phase $G > 0$, whereas for the post-irradiation buildup phase, $G = 0$.

Due to the thermal instability of the holes trapped at the $PO_4^{3-}$ centers (Equations 4.5b and 4.5f), the formation of $Ag^{2+}$-centers is thermally activated, with an activation energy of $E_h$. However, there is no temperature dependence explicitly written for the formation of $Ag_2^+$-centers in the above equations. Since $Ag^+$ ions are the diffusing species in the formation of $Ag_2^+$-pairs, the temperature dependence for the formation of $Ag_2^+$-pairs can be accounted for by introducing a temperature dependence for the diffusion coefficient for the $Ag^+$ ions, $D_c$:

$$D_c = D_0 \exp\left\{-\frac{E_D}{kT}\right\}, \tag{4.6}$$

where $D_c$ and $D_0$ have dimension $cm^2.s^{-1}$ and $E_D$ is the activation energy. $D_c$ can be expressed:

$$D_c = Ks_D, \tag{4.7}$$

where $K$ is a material-dependent constant of dimensions $cm^2$ and is related to the lattice spacing.

In Equation 4.5d, the rate coefficient for $Ag_2^+$ pair formation $A_6$ can be written:

$$A_6 = \sigma\bar{v}, \tag{4.8}$$

where $\sigma$ is the cross-section for pair formation and $\bar{v}$ is the mean thermal velocity of the diffusing $Ag^+$ ions. The hopping frequency may also be written in terms of $\sigma$ and $\bar{v}$:

$$s_D = N\sigma\bar{v}, \tag{4.9}$$

where $N$ is the density of available hopping sites for the $Ag^+$ ions. Combining Equations 4.6 to 4.9:

$$A_6 = \frac{s_D}{N} = \frac{D_0 \exp\left\{-E_D\big/kT\right\}}{KN}. \tag{4.10}$$

Finally, Equation 4.5d is now replaced with:

$$\frac{dn_2}{dt} = n_1 \frac{D_0 \exp\left\{-E_D/kT\right\}}{KN}\left(N_1 - n_1 - m_1 - n_2\right) = n_1 K_D \exp\left\{-E_D/kT\right\}\left(N_1 - n_1 - m_1 - n_2\right), \tag{4.11}$$

where constant $K_D = D_0/KN$ has dimensions $cm^3.s^{-1}$.

Equations 4.5a–4.5f, with Equation 4.11 replacing Equation 4.5d, can be solved numerically. Example solutions are shown in Figure 4.8. Once again, it may be observed that the growth of the RPL centers ($Ag_2^+$-centers) exhibits upward curvature during irradiation (Figure 4.8a), followed by buildup after irradiation (Figure 4.8b). (Note that in Figure 4.8b the RPL center concentration is shown after a very short irradiation period compared to the longer irradiation period shown in Figure 4.8a.)

---

## Exercise 4.2   RPL BuildUp (2)

Consider a model[see footnote 1] for color center growth in a LiF crystal consisting of a pre-existing concentration of anion vacancies (empty $F^-$-ion vacancies known as $V_a$-centers), that can capture free electrons during irradiation to form $F$-centers, thus:

$V_a + e^- = F$.

Allow further $V_a$-centers to cluster with $F$-centers via diffusion:

$V_a + F = F_2^+$,

which can capture further electrons to form $F_2$-centers:

$F_2^+ + e^- = F_2$.

Consider further diffusion of $V_a$-centers to form $F_3{}^+$-centers:

$V_a + F_2 = F_3{}^+$

and further capture of electrons to form $F_3$-centers

$F_3{}^+ + e^- = F_3$.

Ignore any back reactions.

For charge neutrality allow holes to be trapped at deep hole traps.

(1) Write a set of rate equations to describe the above processes.

(2) Consider $F_2$-centers to be the RPL centers. Show the growth of $F_2$-centers (i.e. the growth of RPL) at room temperature (i) during irradiation, and (ii) immediately after irradiation, by inserting appropriate values for the various parameters used in the rate equations, and solving the rate equations numerically.

1 This is an overly simple model for the creation of $F$-centers. $F^-$-ion vacancies and $F$-centers are also produced by the irradiation itself via the creation of excitons and subsequent energy transfer to create stable $F$- and $H$-centers. ($H$-centers are $F^-$ ions in interstitial positions that have captured holes.) The presence of impurity ions, $Mg^{2+}$, $Ti^{4+}$, etc., complicates the process further. These mechanisms are not included in this exercise.

### 4.3.3  More on Buildup Processes

#### 4.3.3.1  *After Irradiation*

The buildup curves illustrated in the simulated results shown in Figures 4.5 and 4.8b have the form of a saturating exponential-like growth, although analytical expressions for the curves are not available. If the buildup curve is described by the function $I_{BU}(t)$, an empirical expression for $I_{BU}(t)$ might be:

$$I_{BU}(t) \approx \sum_{j=1}^{k} K_j (1 - \exp\left\{-\frac{t}{\tau_j}\right\})$$

(4.12)

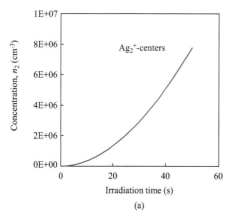
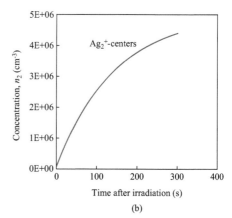

**Figure 4.8**  Examples solutions to Equations 4.5 and 4.11 showing the growth in the concentration of $Ag_2{}^+$ RPL centers, (a) during irradiation and (b) after irradiation. The parameters used in the calculations (see Equations 4.5 and 4.11) are: $N_1 = 10^{18}$ cm$^{-3}$; $M_2 = 10^{21}$ cm$^{-3}$; $A_1 = 10^{-14}$ cm$^3$s$^{-1}$; $A_2 = 10^{-15}$ cm$^3$.s$^{-1}$; $A_3 = 10^{-13}$ cm$^3$.s$^{-1}$; $A_4 = 10^{-12}$ cm$^3$.s$^{-1}$; $A_5 = 10^{-12}$ cm$^3$.s$^{-1}$; $K_D = 5\times10^{-18}$ cm$^3$.s$^{-1}$; $E_D = 0.17$ eV; $E_h = 0.45$ eV; $s_h = 2\times10^9$ s$^{-1}$; along with $k = 8.617\times10^{-5}$ eV.K$^{-1}$. The temperature $T$ was varied from 300 K to 400 K and the irradiation rate was $G = 10^6$ cm$^{-3}$.s$^{-1}$.

where $K_j$ is a constant, $\tau_j$ is the characteristic buildup time constant for each of the $j$ components. Here time $t = 0$ is defined as the end of the irradiation period and for a very short irradiation $I_{BU}(t) \approx 0$ at $t = 0$. In practice, the number of components $k$ should be chosen from fits of Equation (4.12) to the experimental buildup curve. Alternative empirical expressions (e.g. polynomials) may also yield acceptable fits for short buildup times.

### 4.3.3.2 During Irradiation

Once a suitable empirical expression has been obtained to describe the buildup curve, this can be used to determine the shape of the RPL versus irradiation time (e.g. shown in simulations in Figure 4.8a). To understand this upward curvature shape, consider the radiation dose $D$ to be absorbed in elemental dose increments, $dD$, each in a time period $dt$ at a fixed dose rate $dD/dt$. If irradiation time $t_{irr} = t_D$ is the time taken to reach dose $D$, then $dD/dt = D/t_D$. The RPL signal $dI$ due to each dose element builds up and, while doing so, the next dose element $dD$ is deposited and the RPL signal due to it, $dI$, also starts to buildup in turn, delayed by $dt$ with respect to the previous component. If each segment of RPL builds up according to the function $I_{BU}(t)$, then the total RPL as a function of the irradiation time $t$ is given by:

$$I_{RT}(t) = \frac{1}{t_D} \int_0^t I_{BU}(t)\, dt, \tag{4.13}$$

where the expression is normalized to the irradiation time, $t_D$ and the subscript "$RT$" refers to real-time measurement of the RPL intensity during irradiation. In this equation $t = 0$ means the start of the irradiation period. To obtain $I_{RT}(t)$ experimentally, the specimen being irradiated would have to be probed during irradiation using light (e.g. from a laser) of a wavelength that induces excitation of the RPL center from its ground state to its excited state, and the emission during this excitation, due to relaxation from the excited state to the ground state, would have to be continuously monitored.

As an example, consider $k = 2$ in Equation 4.12. With $I_{RT} = 0$ at $t = 0$ the integral in Equation 4.13 becomes:

$$\int_0^t I_{BU}(t)\, dt = K_1 \tau_1 \exp\{-t / \tau_1\} + K_2 \tau_2 \exp\{-t / \tau_2\} + (K_1 + K_2)t + C, \tag{4.14}$$

where $C$ is a constant.

Alternatively, the buildup curve could be integrated numerically. This is illustrated in Figure 4.9 where data from the simulated RPL growth curve from Figure 4.8a are compared with the numerical integration of the simulated buildup curve shown in Figure 4.8b.

### 4.3.3.3 Temperature Dependence

In Equations 4.4 (for electronic processes) and 4.5 with 4.11 (for ionic processes), explicit temperature dependencies were included in the models through terms $s_{ST}\exp\left\{-\dfrac{E_{ST}}{kT}\right\}$, $s_h\exp\left\{-\dfrac{E_h}{kT}\right\}$, and $D_0\exp\left\{-\dfrac{E_D}{kT}\right\}$. All processes speed up as the temperature increases and it can be expected that buildup will occur more rapidly at higher temperatures. Example simulated temperature dependencies for both models are shown in Figure 4.10. Figure 4.10a shows both the irradiation period and the following buildup period for the electronic model using the same parameters as in Figure 4.5c, but at temperatures between 300 K and 400 K. As the buildup process becomes much faster, the growth during irradiation becomes linear and there is

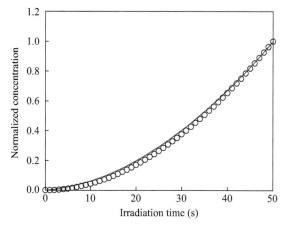

**Figure 4.9**  Comparison of the RPL growth during irradiation (data points) and the numerical integral of the simulated buildup curve from Figure 4.8b (red line).

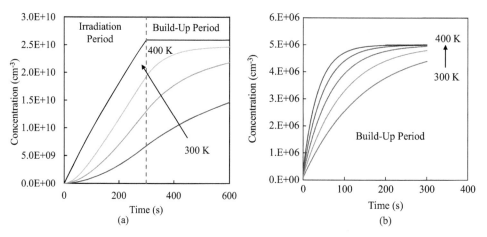

**Figure 4.10**  (a) Dependence of the growth of RPL centers on temperature. Simulations are shown for during and after the irradiation using the 2T1R model of Figure 4.4 with the same parameter values as used for Figure 4.5c, but for temperatures ranging from 300 K to 400 K. (b) Changes in the shape of the buildup curve with temperature using the model of Figure 4.7 and the parameter values listed in the caption to Figure 4.8.

correspondingly no buildup after irradiation since all the buildup occurred during irradiation. Figure 4.10b shows just the buildup period for the ionic process. Here it can also be seen that as the temperature increases the buildup rate also increases. Because of the parameters chosen for the model, there is still some post-irradiation buildup, albeit faster, even at 400 K.

Due to buildup and its temperature dependence, RPL dosimeters have to be post-irradiation annealed to ensure buildup has been completed before RPL measurement can be made. The temperature to which the sample is heated, and the time at that temperature, are both material dependent. No recommendations are available in the published literature for LiF:Mg or $Al_2O_3$:C,Mg, but for Ag-doped alkali phosphate glass the recommendation is to hold the dosimeter at 70 °C for 30 minutes following irradiation, when the irradiation time (at room temperature) is less than or comparable to the buildup time at that temperature. On the other hand, if the irradiation time is much greater than the buildup time such that buildup has sufficient time to be completed during the irradiation process, then no post-irradiation annealing is required. An example of the latter is the use of RPL detectors to measure environmental doses at very low dose rates and correspondingly long irradiation times. Here, the buildup rate is faster than the growth rate due to the natural environmental radiation and post-irradiation annealing is unnecessary.

### Exercise 4.3    RPL Buildup (3)

(1) Include a temperature dependence to the rate equations you developed in Exercise 4.2 by including a temperature-dependent diffusion coefficient for the anion vacancies. Using suitable parameter values, show the changes in the RPL-center ($F_2$-center) growth as a function of temperature.

(2) Now add shallow electron traps to the model and choose $E_t$ and $s$ values that make the shallow traps unstable at room temperature. Re-write the rate equations, and solve again to show the new temperature dependence of the RPL-center ($F_2$-center) growth, both during and after irradiation.

# 5

# Analysis of TL and OSL Curves

*Without analysis, no synthesis.*

– F. Engels 1878

## 5.1 Analysis of TL Glow Curves

One of the most frequently published topics in the field of TL is the analysis of the experimentally observed glow peaks in order to extract values for the trap depth $E_t$ and the frequency factor $s$, and to separate one glow peak from another in a glow curve consisting of a complex overlay of multiple peaks. The reliable use of TL glow curves in dosimetry requires a deep understanding of the measured signal and its expected behavior as a function of absorbed radiation dose, radiation quality, and time. Peak separation is often essential in this regard for some materials, albeit less important for others.

Regrettably, analysis of TL glow curves is also one of the most unreliable aspects of the subject of TL in the sense that many published $E_t$ and $s$ results are, in essence, worthless. Even simple separation of the glow curves into their component parts is often accepted at face value in the absence of the necessary checks and balances that are needed to underline the results' reliability. Unreliability stems from several factors, including unsatisfactory experimental technique in gathering the original data, and the inapplicability of the chosen models to analyze the data. The problem is a complex one and requires, at a minimum, an understanding of the theoretical background for the chosen analytical methods, and the limitations and constraints of those methods.

Similar comments can be made regarding OSL, although in the use of OSL in dosimetry the reliance on the separation of the OSL curves into their component parts is perhaps less needed. Nevertheless, the requirements of accurate analysis of OSL curves (for example, to obtain photoionization cross-sections, $\sigma_p(E)$) are fundamentally the same as those needed to analyze TL curves, including good experimental technique and justifiable use of models and analysis methods.

In this chapter, the various analytical methods most popularly applied in the literature are discussed, along with the supplemental cross-checks and tests that should be carried out to support the results obtained.

*A Course in Luminescence Measurements and Analyses for Radiation Dosimetry,* First Edition.
Stephen W.S. McKeever.
© 2022 John Wiley & Sons Ltd. Published 2022 by John Wiley & Sons Ltd.
Companion Website: www.wiley.com/go/mckeever/luminescence-measurements

## 5.2   Analytical Methods for TL

There are several methods that have been proposed for analysis of TL glow peaks. They may be categorized as:

- Partial-peak methods;
- Whole-peak methods;
- Peak-shape methods;
- Peak-position methods;
- Curve-fitting methods;

and

- Isothermal methods.

Each is reliant upon accurate, reproducible and reliable data. (Otherwise to quote a well-known saying, "garbage in; garbage out."[1])

### 5.2.1   Partial-Peak Methods

#### 5.2.1.1   A Single TL Peak with a Discrete Value for $E_t$

The most important of the partial-peak methods is called the Initial Rise Method (IRM), introduced by Garlick and Gibson (1948). Referring to Figure 5.1a, the "initial rise" can be defined as that initial part of an individual TL peak for which the change in the concentration of trapped electrons ($\Delta n$) is much less that the initial trapped charge concentration ($n_0$, where $n_0 = n$ at the start of the heating). It is that region of the TL peak where $T < T_c$, over which the inequality $\Delta n \ll n_0$ holds true. $T_C$ is a "critical temperature" above which the inequality is no longer valid.

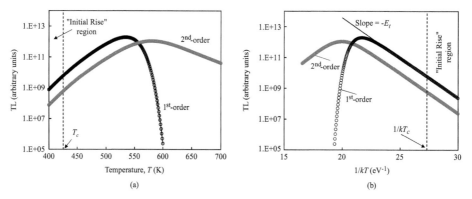

**Figure 5.1**   (a) First-order TL peaks, calculated using the Randall-Wilkins expression (Equation 3.17), and second-order TL peaks, calculated using the Garlick-Gibson expression (Equation 3.21), with $E_t = 1.2$ eV, $s = 1\times10^{12}$ s$^{-1}$, $n_0 = 1\times10^{14}$ cm$^{-3}$, and $N = 1\times10^{15}$ cm$^{-3}$. The peaks are illustrated as $\ln(I_{TL})$ versus $T$. The "initial rise" region is that region for which $T < T_c$ and $\Delta n \ll n_0$. (b) Same data expressed as $\ln(I_{TL})$ versus $1/kT$. The initial rise region is a straight line of slope $= -E_t$.

---

[1] According to *Wikipedia*, the phrase is attributed to Wilf Hey of IBM. Much earlier, Charles Babbage alluded to the same issue when discussing his Analytical Engine: "*On two occasions I have been asked, 'Pray, Mr. Babbage, if you put into the machine wrong figures, will the right answers come out?' ... I am not able rightly to apprehend the kind of confusion of ideas that could provoke such a question.*" Babbage (1864).

In the case of first-order kinetics, the Randall-Wilkins expression for the shape of a TL peak, Equation 3.17, is:

$$I_{TL}(T) = n_0 s \exp\left\{-\frac{E_t}{kT}\right\} \exp\left\{-\frac{s}{\beta_t}\int_{T_0}^{T}\exp\left[-\frac{E_t}{k\theta}\right]d\theta\right\}. \tag{3.17}$$

Under the conditions that $\Delta n \ll n_0$, the term $\exp\left\{-\dfrac{s}{\beta_t}\displaystyle\int_{T_0}^{T}\exp\left[-\dfrac{E_t}{k\theta}\right]d\theta\right\} \approx 1$ and therefore:

$$I_{TL}(T) \approx n_0 s \exp\left\{-\frac{E_t}{kT}\right\} \tag{5.1}$$

Assuming $s$ to be approximately constant up to $T = T_c$, Equation 5.1 becomes a straight line on a plot of $\ln(I_{TL})$ versus $1/kT$ with a slope of $-E_t$, as shown in Figure 5.1b.

Figures 5.1a and 5.1b also illustrate that the IRM is independent of kinetic order, with example data also shown for a second-order TL peak. This can be seen from the Garlick and Gibson (1948) equation for TL, namely:

$$I_{TL}(T) = \frac{n_0^2 s' \exp\left\{-E_t\,/\,kT\right\}}{\left[1+\left(\dfrac{n_0 s'}{\beta_t}\right)\displaystyle\int_{T_0}^{T}\exp\left\{-E_t\,/\,k\theta\right\}d\theta\right]^2}, \tag{3.21}$$

which, for $\Delta n \ll n_0$, becomes:

$$I_{TL}(T) \approx n_0^2 s' \exp\left\{-\frac{E_t}{kT}\right\}, \tag{5.2}$$

and again a plot of $\ln(I_{TL})$ versus $1/kT$ yields a straight line of slope $-E_t$ in the initial rise region. The same holds true for general-order and mixed-order kinetics. Thus, the IRM is independent of kinetic order – that is, it is independent of the relative sizes of the trapping, retrapping, and recombination rates and probabilities – and independent of the degree of trap filling $n_0$, or $n_0$ /$N$ (i.e., dose) and heating rate $\beta_t$. In this sense it is quite general.

The analysis relies on the assumption that $s$ is constant over the temperature range up to $T_c$. From Chapter 2, a temperature dependence for $s$ of the type $s \propto T^a$ might be expected. In this case:

$$I_{TL} \approx CT^a \exp\left\{-\frac{E_t}{kT}\right\} \tag{5.3}$$

where $C$ is a constant. From this:

$$\ln(I_{TL}) \approx a\ln(T) - \frac{E_t}{kT} \tag{5.4}$$

and

$$\frac{d\ln(I_{TL})}{dT} \approx \frac{a}{T} + \frac{E_t}{kT^2}. \tag{5.5}$$

The IRM assumes $a = 0$, and thus:

$$\frac{\mathrm{d}\ln(I_{TL})}{\mathrm{d}T} \approx \frac{E_t^{IR}}{kT^2}. \tag{5.6}$$

$E_t^{IR}$ is that value of $E_t$ obtained assuming $a = 0$. From Equations 5.5 and 5.6:

$$E_t^{IR} \approx akT + E_t. \tag{5.7}$$

Thus, the activation energy measured with the IRM ($E_t^{IR}$) differs from the actual value ($E_t$) by the term $akT$, and is slightly temperature dependent.

An additional correction term can arise from the QE assumption. It was shown in Chapter 3 that:

$$I_{TL}(T) = \frac{I_{RW}(T)}{Q(T)}. \tag{3.86}$$

That is, for first-order kinetics and the NMTS model, the actual TL curve shape differs from the Randall-Wilkins curve ($I_{RW}(T)$) by the term $Q(T)$. Only when $Q(T) = \text{constant} = Q$ does the $I_{RW}(T)$ function describe the real TL curve. If QE holds, $Q = 1$ and no correction is needed. If, however, $Q(T)$ is not constant, a further correction is required.

An additional correction is needed if thermal quenching exists. In Chapter 3, the expression for the change in luminescence efficiency as a function of temperature assuming the Mott-Seitz mechanism was shown to be:

$$\eta(T) = \frac{1}{1 + C\exp\{-W/kT\}}. \tag{3.89}$$

Applying this term to the IRM yields the following expression for the initial rise region:

$$I_{TL} = C\exp\left\{-\frac{(E_t - W)}{kT}\right\}, \tag{5.8}$$

with $C$ a constant so that the slope of $\ln(I_{TL})$ versus $1/kT$ yields ($E_t - W$), not $E_t$.

Independent of the above corrections, a critical question for the IRM relates to the width of the initial rise region. How is the value of $T_c$ chosen? An often-applied "rule of thumb" is to select that value of $T_c$ for which $I_{TL}(T_c)$ is less than 10% of $I_{TL}(T_m)$, where $T_m$ is the temperature of the peak maximum. This rule was developed by trial-and-error studies by several early authors. (See McKeever (1985) and Coleman and Yukihara (2018) for descriptions of this early work.) A key finding is that the appearance of a straight line on a $\ln(I_{TL})$ versus $1/kT$ plot is by itself an insufficient criterion for selection of $T_c$. The larger the value of $T_c$, the more inaccurate the determined value of $E_t$, despite the appearance of a straight line. It is generally wiser to choose the lowest value of $T_c$ that will give enough good data to reliably fit a straight line on a $\ln(I_{TL})$ versus $1/kT$ plot.

Apart from selection of $T_c$, a related issue is how to select the temperature at which to start the initial rise analysis ($T_{min}$). The data of Figure 5.1 were simulated without a background signal and, because of this, it does not matter at what temperature the analysis starts. Only data representing TL emission is present in the data set. However, in a real experiment there is always a background and therefore selection of the start temperature is an important consideration. In principle, the subtraction of the background should solve this problem, but this leads to a further consideration, namely experimental noise in the data such that defining a single value for the background is not so straightforward. Both the background and the TL signal will each have noise associated with them. Coleman and Yukihara (2018) examined this issue by including a background signal and adding Poisson noise to the simulated data. One of the methods chosen by them was to define the start temperature $T_{min}$ as that temperature at which the background-subtracted TL intensity has reached three times the standard deviation of the background noise. They further suggested choosing $T_c$ to be such that $I_{TL}(T_c)$ is no more than 5% of $I_{TL}(T_m)$. These are good practical rules to follow when performing IRM analysis of experimental data.

### 5.2.1.2   Multiple Overlapping Peaks, and Trap Energy Distributions

In the case of overlapping peaks, the situation becomes much less clear because the initial rise portion of any one peak may be obscured by overlap with its neighboring peaks such that the obtained $E_t$ value may be compromised – even to the extent that the obtained value is unusable. The extent to which this occurs will depend upon the degree of overlap and the relative sizes of the TL peaks. This situation is exacerbated when dealing with distributions of activation energies rather than discrete values. Here, the TL curve can be considered to be a superposition of many individual TL curves, weighted by the density of states function that describes the distribution.

Several strategies have been suggested in attempts to overcome these limitations and difficulties. Each of them attempt to separate the TL curve into its component parts and can be categorized as methods that employ step annealing after one irradiation, or multiple irradiations with a different preheat temperature before each measurement, and multiple irradiations but each at a different temperature. The classic papers describing these approaches are Gobrecht and Hofmann (1966), McKeever (1980), and Van den Eeckhout et al. (2013).

#### 5.2.1.2.1   Step-annealing Analysis

The principle of step annealing is illustrated in Figure 5.2. Here two overlapping, 1st-order TL peaks are illustrated. The irradiated sample is first heated, at rate $\beta_t$, to a certain temperature

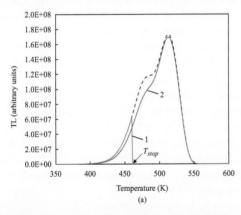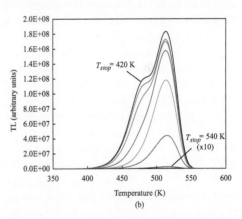

**Figure 5.2**   (a) The principle of step annealing and the definition of $T_{stop}$. Instead of recording the whole glow curve, the heating is stopped at $T_{stop}$ (curve 1) and rapidly cooled. (Instantaneous cooling is assumed in this illustration.) Then the sample is reheated to record the remaining glow curve (curve 2). The dotted line shows the TL glow curve that would have been measured immediately after irradiation without a step heating to $T_{stop}$. (b) The changes to the glow curve as a result of using different $T_{stop}$ values. (The data were obtained from numerical solutions to the rate equations for a 3T1R model, with the third trap assumed to be thermally disconnected. The $E_t$ values for traps 1 and 2 were 1.3 eV and 1.4 eV, respectively, each with $s = 5\times10^{12}$ s$^{-1}$. A new irradiation, starting from all traps empty, was simulated for each curve.).

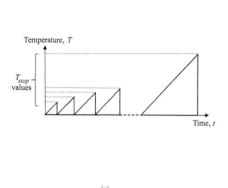

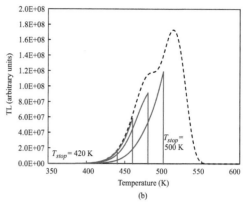

(a)

(b)

**Figure 5.3** (a) "Saw-tooth" heating profile for the method of Gobrecht and Hofmann (1966). (b) Resultant TL profiles (in red). The dotted line shows the glow curve that would have been attained without any pre-heat. In this illustration $T_{stop}$ changes by 20 °C. In an experiment the change in $T_{stop}$ should be no more than 1 °C to 5 °C (the smaller the better). The same $E_t$ and $s$ values as in Figure 5.2 were used for Figure 5.3b.

$T_{stop}$ and then rapidly cooled to the starting temperature. The sample can either then be reheated, at rate $\beta_t$, to record the remaining full TL glow curve followed by a new irradiation for each cycle (as shown in Figure 5.2), or it can be partially heated to a new value of $T_{stop}$ and the sequence repeated multiple times after just one irradiation (as illustrated in Figure 5.3). In each case, the value of $T_{stop}$ can be incremented in small intervals, say 1 °C – 5 °C.

If the full glow curve is recorded for each new value of $T_{stop}$ (Figure 5.2), the sample has to be re-irradiated before another value of $T_{stop}$ can be selected. As a result, the experiment can be rather tedious and lengthy, as well as assuming no fundamental changes to the TL glow curve have occurred due to multiple irradiations. (Alternatively, a fresh sample can be used for each irradiation.) Note that instantaneous cooling from $T_{stop}$ is assumed in Figure 5.2; in reality a finite cooling rate must be used. Additionally, the same heating rate is used for each value of $T_{stop}$ and for recording the final glow curves. (1 K.s$^{-1}$ in these simulations.)

As $T_{stop}$ is increased, the leading edge (i.e. the initial-rise region) of the glow curve changes progressively from that of peak 1 toward that of peak 2. In principle, successive application of the IRM as $T_{stop}$ increases should yield the trap depths of the two peaks, with the analysis yielding values of $E_t$ that change from ~$E_{t1}$ to ~$E_{t2}$ as $T_{stop}$ increases. However, the success of this strategy depends heavily upon the degree of overlap between the two peaks. It may simply be impossible to separate the initial rise region of peak 2 from the effects of peak 1. In this event, the results for peak 1 only may be taken to be reliable. In the example shown in Figure 5.2, the individual peaks are in fact too close for reliable analysis using the IRM for peak 2 and only the first peak ($E_t = 1.3$ eV) can be analyzed from this data set. (See Coleman and Yukihara (2018) for a detailed discussion of this point; also see Exercise 5.1.)

---

**Exercise 5.1    Peak cleaning using step annealing and the IRM**

In this exercise the intent is to show how the degree of overlap between two TL peaks dictates how well the IRM can be applied to extract the $E_t$ values for each trap.

(a) Select two sets of $E_t$ and $s$ values and construct a glow curve consisting of two overlapping first-order TL peaks. (To construct the glow curves use the expressions for the exponential integral developed by Kitis et al. (1998).)

(b) Calculate a set of glow curves corresponding to different values of $T_{stop}$.

(c) Apply the IRM to the leading edge (initial rise region) of each of the glow curves obtained in part (b) and evaluate a value for $E_t$ for each $T_{stop}$.

(d) How does the calculated $E_t$ compare with the input values of $E_t$ for the two peaks?

(e) Vary the separation between the two peaks by changing the $E_t$ value for one of the peaks and repeat the IRM analysis for the various $T_{stop}$ values. Recalculate $E_t$ for each value of $T_{stop}$ using the IRM and compare with the input $E_t$ values.

(f) What do you conclude is a minimum separation of two TL peaks for accurate values of $E_t$ to be obtained for each peak?

### 5.2.1.2.2    Fractional-Glow Analysis

As noted, the alternative to heating to record the full glow curve and irradiating afresh after each measurement is to irradiate the sample just once and progressively apply increasing $T_{stop}$ values in a "saw-tooth" heating profile, as indicated in Figure 5.3a. The method was described by Gobrecht and Hofmann (1966) and is known as the "fractional-glow" technique. The resultant TL is shown in Figure 5.3b. By incrementally increasing $T_{stop}$ in steps of 1 °C to 5 °C (the smaller the better), a succession of partial glow curves is obtained until eventually all traps are emptied. By applying the IRM to each segment the values of $E_t$ can be obtained, changing from the lowest $E_t$ to the highest $E_t$ as each segment passes through the glow curve. Furthermore, by calculating the area under each glow curve segment a distribution of trapping states can be obtained as a function of $E_t$.

Note that instantaneous cooling is again assumed in the conceptual Figure 5.3; in reality there will be a finite cooling rate, and Gobrecht and Hofmann (1966) suggested continuing to record the TL during the cooling phase also. By applying the IRM method to both the heating and the cooling cycle two values of $E_t$ are obtained. The average of the two values can then be taken as $E_t$ for that value of $T_{stop}$.

The same limitations caused by the degree of peak overlap apply to the fractional-glow technique. However, this method of step annealing generally produces better results than the step-annealing methods described in Figure 5.2. The reason is that each heating step successively reduces the lower temperature peak before starting to reduce the higher temperature peak and, thus, there is less interference from peak 1 by the time peak 2 is reached. By extension, a variation on the method is to keep $T_{stop}$ fixed, and to repeatedly heat to the same $T_{stop}$ value in order to continually and incrementally reduce peak 1 without appreciably reducing peak 2. The value chosen for $T_{stop}$, and the number of heating cycles required, depend upon the relative peak sizes and the degree of overlap, but if a point is reached whereby there is hardly any further TL emitted during each heating cycle, then it can be assumed that the lower temperature peak has been removed. In this way a cleaner version of peak 2 is obtained and by continuing the process but with a new (higher) $T_{stop}$ value, peak 2 can now be analyzed. In principle, these techniques will also apply to glow curves consisting of multiple overlapping peaks.

---

**Exercise 5.2    IRM and the fractional-glow method**

Again construct a glow curve consisting of two overlapping, first-order peaks. (Use the same $E_t$ and $s$ values you used in Exercise 5.1.)

(a) Apply the fractional-glow method by simulating successive heating to incrementally increasing values of $T_{stop}$. Use small increments in $T_{stop}$ and produce a set of partial glow curves, each corresponding to a particular $T_{stop}$ value.

(b) Apply the IRM to each partial glow curve, and compare the $E_t$ values with the input values as $T_{stop}$ is increased incrementally. How do the results of this method compare to those obtained in Exercise 5.1?

(c) Plot the areas under each partial glow curve (numerically integrate them using a spreadsheet) as a function of the $E_t$ value obtained from that same partial glow curve. Display the results as a histogram of TL area versus $E_t$.

In the above-described methods, the situation becomes progressively more complicated if in fact the energies are not discrete, but distributed. A glow curve resulting from a distribution of trap energies can be considered to be a sum of many individual glow curves each corresponding to a specific $E_t$ and differing from the next glow curvelet by energy $\delta E_t$ (assuming $s$ is the same for each curvelet.) In the case of the fractional-glow method, the distribution obtained (glow-curve area versus $E_t$) should then reflect the trap-depth distribution, $g(E_t)$. An example is shown in Figure 5.4. In this example the glow curve (Figure 5.4a) is made up of 40 individual glow curves, each corresponding to a different value of $E_t$, according to the distribution $g(E_t)$ (illustrated in Figure 5.4d). The distribution $g(E_t)$ is a sum of two overlapping Gaussians, with mean energies at $E_t = 1.5$ eV and $E_t = 1.75$ eV. Figure 5.4b shows a set of glow curves following heating to various $T_{stop}$ values, from 370 K to 730 K. By applying the IRM for each value of

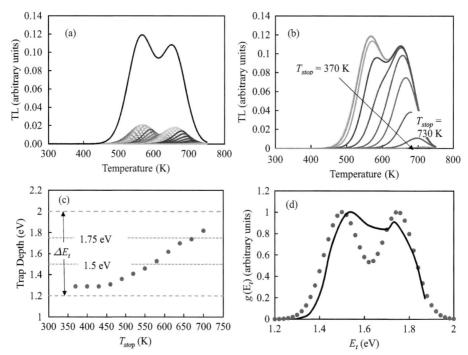

**Figure 5.4**    Results of the IRM using the Gobrecht-Hofmann Fractional Glow technique for two overlapping TL peaks arising from two Gaussian activation energy distributions. (a) The simulated glow curve obtained from the sum of 40 individual glow curves corresponding to the two Gaussian distributions in trap depth, $g(E_t)$, with mean values centered at $E_{tm} = 1.5$ eV and 1.75 eV. Each glow curvelet uses the same $s$ value ($10^{12}$ s$^{-1}$). (b) The changes in the glow curve for selected $T_{stop}$ values. (c) Calculated trap depths using the IRM for each $T_{stop}$ value. (d) The obtained $E_t$ distributions from the glow curve (black line) compared with the input distributions (blue dots).

$T_{stop}$, the energies shown in Figure 5.4c are obtained, as a function of $T_{stop}$. Of note in Figure 5.4c is the fact that the derived energies using the IRM are less than the lowest mean energy (1.5 eV) at low $T_{stop}$, and greater than the highest mean energy (1.75 eV) at high $T_{stop}$ values. The calculated values span the range of $E_t$ values in the distribution $g(E_t)$.

When the Fractional Glow method is applied in order to obtain the $E_t$ distribution, the distribution shown by the black line in Figure 5.4d is obtained. This is obtained by taking the area under the TL curve up to each $T_{stop}$ value and plotting against each $E_t$ value evaluated. The distribution can be seen to approximate the input $g(E_t)$ distribution, shown by the blue dots. In particular, the distribution maxima occur near ~1.5 eV and ~1.75 eV, in approximate agreement with $g(E_t)$. Using small increments in $T_{stop}$ would produce a more accurate result.

More detailed analysis (Coleman and Yukihara 2018) reveals that Gaussian-shaped distributions are reproduced better than uniform distributions using this approach. However, the method critically relies upon the values of $T_{min}$ and $T_c$, and care must be taken to choose these wisely. Choosing $T_{min}$ to be that temperature at which the TL signal rises beyond three times the standard deviation of the background, and $T_c$ to be that temperature at which the TL intensity is no more that 5% of the TL intensity at the peak maximum, are good practical strategies to employ experimentally. Note that as $T_{stop}$ increases, the TL intensity decreases and therefore $T_{min}$ will increase accordingly and $T_c$ will vary as the peaks are depleted sequentially. However, when working with simulated data where neither background nor noise is included, there is no restriction on how low $T_{min}$ can be and there will be a temptation to always use the lowest temperature in the data to start the initial analysis. Noise-free data in a simulation can be obtained down to the lowest temperature used in the simulation, even when the calculated TL intensity is orders of magnitude less than the peak intensity, and there is almost always some remnant of the lowest temperature peaks present, no matter how weak it is.[2] As a result, exercises that use simulated TL curves only, without noise and without a background, tend to give results biased toward the low temperature peak if care regarding $T_{min}$ is not taken.

### 5.2.1.2.3 *High-irradiation-temperatures Analysis*

The third analysis method is to use multiple irradiations, but with each irradiation at a different temperature, as described by Van den Eeckhout et al. (2013). This approach relies on the consideration that as the temperature of the irradiation $T_{irr}$ increases, the traps corresponding to the lower temperature peaks will, progressively, not be filled due to their increasing thermal instability. In this way, a series of glow peaks can be obtained, each with sub-sets of the TL peaks from the overall glow curve. The concept is illustrated in Figure 5.5 for an arbitrary distribution of trapped electrons $n(E) = N(E)f(E)$. Since the system has been irradiated and is no longer in thermal equilibrium (see Chapter 1), the quasi-Fermi level $E_{Fe}$ is used to calculate $f(E)$ with an approximation to post-irradiation, thermal equilibrium. The Fermi-Dirac occupancy function is then:

$$f(E) = \frac{1}{\exp\left\{\dfrac{E - E_{Fe}}{kT}\right\} + 1}. \tag{5.9}$$

---

[2] This is because after heating to $T_{stop}$, there will be electrons in the conduction band. As the system is cooled the concentration of electrons in the conduction band decreases and some of them are re-trapped into the lower temperature traps. Thus, during the subsequent TL measurement, there will always be a low-temperature TL peak, even though it may be much weaker than it was originally. Always using $T_{min}$ to be the lowest temperature used in the calculation will therefore bias the $E_t$ result to that of the lowest temperature peak.

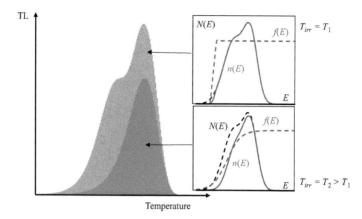

**Figure 5.5**  Schematic illustration of the method of the Van den Eeckhout et al. (2013). In this figure, irradiation at two temperatures, $T_1$ and $T_2$, results in two different trapped electron distributions $n(E)$. This change is reflected in the subsequent glow curves. The difference in the glow-curve areas is proportional to the change in $n(E)$ as $T_{irr}$ is increased from $T_1$ to $T_2$.

As discussed with reference to Figure 1.3 in Chapter 1, as the temperature is increased, the quasi-Fermi level approaches the Fermi level, changing the occupancy of the trap distribution. This is illustrated in the two insets in Figure 5.5 where the black dotted lines in each inset show $N(E)$, and the blue lines display $n(E)$. The Fermi-Dirac occupancy function is also shown (dotted red line), given by Equation 5.9. The Fermi-Dirac function $f(E)$ shifts to higher energies as the temperature increases, and the slope changes. As a result, the occupancy of the available traps decreases as $T_{irr}$ increases. Eventually a value of $T_{irr}$ will be reached at which $E_{Fe} = E_F$ and no traps are filled. (Note in the original work by Van den Eeckhout et al. (2013), $f(E)$ is represented by a vertical line at a given energy, with the traps full for all energies above this line and all traps empty for all energies below the line. However, this would only be true if $T_{irr}$ = 0 K, which is not the case.)

The change in the trapped electron distribution function $n(E)$ due to irradiating at higher temperature results in a change in the subsequent TL glow curve, as illustrated. The change in the area under the glow curve is reflective of the alteration of $n(E)$. An approximation to $n(E)$ can then be found by applying the IRM analysis to the leading edge of each glow curve obtained, for different values of $T_{irr}$. For example, the difference between the area of the glow curve for $T_{irr} = T_1$ and that for $T_{irr} = T_2$ is proportional to that part of $n(E)$ that is lost, i.e. $\Delta n(E)$, due to change $T_{irr}$. If the calculated trap depth changes from $E_{t1}$ to $E_{t2}$, i.e. $\Delta E_t = E_{t2} - E_{t1}$, then a histogram of $\Delta n(E)$ versus $\Delta E_t$ yields an approximation to the original $n(E)$.

Again, the energy resolution and accuracy of the method depend on the size of the increments in $T_{irr}$, with small increments and multiple measurements preferred. Interference from the lower part of the glow curve (see Footnote 2) when calculating $E_t$ should not be a problem with this method if the relaxation of the electrons in the conduction band is allowed to occur while the system is at $T_{irr}$; a "cleaner" analysis of $E_t$ should result. The same rules as noted for selection of $T_{min}$ and $T_c$ still apply.

### 5.2.1.2.4   $T_m$-$T_{stop}$ Analysis

The $T_m$-$T_{stop}$ procedure is related to the methods described above, but is not explicitly used to calculate $E_t$. Instead, its purpose is to identify the number of TL peaks in a complex glow curve consisting of overlapping peaks, and to provide some information on whether the kinetics are first-order or non-first-order (when dealing with discrete traps), or whether the glow curve arises from a distribution of traps. As with the step-annealing method described above, a freshly irradiated sample is first heated to $T_{stop}$, then cooled and reheated to produce the remaining glow curve. The first peak in that glow curve is identified, and its peak temperature $T_m$, is noted. It was demonstrated in Chapter 3 that a first-order peak will remain at the same temperature $T_m$ as $T_{stop}$ increases (Figure 3.3c). Therefore, a plot of $T_m$ as a function of $T_{stop}$ for a single, first-order peak will a produce flat, horizontal line. An overlay of several first-order peaks will yield a staircase-like plot of $T_m$ against $T_{stop}$. In contrast, a non-first-order TL peak will shift slightly to higher temperatures as $T_{stop}$ increases, as was demonstrated in Figure 3.5c for the second-order case. However, since the change in $T_m$ as a function of $T_{stop}$ is only pronounced at the higher values of $T_{stop}$, i.e. as the trap becomes almost depleted, an overlap of several non-first-order peaks would also yield a staircase-like plot of $T_m$ against $T_{stop}$, but with the steps much less pronounced.

In each of these cases, whether the glow curve is comprised of first-order or non-first-order peaks, the number of individual peaks in the glow curve should become visible from the horizontal, or near-horizontal "steps" in the $T_m$-$T_{stop}$ plot, and this information becomes particularly useful when applying peak-fitting methods, described in a later section.

The principle is illustrated in Figures 5.6a, 5.6b, and 5.6c. A series of arbitrary TL glow curves consisting of four overlapping second-order peaks is shown in Figure 5.6a for several different $T_{stop}$ values. In Figure 5.6b, the glow curve for $T_{stop} = 300$ K (i.e. without a pre-heat) is shown along with the differential of the glow curve ($dI_{TL}/dT$). Comparing the TL curve with its differential identifies the positions of the peak maxima, $T_{m1}$, $T_{m2}$, $T_{m3}$, and $T_{m4}$. In the $T_m$-$T_{stop}$ method, the position of the first maximum in the overall glow curve is recorded for each value of $T_{stop}$, and a plot of the first $T_m$ value versus $T_{stop}$ is produced (Figure 5.6c). For a set of TL peaks arising from discrete traps a staircase-like plot is obtained, each horizontal/near-horizontal section indicating the presence of a peak. In this way the minimum number of individual peaks in the glow curve is identified. The word "minimum" is emphasized here since small peaks may not be clearly indicated.

If the glow curve consists of a sum of first-order peaks, a clear staircase structure is obtained. A limitation of the method arises when the individual peaks are closely overlapping. In this case the individual peak maxima may again be difficult to identify. For the extreme case of a distribution of trap energies, the $T_m$ value will display a continuous shift to higher values as $T_{stop}$ increases. Conceptual examples are illustrated schematically in Figure 5.7.

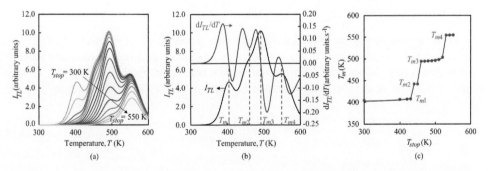

**Figure 5.6** Concept for the $T_m$-$T_{stop}$ method. (a) Changes to the glow-curve shape as $T_{stop}$ increases. The glow curve consists of four overlapping second-order peaks. (b)The first glow curve plotted with the numerical differential of the glow curve to identify the peak positions, $T_{m1}$-to-$T_{m4}$. (c) The first maximum in the glow curve $T_m$ versus $T_{stop}$, showing the characteristic staircase structure.

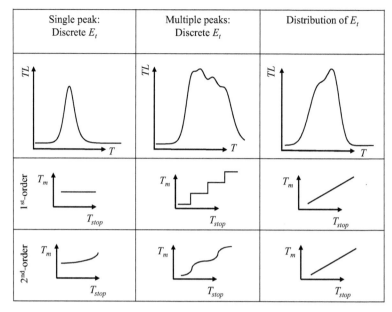

**Figure 5.7** A schematic representation of the $T_m$-$T_{stop}$ shapes that may be expected for different conditions.

---

**Exercise 5.3** $T_m$-$T_{stop}$ **analysis**

Construct a glow curve consisting of four overlapping first-order TL peaks using the Randall-Wilkins expression for first-order curves (Equation 3.17) and the approximations of Kitis et al. (1998).

(a) Simulate heating to $T_{stop}$, and then recording the rest of the glow curve and evaluate the position of the first maximum $T_m$. Increase $T_{stop}$ in steps of 5 K and plot $T_m$ against $T_{stop}$. Do your results show that there are four peaks in the glow curve?

(b) Repeat, but varying the height of the individual peaks and the separation between them. Do your results always produce clear "steps" for each peak in the glow curve?

(c) Repeat, but using second-order kinetics.

(d) What do you conclude about the limitations of the method?

---

### 5.2.2 Whole-Peak Methods

One of the weaknesses of the IRM is that it uses only a portion of the TL peak; the rest of the peak is ignored and this means that a considerable amount of data is discarded and potential information is lost. To overcome this limitation, a "whole-peak" method of analysis was proposed by Halperin et al. (1960), relying upon Equation 3.16, for the first-order case:

$$I_{TL}(t) = ns\exp\left\{-\frac{E_t}{kT}\right\},$$ (3.16)

or, Equation 3.24, for the general-order case:

$$I_{TL}(t) = n^b s'\exp\left\{-\frac{E_t}{kT}\right\}.$$ (3.24)

Expressed as functions of $T$, these become:

$$I_{TL}(T) = \left(\frac{ns}{\beta_t}\right)\exp\left\{-\frac{E_t}{kT}\right\},\tag{5.10}$$

and

$$I_{TL}(T) = \left(\frac{n^b s'}{\beta_t}\right)\exp\left\{-\frac{E_t}{kT}\right\}.\tag{5.11}$$

From Equation 5.10, a plot of $\ln(I(T)/n)$ versus $1/T$ will yield a straight line of slope $(-E_t/k)$ and an intercept of $\ln(s/\beta_t)$. Noting that $I(t) = dn/dt$, the value of $n$ is proportional to the integral under the TL peak from temperature $T$ to the end of the TL peak. Similarly, for non-first-order kinetics, Equation 5.11 shows that a plot of $\ln(I(T)/n^b)$ versus $1/T$ will also give a slope $(-E_t/k)$ and an intercept of $\ln\left(s'/\beta_t\right)$. The unknown kinetic-order parameter $b$ can be varied and the correct choice of $b$ will be that value that gives a straight line on the $\ln(I(T)/n^b)$ versus $1/T$ plot. The concept is illustrated in Figure 5.8 for the case of first-order kinetics ($b = 1$).

The obvious difficulty with this procedure is the need to isolate individual peaks. One potential solution (for first-order peaks only) is a method wherein the last peak in the glow curve is first isolated by removing all the lower temperature peaks. Then, by a process of successive subtraction, the individual peaks are isolated one by one, starting at the highest temperature peak and working toward the peak at the lowest temperature. The method is described by Taylor and Lilley (1978) and is shown in Figure 5.9. The procedure requires a clean trailing edge of the glow peak, without interference of the black-body emission background (from the sample, heater and surroundings) that occurs at high temperatures. Subtraction of the background may, of course, be performed, but extra errors will be introduced.

The procedure is to heat the specimen to a temperature $T_{stop}$ beyond the temperatures at which the lower-temperatures peaks contribute to the glow curve, as illustrated in Figure 5.9a. A second heating (Figure 5.9b) then reveals a clean remnant of the last peak (peak 3 in this example). By scaling (by a factor $f$) the remnant peak (curve 3 in Figure 5.9b) to match the trailing edge of the overall glow curve, the scaled peak 3 ($f \times 3$) may be subtracted from the glow curve to reveal only the sum of the lower temperature peaks. By repeating the procedure,

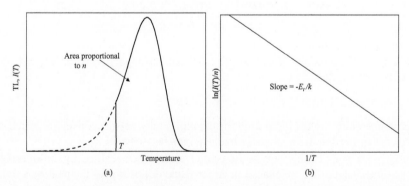

**Figure 5.8**   (a) By choosing a particular temperature $T$ in the glow curve, the area under the rest of the glow curve, for temperatures greater than $T$, is proportional to the value of $n$, the concentration of trapped electrons remaining. (b) Plotting $\ln(I(T)/n)$ against $1/T$ reveals a straight line, of slope $E_t/k$, from which $E_t$ is determined. For non-first-order kinetics, a plot of $\ln(I(T)/n^b)$ against $1/T$ also gives a straight line of slope $E_t/k$ when the correct value of $b$ is inserted.

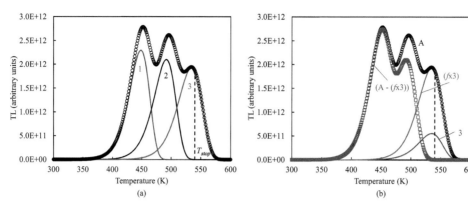

**Figure 5.9**  (a) TL glow curve consisting of three first-order peaks. If the specimen is heated to $T_{stop}$, beyond the maximum of the last peak, the lower-temperature peaks (1 and 2) are removed but some of peak 3 remains. (b) The remainder of peak 3, after heating to $T_{stop}$ (in figure (a)). This is then multiplied by a factor $f$ such that the trailing edge of the peak matches the original glow curve, A. By subtraction of ($f \times 3$) (green curve) from A, the remaining glow curve now consists of just 2 peaks (red squares). Repeating the exercise then isolates peaks 2 and 1.

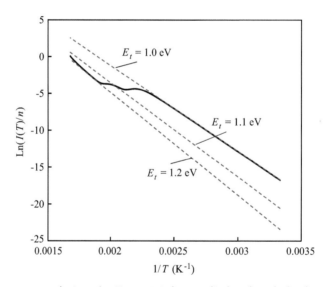

**Figure 5.10**  The same analysis as for Figure 5.8, but applied to the whole glow curve. Breaks in the slope indicate the presence of individual peaks, but the slope analysis is only reliable where there is weak peak overlap.

the lower temperature peaks (1 and 2 in this example) may be successively revealed. The net result is clean, isolated, individual peaks that can then be analyzed.

This procedure is applicable only to first-order peaks since scaling, by multiplying by the factor $f$, assumes that there is no change in either shape or position of the peak as a function of $T_{stop}$, an assumption valid only for first-order kinetics. Furthermore, the method is useful for, say, two or three peaks but becomes progressively more difficult the more peaks there are in the glow curve, and the closer they overlap. Errors accrued in separating the last peak are transferred and compounded as the procedure moves to the lower-temperature peaks.

Rather than separating the individual TL peaks using the above method, one could apply the whole-peak method to the complete glow curve. This is shown in Figure 5.10 where $\ln(I(T)/n)$ is plotted against $1/T$. The input $E_t$ values for the data were 1.0 eV, 1.1 eV, and 1.2 eV and the slopes of $E_t/k$ are shown for each $E_t$ value (dotted lines). This analysis assumes that the proportionality constant between the individual peak areas and the individual trap occupancies $n$ is the same throughout the glow curve, which may not be the case. Closely overlapping peaks (such as those shown) produce small regions of different slopes and only those regions where the overlap is weak yield good data. However, breaks in the slope indicate the presence of individual peaks and this may be useful for peak fitting procedures and for comparing with $T_m$-$T_{stop}$ analysis.

---

### Exercise 5.4   Whole-peak analysis

Construct a glow curve consisting of four overlapping, first-order TL peaks, using either the Randall-Wilkins formula or numerically solving a set of differential equations. (As usual, you have the freedom to select the $E_t$ and $s$ values.)

(a) Using the constructed glow curve, apply the peak-separation procedure described in the text using a first $T_{stop}$ value which is just beyond the temperature of the last peak. Progressively remove the highest temperature peaks until the four peaks have been isolated.

(b) Apply the whole curve analysis to each peak to determine $E_t$ and $s$ for each peak. How do the values obtained compare to your selected values?

(c) Apply the initial-rise method (IRM) to each peak and compare the results.

(d) Repeat using peaks that overlap more closely, and peaks that overlap less closely.

---

#### 5.2.3   Peak-Shape Methods

Peak-shape methods require only two or three points in the glow curve, thereby rejecting most of the data and possibly losing potential information. The key parameters are the temperatures of the peak maximum $T_m$, the low-temperature half-height $T_1$, and the high-temperature half-height $T_2$, as illustrated in Figure 5.11. Various peak-shape formulae have been developed using these parameters and have been analyzed in depth by Chen (1969a). Three width-parameters are defined based on the three temperatures. They are the full-width at half maximum:

$$\omega = T_2 - T_1; \tag{5.12a}$$

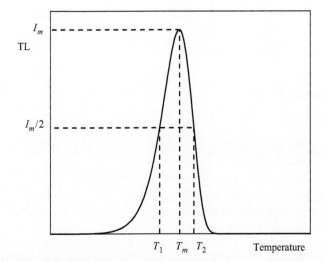

**Figure 5.11**   Definitions of the temperatures of the peak maximum $T_m$, the low-temperature half-height $T_1$, and the high-temperature half-height $T_2$.

the low-temperature half-width:

$$\tau = T_m - T_1; \tag{5.12b}$$

and the high-temperature half-width:

$$\delta = T_2 - T_m. \tag{5.12c}$$

In addition, a "shape factor" $\mu_g$ is introduced, where:

$$\mu_g = \frac{\delta}{\omega}. \tag{5.12d}$$

A general formula can then be written relating the trap depth $E_t$ to the shape parameters, thus:

$$E_t = c_\gamma \left( \frac{kT_m^2}{\gamma} \right) - b_\gamma \left( 2kT_m \right), \tag{5.13}$$

where $\gamma$ is either $\omega$, $\tau$, or $\delta$. The values for the constants $c_\gamma$ and $b_\gamma$ depend on the order of kinetics and are listed in Table 5.1. It can seen that when $\mu_g = 0.42$, the $b_\gamma$ and $c_\gamma$ values for the general-order case become equal to those for the first-order case, whereas if $\mu_g = 0.52$ the $b_\gamma$ and $c_\gamma$ values become the same as those for the second-order case. Thus, a first-order peak is defined by a shape factor $\mu_g = 0.42$, whereas a second-order peak is defined by a shape factor $\mu_g = 0.52$. If a temperature dependence for $s$ of the form $s \propto T^a$ is included, a term equal to $a/2$ must be added to the constant $b_\gamma$. Chen (1969b) described how $\mu_g$ varies with the kinetic-order parameter $b$. The result is shown in Figure 5.12.

The peak-shape method of analysis is also dependent on the ability to produce a clean peak and the same peak separation methods that have been discussed need to be applied when the glow curve consists of a series of overlapping peaks.

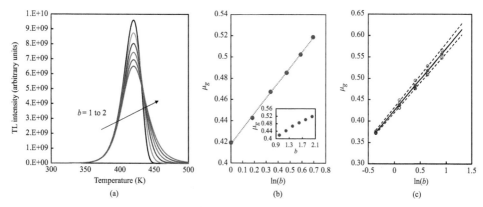

**Figure 5.12**   Variation of the shape factor $\mu_g$ with kinetic-order parameter $b$. (a) TL glow peaks for the same $E_t$ and $s$, but with different kinetic order. (See also Figure 3.6.) (b) Geometric factor as $\mu_g$ versus $\ln(b)$, from the data of figure (a). (The inset shows $\mu_g$ against $b$.) (c) Data from Chen (1969b), also plotted as $\mu_g$ versus $\ln(b)$. The black data points indicate the average of all the $\mu_g$ values for various $E_t$ and $s$ combinations, at each value of $b$, and the black line is a linear fit. The open circles are the upper and lower bounds of the $\mu_g$ values, while the dashed lines are linear fits to the data. (All data from Table II in Chen 1969b).

**Table 5.1** Values for constant $b_\gamma$ and $c_\gamma$ for use in Equation 5.13. Data from Chen (1969a, 1969b).

|  | First Order | | | Second Order | | | General Order | | |
|---|---|---|---|---|---|---|---|---|---|
|  | $\omega$ | $\tau$ | $\delta$ | $\omega$ | $\tau$ | $\delta$ | $\omega$ | $\tau$ | $\delta$ |
| $b_\gamma$ | 1.0 | 1.58 | 0 | 1.0 | 2.0 | 0 | 1.0 | $1.58 + 4.2(\mu_g - 0.42)$ | 0 |
| $c_\gamma$ | 2.52 | 1.51 | 0.976 | 3.54 | 1.81 | 1.72 | $2.52 + 10.2(\mu_g - 0.42)$ | $1.51 + 3.0(\mu_g - 0.42)$ | $0.976 + 7.3(\mu_g - 0.42)$ |

---

**Exercise 5.5    Peak-shape analysis (I)**

Using the glow curve(s) constructed in Exercise 5.4, apply the peak shape methods (Equation 5.13 with Table 5.1) to calculate the $E_t$ values for the four traps. Compare the results with the results from Exercise 5.4.

---

**Exercise 5.6    Peak-shape analysis (II)**

Calculate a series of glow peaks, for fixed $E_t$ and $s$, but using different values of $b$ (e.g. as in Figure 5.12).
(a) Determine the shape factor $\mu_g$ and plot as a function of $b$.
(b) Repeat, using a variety of $E_t$ and $s$ values. How much variation in $\mu_g$ is observed, for the same $b$, as $E_t$ and $s$ vary?

---

### 5.2.4  Peak-Position Methods

It was shown in Chapter 3 that the temperature of a TL maximum, $T_m$, varies with $E_t$ (for fixed $s$) and $s$ (for fixed $E_t$). One of the earliest estimates of the trap depth assumed that $s\exp\{-E_t/kT\}=1$ at $T = T_m$, or $E_t = kT_m\ln(s)$. The estimate requires a known or assumed value for $s$, which typically may range from, say, $10^8$ s$^{-1}$ to $10^{14}$ s$^{-1}$, and so $E_t$ estimates could range correspondingly from ~$18kT_m$ to ~$32kT_m$. Early researchers suggested $25kT_m$ (Randall and Wilkins 1945a, 1945b).

Apart from the obvious large uncertainty with such estimates due to a lack of knowledge of $s$, the estimates also ignore the dependence of $T_m$ on heating rate $\beta_t$, for a fixed $E_t$ and $s$. The peak maximum occurs when $dI_{TL}/dT = 0$ and from the Randall-Wilkins, first-order TL equation (Equation 3.17) at peak maximum:

$$\frac{\beta_t E_t}{kT_m^2} = s\exp\left\{-\frac{E_t}{kT_m}\right\}, \tag{5.14}$$

while from the Garlick-Gibson, second-order TL equation (Equation 3.21):

$$\frac{\beta_t E_t}{kT_m^2} = s\exp\left\{-\frac{E_t}{kT_m}\right\}\left[1+\left(\frac{2kT_m}{E_t}\right)\right], \tag{5.15}$$

and for general-order kinetics (Equation 3.25):

$$\frac{\beta_t E_t}{kT_m^2} = s\exp\left\{-\frac{E_t}{kT_m}\right\}\left[1+(b-1)\left(\frac{2kT_m}{E_t}\right)\right]. \tag{5.16}$$

Equations 5.14–5.16 demonstrate the reliance of $T_m$ on $\beta_t$, $E_t$, and $s$ and show that the shift in $T_m$ as $\beta_t$ changes is a function of $E_t$ and $s$. Using the first-order equation (Equation 5.14) as an example, it can be seen that for two heating rates $\beta_{t1}$ and $\beta_{t2}$ with corresponding peak maximum temperatures of $T_{m1}$ and $T_{m2}$:

$$E_t = k\left(\frac{T_{m1}T_{m2}}{T_{m1} - T_{m2}}\right)\ln\left[\frac{\beta_{t1}}{\beta_{t2}}\left(\frac{T_{m2}}{T_{m1}}\right)^2\right] \tag{5.17}$$

This can be generalized (Hoogenstraaten 1958) for several different heating rates:

$$\ln\left(\frac{T_m^2}{\beta_t}\right) = \frac{E_t}{kT_m} + \ln\left(\frac{E_t}{sk}\right), \tag{5.18}$$

from which a plot of $\ln\left(T_m^2 / \beta_t\right)$ against $1/kT_m$ yields a straight line of slope $E_t$ and intercept $\ln(E_t/sk)$. An example for first-order kinetics is shown in Figure 5.13.

Experimentally, there is still the issue of separating the peaks by thermal cleaning in order to identify accurate values for $T_m$. The problem is the same as that seen with the $T_m$-$T_{stop}$ method where the $T_m$ value changes as the peak is thermally cleaned when the kinetics are non-first-order.

---

**Exercise 5.7    Heating-rate analysis (I)**

Again use the same glow curve as used in Exercise 5.4, but do so for 4 or 5 different heating rates.

(a) Note how sensitive the glow curve is to the heating rate. How large does the change in $\beta_t$ have to be to induce significant changes in the $T_m$ values?

(b) Separate the peaks using the whole-peak analysis method of Section 5.2.2. For each peak plot $\ln\left(T_m^2 / \beta_t\right)$ against $1/kT_m$. How do the evaluated $E_t$ values compare with the input values and the values obtained by the other analysis methods?

---

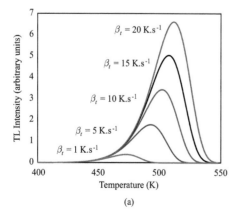

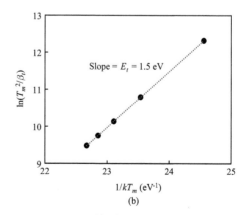

**Figure 5.13**    (a) First-order TL peaks with $E_t = 1.5$ eV and $s = 1\times10^{14}$ s$^{-1}$, for five heating rates, as shown. (b) $\ln(T_m^2 / \beta_t)$ versus $1/kT_m$, showing a slope of 1.5 eV.

The heating rate method as described so far requires separated peaks so that the value of $T_m$ can be accurately obtained. This in turn requires a method of peak cleaning, which itself can introduce errors into the analysis. A variant of the method, suggested by Sweet and Urquhart (1981), does not require peak cleaning. To explain the method, first consider a single peak recorded at several different heating rates, such as shown in Figure 5.13a. The core of the method can be understood by examining Equation 3.16 for the first-order case (or Equation 3.24 for the general-order case). Consider a temperature $T_x$ for which the remaining trapped charge is proportional to the remaining area under the glow curve, that is:

$$n(T_x) = \left(\frac{1}{\beta_t}\right) \int_{T_x}^{\infty} I(T) dT,$$    (5.19)

where the variable is changed from $t$ to $T$ using heating rate $\beta_t$. For a set of glow curves at different $\beta_t$, a temperature $T_x$ can be found for each peak for which $n(T_x)$ is the same value in each case. That is, $T_x$ is chosen for each glow curve such that $\left(\frac{1}{\beta_t}\right) \int_{T_x}^{\infty} I(T) dT$ is a constant (and therefore, $n(T_x)$ is a constant). From Equation 3.16, this means:

$$I(T_x) = K \exp\left\{\frac{-E_t}{kT_x}\right\},$$    (5.20)

where $K$ is a constant and assuming $s$ is independent of $T$. A similar expression can be written for non-first-order kinetics.

A plot of $\ln(I(T_x))$ versus $1/kT_x$ will produce a straight line of slope $-E_t$. The method is shown schematically in Figure 5.14a where $T_{x1}$ and $T_{x2}$ are chosen such that $\left(\frac{1}{\beta_{t1}}\right) \int_{T_{x1}}^{\infty} I(T) dT = \left(\frac{1}{\beta_{t2}}\right) \int_{T_{x2}}^{\infty} I(T) dT$. An example result applied to the data set of Figure 5.13a is shown in Figure 5.14b, where a value of $T_x$ for $\beta_t = 1$ K.s$^{-1}$ was first randomly chosen,

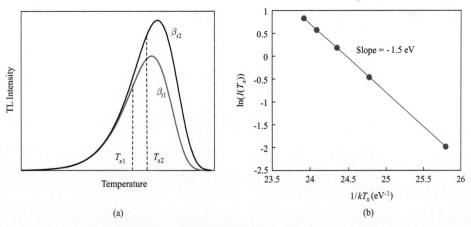

(a)                                    (b)

**Figure 5.14**    (a) Schematic of the method described by Sweet and Urquhart (1981). Glow curves are recorded at two different heating rates, $\beta_{t1}$ and $\beta_{t2}$, and temperatures $T_{x1}$ (for $\beta_{t1}$) and $T_{x2}$ (for $\beta_{t2}$) are chosen such that $\left(\frac{1}{\beta_{t1}}\right) \int_{T_{x1}}^{\infty} I_1(T) dT = \left(\frac{1}{\beta_{t1}}\right) \int_{T_{x2}}^{\infty} I_1(T) dT$, where $I_1(T)$ is the glow curve recorded at $\beta_{t1}$, and $I_2(T)$ is the glow curve recorded at $\beta_{t2}$. (b) For several different $\beta_t$, a plot of $\ln(I(T_x))$ against $1/kT_x$ gives a straight line of slope $-E_t$. The example shown is for a single, first-order peak with $E_t = 1.5$ eV, recorded at five different heating rates ranging from $\beta_t = 1$ K.s$^{-1}$ to 20 K.s$^{-1}$.

and $\left(\dfrac{1}{\beta_t}\right)\displaystyle\int_{T_x}^{\infty} I(T)\,dT$ calculated. For each of the other curves (for $\beta_t = 5$ K.s$^{-1}$ to 10 K.s$^{-1}$), $T_x$ values were found such that $\left(\dfrac{1}{\beta_t}\right)\displaystyle\int_{T_x}^{\infty} I(T)\,dT$ was the same value for each heating rate. A plot of $\ln(I(T_x))$ versus $1/kT_x$ shows a straight line with the slope equal to $-1.5$ eV, equal to the input value of $E_t$.

In principle, the advantage of this technique is that the procedure can be used when there are several overlapping peaks, without the need for thermal cleaning to separate them. A simulated example is shown in Figure 5.15 where figure (a) shows the glow curve formed from two overlapping first-order peaks, with $E_t$ values of 1.42 and 1.5 eV and $\beta_t$ ranging from 1 K.s$^{-1}$ to 20 K.s$^{-1}$. (For reference, the figure shows the two individual peaks (dashed lines) at $\beta_t = 20$ K.s$^{-1}$.) Figures 5.15b and 5.15c shows the successive slopes of $\ln(I(T_x))$ versus $1/kT_x$ for increasing or decreasing $T_x$ values. The procedure begins by either selecting a $T_x$ value for $\beta_t = 1$ K.s$^{-1}$ (call this $T_{x1}$) and finding by inspection those values of $T_x$ for the other heating rates that give the same value for $\left(\dfrac{1}{\beta_t}\right)\displaystyle\int_{T_x}^{\infty} I(T)\,dT$. The resultant plots are shown in Figure 5.15b for increasing $T_{x1}$.

Alternatively, one could choose a $T_x$ value for the $\beta_t = 20$ K.s$^{-1}$ glow curve (call this $T_{x20}$) and successively decrease the $T_{x20}$ value, again obtaining $\ln(I(T_x))$ versus $1/kT_x$ plots. The resultant plots are shown in Figure 5.15c. In either case, values of $E_t$ as a function of $T_x$ can be obtained, as shown in Figure 5.15d. A "staircase" $E_t$-versus-$T_{x1}$ (or versus $T_{x2}$) plot is observed in each case, but displaced along the temperature axis and in each case the initial and final values for $E_t$ are too high by ~0.10 eV to 0.15 eV, or up to ~10%. Since each value of $T_x$ defines a fraction

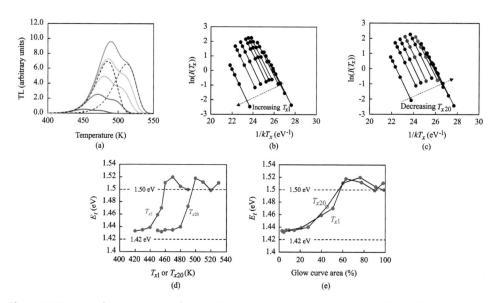

**Figure 5.15**    (a) Glow curves made up of two, overlapping, first-order peaks, with $E_t = 1.42$ eV and 1.5 eV, respectively, at five different heating rates, from $\beta_t = 1$ K.s$^{-1}$ to 20 K.s$^{-1}$. The individual peaks are shown as dashed lines for $\beta_t = 20$ K.s$^{-1}$. In (b) and (c), $\ln(I(T_x))$ against $1/kT_x$ plots for different chosen values for $T_{x1}$ or $T_{x20}$ are shown, where $T_{x1}$ and $T_{x20}$ are the values of $T_x$ for $\beta_t = 1$ K.s$^{-1}$ and $\beta_t = 20$ K.s$^{-1}$, respectively. (d) Evaluated $E_t$ values as a function of $T_{x1}$ or $T_{x20}$, as $T_x$ values are changed to pass through the glow curve. (e) By expressing the $E_t$ values as a function of area under the respective glow curves, the plots coincide. More detail of the procedure is described in the accompanying web site (see Exercises and Notes, Chapter 5, Figure 5.15).

of the area under the respective glow curve (e.g. $T_{x1}$ defines an area under the glow curve obtained with $\beta_t = 1$ K.s$^{-1}$) the data can also be expressed as $E_t$ versus area under the glow curve. When visualized in this way, the $E_t$ plots coincide on the horizontal axis (Figure 5.15e).

The method is experimentally easy to perform, since no thermal pre-heats are involved, but it is analytically tedious. Also, the process is highly dependent on the degree of overlap between the peaks. Further, since the areas under the TL curve are compared at different heating rates, the method requires either separate samples each of the same sensitivity, or no changes to the sensitivity if repeated irradiation and heating of one sample is performed.

---

**Exercise 5.8   Heating-rate analysis (II)**

Again use the same glow curve as used in Exercise 5.4, for 4 or 5 different heating rates. Apply the method of Sweet and Urquhart (1981) to the glow curve, as described in the text. Compare the evaluated $E_t$ values with the input values and the values obtained by the previous methods.

---

### 5.2.5   Peak-Fitting Methods

#### 5.2.5.1   Principles

With the advent of powerful desktop computers and modern luminescence detection equipment that is able to digitally record luminescence intensity and temperature data, fitting complex glow curves with a series of TL peaks has become a straightforward analytical procedure, albeit one that requires critical assessment of the results obtained. The basic concept rests on the notion of fitting a weighted sum of TL equations to the experimental glow curves, as already outlined in Chapter 3. Specifically, the overall TL glow curve is fitted to an expression of general form:

$$I_{TL}(T) = \int_0^\infty n(E_t) I(T) dE_t, \tag{5.21}$$

where a distribution function, or weighting function, $n(E_t)$ is assumed to describe the density of filled trapping states in the material after irradiation. As previously discussed in Chapter 3, Equation 5.21 assumes a distribution in "$E_t$-space" only, whereas a more-accurate expression would include the possibility of a distribution in "$s$-space" also:

$$I_{TL}(T,\beta_t) = \int_{s_1}^{s_2} \int_{E_{t1}}^{E_{t2}} n(E_t,s) I(T,\beta_t) dE_t ds. \tag{5.22}$$

Since the right-hand side of this equation is a two-dimensional expression, in $E_t$ and $s$, the right-hand-side likewise needs to describe a corresponding two-dimensional data set, and thus the TL intensity $I_{TL}$ is written as a function of both temperature $T$ and heating rate $\beta_t$. That is, multiple data sets of $I_{TL}(T)$ at different $\beta_t$ are required, and all $I_{TL}(T)$ curves, for each value of $\beta_t$, should be fitted simultaneously. However, this is rarely done and a single measurement at a fixed $\beta_t$ is the norm.

Fitting the experimental data to extract the two convolved functions $n(E_t,s)$ and $I(T,\beta_t)$ is the overall goal. In general practice, the double integral in Equation 5.22 is normally replaced by a single summation, thus:

$$I_{TL}(T) = \sum_{j=1}^{j=k} n_j I_j(T) \tag{5.23}$$

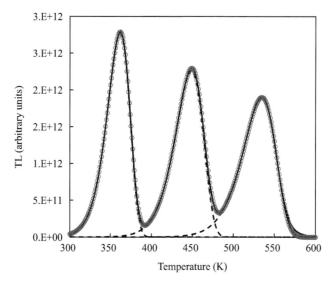

**Figure 5.16**   An example fit (red circles) of a glow curve consisting of three well-separated, first-order TL peaks calculated using a numerical solution to the appropriate rate equations (black line). The three individual peaks are shown as dashed lines. The input values were $E_{t1}$ = 0.8 eV, $E_{t2}$ = 1.0 eV, $E_{t3}$ = 1.2 eV, and $s_1 = s_2 = s_3 = 10^{10}$ s$^{-1}$. The fitted values were $E_{t1}$ = 0.8 eV, $E_{t2}$ = 1.0 eV, $E_{t3}$ = 1.2 eV and $s_1 = 9.8 \times 10^9$ s$^{-1}$, $s_2 = 9.1 \times 10^9$ s$^{-1}$, and $s_3 = 9.5 \times 10^9$ s$^{-1}$.

where it is assumed that the measured glow curve $I_{TL}(T)$ consists of $k$ individual glow peaks, each of intensity $n_j$, and $I_j(T)$ represents an individual glow peak shape. Equation 5.23 represents the superposition principle in that the measured glow curve can be considered to be a weighted sum of individual glow peaks. In this sense, the word "deconvolution" is not relevant since the function $n(E_t,s)$ has been replaced by a set of numbers $n_j$. Nevertheless, the act of fitting experimental glow curves with a summation of individual peaks is most popularly referred to as Computerized Glow Curve Deconvolution (CGCD; e.g. Horowitz et al. 1986).

A complication with this approach exists for non-first-order kinetics. It was discussed in Chapter 3 how only first-order kinetics conform to the principle of superposition since representing each individual peak by a single intensity $n_j$ becomes invalid for non-first-order kinetics. (See Chapter 3, Section 3.4.1 (Figure 3.16 in particular) and Section 3.5.) Thus, Equation 5.23 is only valid if:

$$I_{TL}(T) = \sum_{j=1}^{j=k} n_j I_{RW_j}(T), \tag{5.24}$$

where the Randall-Wilkins equation ($I_{RW}(T)$, Equation 3.17) is substituted for the previously unspecified peak shape $I(T)$. Attempts to use peak fitting/CGCD with anything other than first-order TL are not worthwhile.

The fitting procedure requires the selection of an approximation to the exponential integral function $F(E_t,T) = \int_{T_0}^{T} \exp\left[-\dfrac{E_t}{k\theta}\right] d\theta$. As discussed in Chapter 3, several possibilities have been used in the literature (e.g. Bos et al. 1993; Chen and McKeever 1997). However, modern software often includes built-in routines to implement the exponential integral such that it is possible to use:

$$F(E_t,T) = \int_{T_0}^{T} \exp\left[-\frac{E_t}{k\theta}\right] d\theta = T\exp\left\{-\frac{E_t}{kT}\right\} + \frac{E_t}{k}\,\mathrm{Ei}\left[-\frac{E_t}{kT}\right], \tag{5.25}$$

where Ei[..] is the exponential integral function.

Once a method to evaluate $F(E_t,T)$ has been selected, the next step is to estimate how many glow peaks, at a minimum, are present in the glow curve. This is non-trivial. An essential step is to conduct a $T_m$-$T_{stop}$ analysis and to identify the minimum number of peaks, $k$, in the glow curve. A second critical step is to select starting values for $E_t$ and $s$ for each of the $k$ peaks. To do so, it is best to apply any of the above-mentioned techniques for peak cleaning, followed by one of the analytical approaches discussed in the previous sections. After these preliminary efforts, peak fitting may then proceed from a reliable, or at least justifiable, starting point.

Several peak fitting routines are commercially available, perhaps the most popular of which is the Levenberg-Marquardt method for non-linear function minimization. The routine determines a parameter $\chi^2$, given by:

$$\chi^2 = \sum_i \left[ I_{RW}(T_i) - F(T_i, E_{t1}...E_{tk}, s_1...s_k, n_1...n_k) \right]^2 \tag{5.26}$$

and determines those values of the parameters $E_{tj}$, $s_j$, and $n_j$ that minimize $\chi^2$. For $k$ peaks, there are $3k$ parameters to determine, and the $3k$-dimensional parameter space can include many local minima apart from the optimum minimum. Although the local minima in parameter space can provide mathematically acceptable fits, the values obtained for $E_t$ and $s$ may be unrealistic (or incorrect, at least) and therefore additional analysis methods, as described in the previous sections, are important to define the ranges of $E_t$ and $s$ that are acceptable/probable. User input may be required to force the computation away from a misleading local minimum. In this context, selecting the initial input values is essential.

A parameter describing the final quality of the fit is the Figure of Merit, FOM, defined as:

$$\text{FOM}(\%) = \frac{100 \sum_i \left| I_{TL}(T_i) - I_{RW}(T_i) \right|}{\sum_i I_{TL}(T_i)}, \tag{5.27}$$

where $I_{TL}(T_i)$ is the measured TL intensity at temperature $T_i$, and $I_{RW}(T_i)$ is the calculated TL value at the same temperature. The FOM is normally saved for each fit, as are the residuals $I_{TL}(T_i) - I_{RW}(T_i)$. Ideally, the residuals should be a random set of small numbers (compared to $I_{TL}(T_i)$) and centered around $I_{TL}(T_i) - I_{RW}(T_i) = 0$. (The Randall-Wilkins equation for first-order TL is used in all expressions.)

When applied to simulated data where there is no noise, a straightforward minimization of the FOM is the goal. However, for real experimental data, where there is noise of $\pm\sigma\%$, any FOM value that is less than $\sigma$ should be considered as an acceptable fit. There may be more-than-one combinations of the various $E_t$ and $s$ values that meet this criterion and this helps to define the range of uncertainties in the determined $E_t$ and $s$ values.

A straightforward example of fitting three first-order TL peaks to a numerically generated data set is shown in Figure 5.16. The input parameter values for the numerical simulation (numerically solving the relevant rate equations) are $E_{t1} = 0.8$ eV, $E_{t2} = 1.0$ eV, $E_{t3} = 1.2$ eV, and $s_1 = s_2 = s_3 = 10^{10}$ s$^{-1}$. The fitted values are $E_{t1} = 0.8$ eV, $E_{t2} = 1.0$ eV, $E_{t3} = 1.2$ eV and $s_1 = 9.8 \times 10^9$ s$^{-1}$, $s_2 = 9.1 \times 10^9$ s$^{-1}$, and $s_3 = 9.5 \times 10^9$ s$^{-1}$.

### Exercise 5.9   Peak-fitting analysis (I)

In 1993 several international laboratories took part in an inter-laboratory comparison of peak-fitting routines using the same reference TL glow curves. The project was called GLOCANIN (**GLO**w Curve **AN**alysis **IN**tercomparison) and the results were subsequently published by the participating laboratories (Bos et al. 1993, 1994). The nine reference curves used in the exercise have been used extensively since then to test various peak-fitting routines (Kitis et al. 1998; Puchalska and Bilski 2006; van Dijk 2006; Kiisk 2013; Sadek et al. 2015; Sature et al. 2017; El-Kinawy et al. 2019, and several others).

Included in the reference glow curves were two simulated curves (Refglow001 and Refglow002). The data for these reference curves can be found in an Excel file on the accompanying web site (Exercises and Notes, Chapter 5, Exercise 5.9).

Download these reference glow curves and fit them using whatever peak-fitting routine/software package you wish. (For example, using Excel construct a glow curve using the expressions derived by Kitis et al. (1998) to approximate the glow curve, and compare with the Refglow001 and 002 curves. Vary the $E_t$ and $s$ parameters until best fits to the reference glow curves are obtained, using Solver. Note, however, that you may use whatever software package you prefer.)

Compare your obtained values and FOM values with those obtained by the GLOCANIN participating laboratories.

### 5.2.5.2    Peak Resolution

As with all analytical techniques described in this chapter, overlap of the individual TL peaks is a significant concern, such that closely overlapping peaks may be identified as one peak. The resolution $R_s$ between two TL peaks can be defined using the shape parameters defined in Section 5.2.3, namely $\tau$ and $\delta$, and the positions of the maxima of the two peaks, $T_{m1}$ and $T_{m2}$:

$$R_s = \frac{T_{m2} - T_{m1}}{\delta_1 + \tau_2}. \tag{5.28}$$

The parameters are illustrated schematically in Figure 5.17. The gap between the two peaks at the half-maximum height is $x$, and thus:

$$T_{m2} - T_{m1} = x + \delta_1 + \tau_2. \tag{5.29}$$

When $x = 0$, $R_s = 1$; when $x < 0$, $R_s < 1$, and when $x > 0$, $R_s > 1$. When the two peaks coincide ($T_{m2} = T_{m1}$) then $R_s = 0$.

Kitis and Pagonis (2019), calculated $R_s$ for several first-order and second-order simulated TL peaks. Considering that the superposition principle applies only to first-order TL peaks, the calculated mean $R_s$ values, for 50–60 $E_t/s$ pairs and first-order peaks only, are shown in Table 5.2.

The resolution clearly depends on the shape parameters for the peaks, which are smaller (larger $R_s$) for first-order compared to non-first-order TL peaks. Kitis and Pagonis show that when $R_s > 0.5$, CGCD methods reliably yield $E_t$ and $s$ values for the individual peaks. For $0.25 < R_s < 0.5$, the CGCD methods can give reliable results, but only if supported by other analytical methods. For $R_s < 0.25$, the CGCD results are not reliable. Note that this analysis assumes similarly sized peaks; the conclusions when the peaks differ in size may be different.

Although this analysis of resolution has only been carried out for CGCD methods, it is probably safe to assume that similar resolution between peaks is also relevant to the other analysis methods.

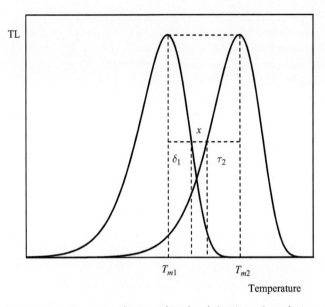

**Figure 5.17**    Parameters $T_{m1}$, $T_{m2}$, $\delta_1$, and $\tau_2$, used in the definition of resolution, $R_s$.

**Table 5.2** Values of resolution $R_s$ for various peak separations $T_{m2} - T_{m1}$ for first-order kinetics. Data from Kitis and Pagonis (2019).

| $T_{m2} - T_{m1}$ | $R_s$ |
|---|---|
| 10 | 0.34±0.07 |
| 20 | 0.67±0.10 |
| 30 | 1.02±0.16 |
| 40 | 1.31±0.21 |
| 50 | 1.61±0.27 |

### 5.2.5.3    CGCD Using More-Than-One Heating Rate

It was noted above with respect to Equation 5.22 that evaluation of $E_t$ and $s$ can be considered more reliable if a two-dimensional data set is used. Thus, measurement of several glow curves using different heating rates should be collected and the glow curves fitted simultaneously. This is because several combinations of $E_t$ and $s$ may give a TL peak at the same temperature $T_m$, but only one unique set of $E_t$ and $s$ will give the correct $T_m$ at two or more different heating rates. This principle is illustrated in Figure 5.18 where two sets of $E_t/s$ are shown to give the same $T_m$ value at one heating rate (1 K.s$^{-1}$), but the same parameters give different $T_m$ values at a different heating rate (5 K.s$^{-1}$).

This behavior can be understood from differentiation of the Randall-Wilkins expression, Equation 3.17. At the peak maximum, at temperature $T_m$, the derivative is zero, from which condition Equation 5.14 applies. From this equation it can be seen that for a fixed $\beta_t$, there are several combinations of $E_t$ and $s$ that will give the same $T_m$. However, if two different combinations of $E_t$ and $s$ give the same $T_m$ at one heating rate, they will yield different $T_m$ values at a different heating rate.

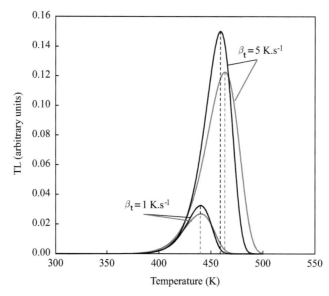

**Figure 5.18**    Two TL peaks for parameters $E_{t1} = 1.15$ eV and $s_1 = 10^{12}$ s$^{-1}$, and $E_{t2} = 1.4$ eV and $s_2 = 9 \times 10^{14}$ s$^{-1}$. At $\beta_t = 1$ K.s$^{-1}$ both sets of parameters give a TL peak at the same temperature, $T_{m1} = T_{m2} = 440$ K, but at $\beta_t = 5$ K.s$^{-1}$, the peaks were at $T_{m1} = 463.5$ K and $T_{m2} = 459$ K, respectively.

It is also clear that the resolution of the peaks may be changed by changing the heating rate. In the example shown in Figure 5.18 the resolution changes from $R_s = 0$ at $\beta_t = 1$ K.s$^{-1}$, to $R_s = 0.13$ at $\beta_t = 5$ K.s$^{-1}$. Thus, changing the heating rate, especially over a wide range, can help to determine if the glow curve being studied in the result of closely overlapping peaks that are too close to be resolved by eye.

An example of a data set simulated at different heating rates is shown in Figure 5.19a. Here, two first-order TL peaks are displayed at three different heating rates. The peaks barely overlap in this example and the resolution of the peaks is not an issue. The black line shows the sum of the three glow curves to obtain a "pseudo" glow curve made up of the sum of the glow curves obtained for the three different heating rates. The procedure is then to fit the "pseudo" TL curve to the sum of three glow curves at three different $\beta_t$ values, each glow curve consisting of two RW peak shapes formed using the same two sets of $E_t$ and $s$, namely $E_{t1}/s_1$ and $E_{t2}/s_2$. The input values are $E_{t1} = 1.15$ eV and $s_1 = 7.0 \times 10^{13}$ s$^{-1}$, and $E_{t2} = 1.40$ eV and $s_2 = 1.4 \times 10^{14}$ s$^{-1}$. The fitted values are $E_{t1} = 1.15$ eV and $s_1 = 6.9 \times 10^{13}$ s$^{-1}$, and $E_{t2} = 1.40$ eV and $s_2 = 1.39 \times 10^{14}$ s$^{-1}$. Forming a "pseudo" glow curve in this way is a simple way to fit all three glow curves, obtained at three different heating rates, simultaneously.

A more challenging situation is shown in Figure 5.19b. Here the input parameters are $E_{t1} = 1.15$ eV and $s_1 = 10^{12}$ s$^{-1}$, and $E_{t2} = 1.40$ eV and $s_2 = 9 \times 10^{14}$ s$^{-1}$. As was observed in Figure 5.18, which uses the same parameters, the resolution of these peaks is zero at $\beta_t = 1$ K.s$^{-1}$ and improves slightly as $\beta_t$ increases. The two individual peaks (shown as dotted lines in Figure 5.19b) still strongly overlap, even at $\beta_t = 15$ K.s$^{-1}$. By forming the "pseudo" glow curve by combining all data at the three different heating rates (1 K.s$^{-1}$, 5 K.s$^{-1}$, and 15 K.s$^{-1}$) and fitting the "pseudo" TL curve using two sets of $E_t$, $s$, and $n_0$ values, fitted values of $E_{t1} = 1.16$ eV and $s_1 = 1.2 \times 10^{12}$ s$^{-1}$, and $E_{t2} = 1.40$ eV and $s_2 = 9.6 \times 10^{14}$ s$^{-1}$ are obtained. In this extreme case the $R_s$ value is less than the limit of 0.25 as defined by Kitis and Pagonis (2019) but the performance of the fitting is still as good as in Figure 5.19a where the peaks are well resolved. The use of multiple heating rates appears able to overcome the stated limitation on TL peak resolution ($R_s < 0.25$).

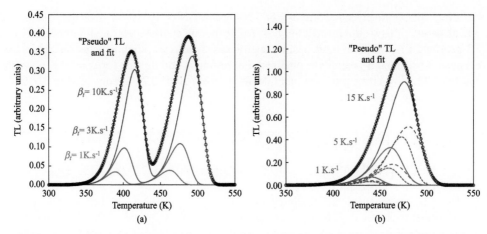

**Figure 5.19** Fitting of "pseudo" glow curves made up of the sum of three separate glow curves, each consisting of two peaks, obtained at three different heating rates. In (a) and (b) the "pseudo" glow curve is shown as a black line and is the sum of the three glow curves obtained at three different heating rates, shown as orange, red, and blue lines. The fit to the "pseudo" curve is shown as open circles. In (a) the input values were $E_{t1}$ = 1.15 eV, $s_1$ = 7.00x10$^{13}$s$^{-1}$ and $n_{01}$ = 1.0 arbitrary unit; and $E_{t2}$ = 1.40 eV, $s_2$ = 1.40x10$^{14}$ s$^{-1}$ and $n_{02}$ = 1.30 arbitrary units. The fitted values were $E_{t1}$ = 1.15 eV, $s_1$ = 6.91x10$^{13}$s$^{-1}$, and $n_{01}$ = 1.0 arbitrary unit; and $E_{t2}$ = 1.40 eV, $s_2$ = 1.42x10$^{14}$ s$^{-1}$ and $n_{02}$ = 1.30 arbitrary units. In (b) the input values were $E_{t1}$ = 1.15 eV, $s_1$ = 10$^{12}$ s$^{-1}$ and $n_{01}$ = 1.0 arbitrary unit; and $E_{t2}$ = 1.40 eV and $s_2$ = 9x10$^{14}$ s$^{-1}$ and $n_{02}$ = 1.00 arbitrary unit. The fitted values were $E_{t1}$ = 1.16 eV, $s_1$ = 1.6x10$^{12}$ s$^{-1}$ and $n_{01}$ = 1.53 arbitrary units; and $E_{t2}$ = 1.40 eV, $s_2$ = 9.6x10$^{14}$ s$^{-1}$ and $n_{02}$ = 0.97 arbitrary units. The full lines are the composite peaks at the three heating rates. The dotted lines are peak 1 and the dashed lines are peak 2, at the three heating rates.

In the fitting of the glow curve in Figure 5.19b, two peaks were used since it was known that the simulated glow curves were made up of two peaks. With experimental data, this degree of prescience is absent and the number of peaks may be unknown. Indeed, from the shapes of each glow curve at the three heating rates a single peak of non-first-order kinetics could be suspected. Therefore, in a real experiment it is essential to carry out ancillary measurements to determine, first of all, if the glow curve displays first-order kinetics. The easiest way to do this is to examine how the peak shape changes with dose for a fixed $\beta_t$. In Section 3.2.1 and Figure 3.3, it was demonstrated that the position of a first-order peak is invariant with dose. Thus, a critical measurement is to determine if the glow curve changes shape and position with dose, or if it simply scales with dose, indicating first-order kinetics. Once it is confirmed that first-order kinetics are prevalent, then it is clear that the overall, non-first-order shape of the glow curve must be due to a superposition of more-than-one first-order TL peaks and fitting using the superposition principle can proceed. Of course, the actual number of peaks present may still be uncertain. $T_m$-$T_{stop}$ analysis may help but, in a case like Figure 5.19b, it will be unlikely that the values of the individual $T_m$ values will emerge due to the extreme overlap of the peaks (especially at 1 K.s$^{-1}$). At higher (or lower) heating rates where separation of the peaks is a little greater, a gradual rise in $T_m$ with increasing $T_{stop}$ may be observed, but this would just confirm the presence of more-than-one peak. A final recourse will be to fit the data using one, two, or more peaks and checking for consistency and the quality of the fit. In principle, only the correct number of peaks will predict the curve shapes obtained at multiple heating rates.

For example, by trying to fit the data of Figure 5.19b to a single, first-order peak, a poor fit is obtained (particularly on the high-temperature side) and a large FOM is calculated (2.6% compared to 0.07% for the 2-peak fit shown in the figure). When three peaks are used, however, comparatively good fits can be obtained (FOM = 0.07% in both the 2-peak and 3-peak cases). This would make it very difficult experimentally to tell the difference between a 2-peak and a 3-peak fit. If more heating rates are used, however, slightly better discrimination between 2-peak and 3-peak fits is obtained. For example, adding data for $\beta_t = 10$ K.s$^{-1}$ to the "pseudo" TL glow curve produces an FOM for a 3-peak fit of 0.26%, compared to 0.03% for the 2-peak case.

---

**Exercise 5.10  Peak-fitting analysis (II)**

On the accompanying web site, go to Chapter 5 Exercises and Notes, and find the folder Exercise 5.10. Open the Excel spreadsheet and plot the data. Each TL column is a TL glow curve recorded at a different heating rate (0.1 K.s$^{-1}$, 0.3 K.s$^{-1}$, and 1.0 K.s$^{-1}$), as indicated in the column title. Each TL curve is a sum of three overlapping first-order TL peaks.

Use a peak fitting routine of your choice to simultaneously fit the three glow curves in order to extract the values of $E_t$, $s$, and $n_0$ for each peak.

What values of $E_t$, $s$, and $n_0$ do you get? How do they compare with the input values for the simulated glow peaks, which were:

- $E_{t1} = 1.05$ eV
- $E_{t2} = 1.15$ eV
- $E_{t3} = 1.25$ eV
- $s_1 = s_2 = s_3 = 10^{12}$ s$^{-1}$
- $n_{01} = n_{02} = n_{03} = 10^{14}$ cm$^{-3}$

What is the FOM for the best fit?

---

### 5.2.5.4  Continuous Trap Distributions

The discussion of overlapping peaks leads directly to a discussion of trap distributions. In this case, the simple expression Equation 5.24 cannot be used, and it is necessary to revert to Equation 5.22, with $I_{RW}(T,\beta_t)$ inserted for $I(T,\beta_t)$. Several suggested procedures have been published to account for a trap-depth distribution when fitting glow curves. One approach is described by Benavente et al. (2019) who combined the Randall-Wilkins expression (Equation 3.17) and the equation at the peak maximum, Equation 5.14, to give:

$$I_{RW}(T) = I_m \exp\left\{\frac{E_t}{kT_m} - \frac{E_t}{kT}\right\} \exp\left\{-\frac{E_t}{kT_m^2}\int_{T_m}^{T}\exp\left\{\frac{E_t}{kT_m} - \frac{E_t}{k\theta}\right\}d\theta\right\}, \tag{5.30}$$

where $I_m$ is the peak maximum at $T = T_m$. Here, parameters $E_t$, $T_m$, and $I_m$ are used to define the glow peak instead of $E_t$, $s$, and $n_0$. Benavente et al. (2019) used a rational approximation to the exponential integral in Equation 5.30, to get:

$$I_{RW}(T) = I_m \exp\left\{\frac{E_t}{kT_m} - \frac{E_t}{kT}\right\} \exp\left\{-\frac{E_t}{kT_m}\left[R\left(\frac{E_t}{kT_m}\right) - \frac{T}{T_m}\exp\left\{\frac{E_t}{kT_m} - \frac{E_t}{kT}\right\}R\left(\frac{E_t}{kT}\right)\right]\right\} \tag{5.31}$$

where:

$$R(x) = 1 - \frac{0.250621 + 2.334733x + x^2}{1.681534 + 3.330657x + x^2}, \tag{5.32}$$

and $x = E_t/kT_m$.

For a Gaussian distribution of traps:

$$N(E_t) = N_m \exp\left\{-a(E_t - E_{tm})^2\right\}, \tag{5.33}$$

where the maximum concentration $N_m$ is found at $E_t = E_{tm}$, and $a$ is a constant. The distribution of filled traps is then:

$$n(E_t) = N(E_t) f(E_t). \tag{5.34}$$

With the approximation that $s$ is not distributed, and with a fixed heating rate $\beta_t$, Equation 5.22 becomes:

$$I_{TL}(T, \beta_t) = \int_{E_{t1}}^{E_{t2}} n(E_t) I_{RW}(T) \mathrm{d}E_t, \tag{5.35}$$

with $s$ now part of the RW expression. For a continuous distribution such as this, Benavente et al. (2019) define parameters $T_N$ and $I_N$ (instead of $T_m$ and $I_m$) thus:

$$\frac{E_t}{kT_N^2} = \frac{s}{\beta_t} \exp\left\{-\frac{E_t}{kT_N}\right\} \tag{5.36}$$

and $I_{RW}(T_N) = I_N$.

With $I_{RW}(T)$ written in the form of Equation 5.31, Equation 5.35 now becomes:

$$I_{TL}(T) = I_N \frac{\displaystyle\int_{E_{t1}}^{E_{t2}} f(E_t) \exp\left\{-\frac{E_t}{kT}\right\} \exp\left\{-\frac{E_{tm}}{kT_N}\left[\frac{T}{T_N}\right]\right\} \exp\left(\frac{E_{tm}}{kT_N} - \frac{E_t}{kT}\right) R\left(\frac{E_t}{kT}\right)\right\} \mathrm{d}E_t}{\displaystyle\int_{E_{t1}}^{E_{t2}} f(E_t) \exp\left\{-\frac{E_t}{kT}\right\} \exp\left\{-\frac{E_{tm}}{kT_N} \exp\left(\frac{E_{tm}}{kT_N} - \frac{E_t}{kT_N}\right) R\left(\frac{E_t}{kT_N}\right)\right\} \mathrm{d}E_t}. \tag{5.37}$$

The integrals need to be solved numerically. The method is computationally intensive and requires evaluating Equation 5.37 numerically for each value to $E_t$, $T_N$, and $I_N$, and varying the latter three parameters to minimize $\chi^2$.

Recall from Chapter 3 (Section 3.5 and Figure 3.21b) that a Gaussian distribution of thermal trap depths, centered at $E_{tm}$, will produce a near-Gaussian-shaped final TL peak. Conversely, therefore, if a Gaussian-shaped TL peak is observed, and if its shape is invariant with dose (indicating first-order kinetics), then it can be inferred that the TL peak is a sum of first-order, RW-equations, with a Gaussian distribution in $E_t$. To estimate $E_{tm}$, the variable heating rate analysis described by Equation 5.18 can be applied by noting the change in the temperature of the maximum of the Gaussian-shaped TL peak ($T_m$) as a function of heating rate. From a plot of $\ln(T_m^2/\beta_t)$ against $1/T_m$ the slope will give $E_{tm}$. The method is explored in Exercise 5.11.

An alternative approach is not to assume a particular form for the distribution $n(E_t,s)$, but to evaluate the optimum distribution function that fits simultaneously a series of glow curves recorded at multiple heating rates. This is an extension of the procedure noted above but here the single values for $n_0$ are replaced by $n(E_t,s)$ and many individual TL glow curves are constructed with parameters separated by $\delta E_t$ and $\delta s$. The procedure was outlined by Whitley et al. (2002) who demonstrated the feasibility of the method by simulating four Gaussian distributions in $E_t$ and $s$ space, along with the resultant TL curves (actually, Whitley et al. used thermally stimulated

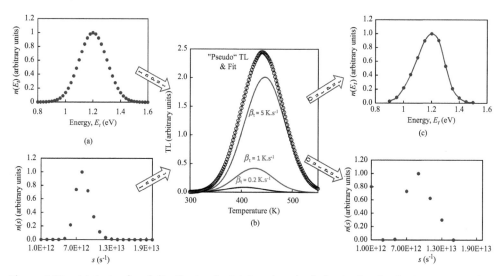

**Figure 5.20**    (a) A simulated distribution in $E_t$ (above) and $s$ (below). The distributions are used to give the simulated TL curves shown on Figure 5.20b, at three heating rates, as shown. The sum of these glow curves gives the "pseudo" TL curve, which is then fitted using Randall-Wilkins, first-order equations at the three heating rates, and almost-random $E_t$- and $s$-distributions. The result of the fit is shown by the circles in Figure 5.20b, and the output distributions are shown in Figure 5.20c (with $E_t$ above, and $s$ below).

conductivity curves, but the principle remains exactly the same). As emphasized by Whitley et al. (2002), a wide range of heating rates is preferred in order to achieve the best results. Applying the method experimentally to $Al_2O_3$:C produced the results already shown in Figure 3.23.

Here the method is demonstrated with the distributions shown in Figure 5.20a for $E_t$ (above) and $s$ (below). The resultant glow curves simulated using these distributions are shown in Figure 5.20b, for $\beta_t = 0.2$ K.s$^{-1}$, 1.0 K.s$^{-1}$, and 5.0 K.s$^{-1}$. The resultant "pseudo" TL curve, obtained by adding together the glow curves obtained at the three heating rates, is then fitted using the Randall-Wilkins function weighted by almost-random $E_t$- and $s$-distributions.[3] The resultant output distributions obtained from the analysis are shown in Figure 5.20c, with the $E_t$-distribution, and the $s$-distribution.[4]

---

### Exercise 5.11    Gaussian distribution of traps

Part A
Assuming first-order kinetics, construct a TL glow peak consisting of a Gaussian distribution of traps, $n(E_t)$. Choose a mean value of the trap depth $E_{tm}$, a width for the distribution, and a value for $s$; for the calculation assume the same value of $s$ for each of the values of $E_t$. (Note, use at least 30 values for $E_t$ in the $n(E_t)$ distribution so that the net TL curve is made up of at least 30 individual glow curves, each with its own value of $E_t$.)

---

[3] Although in principle random distributions for $E_t$ and $s$ can be used, the fitting program converges more quickly and reliably if an outline of the distributions is known beforehand. Therefore, preliminary $T_m$-$T_{stop}$ and IRM analysis is highly recommended to give a guide to the possible distribution shapes and assist the computation.

[4] The simulated glow curves are a sum of 820 individual glow curves, obtained using a matrix of 41 $E_t$-values and 20 $s$-values, each with a Gaussian distribution, as shown in Figure 5.20a. For faster computation the fitting uses a coarser matrix, of 104 glow curves, from 13 $E_t$-values and 8 $s$-values. The resulting fitted distributions, therefore, do not have the same resolution as the input distributions.

(a) Verify that the net glow peak (the sum of all the individual glow curves) can be approximated by a Gaussian shape.

Part B

(b) Construct several TL curves, each for a different value of heating rate $\beta_t$. Choose 4 or 5 heating rates.

(c) Plot $\ln(T_m^2/\beta_t)$ against $1/T_m$, where $T_m$ is the temperature of the maximum of the net TL peak and determine the mean trap depth $E_{tm}'$ from the slope. How does the value obtained for $E_{tm}'$ compare to the chosen mean value $E_{tm}$ used in your initial Gaussian distribution? Similarly, calculate the value for $s'$ from the intercept of the $\ln(T_m^2/\beta_t)$-versus-$1/T_m$ plot. Compare the value of $s'$ with your initial chosen value of $s$.

(d) Using the $E_{tm}'$ and $s'$ values calculated from the $\ln(T_m^2/\beta_t)$-versus-$1/T_m$ plot, fit the net glow curve that you initially constructed using an equation of the form:

$$I_{TL}(T) = \sum_{E_t} n'(E_t) I_{RW}(E_t, T)$$

where $I_{RW}(E_t, T)$ is the Randall-Wilkins, first-order TL equation; $n'(E_t)$ is a fitted weighting function and is a Gaussian centered at $E = E_{tm}'$. How does your fitted function $n'(E_t)$ compare with your input Gaussian function, $n(E_t)$?

### 5.2.6 Calculation of *s*

Almost all of the discussion so far has concerned the analysis of TL glow curves to obtain values for $E_t$, and the assessment of $s$ has been mentioned only in passing. $E_t$ and $s$ are not independent parameters and, as has already been mentioned several times, only certain pairs of $E_t$ and $s$ can give a TL peak at a given value of $T_m$ for a given heating rate. The key equation is Equation 5.14, re-written here as:

$$s = \frac{\beta_t E_t}{k T_m^2} \exp\left\{ \frac{E_t}{k T_m} \right\}. \tag{5.38}$$

Thus, if $T_m$ is known (from the measurement of the glow curve) and $E_t$ is evaluated (using one of the methods so far described), $s$ can be evaluated.

Some of the methods described are able to compute $s$ directly. For example, the whole-peak method of Equation 5.10 leads to a plot of $\ln(I(T)/n)$ versus $1/T$ and an intercept of $\ln(s/\beta_t)$. Knowing the value of $\beta_t$ enables $s$ to be obtained from the intercept. Similarly, the heating rate method of Equation 5.18 yields a plot of $\ln\left(T_m^2 / \beta_t\right)$ against $1/kT_m$ with an intercept $\ln(E_t/sk)$. Knowing $E_t$ from the slope, $s$ can be calculated from the intercept.

One important point, which is made clear in Equation 5.14, is that the peak position $T_m$ is far less sensitive to changes in $s$ than it is to changes in $E_t$, since $E_t$ appears inside the exponential term while $s$ is outside. A consequence of this is that peak fitting is far less sensitive to changes in $s$ than to changes in $E_t$ and, as a result, the obtained $E_t$ value has less uncertainty than the obtained $s$ value. Errors of ~5% in $E_t$ lead to similar errors in the exponent of $s$. As a result, calculated $s$ values are often listed in the form $a10^{x \pm y}$. This uncertainty can easily be seen by comparing the input and output $s$ distributions in Figure 5.20.

### 5.2.7 Potential Distortions to TL Glow Curves

In any of the above analysis methods to determine $E_t$ and $s$ there are certain inferred, but often unstated, assumptions that are inherent to all the methods. The first is that the temperature recorded is the actual temperature of the sample and, related to this, that all of the sample is at the same temperature. Another relates to Equation 3.7 in which it is stated that the luminescence intensity is given by the product, $\eta |dm/dt|$, where $m$ is the concentration of recombination centers and $\eta$ is an efficiency term. In all of the analyses presented, it is assumed that $\eta$ remains constant throughout the glow curve. If the efficiency term is a function of temperature, this must be accounted for before analysis can begin.

These issues are discussed in the sections that follow.

### 5.2.7.1    *Thermal Contact*

In most TL experiments, the sample is heated though contact heating with a hot plate or heating strip. Some TL systems use hot gas and some laser heating, but the majority use contact heating. In these cases, the temperature recorded is the temperature of the heating strip and not the temperature of the sample. Unless the two are the same, all analysis methods are fruitless since the uncertainties are very sensitive to $T$ due to the exponential term $\exp\{-E_t/kT\}$.

Betts et al. (1993a, 1993b) describe the possible temperature distributions that can occur when heating a sample. The practical issues are highlighted in the schematic diagram, Figure 5.21. Although many variations on heater strip design are available, especially in commercially available TL/OSL equipment, the physical principles outlined here apply to most. Note, that as long as the heating of the actual sample is reproducible, the system may be perfectly acceptable for dosimetry, but not necessarily acceptable for $E_t$ and $s$ analysis.

Three temperature gradients are of concern in which $T$ may vary with position $(x, y, z)$ as indicated in the figure. Strip heaters tend to be hottest in the center due to thermal conduction of heat away from the ends. The hottest spot can be confined somewhat by notching the heater strip and making it narrower in the center, as indicated in the figure. The current density is then highest at the mid-point of the heater where the sample (the dosimeter) is placed. This helps to reduce temperature gradients in the $x$- and $z$-directions. Other heater designs to minimize gradients have been suggested (Yin et al. 2020).

The temperature is normally monitored via a thermocouple spot-welded to the underside, therefore it is essential that there is excellent thermal contact between the strip and the sample.[5]

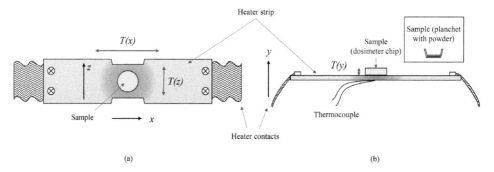

**Figure 5.21**    (a) Top view of a typical heater strip with dosimeter chip. (b) Side view of the same heater strip. The sample can be either a chip, or powder contained in a metal planchet. Potential thermal gradients and non-uniform temperatures ($T(x)$, $T(y)$, and $T(z)$) can be expected in the $x$-, $y$-, and $z$-directions.

---

[5] To achieve this, a small amount of thermal paste or silicone grease may be applied to the underside of the sample, and also the atmosphere in which the heater/sample combination is placed should be filled with an atmosphere of an inert gas with good thermal conductivity properties. Nitrogen is commonly used; helium is also good. (A vacuum should not be used, and the pressure of the gas should be just below atmospheric.)

This becomes more problematic if the sample is in powder form inside a metal planchet, which itself is placed on to the heater strip.

Even if good thermal contact is achieved, thermal gradients across the thickness of the sample, in the *y*-direction, can be problematic. This is especially so at high heating rates. The thicker the sample/powder layer, the worse the effect. Dosimeter chips are typically of dimension 4 mm to 5 mm across and up to 1 mm thick. They may be round or square, but in either case 1 mm is too thick to ensure no thermal gradients in the *y*-direction. To mitigate this problem, thinned samples (e.g. 0.1 mm to 0.2 mm) are recommended when performing $E_t/s$ analysis. When using powder samples, a thin mono-layer of grains placed directly in the heater strip (with, say, silicone grease) should ensure that there are minimal or no thermal gradients. The importance of this can sometimes be seen as changes in the peak position and intensities when comparing TL from powders with that from crystals (Yin et al. 2020). Finally, whenever possible, lower rather than higher heating rates should be preferred. Although heating rates up to 20 K.s$^{-1}$ have been used in some of the simulations shown in this text in order to illustrate the effects, actual heating rates should be kept below ~5 K.s$^{-1}$ for glow-curve analysis when performing experiments (Betts et al. 1993a). This means that when a wide range of heating rates is needed for the analysis, some of the heating rates chosen will be very low (e.g. 0.1 K.s$^{-1}$, or less) in order to produce large changes in $T_m$.

### 5.2.7.2   Thermal Quenching

In Section 3.7.1 the loss of luminescence efficiency with increasing temperature was discussed. Known as thermal quenching, two major models – the Mott-Seitz model and the Schön-Klasens model – were discussed in some detail. A mathematical form of the temperature dependence of the luminescence efficiency for the Mott-Seitz model was given as:

$$\eta(T) = \frac{1}{1 + C\exp\{-W/kT\}},$$

(3.89)

where $C$ is a constant and $W$ is the activation energy for quenching. A similar expression exists for the Schön-Klasens model (Equation 3.91).

Thermal quenching distorts the shape of the glow curve such that analysis, without correction for $\eta(T)$, will be unreliable and meaningless. It is essential that a test for thermal quenching be performed and, if it exists, parameters $C$ and $W$ need to be established and the appropriate correction made.

### 5.2.7.3   Emission Spectra

One aspect of TL analysis that has not yet been discussed in this text, and is often, even usually, overlooked is that of the spectrum of the luminescence emission. When writing equations for the TL intensity it is assumed that:

$$I_{TL}(t) = \left|\frac{dm}{dt}\right|$$

(5.39)

from which are derived the usual expressions for the TL glow-curve shape. Writing it as an equality it this way implies that every recombination event results in a measured TL photon (or an OSL photon if OSL is being measured). In reality, this is not the case and many recombination events occur without producing a detected photon. (This ignores thermal quenching, for which a recombination event might not emit a photon at all.) Emphasis is placed upon the word *detected* here since photons may be emitted but, because of the efficiency of the detection system, may not be measured. (Furthermore, the geometry of the detection system may mean that the emitted photon misses the detector entirely.)

Conventional TL measurements record the TL emission using a broad-band detector, such as a photomultiplier tube with the TL emission usually filtered through a broad-band optical filter. The detector has an inherent efficiency (expressed as its quantum efficiency) as a function of the wavelength of the incident photon. Furthermore, the filter through which the photons pass before impinging on the detector has a transmission band and transmission efficiency which means that not all photons emitted will be detected with the same efficiency. If the spectrum of the emitted light does not change during the TL measurement then a more accurate statement of Equation 5.39 would be:

$$I_{TL}(t) \propto \left| \frac{dm}{dt} \right| = K \left| \frac{dm}{dt} \right| \tag{5.40}$$

with a fixed proportionality constant, $K$. On the other hand, if the wavelength emitted during TL shifts with temperature, then the proportionality will change and the expression must then become:

$$I_{TL}(t) = K(t) \left| \frac{dm}{dt} \right| \tag{5.41}$$

or

$$I_{TL}(T) = K(T) \beta_t \left| \frac{dm}{dT} \right|. \tag{5.42}$$

Figure 5.22a demonstrates a simulated isometric plot of TL intensity-versus-temperature-versus photon energy. In this 2T2R model, two traps are present ($E_{t1} = 1.1$ eV and $E_{t2} = 1.2$ eV, with $s_1 = s_2 = 10^{12}$ s$^{-1}$) but with each exclusively using its own recombination center. The emission spectra are modeled as Gaussians, with electrons from trap 1 recombining with one recombination center producing emission at 3.5 eV, and electrons from trap 2 recombining at

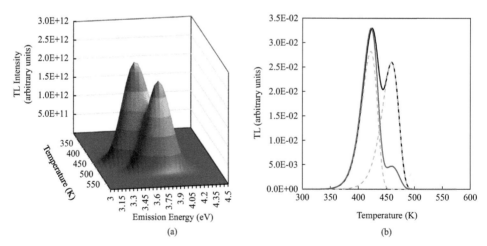

(a)                                                          (b)

**Figure 5.22**   Calculated TL glow curve for two traps, each using a different recombination center. In this case $E_{t1} = 1.1$ eV, $E_{t2} = 1.2$ eV, and $s_1 = s_2 = 10^{12}$ s$^{-1}$. (a) The expected isometric plot of TL-versus-temperature-versus photon energy for Gaussian emissions, with one recombination center peaking at 3.5 eV and the other at 3.75 eV. (b) If the glow curve was measured with a perfect detector (i.e. efficiency independent of photon energy) and no filter, the glow curve shown by the black line would result. When measured with a real detector and filter, a distorted glow curve results. (In this case the calculation was for a bi-alkali PMT with a BG-39 filter.).

the other center and producing emission at 3.75 eV. If the TL is integrated over all photon energies, the glow curve illustrated by the black line in Figure 5.22b is obtained, where the individual TL peaks are illustrated in gray dotted lines. However, if such an emission was detected using a bi-alkali photomultiplier and, say, a commonly used BG-39 broadband filter, the glow curve indicated by the orange line would actually be measured. (All curves are normalized to the same maximum in this figure for the sake of comparison). In this situation, TL due to trap 1 would be given by:

$$I_{TL_1}(T) = K_1 \beta_t \left| \frac{dm_1}{dT} \right| \tag{5.43}$$

and TL due to trap 2 would be given by:

$$I_{TL_2}(T) = K_2 \beta_t \left| \frac{dm_2}{dT} \right| \tag{5.44}$$

where each emission band is characterized by its own proportionality constant ($K_1$ and $K_2$).

Although the TL glow curve measured (red line) is different from the luminescence actually emitted (black line), analysis of the glow curve should still produce the correct values for $E_{t1}$ and $s_1$, and $E_{t2}$ and $s_2$.

A different situation is shown in Figure 5.23. Here the model is also 2T2R, but the second trap is deep and thermally disconnected, therefore there should be only one TL peak. However, there are two recombination centers, of concentrations $m_1$ and $m_2$, both of which are radiative. Again, Gaussian emissions are assumed, with recombination at $m_1$ peaking at 3.75 eV and at $m_2$ peaking at 3.5 eV. For this situation, the appropriate rate equations were solved numerically and the parameters (listed in the figure caption) were chosen such that $m_1$ is depleted before $m_2$. Since:

$$I_{TL}(t) = K_1 \left| \frac{dm_1}{dt} \right| + K_2 \left| \frac{dm_2}{dt} \right|. \tag{5.45}$$

At first sight, this may appear as the same situation as described above, but in the first example there were two independent traps, each with its own recombination center. That is, it was as if there were two independent OTOR systems acting simultaneously with two TL peaks expected, each at a different wavelength. In the second example, however, there is only one trap emptying and the two recombination centers are competing for electrons released from the trap.

Figure 5.23a shows the time dependencies of $n$, $m_1$, and $m_2$, where it can be seen that over the region where the trap empties ($n$; approximately between the vertical dashed lines), first $m_1$, then $m_2$ decays. The net result is that during the emptying of trap 1, the emission shifts from the peak at 3.5 eV to that at 3.75 eV. The overall situation is seen in the isometric plot of Figure 5.23b where not only are two different emissions observed, but the appearance is given of two separate TL peaks – for just one trap! If the TL was recorded at the monochromatic energy of 3.5 eV, the black TL curve seen in Figure 5.23c would result ($I_{TL_1}(T) = K_1 \beta_t \left| \frac{dm_1}{dT} \right|$), whereas if it was recorded at 3.75 eV, the red TL peak would be recorded ($I_{TL_2}(T) = K_2 \beta_t \left| \frac{dm_2}{dT} \right|$). In other words, the TL peak would appear in an entirely different position depending upon which emission energy is monitored. Additionally, since the detector/filter response is dependent on wavelength (photon energy), it means that the efficiency of detection varies with temperature for purely equipment-based reasons, in addition to the kinetic reasons illustrated here. The net result is that the overall TL is not analyzable by any of the equations and methods discussed in this chapter.

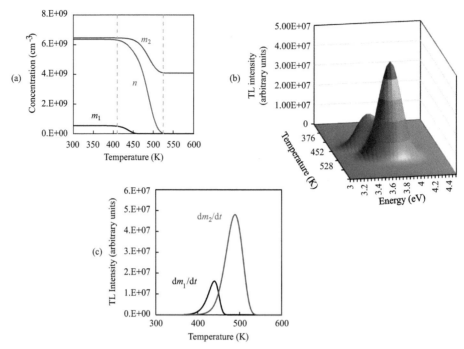

**Figure 5.23** Calculated data for a model in which one trap empties, but uses two recombination centers, of concentrations $m_1$ and $m_2$. The data were calculated using numerical solution to the rate equations, using the parameters listed below. (a) Temperature dependencies during heating of the trapped electron concentration $n$, and the trapped hole concentrations $m_1$ and $m_2$. The parameters are such that one recombination center (concentration $m_1$) empties before the second (concentration $m_2$). Both recombination centers are characterized by Gaussian emissions with the first peaking at 3.5 eV and the second at 3.75 eV. (b) Isometric plot of TL-versus-temperature-versus emission energy. (c) If measured with a monochromatic filter (at 3.75 eV) the orange TL peak would result. If measured at 3.5 eV, the black TL peak would result. The parameters used in the calculation were: $N_1 = 10^{11}$ cm$^{-3}$, $E_{t1} = 1.2$ eV, $s_1 = 10^{12}$ s$^{-1}$, $N_2 = 10^{10}$ cm$^{-3}$, $E_{t2} = 5.0$ eV, $s_3 = 10^{12}$ s$^{-1}$, $M_1 = 5 \times 10^{11}$ cm$^{-3}$, $M_2 = 10^{13}$ cm$^{-3}$, $A_1 = A_2 = 10^{-10}$ cm$^3$.s$^{-1}$, $B_1 = 10^{-8}$ cm$^3$.s$^{-1}$, $B_2 = 10^{-10}$ cm$^3$.s$^{-1}$, $A_{m1} = 10^{-9}$ cm$^3$.s$^{-1}$ and $A_{m2} = 10^{-10}$ cm$^3$.s$^{-1}$.

This example was deliberately chosen with parameters that yield two TL peaks for only one trap and, as a result, conventional analysis methods are inapplicable. Additional recombination centers with multiple emission wavelengths can also be imagined. In fact, for Figure 5.23, the value of $P$ (see Chapter 3) is >1 and the kinetics are definitely non-first-order. This is known in this case only because the selected parameters are known. In a real case, and presented with experimental data similar to those shown in Figure 5.23, two traps would be suspected and analysis of each peak would be undertaken, yielding meaningless results.

Other examples using this model are discussed more fully on the accompanying web site, (see Discussion of Figure 5.23 in Chapter 5 Exercises and Notes) and cases where only one TL peak is observed with this same model are illustrated. The important message here is that recording the emission spectrum is a vital part of the overall glow-curve characteristics. An isometric plot is preferred, although such equipment is not commonly available. (Note, it is essential when recording isometric plots to correct the intensity data for the wavelength response of the light detector and the energy-dispersive system used (usually a grating), and to correct for constant-wavelength bandwidth to constant-energy bandwidth since spectra are recorded as a function of photon wavelength, not energy (e.g. see Wang and Townsend 2013)).

**Exercise 5.12   TL emission spectra issues**

(a)  Write the rate equations describing the model shown below. (This is the same model used to obtain the data in Figure 5.23.)

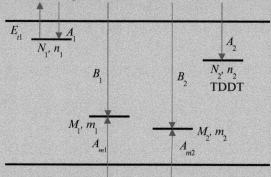

(b)  Using suitable software, solve the equations numerically using the parameters shown in the figure caption of Figure 5.23, for three different doses. (Use doses of your choice.)

(c)  Plot the overall TL, the TL due to recombination at recombination center 1 ($m_1$) only, and the TL due to recombination at center 2 ($m_2$) only. Also plot the variation of $n_c$ (electrons in the conduction band) against temperature and compare with the TL curves. Do this for each separate dose.

(d)  Explain the features that you see. Using your observations, would you conclude that this glow curve can be fitted using peak fitting methods, or not? If not, why not?

(e)  Evaluate $P$ and $Q$ for this model (refer to Chapter 3). Do the values for $P$ and $Q$ support your conclusions in part (d) or not?

### 5.2.7.4   Self-absorption

Many TL and OSL materials emit luminescence in a range that overlaps with optical absorption features that are themselves both dose- and temperature-dependent. The absorption features may or may not be directly associated with the TL or OSL signals – that is, the absorption peak may or may not be due to either the trap or the recombination center, but nevertheless the absorption is expected to increase with dose and may display thermal annealing characteristics that overlap with the temperature range of, say, a particular TL peak. The height of the TL peak increases as the dose increases, but so too does the optical absorption. Thus, for higher doses, there will be greater absorption of the TL emitted light, and the extent of the absorption will change as the temperature changes during TL readout. Since the TL signal measured is the light that reaches the detector, self-absorption of the TL light will lead to a distortion to the measured shape of the TL peak.

A simulated example is shown in Figure 5.24. Here the optical density (*OD*) due to an absorption feature is shown as a function of temperature, along with the transmittance, expressed as a fraction (rather than the usual percentage). The wavelength range of the absorption is assumed to overlap with the TL emission spectrum. Also shown in the figure is the TL peak that would be obtained without self-absorption. Clearly the emission from higher temperature regions of the TL peak has a greater probability of escaping the crystal than emission from the lower temperature region. For the *OD* shown, the measured TL peak is also shown, indicating a clear distortion on the lower temperature side of the peak compared with the no-absorption case. The apparent peak position has also changed.

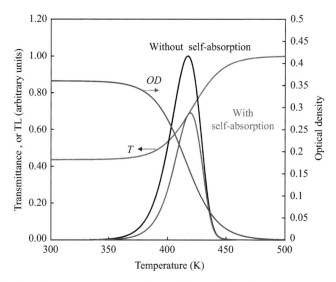

**Figure 5.24**  Simulated temperature dependence of optical density (*OD*) and transmittance (*T*) as a function of temperature for an optical absorption band that overlaps with the emission band of a TL peak, causing self-absorption of the TL light. The shape of the TL peak without absorption is shown in comparison with the peak with self-absorption, showing the distortion to the peak shape that results.

Since both the absorption and the TL peak vary with dose, but likely in different ways, the distortion will depend on dose, and since the annealing of the absorption curve and the position of the TL peak may also depend on heating rate, the effect will also be dependent on $\beta_t$.

Regarding the dependence of the measured TL peak height on dose, the net effect could be a complex dose-response behavior, with perhaps the TL signal initially increasing, then decreasing and shifting as the absorption begins to dominate. The exact behavior will depend upon the dose-response characteristics of all of the defects involved in producing the TL peak and the absorption band, the extent of the overlap between the absorption band and the TL emission band, and the thermal annealing behavior of the absorption band.

### 5.2.8  Summary of Steps to Take Using TL Curve Fitting

From the previous sections it is clear that several steps must be taken before peak fitting can be undertaken to ensure that the results of the peak fitting exercise can be accepted with a satisfactory degree of confidence. The steps should include:

1) *TL versus dose*: Do the TL peaks shift as a function of dose, or do they remain constant? If they shift, do they move in a way consistent with non-first-order kinetics, or simply first-order kinetics with peak overlap and differential growth of the individual peaks?
2) *Perform a $T_m$-$T_{stop}$ analysis*. Although laborious, this test provides essential information regarding the positions and number of the peaks within the glow curve. Are the data consistent with first-order kinetics?
3) *Measure emission spectra*: Could the emission be distorted by the optical filtration and detector system used in the measurement of the TL glow curve? Ensure that the spectra are corrected for the response of the detection system. Are there wavelength shifts with temperature? Ideally, use an isometric measurement if the required equipment is available. If possible, collect and sum data over all emitted wavelengths.

4) *Use several different heating rates.* Do the peaks shift in temperature and size in a manner consistent with first-order TL? To ensure accurate temperature readings, use small amounts of material (thin samples, or mono-layers of grains) and slow heating rates.

5) *Test for thermal quenching*: Is thermal quenching of the measured emission occurring? Test using one or more of the methods discussed in Chapter 3.

6) *Perform an Initial Rise analysis.*Using the data from the $T_m$-$T_{stop}$ experiment, perform an initial rise analysis to estimate values for $E_t$ and $s$ for each peak.

7) *Perform a heating rate analysis*: Using the glow curves measured at different heating rates, estimate $E_t$ and $s$ based on these results using one of the heating rate methods discussed above.

8) *Peak fit.* Only if the results of the above experiments are consistent with first-order glow curves, attempt a peak fit using the estimated $E_t$ and $s$ values from the Initial Rise and Heating Rate methods and using the estimated numbers of peaks within the glow curve from the $T_m$-$T_{stop}$ data. (*Note 1*: the smallest number of individual peaks that gives consistent results should be used.) Fit the curves using individual heating rates, and also fit using all heating rates simultaneously. (*Note 2*: The data to be fitted should have the background signal subtracted. The background is the data set obtained after reading the TL, with the sample still in place to ensure that the background black-body emission is the same as when the TL was measured. The TL measurement minus background is the data set to be fitted, after corrections for thermal quenching, etc.)

9) *Check the FOM against the noise in the data*: How does the FOM compare to the standard deviation of the noise in the data? Any FOM which is less than the noise should be considered to be an acceptable fit. There may be a range of $E_t$ and $s$ values that meet this criterion.

10) *Check for self-consistency*: Are all of the data consistent with first-order kinetics? Are the $E_t$ and $s$ values consistent?

Only if the results are self-consistent can there be a reasonable degree of confidence that the $E_t$ and $s$ values are correct.

### 5.2.9  Isothermal Analysis

A final analytical procedure, but one that is not often used, is to measure the time-dependent decay of the TL signal while maintaining the sample at a fixed temperature. For a single, first-order peak, Equation 3.16 can be used to give:

$$I(t) = I(0)\exp\left\{-s\exp\left\{-\frac{E_t}{kT}\right\}t\right\}, \tag{5.46}$$

where $I(0)$ is the intensity at time $t = 0$ (i.e. $I(0) = n_0 s\exp\left\{-\dfrac{E_t}{kT}\right\}$). From this it can be seen that the luminescence will decay exponentially and a plot of $\ln(I(t)/I(0))$ versus $t$ will give a straight line of slope $S = s\exp\left\{-E_t / kT\right\}$. By measuring the decay at several different temperatures a plot of $\ln(S)$ versus $1/kT$ will yield a slope of $E_t$, with an intercept of $\ln(s)$. The procedure was first described by Randall and Wilkins (1945b).

Experimentally, an isothermal decay analysis can be performed by holding the sample at a fixed temperature for a defined time after irradiation, and then recording the glow curve and noting the intensity of the subsequent TL peaks. A new irradiation of the sample is then needed for each subsequent temperature and time combination. Alternatively, the luminescence emitted during the storage period can be followed (phosphorescence, or after-glow) as a function of time. The experiment may then be repeated at different storage temperatures. The choice of temperature and time depends upon the position (temperature of the peak maximum) of the TL peak, with slower decays being observed at lower temperatures.

There are several difficulties when trying to apply these methods. Perhaps the most obvious one is that more-than-one trap can be emptying at any given temperature, with shallower traps emptying more quickly than deeper traps. The result is a curved $\ln(I(t)/I(0))$ versus $t$ plot, which in principle may be fitted to the sum of first-order exponential decays to extract information about each trap. The half-life $\tau_{1/2}$ of electrons in a trap that empties according to first-order kinetics was given in Chapter 3 as:

$$\tau_{1/2} = \ln(2)\tau_{lifetime} = \ln(2)s^{-1}\exp\left\{\frac{E_t}{kT}\right\}. \tag{3.101}$$

With two, overlapping first-order TL peaks one can expect that the half-life of electrons in the deeper trap is longer than that in the shallower trap, and in most cases this is generally true. Nevertheless, consider two traps, one with $E_t = 1.15$ eV and $s = 10^{12}$ s$^{-1}$, and the other with $E_t = 1.30$ eV and $s = 10^{14}$ s$^{-1}$. The first would yield a first-order TL peak at 411 K, and the second at 434 K. However, for isothermal decay at 378 K both would have the same half-life of ~25 minutes and would therefore be indistinguishable. Below this temperature Peak 1 would have the shorter half-life while above this temperature Peak 2 would have the shorter half-life. The change in half-life with temperature is illustrated in Figure 5.25 for this case. The figure illustrates that there can be significant difficulties with resolution between the two or more traps during isothermal decay.

Non-first-order kinetics will also yield a non-linear $\ln(I(t)/I(0))$ versus $t$ plot. Starting from Equation 3.24, the equivalent expression to Equation 5.46 can be obtained for general-order kinetics, namely:

$$\left(\frac{I(t)}{I(0)}\right)^{\frac{1-b}{b}} = 1 + n_0^{b-1}(b-1)s'\exp\left\{-\frac{E_t}{kT}\right\}t. \tag{5.47}$$

For the correct value of the kinetic-order parameter $b$, a plot of $\ln\left(\frac{I(t)}{I(0)}\right)^{\frac{1-b}{b}}$ versus $t$ will be

exponential, with a slope of $S = n_0^{b-1}(b-1)s'\exp\{-E_t / kT\}$, from which it can be observed

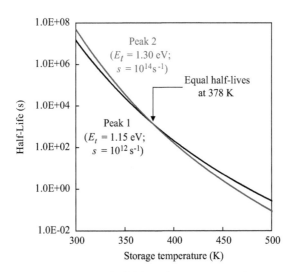

**Figure 5.25** Illustration of the change in half-life with temperature, for two $E_t/s$ combinations. For the $E_t/s$ values chosen, the half-lives are equal at 378 K, whereas at a heating rate of 1 K.s$^{-1}$, the corresponding TL peaks would appear at 411 K and 434 K.

that the slope $S$ will change with the initial level of trap filling $n_0$. This does not occur with first-order kinetics and is another test of the order of kinetics of the system under study.

In many cases the decay is found to follow neither Equation 5.46 nor Equation 5.47, but rather displays a $t^{-k}$ decay, with $k$ close to a value of 1. Randall and Wilkins (1945b) showed that in the case of a uniform distribution of traps of concentration $N_E$ between $E_t$ and $E_t + dE_t$, then the luminescence decay becomes:

$$I(t) = \int_0^\infty N_E \exp\left\{-\frac{E_t}{kT}\right\} \exp\left\{-st\exp\left\{-\frac{E_t}{kT}\right\}\right\} dE_t. \tag{5.48}$$

After integration and assuming $\exp\{-st\} \ll 1$:

$$I(t) = \frac{n_0 kT}{t}, \tag{5.49}$$

from which the $t^{-1}$ law is obtained.

It is important to recognize, however, that the decay described by Equations 5.48 and 5.49 is strongly temperature-dependent. In many systems, a temperature-independent $t^{-1}$ decay is observed. This situation has already been discussed in depth in Chapter 3 where it was shown to be the result of quantum mechanical tunneling between the electron trap ground state and the recombination center. The detailed equations are described in Chapter 3, Section 3.8.1 and are not repeated here. Such temperature-independent decay is described as "anomalous fading," as distinct from the temperature-dependent "thermal fading."

A final complexity with interpreting isothermal decay curves is the possibility of the traps and centers themselves being unstable and undergoing defect reactions during the decay period. Thus, the electrons may be stable at the temperature at which the sample is held, but the trap may undergo reactions with other defects such that the luminescence emitted during isothermal decay and the subsequent TL signal measured after storage may be considerably changed. An example system important for dosimetry is LiF:Mg,Ti (TLD-100) in which clustering reactions of Mg-defects, including precipitation of Mg, can occur at certain temperatures and over time scales that can interfere with the isothermal observations. This situation is summarized by McKeever (1985) and analyzed by Bos and Piters (1993) where it is illustrated that false $E_t$ values can emerge due to clustering and precipitation reactions occurring during storage at specific temperatures. Such reactions may be unknown or unsuspected, leading to false conclusions. This again emphasizes the need to apply multiple analytical methods for $E_t$ and $s$ analysis before conclusions can be reached.

---

### Exercise 5.13   Isothermal analysis

The purpose of this exercise is to give you a feel for the value of storage temperature and time on the stability of TL peaks, for given values of $E_t$ and $s$. Part A asks you to vary the storage temperature and time, and the $E_t$ and $s$ combinations and requires you to observe how the subsequent glow curve changes shape. Part B asks for some calculations of the half-lives at different temperatures.

Part A
Using a 2T2R model (see Exercise 5.12) select two pairs of values for $E_t$ and $s$ ($E_{t1}/s_1$, $E_{t2}/s_2$) and solve the numerical equations governing trap filling, storage at temperature $T_{store}$ for times $t_{store}$, and subsequent TL after the storage period. You choose the values for the trap concentrations but choose values that represent first-order kinetics. Also choose values for the irradiation rate (electron-hole generation rate), and select $E_{t1}/s_1$,

$E_{t2}/s_2$, $T_{store}$, and $t_{store}$ values such that the subsequent TL peaks fade thermally over a time period from minutes to hours.

(a) Plot the TL glow curves for different $T_{store}$ and $t_{store}$ values.

Part B

(b) For your selected values of $E_{t1}/s_1$ and $E_{t2}/s_2$, estimate the appropriate half-lives using Equation 3.101.

(c) Plot the TL peak heights against $t_{store}$ at different values of $T_{store}$.

(d) Are there any of your chosen $T_{store}$ values for which the half-lives of the two separate traps are the same? If not, change the $E_t/s$ pairs so that the half-lives are the same for one of your selected $T_{store}$ values and repeat the calculations.

## 5.3    Analytical Methods for OSL

The goal of analyzing TL glow curves is to identify the number of different traps contributing to the TL glow curve and to estimate values for the parameters $E_t$ and $s$ for each trap. The $E_t$ and $s$ parameters govern the rate of thermal detrapping, $p = s\exp\{-E_t/kT\}$, at a given temperature. Similarly, the purpose of analyzing OSL curves is to arrive at the number of different traps contributing to the OSL signal, and to evaluate those parameters that govern the rate of optical detrapping for each trap, $p = \sigma_p\Phi$, for a given wavelength and stimulation power. The stimulation intensity $\Phi$ is an experimental parameter dependent upon the source of optical excitation used in the experiment. It can be measured using a calibrated power meter, usually giving a result in units of W.cm$^{-2}$. If the stimulation source is monochromatic (laser) or near-monochromatic (an LED) the energy (usually in eV) carried by each photon can be estimated from its known peak wavelength, in nm ($E = h\nu / \lambda$) and the intensity $\Phi$, i.e. number of photons impinging on the sample per cm$^2$ per second, can then be calculated (in units of number of photons.cm$^{-2}$.s$^{-1}$). Thus, if $p$ is determined from the OSL experiment, $\sigma_p$ can then be calculated.

### 5.3.1    Curve Shape Methods

#### 5.3.1.1    CW-OSL

For first-order kinetics a CW-OSL curve follows Equation 3.14, re-written here as:

$$I_{CW-OSL}(t) = n_0\sigma_p\Phi\exp\{-t\sigma_p\Phi\}. \tag{3.14}$$

Here, $t$ is the stimulation time, and the intensity of the OSL at $t = 0$ is $I_{CW-OSL}(0) = n_0\sigma_p\Phi$.

Although equivalent curves for non-first-order CW-OSL can be written mathematically, the discussions concerning TL glow curves indicated that first-order detrapping kinetics dominate in nature. This is because there are always deeper traps and large numbers of trapped hole sites, making detrapping first-order. Also quasi-equilibrium (a critical assumption in the development of the TL and OSL equations) is most consistent with first-order kinetics. Although mathematically QE and non-first-order are possible, logical arguments favor first-order kinetics. (See accompanying web site, Exercises and Notes, Chapter 3, Discussion of Section 3.6.2.) Finally, similar to TL, a description of an experimental OSL curve as the sum of individual OSL curves requires the principle of superposition, which only applies in the case of first-order kinetics. Therefore, from this point on in the description of OSL analysis, only first-order expressions are used.

Using Equation 3.14, Huntley et al. (1996) calculated the ratio of the rate at which the CW-OSL signal decreases, $S_{CW-OSL}$, to the intensity of the CW-OSL at a given stimulation time:

$$\frac{dI_{CW-OSL}}{dt} = S_{CW-OSL} = -\sigma_p \Phi I_{CW-OSL}, \tag{5.50}$$

or,

$$\sigma_p = -\frac{1}{\Phi}\left(\frac{S_{CW-OSL}}{I_{CW-OSL}}\right) \tag{5.51}$$

and the value of $\sigma_p$ can be evaluated from the shape of the CW-OSL decay curve.

An alternative strategy is to employ a very low stimulation rate (weak $\Phi$) for which the change in the trapped electron concentration $\Delta n$ is very small such that $\Delta n \ll n$ and therefore $n$ and $m$ (the concentration of trapped holes in the recombination center) are approximately constant. Under quasi-equilibrium $np = n\sigma_p \Phi = dn/dt = dm/dt$, which gives:

$$\sigma_p = \frac{n_c B m}{\Phi} = n_c\left(\frac{K}{\Phi}\right) \tag{5.52}$$

where $B$ is the recombination probability and $n_c$ is the concentration of free carriers. Since $m$ is approximately constant under weak stimulation, $Bm = K$ is a constant and the photoionization cross-section is proportional to the free-carrier concentration. The latter can be monitored using photoconductivity, but OSL can also be used since $I_{CW-OSL} = dm/dt = n_c Bm$, and so $\sigma_p \approx I_{CW-OSL}/\Phi$, for weak stimulation, i.e. very short stimulation times/low stimulation powers.

### 5.3.1.2   LM-OSL

Kitis and Pagonis (2008) followed the methodology of Chen and developed a series of shape parameters for LM-OSL curves that mimicked the shape parameters for TL peaks, developed by Chen (1969a, 1969b). Defining the times $t_1$, $t_{max}$, and $t_2$ corresponding to the points at which the LM-OSL curve reaches its half-maximum ($I_{max}/2$) at low-$t$, maximum ($I_{max}$), and half-maximum ($I_{max}/2$) at high-$t$, respectively, these authors defined the parameters $\tau = t_{max} - t_1$, $\delta = t_2 - t_{max}$, and $\omega = t_2 - t_1$. Although the values of these terms can be mathematically defined for non-first-order kinetics, the values for first-order kinetics only are displayed here, in Table 5.3, where they are each normalized by the value for $t_{max}$.

However, the normalized values for the parameters listed in Table 5.3 are independent of both $\beta_\Phi$ and, importantly, $\sigma_p$, and therefore not useful for analysis purposes.

**Table 5.3**   Values for $\omega/t_{max}$, $\delta/t_{max}$, and $\tau/t_{max}$ for first-order LM-OSL curves. Data from Kitis and Pagonis (2008)

| $\omega/t_{max}$ | $\delta/t_{max}$ | $\tau/t_{max}$ |
|---|---|---|
| $1.6023 \pm(1 \times 10^{-4})$ | $0.9214 \pm(1 \times 10^{-4})$ | $0.6809 \pm(5 \times 10^{-5})$ |

### 5.3.2   Variable Stimulation Rate Methods: LM-OSL

Although the normalized values for the parameters listed in Table 5.3 are independent of both $\beta_\Phi$ and, importantly, $\sigma_p$, the non-normalized values, $\omega$, $\delta$, and $\tau$ are related to $\sigma_p$ by a power law.

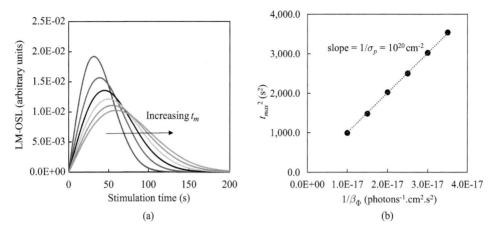

**Figure 5.26** (a) Variation in the shape of an LM-OSL curve as a function of $\beta_\Phi = \Phi_m / t_m$. (b) Plot of $t_{max}^2$ against $1/\beta_\Phi$, with a slope of $\sigma_p^{-1}$. The values used in the simulation were: $\Phi_m = 10^{18}$ photons.cm$^{-2}$.s$^{-1}$, $\sigma_p = 10^{-20}$ cm$^2$, and $t_m$ varying from 10 s to 35 s.

This can be seen from Exercise 3.5 in Chapter 3, where the expression for the time at which an LM-OSL peak reaches its maximum for a given $\beta_\Phi$ was given as:

$$t_{max} = \sqrt{\frac{t_m}{\sigma_p(E)\Phi_m}} = \sqrt{\frac{1}{\sigma_p\beta_\Phi}}. \tag{5.53}$$

In Figure 5.26a several LM-OSL curves are simulated for several ramp times, as noted in the figure caption. From Equation 5.53 it is seen that a plot of $t_{max}^2$ against $1/\beta_\Phi$ will produce a straight line of slope $1/\sigma_p$ (Figure 5.26b). Thus, by varying $\beta_\Phi$ and noting the value of $t_{max}$, the value for $\sigma_p$ can be obtained.

From the form of Equation 5.53, an empirical equation can be written

$$t_x = K_x\sqrt{\frac{1}{\sigma_p}}, \tag{5.54}$$

where $t_x$ is an arbitrary time during the LM-OSL stimulation and $K_x$ has the same dimensions as $\sqrt{1/\beta_\Phi}$. Choosing $t_x = t_1$ gives:

$$\tau = t_{max} - t_1 = \sqrt{\frac{1}{\sigma_p}}\left(\sqrt{\frac{1}{\beta_\Phi}} - K_1\right) = K\sqrt{\frac{1}{\sigma_p}}. \tag{5.55}$$

Equally, similar expressions for $\delta$ and $\omega$ can be written. Whereas Equation 5.53 shows that $t_{max}$ changes with $\sigma_p$ for a fixed $\beta_\Phi$, Equation 5.55 emphasizes that the width $\tau$ (or $\delta$ or $\omega$) also varies, with the peak getting narrower as $\sigma_p$ increases.

Example data are shown in Figure 5.27a where the shift and narrowing of the LM-OSL peak is clearly seen as $\sigma_p$ increases, for a fixed value of $\beta_\Phi$. Figure 5.27b illustrates the behavior of $\omega = t_2 - t_1$ as a function of $\sigma_p$, as an example, displaying the predicted $\sigma_p^{-0.5}$ relationship.

The interplay between $\sigma_p$ and $\beta_\Phi$ upon the determination of $t_{max}$ is summarized in the isometric plot of $t_{max}$-versus-$\sigma_p$-versus-$t_m$, shown in Figure 5.28. For known values of $\beta_\Phi = \Phi_m / t_m$, $t_{max}$ can be measured and $\sigma_p$ determined.

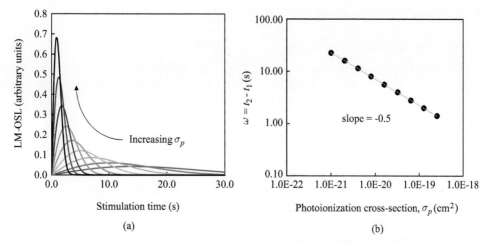

(a)                                                                    (b)

**Figure 5.27**    (a) Variation in the shape of an LM-OSL curve as a function of $\sigma_p$. (b) Dependence on the LM-OSL peak width defined by $\omega = t_2 - t_1$, as a function of the photoionization cross-section. The parameters used in the simulation were: $\Phi_m = 10^{20}$ photons.cm$^{-2}$.s$^{-1}$, $t_m = 10$ s, and $\sigma_p$ varying from $2.00 \times 10^{-21}$ cm$^2$ to $2.56 \times 10^{-19}$ cm$^2$.

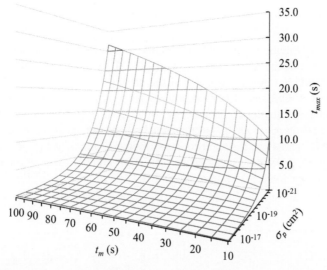

**Figure 5.28**    Variation in $t_{max}$ as functions of $\sigma_p$ and $t_m$, for values of $\sigma_p$ values ranging from $10^{-21}$ to $10^{-17}$ cm$^2$, and for $t_m$ values from 10 s to 100 s, for $\Phi_m = 10^{20}$ photons.cm$^{-2}$.s$^{-1}$.

---

### Exercise 5.14    Determination of $\sigma_p$ from single CW-OSL or LM-OSL curves

Select a fixed value of $\Phi$ for CW-OSL, or $\Phi_m$ for LM-OSL, and generate a series of CW-OSL and LM-OSL curves for a range of $\sigma_p$ and a range of $t_m$, where $\beta_\Phi = \Phi_m / t_m$. Examine how the CW-OSL curve shape, and the LM-OSL peak shape (width and position) vary with $\sigma_p$ and a range of $t_m$.

From the curves generated, apply the methods described in the text to determine $\sigma_p$ from the CW-OSL and the LM-OSL data. Do the values obtained match the values you used to generate the curves?

### 5.3.3   Curve-Fitting Methods

#### 5.3.3.1   The Curve Overlap Problem

For CW-OSL signals due to one trap emptying it is a trivial procedure to fit to the single exponential decay described by Equation 3.14 and, knowing $\Phi$, determine $\sigma_p$ from $p = \sigma_p \Phi$. Difficulties arise when there are several traps contributing to the OSL signal such that:

$$I_{CW-OSL}(t) = \sum_{i=1}^{k} n_{0i} \sigma_{pi} \Phi \exp\{-t\sigma_{pi}\Phi\}, \tag{5.56}$$

where $k$ is the total number of separate traps contributing to the overall OSL signal. An important feature of Equation 5.56 is that all traps contribute to the CW-OSL signal at the same time, beginning at $t = 0$. Thus, whereas in TL each trap can be considered to empty somewhat sequentially, from the shallowest to the deepest, subject to the $E_t$ and $s$ values, in CW-OSL all traps empty at once at rates dependent upon the values of their photoionization cross-sections $\sigma_p$, but each starting from $t = 0$. Deconvolution of the separate signals becomes more problematic as a result.

The same problem also applies to LM-OSL when multiple traps are emptying for which the net LM-OSL signal is:

$$I_{LM-OSL}(t) = \sum_{i=1}^{k} n_{0i} \sigma_{pi} \frac{\Phi_m}{t_m} t \exp\left\{-\sigma_{pi} \frac{\Phi_m}{t_m} t^2 / 2t_m\right\}. \tag{5.57}$$

Again, all traps contributing to the LM-OSL signal do so from $t = 0$ and empty simultaneously. This is emphasized by the realization that the LM-OSL curve is just a mathematical transformation of the CW-OSL curve (as was demonstrated in Exercise 3.5).

Examples of CW-OSL and LM-OSL curves for $k = 3$ traps, emptying under first-order kinetics according to Equations 5.56 and 5.57, are given in Figure 5.29. The parameters used in these simulated curves are given in the figure caption and the linear-modulated intensities used for the LM-OSL curves are assumed to reach their maximum values $\Phi_m$ after $t_m = 100$ s. Furthermore, the $\Phi_m$ values are such that the maximum value of $p = \sigma_p \Phi$ (denoted $p_m$ in Figure 5.29b) equals the fixed value for $p$ used in the CW-OSL simulations of Figure 5.29a.

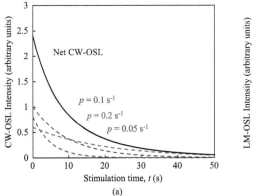
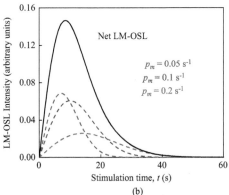

**Figure 5.29**   (a) CW-OSL and (b) LM-OSL curves for three traps exhibiting first-order kinetics. Each signal starts from the beginning of the stimulation process. The parameters used are shown in the figure, plus: $n_{01} = 1.0$ concentration units, $n_{02} = 0.8$ concentration units, $n_{03} = 0.6$ concentration units, and for LM-OSL $t_m = 10$ s.

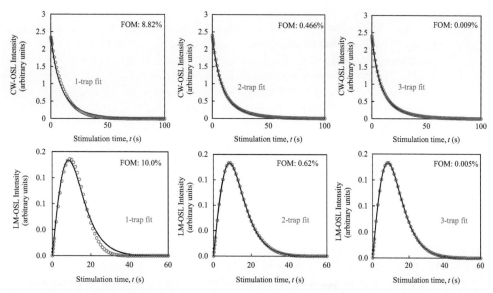

**Figure 5.30** Top row: fits to the net CW-OSL from Figure 5.29, using the assumptions of one trap, two traps, or three traps, along with the respective FOM values. Bottom row, same, but for LM-OSL.

The unavoidable overlap of OSL signals (CW-OSL or LM-OSL) means that curve fitting is almost exclusively used in analysis of OSL data, and resolution of the signals is moot in most cases. Occasionally the differences in $n_0$ and $\sigma_p$ for the traps are such that distinct LM-OSL peaks can sometimes be observed. Nevertheless, since all LM-OSL curves start a $t = 0$, and since small values of $\sigma_p$ produce stretched-out, low, broad peaks at longer times, while high $\sigma_p$ values produce narrower, sharper peaks at short times, significant resolution between processes is still difficult.

Under such conditions fitting experimental data in a definitive fashion is problematic. This is illustrated in Figure 5.30 where the data of Figure 5.29 are fitted assuming one, two or three traps. (Of course, Figure 5.29 shows simulated, noise-free data and it is known that there are three traps emptying. Such knowledge and a smooth data set is not possible experimentally.) On the top row of plots in Figure 5.30 are shown the corresponding 1-trap, 2-trap, and 3-trap fits, to the CW-OSL curves, along with the respective FOM values for the fits. On the bottom row are shown the similar data for the LM-OSL curves.

Clearly, fitting either the CW-OSL or the LM-OSL curve with the assumption of just one trap emptying gives unacceptable fits. Adding a trap, however, yields fits that are difficult to reject. In fact, there is very little difference between the 2-trap fit and the 3-trap fit, even though the data sets were simulated assuming three traps. For this latter reason the 3-trap case gives the much better FOM (and the right values for $n_0$ and $\sigma_p$) but for experimental data this is not known a priori, and there will be noise in the experimental data, such that if the FOM $< \sigma_{noise}$, distinction between a two-trap fit and a three-trap fit may be impossible. Of course, one could continue adding traps and getting better and better fits, but as long as FOM $< \sigma_{noise}$, all fits must be deemed statistically acceptable.

---

### Exercise 5.15    CW-OSL and LM-OSL curve fitting

Simulate the simultaneous optical detrapping of four traps, each with their own values for $n_0$ and $p$. Calculate CW-OSL and LM-OSL curves. For the latter make the maximum value of $p = \sigma_p \Phi$ (i.e. $p_m = \sigma_p \Phi_m$) equal the values you chose for $p$ during CW-OSL for each trap. (Note: The rate of increase in stimulation intensity is $\beta_\Phi = \Phi_m / t_m$ is the same for each trap, but $p_m/t_m$ differs in proportion to the value of $\sigma_p$ for each trap.)

Use a suitable peak fitting routine to fit the CW-OSL and LM-OSL curve so obtained with 1, 2, 3, and 4 traps. Observe the changes in the FOM for each fit.

Repeat the exercise using different initial values for $n_0$ and $\sigma_p$ for the four traps.

What do you conclude regarding the efficacy of peak fitting CW-OSL and LM-OSL curves without prior knowledge of how many traps contribute to the OSL curves?

(A four-trap model was used by Kitis and Pagonis (2008). Compare your conclusions with those of these authors.)

### 5.3.3.2 Simultaneous Fitting of LM-OSL Peaks Generated by Varying the Stimulation Rate

Figures 5.26–5.29 demonstrate the interplay between $\sigma_p$ and $\beta_\Phi$ in determining the position of the LM-OSL peak $t_{max}$. For a given value of $\beta_\Phi$ there is only one value of $\sigma_p$ that will produce the peak in the correct position. As a result, an advantage may be gained by simultaneously fitting multiple LM-OSL curves, each obtained at a known but different stimulation rate (by choosing different $t_m$ values and keeping $\Phi_m$ fixed). Only one value of $\sigma_p$ will give the LM-OSL peak in the right positions. An example is shown in Figure 5.31 where two-traps were used to simulate the LM-OSL curve and four different values of $\beta_\Phi$ were used to simulate the results. The results were then added together to form a "pseudo-LM-OSL" curve (not to be confused to the "pseudo-LM-OSL" curve referred to by Bulur (2000)). The pseudo-LM-OSL curve was then fitted using just two values of $\sigma_p$, but with four different values of $\beta_\Phi$. The theoretical best FOM (using Solver software) was found to be $7.1 \times 10^{-15}\%$, and by initiating the calculation using different sets of starting values, the program converged quickly to produce values of FOM $< 10^{-3}\%$ and with $\sigma_p$ and $n_0$ values around $\pm 10\%$ of the input values. This strategy may produce an increase in confidence for the values obtained, but the basic problem of an unknown number of overlapping components is ever-present.

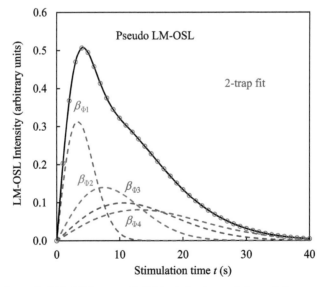

**Figure 5.31** A "pseudo-LM-OSL" curve (full black line) consisting of the sum of four LM-OSL curves recorded at different $\beta_\Phi$ values (dotted lines), and fitted using two traps (with two values for $\sigma_p$ and $n_0$).

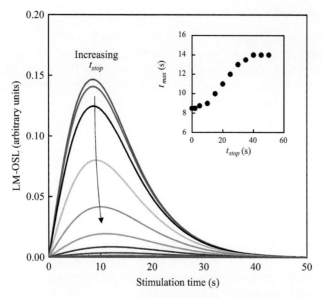

**Figure 5.32** Variation in the LM-OSL curve after stopping the stimulation at certain times ($t_{stop}$) and monitoring the position of the first peak in the LM-OSL curve ($t_{max}$). A $t_{max}$-$t_{stop}$ plot is shown in the inset.

### 5.3.4 How Can the Number of Traps Contributing to OSL Be Determined?

What is missing in OSL peak fitting is a way of determining beforehand how many traps are emptying at the stimulation wavelength chosen. Two possible approaches are now described.

#### 5.3.4.1 $t_{max}$-$t_{stop}$ Analysis

An equivalent to the $T_m$-$T_{stop}$ procedure for TL can be imagined. For the LM-OSL procedure the stimulation could be stopped at time $t_{stop}$, and the complete, remaining LM-OSL then measured and the position of the first maximum $t_{max}$, noted. By re-irradiating the specimen and repeating the exercise, but for a different $t_{stop}$, the variation in the $t_{max}$ with $t_{stop}$ can be plotted. This is the logical equivalent of the $T_m$-$T_{stop}$ procedure in TL and was demonstrated by Kitis and Pagonis (2008) for a complex model devised by Bailey (2001) to describe OSL from quartz. An example is shown here in Figure 5.32 for the same data set as used in Figure 5.29. As the LM-OSL intensity decreases, the position of the first peak shifts to longer time as the traps with the largest $\sigma_p$ preferentially empty. However, the same problem – namely the simultaneous emptying of all traps – leads not to a stair-step $t_{max}$-$t_{stop}$ plot, but a gradually increasing curve as all traps empty at the same time. Different concentrations of the various traps and different $\sigma_p$ values may yield different $t_{max}$-$t_{stop}$ plots, but the basic problem remains.

---

### Exercise 5.16 $t_{max}$-$t_{stop}$ plots

Using similar parameters to those you chose in Exercise 5.15, simulate a $t_{max}$-$t_{stop}$ experiment and plot $t_{max}$ against $t_{stop}$. Are there values of $n_0$ and $\sigma_p$ for the different traps that produce a stairstep $t_{max}$-$t_{stop}$ structure? What do you conclude about the efficacy of a $t_{max}$-$t_{stop}$ experiment?

Try using the Bailey (2001) model and repeating the exercise. Can you replicate the results of Kitis and Pagonis (2008)?

### 5.3.4.2    Comparison with TL

It is not always guaranteed that a trap active in TL production is also active in OSL production, and vice-versa, but many times (perhaps even the majority of times) it is found that stimulating with light to produce OSL also has the effect of bleaching TL. This was described comprehensively in Section 3.7.4, which dealt with potential optical effects on the shape of TL glow curves. In principle, this may provide a clue as to how many traps contribute to the OSL signal. For example, if stimulating an irradiated sample during OSL, at a particular wavelength and for a particular time, also causes a reduction in several different TL peaks, then it might be inferred that the traps governing those TL peaks are the ones that contribute to the OSL signal. This is especially so if the emission wavelengths observed in OSL and TL are the same. This is the simplest explanation, and in most cases it is perhaps the right one, but, as always, there are caveats with this straightforward interpretation. For example, the OSL measurement might reduce the recombination center concentration such that the TL signal is reduced accordingly but the "TL traps" themselves might not actually empty during the optical stimulation phase. Such conditions require a sophisticated multiple-trap/multiple-recombination-site model and, as was described in Section 3.7.4, such models are certainly possible. Rather than repeating here those calculations and discussions concerning all the relevant possibilities, the reader is advised to refer back to this section in Chapter 3.

If it is observed that the emission spectra for TL and OSL are the same, and if it is observed that only some of the TL peaks are reduced in height and not all, then it can be reasonably concluded that the traps responsible for those TL peaks are also responsible for the OSL signal. In such a case an immediate estimate of the number of traps contributing to the signal can be made.

In like manner, it is also possible to monitor the changes in the OSL signal during post-irradiation, isothermal storage at different temperatures, and to compare these with the corresponding changes to the TL glow curve under similar conditions. Section 5.2.9 described the kinetics of isothermal analysis, and exactly the same procedures can be applied to OSL. Direct comparison with the changes in the TL glow curve can indicate which TL peaks, and therefore how many traps, are affected during isothermal storage, from which the number of traps contributing to the OSL signal may be estimated.

Comparisons such as these are essential if an analysis of OSL is to produce results that are supported by other data and can be treated with a degree of confidence.

### 5.3.5    Variation with Stimulation Wavelength

Figure 3.18 in Chapter 3 illustrated how changing the stimulation wavelength (energy) can change not just the probability of detrapping, but the relative detrapping probabilities of the different traps, such that $\sigma_{p1} > \sigma_{p2}$ (and therefore $p_1 > p_2$) at one wavelength, while $\sigma_{p2} > \sigma_{p1}$ (and therefore $p_2 > p_1$) at another. It is instructive to examine the changes to the CW-OSL and LM-OSL curves that can result for such changes in stimulation wavelength. For clarity, just two traps (traps 1 and 2) are considered, with $p_1 > p_2$ at $\lambda_1$, and $p_2 > p_1$ at $\lambda_2$ (where $\lambda_1 > \lambda_2$). Figure 5.33 illustrates the changes expected for this case, when $n_{02}$ is assumed to be just 75% of $n_{01}$. Using Figure 3.18 and selecting stimulation energies of 1.6 eV (774.9 nm) and 3.0 eV (413.3 nm), and assuming the stimulation intensity $\Phi$ to be the same at both wavelengths,[6] the CW-OSL and LM-OSL curves of Figure 5.33 are obtained.

Unfortunately, the data of Figure 5.33 do not help much with curve fitting. The problem is exactly the same as that already discussed in each data set – namely, curve overlap starting

---

[6]This is unlikely in a practical experiment. The intensities of, say, light emitting diodes or laser sources at these two wavelengths are likely to be different, and this too would have to be taken into account to interpret the differences seen in the OSL curve shapes.

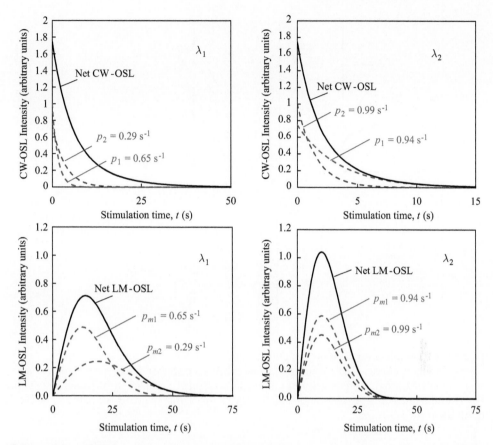

**Figure 5.33** CW-OSL curves (top row) and LM-OSL curves (bottom row) for two different stimulation wavelengths, $\lambda_1$ and $\lambda_2$. The values of $\lambda_1$ and $\lambda_2$ are chosen from Figure 3.18 such that $\sigma_{p1} > \sigma_{p2}$ (and therefore $p_1 > p_2$) at one wavelength, while $\sigma_{p2} > \sigma_{p1}$ (and therefore $p_2 > p_1$) at the other.

from $t = 0$, and there is no advantage to be gained from fitting the curves simultaneously. By changing the wavelength, the detrapping parameter $\sigma_p$ is changed and fitting the curves simultaneously is now an exercise in fitting four (instead of two) $\sigma_p$ and $n_0$ pairs of values to the data set. All that has been done is to increase the number of fitting parameters to six, rather than four, to describe the data (two $n_0$ values and four $\sigma_p$ values). No major advantage is gained.

### 5.3.6   Trap Distributions

Equations 5.56 and 5.57 describe the OSL as a sum of several detrapping processes from a set of discrete traps. The extension of this is to assume a continuous distribution of trapping states and write:

$$I_{OSL}(t) = \int_0^\infty g(E_o) I'_{OSL}(t) \, dE_0, \tag{5.58}$$

where $I'_{OSL}(t)$ is either the CW-OSL or the LM-OSL expression for a single value of the optical trap-depth $E_o$, and $g(E_o)$ is the distribution of optical trap depths. At a given stimulation

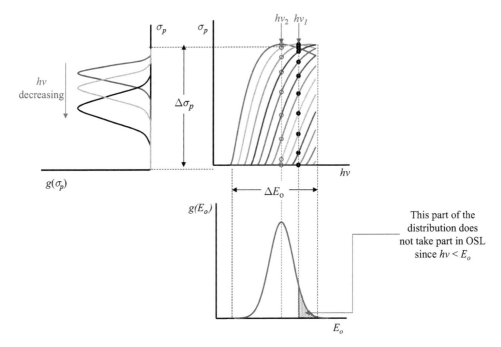

**Figure 5.34**  In the lower right of the figure a distribution in $E_o$ is shown, over the range $\Delta E_o$. The corresponding $\sigma_p$-versus-$E_o$ curves (using Equation 2.16) are shown in the upper right. For a fixed stimulation energy $h\nu_1$, there is a corresponding distribution of $\sigma_p$ values, $g(\sigma_p)$, shown in the upper left, over the range $\Delta\sigma_p$. As the stimulation energy changes, the range $\Delta\sigma_p$ stays the same but the distribution $g(\sigma_p)$ changes.

wavelength (energy), the optical trap depth defines the value for the photoionization cross-section $\sigma_p$, and therefore a distribution in $E_o$ is also a distribution in $\sigma_p$. The distribution in $\sigma_p$, however, will depend upon the choice of stimulation wavelength $\lambda$ (or stimulation energy $h\nu$). This is illustrated in Figure 5.34. In the lower-right of this figure an arbitrary distribution of traps $g(E_o)$ is indicated in terms of the available optical trap depths $E_o$. For each value of $E_o$, the photoionization cross-section will vary as a function of stimulation energy $h\nu$, as shown in the upper-right figure, where each curve corresponds to a different value of $E_o$. The range of available $E_o$ values is indicated by $\Delta E_o$. (The $\sigma_p$-versus-$h\nu$ curves in this figure are calculated using Equation 2.16.)

For a stimulation energy $h\nu_1$, only that sub-set of traps with $E_o < h\nu_1$ will participate in OSL. Thus, the part of the distribution that is shown by the red shaded area does not participate in production of the OSL signal. The range of $\sigma_p$ values ($\Delta\sigma_p$) at a stimulation energy $h\nu_1$ is indicated by the black dots, and the corresponding distribution $g(\sigma_p)$ of $\sigma_p$ values is indicated in the upper-left figure.

If the stimulation energy is reduced to $h\nu_2$, the range $\Delta\sigma_p$ remains the same, but the distribution within that range changes. The distribution shifts to lower values of $\sigma_p$ and broadens, as indicated in the figure. It is clear then, that the $\sigma_p$ distribution for the particular OSL signal being measured depends heavily on the choice of stimulation wavelength. (Note, however, that the original $E_o$ distribution does not change and is independent of the choice of stimulation wavelength, or energy $h\nu$.)

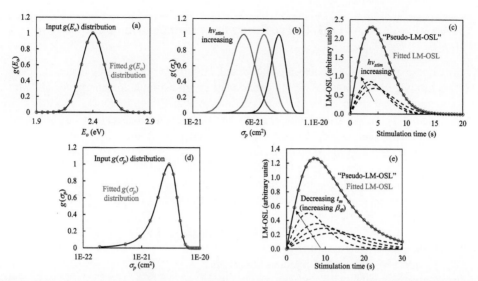

**Figure 5.35** (a) Assumed Gaussian distribution $g(E_o)$ of optical trap depths. (b) Corresponding distributions $g(\sigma_p)$ for three different stimulation energies. (c) Calculated LM-OSL curves at different stimulation energies ($hv_{stim}$) and their sum ("pseudo-LM-OSL"). The fits to the LM-OSL curves and to the $g(E_o)$ distribution are shown by red circles when a Gaussian distribution is assumed for the fitting. (d) Alternatively, a single stimulation energy may be assumed, giving rise to a single $g(\sigma_p)$ distribution. (e) The corresponding LM-OSL curves calculated for four different values of $\beta_\Phi$, and their sum to give a "pseudo-LM-OSL" curve. The $g(\sigma_p)$ distribution necessary to fit the LM-OSL data is shown by red circles in Figure (d), when a Gaussian distribution is assumed.

In the example of LM-OSL, the time at which the peak position occurs is dependent upon both $\beta_\Phi$ and $\sigma_p$. This suggests two approaches that might be adopted to determine the distribution of traps that contribute to the OSL signal – namely, a fixed $\beta_\Phi$ and a selection of several different stimulation wavelengths (and therefore different $\sigma_p$ distributions), or a fixed stimulation wavelength (and therefore a fixed $\sigma_p$ distribution) and variable $\beta_\Phi$. Figure 5.35 illustrates the two approaches. In the top row a distribution in $E_o$, $g(E_o)$, is first assumed (Figure 5.35a, black line); in this example a Gaussian distribution centered at $E_o = 2.4$ eV is used. Three different stimulation wavelengths are assumed ($hv_{stim} = 3.0$ eV, 3.25 eV, and 3.5 eV) producing the corresponding distributions in $\sigma_p$ shown in Figure 3.54b, and the corresponding LM-OSL curves are calculated (Figure 5.35c, dashed lines). By adding the three LM-OSL curves together to get a "pseudo-LM-OSL" curve, the pseudo curve may be fitted and that distribution in $E_o$ that produces the distributions in $\sigma_p$ and fits the LM-OSL curves is obtained. By assuming that the required distribution is a Gaussian, an excellent fit, with FOM values near the idealized number may be obtained. The obtained fitted $g(E_o)$ distribution is shown by the red circles in Figure 5.35a. However, if an arbitrary distribution is assumed, good fits can still be obtained to the LM-OSL data (FOM values < 0.1%), but the obtained distributions only vaguely approximate the actual $g(E_o)$ distribution.

In the second approach, a fixed stimulation wavelength is assumed and therefore only one $g(\sigma_p)$ distribution is used (Figure 5.35d). In this example, four values of $\beta_\Phi$ are used, and a "pseudo-LM-OSL" curve obtained (Figure 5.35e). This is then fitted by assuming a Gaussian distribution in $\sigma_p$, and the fitting converges to produce the correct $g(\sigma_p)$ distribution (red circles in Figure 5.35d). Again, however, if an arbitrary distribution is assumed, multiple acceptable fits can be obtained, making final conclusions difficult.

---

**Exercise 5.17    LM-OSL Curve fitting for a distribution of optical trap depths**

(a) Choose a distribution function $g(E_o)$ for a distribution of optical trap depths. Choose a formula from Chapter 2 for the variation of $\sigma_p$ with stimulation energy $h\nu$, and plot several $\sigma_p$-versus-$h\nu$ curves for your chosen range of $E_o$.

(b) Select three or four stimulation energies and calculate the distributions in photoionization cross-section $g(\sigma_p)$ for each stimulation energy chosen.

(c) Construct several LM-OSL curves for each of the three or four $g(\sigma_p)$ distributions.

(d) Add them together to form a "pseudo-LM-OSL" curve.

(e) Devise another guessed optical energy distribution $h(E_o)$, different from the original $g(E_o)$, and calculate new cross-section distributions $h(\sigma_p)$ for each of the same stimulation energies selected earlier.

(f) Calculate new "pseudo-LM-OSL" curves using these new data.

(g) Using a least-squares-fitting routine, vary the new $h(E_o)$ distribution until the new "pseudo-LM-OSL curve" matches the original.

(h) Compare the final $h(E_o)$ with the original $g(E_o)$.

(i) Repeat, but using different input distributions $g(E_o)$ and guessed distribution $h(E_o)$.

(j) What do you conclude?

---

### 5.3.7    Emission Wavelength

As with TL, it is important to determine if the emission wavelength is changing during the course of the OSL measurement and, in particular, if it is shifting within the detection window used to record the data. For a system with two recombination centers an equivalent expression to Equation 5.45 can be written, shown here re-stated for OSL:

$$I_{OSL}(t) = K_1 \left| \frac{dm_1}{dt} \right| + K_2 \left| \frac{dm_2}{dt} \right| \tag{5.45}$$

A similar calculation as that performed for TL in Figure 5.23 can be done for CW-OSL using the same parameters as in the calculation of Figure 5.23. If this is done, two emission bands (peaking at 3.5 eV and 3.75 eV) would be observed in OSL, but the 3.5 eV signal would be much weaker and the CW-OSL would decay very quickly (less than 1 s using the parameters shown). In contrast, the 3.75 eV emission would show a much stronger signal, with slower decay over several 10s of seconds. CW-OSL experiments would show entirely different signals if measured at these two different wavelengths, and yet one trap only is emptying in both cases. Again, the kinetics are non-first-order and the QE is not found. As a result, the observed CW-OSL could not be fitted with a single exponential with the result that there may be the temptation to erroneously attempt a fit using a sum of exponentials. Meaningless answers would result.

---

**Exercise 5.18    OSL emission spectra issues**

Using the same model as in Exercise 5.12, calculate the expected CW-OSL emission, assuming just one trap is emptying. Use the same parameters as in Exercise 5.12.

(a) Confirm that the CW-OSL intensity for one of the two emission bands is much weaker than that of the other, and that its CW-OSL curve decays much faster than that of the second emission band.

(b) Confirm that the kinetics are non-first-order and that the system is not in QE.

(c) What changes might be made to the parameter values to obtain a result in which the CW-OSL signal from each emission component is similar in intensity and with the CW-OSL curves for two emission bands decaying at the same rate?

(d) What are the kinetics in this case, and is the system in QE?

### 5.3.8   Summary of Steps to Take Using OSL Curve Fitting

Many aspects of fitting OSL curves are the same as for TL, but with one important difference. All OSL signals start at stimulation time zero and are therefore inextricably overlapped. It is not possible to resolve one signal from another. Nevertheless, there are some steps that can be undertaken to increase the confidence of any OSL curve fitting routine, whether it be CW-OSL or LM-OSL.

1) *Estimate the number of traps contributing to the OSL signal.* This is best done by examining the changes to the TL glow curve after the OSL signal has been measured. How many TL peaks are affected? Although not fool-proof, this gives a basis for an assumed number of traps that may be contributing to the OSL signal.

2) *OSL versus dose*: Independent of how many traps contribute to the OSL signal, if each component grows linearly with dose, then there will be no change in the shape of the OSL curve (CW-OSL or LM-OSL) with dose, if the kinetics are first-order. (As with TL, first-order processes are essential for peak fitting).

3) *Perform a $t_{max}$-$t_{stop}$ experiment.* Are the data consistent with the number of contributing components inferred from the comparison with the TL glow curve?

4) *Choose three or four different stimulation wavelengths (stimulation energies) if possible.* Monitor how the OSL curve shape changes at different stimulation wavelengths.

5) *For LM-OSL, measure the LM-OSL curve using three or four different intensity ramp rates, $\beta_{\Phi}$.* What changes occur to the curve shape?

6) *Fit the CW-OSL and/or LM-OSL.* Use the inferred number of components and simultaneously fit the data measured at different stimulation wavelengths and/or $\beta_{\Phi}$.

7) *Check the FOM against the noise in the data*: How does the FOM compare to the standard deviation of the noise in the data? Any FOM which is less than the noise should be considered to be an acceptable fit. There may be a range of fitted parameters $\sigma_p$ and/or $E_o$ that meet this criterion.

8) *Check for self-consistency*: Are all of the data consistent?

### 5.3.9   OSL Due to Optically Assisted Tunneling

In Chapter 3 (Section 3.8.1.3) the notion that tunneling of an electron from the excited state of a trap to the recombination center was introduced. Here, the electron is not excited to the delocalized band and yet nevertheless OSL may result from the tunneling transition. As indicated from the analysis in Section 3.8.1.3, in such circumstances the normal equations describing OSL will not apply and, for example, the CW-OSL curve will follow a $t^{-k}$ law at longer times, with $k \approx 1$. The analysis in Section 3.8.1.3 centered on thermal excitation to the excited state but optical excitation will produce the same result, as examined in Exercise 3.16.

Mathematically, the situation under optical excitation is the same as isothermal decay of luminescence under thermal excitation, but with $\sigma_p\Phi$ replacing $s\exp\{-\Delta E / kT\}$ (see Equation 3.119). Thus, the CW-OSL curve shape will follow $t^{-k}$ ($k \approx 1$) rather than the expected exponential decay (for first-order kinetics and a single trap contributing to the OSL). Expressing the CW-OSL as $\ln(I_{CW\text{-}OSL})$-versus-$\ln(t)$ will show a straight line of slope $\approx -k$. For multiple traps, or for a distribution of traps, a straight line may not be so-obvious and a curved $\ln(I_{CW\text{-}OSL})$-versus-$\ln(t)$ plot may result. If thermal excitation to the excited state is weak, the CW-OSL decay will be independent of temperature.

Huntley (2006) examined luminescence decay due to tunneling after irradiation and derived a simple expression that can be applied to optically or thermally excited tunneling during OSL or TL readout. Huntley used an approximation of a vertical tunneling front (see Figure 3.43) and derived an expression for the decay of OSL, namely:

$$I_{OSL}(t) = I_{OSL}(0)\exp\left\{-\rho'\left(\ln\left[P_0 t\right]\right)^3\right\}, \tag{5.59}$$

where $\rho'$ is a dimensional number density of acceptors (hole traps/recombination centers) equal to $4\pi\rho/3\alpha^3$.

Analytical expressions for CW-OSL and LM-OSL when optically assisted excited-state tunneling is the source of the OSL emission were also derived by Kitis and Pagonis (2013).

$$I_{OSL}(t) = 3n_0\rho'F(t)^2\left(\frac{1}{t}\right)\exp\left\{-\rho'F(t)^3\right\}, \tag{5.60a}$$

where

$$F(t) = \ln\left(1 + z\sigma_p\Phi t\right) \tag{5.60b}$$

for CW-OSL, and

$$F(t) = \ln(1 + \frac{z\sigma_p\beta_\Phi}{2}t^2) \tag{5.61}$$

for LM-OSL. Here, $n_0$ is the original concentration of electron donors and $z$ is a dimensionless constant ($\approx 1.8$).

---

**Exercise 5.19    Excited-state tunneling and LM-OSL**

(a) Examine the papers by Huntley (2006) and Kitis and Pagonis (2013) and follow the arguments in the derivations of Equation 5.59 (Equation 7 in Huntley 2006), and Equations 5.60a and 5.60b (Equations 23 and 24 in Kitis and Pagonis 2013). These equations give approximate analytical expressions for CW-OSL when optically stimulated, excited state tunneling is occurring.

(b) Plot the shapes of CW-OSL using these functions, for assumed values of the relevant parameters, and assuming one trap only contributing to the OSL signal. (Normalize the curves to an initial value of 1.)

(c) Plot CW-OSL as $\ln(I_{CW\text{-}OSL})$-versus-$\ln(t)$.

(d) Repeat (b) and (c), but assuming two or more traps contribute to the OSL signals.

(e) How do the shapes vary? Over what time regions is the $t^{-k}$ law observed?

### 5.3.10  VE-OSL

In Chapter 3, the expression for VE-OSL was given by Equation 3.50, re-written here as:

$$I_{VE-OSL}(E) = n_0 \Phi \sigma_p(E) \exp\left\{-\frac{1}{\beta_E} \int\limits_{hv_0}^{hv} \Phi \sigma_p(\epsilon)\,d\epsilon\right\} \tag{3.50}$$

where all terms have been defined previously. The advantage of using VE-OSL instead of CW-OSL or LM-OSL is that the "TL like" representation for VE-OSL means that the peaks are separated and do not all start together at the beginning of the stimulation process. The overlap problems found with CW-OSL and LM-OSL are not factors, but are similar to those found for TL.

However, the disadvantage (apart from the experimental difficulty of obtaining VE-OSL curves) lies with the integral in Equation 3.50, $\int\limits_{hv_0}^{hv} \Phi \sigma_p(\epsilon)\,d\epsilon$. Evaluation of this integral requires knowledge of the full shape of the $\sigma_p(E)$ curve. As noted with respect to the VE-OSL curve shapes shown in Figure 3.12, the shape of the VE-OSL curve depends upon the shape of the $\sigma_p(E)$ curve, which is what curve fitting is trying to determine. Thus, fitting a VE-OSL experimental curve to Equation 3.50 requires an a priori assumption of $\sigma_p(E)$. Analysis of CW-OSL and LM-OSL curves does not need such an assumption since each of those experiments use a fixed value for the stimulation energy (wavelength), whereas VE-OSL scans the stimulation energy.

# 6

# Dependence on Dose

*Tout......est une question de dose.*
*[All....is a question of dose.]*

C. Bernard 1872

## 6.1 TL, OSL, or RPL versus Dose

The above quotation from French physiologist Claude Bernard actually refers to whether or not substances can be harmful to the human body. In a parallel fashion, the value of luminescence materials as radiation dosimeters also critically depends on the radiation dose. For doses that are too small, many materials and method combinations are unhelpful at measuring the dose since they lack the required sensitivity. If the dose is too large, the luminescence properties may become highly non-linear and the intensity may even decrease for doses that are high enough. The dependence of the luminescence properties on dose, whether the properties be TL, OSL, or RPL, is an essential area of study. The relationship between the intensity of the light (luminescence) output and the energy absorbed must be understood if a material and method combination is to be accepted as useful (or not) in a given dosimetry situation.

This chapter explores the dependence of the luminescence output and the energy absorbed from the radiation field (i.e. the dose).

## 6.2 Dependence on Dose

### 6.2.1 OTOR Model

The OTOR model enables the introduction of some fundamental dose-dependence concepts. From Figure 6.1a the rate equations during trap filling for the OTOR model may be written:

$$\frac{\mathrm{d}n_c}{\mathrm{d}t} = G - n_c\left(N - n\right)A_n - n_c m A_r, \tag{6.1a}$$

*A Course in Luminescence Measurements and Analyses for Radiation Dosimetry,* First Edition.
Stephen W.S. McKeever.
© 2022 John Wiley & Sons Ltd. Published 2022 by John Wiley & Sons Ltd.
Companion Website: www.wiley.com/go/mckeever/luminescence-measurements

$$\frac{dm_v}{dt} = G - m_v (M - m) A_m, \tag{6.1b}$$

$$\frac{dn}{dt} = n_c (N - n) A_n, \tag{6.1c}$$

and

$$\frac{dm}{dt} = m_v (M - m) A_m - n_c mB \tag{6.1d}$$

where the terms have their usual meaning. As usual, there is one equation for each of the energy levels involved (conduction band, valence band, electron trap, and hole trap) and each term appears twice in the equations.

If the system is irradiated for a time $t_{irr}$, then the dose delivered is $\gamma G t_{irr}$, where $G$ is the ionization rate (rate of production of free electrons and holes (e-h pairs) during irradiation) and $\gamma$ is a constant. In SI units the dose $D$ is given in gray = 1 joule/kg, so the constant $\gamma$ has dimensions of J/kg/e-h pair, and measures how much energy in Joules needs to be absorbed per kg of material to produce a given degree of ionization.

During irradiation, the charge neutrality condition is $n_c + n = m_v + m$, but once the irradiation ceases ($G = 0$ at $t = t_{irr}$) the concentrations of electrons and holes in the conduction and valence bands must decay to zero and become trapped in the available electron and hole traps, respectively. Band-to-band recombination is ignored and only Hall-Shockley-Read recombination at the hole trap is considered. The charge neutrality condition then becomes $n = m$ at $t = t_{irr} + \Delta t$, where $\Delta t$ is the extra relaxation time needed for $n_c$ and $m_v$ to decay to zero.

Figure 6.1b shows a typical solution to this model for the parameter values indicated in the figure caption. The growth of the concentration of trapped electrons $n$ is a saturating exponential-like function of irradiation time (plus relaxation time; $t_{irr} + \Delta t$). Such curves are often assumed to follow a law such as:

$$n = N \left(1 - \exp\{-\alpha D\}\right) \tag{6.2}$$

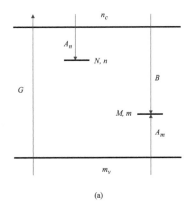

(a)

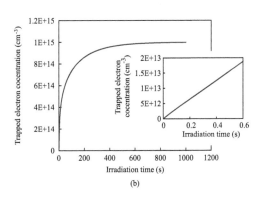

(b)

**Figure 6.1**   (a) The OTOR model used to calculate the growth of $n$ with irradiation time, shown in (b). Over short times, the growth can be approximated to a linear growth (inset). For this figure, $N = 10^{15}$ cm$^{-3}$, $M = 3 \times 10^{14}$ cm$^{-3}$, $A_n = A_m = 10^{-15}$ cm$^3$.s$^{-1}$, $B = 10^{-13}$ cm$^3$.s$^{-1}$, and $G = 10^{14}$ cm$^{-3}$s$^{-1}$.

where $\alpha$ is a constant. Assuming quasi-equilibrium and the special condition that $A_n = B$, Aramu et al. (1975) derived Equation 6.2 from Equations 6.1a–6.1d, with $\alpha = 1/N$. However, in general the growth function is not so simple. (In fact, it is not the case in the example shown in Figure 6.1b.) As a result, care should be taken when using this simple expression for $n(D)$. For example, Lawless et al. (2009) showed that $n$ can grow as $D^{1/2}$ under certain circumstances, even with the OTOR model.

At very low doses, when $n \ll N$, the growth curve can be approximated to a linear function as illustrated in the inset to Figure 6.1b. A linear relationship between $n$ and $D$ is a desired property of a dosimeter. However, the growth curves for all TLDs, OSLDs, or RPLDs, even if they are linear at low doses, must become non-linear, specifically sublinear, at high-enough doses due to inevitable saturation. The goal, therefore, is to find a dosimeter and method for which the luminescence signal remains linear over several orders of magnitude in the dose range of interest (termed the dosimeter's "dynamic range").

---

### Exercise 6.1   The OTOR Model

(a) Lawless et al. (2009) solved Equations 6.1a–6.1d for a variety of parameters values. Using a software package of your choice, repeat the calculations of Lawless et al. (2009) and confirm that you obtain the same results.
(b) Under what circumstances does a $I \propto D^{1/2}$ dose-response relationship emerge?
(c) Under what circumstances does a $I \propto D$ dose-response relationship emerge?

---

Note that with RPL, the RPL intensity is directly proportional to $n$ since the luminescence emitted is due to the relaxation of the RPL centers, which have a concentration $n$ in this simple model. For TL and OSL, however, the luminescence signal is proportional to $dm/dt$, where $m$ is the recombination center concentration. It is only in some models, the OTOR model in Figure 6.1 being an example, that $dm/dt$ is proportional to $n$. In these cases the TL or OSL intensity is also proportional to $n$ in which cases it may be stated that for RPL, OSL, or TL, the intensity $I \propto n$. As a result, the growth curve in Figure 6.1b represents the expected growth curve for luminescence as a function of dose.

*6.2.1.1   Dose-Response Relationships: Linear, Supralinear, Superlinear, and Sublinear*

It is possible to define a function $f(D)$, where:

$$f(D) = \frac{I(D)/D}{I(D_l)/D_l}.$$

(6.3)

Here, $D_l$ is a dose on the linear part of the dose-response curve and $I$ is the intensity of the luminescence signal. If the dosimeter displays a linear dose-response relationship then $f(D) = 1$, by definition. In practice, for many dosimetry material-method combinations it is often found that $f(D) < 1$ or $f(D) > 1$.

Some schematic examples are illustrated in Figure 6.2. In Figures 6.2a to 6.2d, the growth of luminescence intensity (TL, OSL, or RPL) is illustrated as a function of dose. Figures 6.2e to 6.2h show the same data but displayed on a log-log scale. Figures 6.2a and 6.2e show a linear growth ($f(D) = 1$) and when displayed on a log-log scale, linear growth yields a slope = 1. An example of a polynomial, non-linear growth is shown in Figures 6.2b and 6.2f, with $f(D) > 1$. Note that if the growth is of the form $I \propto D^k$, with $k > 1$, then a linear log-log plot is still obtained, but now the slope = $k > 1$.

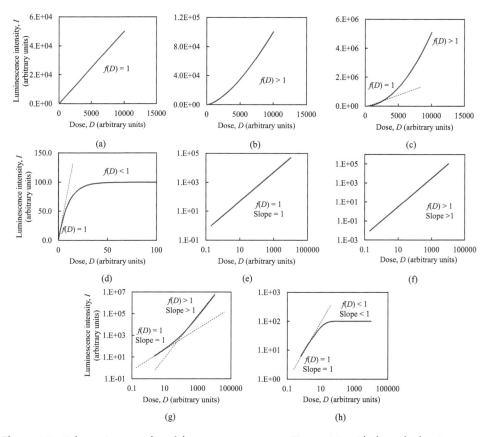

**Figure 6.2**    Schematic examples of dose-response curves. Figures (a) to (d) show the luminescence intensity $I$ against absorbed dose $D$ on linear-linear scales, while (e) to (h) show the same data on log-log scales. (a) and (e) linear; (b) and (f) polynomial; (c) and (g) linear-polynomial; (d) and (h) saturating exponential.

Figures 6.2c and 6.2g show an example where the growth starts linear ($f(D) = 1$), but becomes non-linear as the dose increases, $f(D) > 1$. On a log-log plot, two straight lines are obtained, with slope equal to 1 at low doses and a constant slope, greater than 1, at high doses. Finally, Figures 6.2d and 6.2h demonstrate a saturating exponential-type growth curve in which $f(D) = 1$ at low doses and $f(D) < 1$ as saturation is approached.

A common type of dose-response relationship is shown in Figure 6.3 on (a) linear and (b) logarithmic scales. Here, dose-response relationships that are linear/more-than-linear/saturated are indicated. Using definition (6.3), three regions, $f(D) = 1$, $f(D) > 1$, and $f(D) < 1$, can be identified, indicated by the vertical dashed lines. The three regions are seen most clearly in Figure 6.3b, where $f(D)$ is also plotted as a function of dose.

In the region in which $f(D) > 1$ the luminescence intensity is greater than that which would have been obtained had the system continued with a linear response. The response is then said to be "more-than-linear," or "supralinear," and $f(D)$ is commonly called the "supralinearity index, or the "dose response function."

Examination of Figure 6.3 shows that although $f(D) > 1$ in the middle dose region, within this region the slope of the $I(D)$ relationship can change from > 1 to < 1. This leads to another

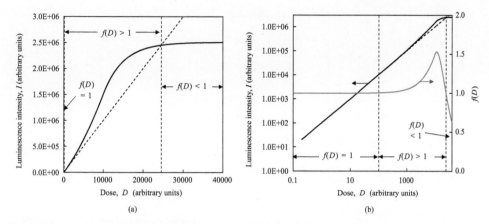

**Figure 6.3** Linear-supralinear-saturating dose-response relationship. (a) Linear-linear scales, and (b) log-log scales. The dose regions in which the supralinearity index $f(D) = 1$, >1 or < 1 are indicated in each plot; $f(D)$ is also shown in Figure (b).

characteristic of non-linear dose-response curves, namely the "superlinearity index," $g(D)$, defined as:

$$g(D) = \left[ \frac{D d^2 I / dD^2}{dI / dD} \right] + 1, \tag{6.4}$$

(see Chen and McKeever 1994).

For an increasing dose dependence, $dI / dD > 0$ and $g(D) > 1$ and this is known as "super-linearity." The relationships between $f(D)$ and $g(D)$ are demonstrated in Figure 6.4. This is the same dose-response relationship as in Figure 6.3, but the plot is divided into regions I, II, III,

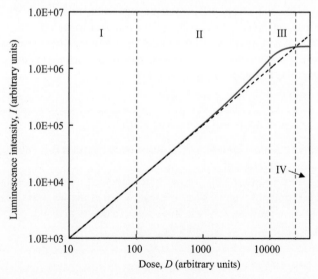

**Figure 6.4** Same plot as Figure 6.3 showing regions of different values of the supralinearity index $f(D)$ and the superlinearity index $g(D)$. Region I: $f(D) = g(D) = 1$; Region II: $f(D) > 1$ and $g(D) > 1$; Region III $f(D) > 1$, $g(D) < 0$; Region IV; $f(D) < 1$, $g(D) < 0$.

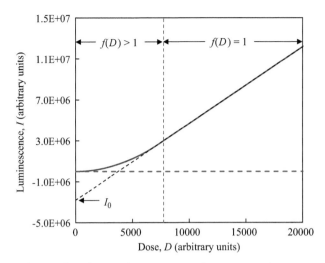

**Figure 6.5**  A supralinear then linear dose-response function, with a negative value for the intercept $I_0$ at zero dose.

and IV. In region I, $f(D) = g(D) = 1$; in region II, $f(D) > 1$ and $g(D) > 1$; in region III, $f(D) > 1$ and $g(D) < 0$; and in region IV, $f(D) < 1$ and $g(D) < 0$.

A more general form of Equation 6.3 may be written:

$$f(D) = \frac{\left[ I(D) - I_0 \right] / D}{\left[ I(D_l) - I_0 \right] / D_l} \tag{6.5}$$

where $I_0$ is the intercept on the luminescence axis when the linear part of the response is extrapolated to $D = 0$. (Note that $f(D)$ cannot be defined at $D = 0$.) For most cases, it can be expected that $I_0 = 0$; however, sometimes a dose-response curve is seen similar to that shown schematically in Figure 6.5. Here $I_0 < 0$ and $f(D)$ would be impossible to define using only Equation 6.3. Using Equation 6.5, however, allows $f(D)$ to be evaluated for all $D > 0$. In the initial region $f(D) > 1$ (i.e. a supralinear region) and the luminescence intensity is greater than that inferred from the backwards extrapolation of the linear part of the dose-response curve. As the dose increases, the response becomes linear, $f(D) = 1$.

---

**Exercise 6.2    Supralinearity and Superlinearity**

(a) Using the solutions to Equations 6.1a–6.1d from Exercise 6.1, calculate and plot $f(D)$ and $g(D)$ as a function of $D$.

(b) Imagine a dose-response function like that shown in Figures 6.2c and 6.2g – i.e. linear at low doses and polynomial ($I \propto D^k$, with $k > 1$) at higher doses. Use a spreadsheet to simulate such a dose-response function. Calculate and plot $f(D)$ and $g(D)$ as a function of $D$ for your simulated dose-response curve.

---

None of the cases where $f(D) > 1$ or $g(D) > 1$ are predicted from the simple OTOR model. To obtain supralinear and superlinear growth of luminescence with dose, more complex, interactive kinetics are needed.

## 6.2.2   Interactive Models: Competition Effects

### 6.2.2.1   *Competition during Irradiation*

Consider an interactive, 2T1R model, as shown in Figure 6.6, with the RPL, TL, or OSL signal resulting from trap 1 – either directly (RPL), or as trap 1 empties and the electrons recombine at the recombination center, concentration $m$ (TL or OSL). During irradiation, trap 1 and trap 2 fill with electrons such that trap 2 can be said to be acting as a "competitor" to trap 1.

Assuming $n_c$, $m_v \ll n_1$, $n_2$, $m$, then charge neutrality dictates that:

$$n_1 + n_2 = m. \tag{6.6}$$

Since trap 1 is the "active" trap, $I \propto n_1$ for RPL, TL, or OSL. Here, $n_1$ is the concentration of electrons in trap 1 at the end of the irradiation period. Consider further that $N_2 < N_1$ and/or $A_{n2} > A_{n1}$; this means that trap 2 will fill faster than trap 1 during irradiation. The result of this is that as trap 2 approaches saturation there will be more electrons available for trap 1. Under such circumstances, and if all traps fill linearly, the luminescence versus dose curve will transition from one linear region initially (as trap 1 and trap 2 both fill) to another, but steeper, linear region as trap 2 saturates. The transition from one linear region to another will be continuous and therefore the dose-response curve will appear to be supralinear (and superlinear) during the transition.

As illustrated by Bowman and Chen (1979), the above qualitative argument can be described mathematically using the rate equations describing the 2T1R model of Figure 6.6 during trap filling. The equations are:

$$\frac{dn_1}{dt} = n_c \left( N_1 - n_1 \right) A_{n1}, \tag{6.7a}$$

$$\frac{dn_2}{dt} = n_c \left( N_2 - n_2 \right) A_{n2}, \tag{6.7b}$$

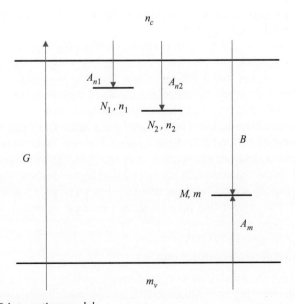

**Figure 6.6**   A 2T1R interactive model.

$$\frac{dm}{dt} = m_v \left( M - m \right) A_m - n_c m B, \tag{6.7c}$$

$$\frac{dn_c}{dt} = G - n_c \left( N_1 - n_1 \right) A_{n1} - n_c \left( N_2 - n_2 \right) A_{n2} - n_c m B, \tag{6.7d}$$

and

$$\frac{dm_v}{dt} = G - m_v \left( M - m \right) A_m. \tag{6.7e}$$

Applying the quasi-equilibrium QE assumption, integrating and eliminating $n_c$, gives:

$$\int_0^{t_{irr}} G dt = D = n_1 - n_{10} + N_2 - n_{20} - \left( N_2 - n_{20} \right) \left[ \frac{N_1 - n_1}{N_1 - n_{10}} \right]^{A_{n2}/A_{n1}}, \tag{6.8}$$

where $n_{10}$ and $n_{20}$ are the concentrations at $t = 0$ and $t_{irr}$ is the irradiation time (and, using the earlier nomenclature, the constant $\gamma = 1$ such that $D = Gt_{irr}$).

Normally, $n_1$ would be expressed as a function of dose, but here dose $D$ is expressed as a function of $n_1$. Since $I \propto n_1$, the superlinearity index (Equation 6.4) can be re-written:

$$g(D) = \left[ \frac{D d^2 n_1 / dD^2}{dn_1 / dD} \right] + 1 = \left[ -D \left( \left( \frac{d^2 D}{dn_1^2} \right) / \left( \frac{dD}{dn_1} \right)^3 \right) / dn_1 / dD \right] + 1. \tag{6.9}$$

Since $dn_1 / dD > 0$, then $dD / dn_1 > 0$; $d^2 n_1 / dD^2 > 0$ means that $d^2 D / dn_1^2 < 0$. The condition for superlinearity ($g(D) > 1$) now becomes $d^2 D / dn_1^2 < 0$.

From Equation (6.8):

$$\frac{d^2 D}{dn_1^2} = -\frac{A_{n2}}{A_{n1}} \left( \frac{A_{n2}}{A_{n1}} - 1 \right) \frac{N_2 - n_{20}}{\left( N_1 - n_{10} \right)^2} \left( \frac{N_1 - n_1}{N_1 - n_{10}} \right)^{(A_{n2}/A_{n2})-2}. \tag{6.10}$$

The condition for this expression to be negative is $A_{n2} > A_{n1}$. This was an explicit condition, previously stated, and therefore $g(D) > 1$.

Note that at low doses, $d^2 D / dn_1^2 \approx 0$, but positive. Thus the dose-response function $n_1(D)$ is linear. At high doses, when trap 2 is close to saturation, $N_2 - n_{20} \approx 0$ and again $d^2 D / dn_1^2 \approx 0$ and $n_1(D)$ is linear again, but with a different slope. Thus, the growth of $n_1$ with dose will consist of two linear regions. The regions are continuous and a superlinear (and supralinear) region will be observed as the curve transitions from one slope to the next.

Figure 6.7 illustrates this situation. The data are from a numerical solution to Equations 6.7a to 6.7e, using the parameters listed in the figure caption. Figure 6.7a shows that the concentration $n_1$ grows linearly at low dose but transitions to a steeper linear growth as concentration $n_2$ saturates. Figure 6.7b shows the same data but extended to higher doses to show saturation of $n_1$, and is plotted on a log-log scale. The shape of $n_1(D)$ shows a slope = 1 on a log-log plot at low $D$ and at higher $D$, with a transition between them where the slope > 1, before saturation is reached.

### 6.2.2.2   Competition during Trap Emptying

If competition between traps and centers occurs during trap filling it can be reasonably expected to occur during trap emptying also. Although RPL can be immediately measured after irradiation, TL and OSL each require that the traps be emptied in order to record the luminescence

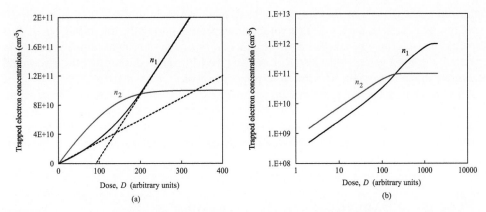

**Figure 6.7** Numerical solutions to Equations 6.7a–6.7e, using the following parameters: $N_1 = 10^{12}$ cm$^{-3}$, $N_2 = 10^{11}$ cm$^{-3}$, $A_{n1} = 10^{-9}$ cm$^3$.s$^{-1}$, $A_{n2} = 3\times10^{-8}$ cm$^3$.s$^{-1}$, $B = 10^{-9}$ cm$^3$.s$^{-1}$, and $G = 10^9$ cm$^{-3}$.s$^{-1}$. (a) Linear-linear plot showing the transition between linear growths of different slopes; (b) same data but on a log-log plot and extended to high dose showing saturation.

signal. Thus, competition processes occurring during the heating stage (for TL) or optical stimulation stage (for OSL) also need to be considered.

To examine the importance of competition during the trap emptying phase the rate equations describing trap emptying using the 2T1R interactive model can be solved. Using TL as an example and the model of Figure 6.6, trap filling is first modeled (using the same parameters as listed in the caption to Figure 6.7), followed by trap emptying with $\beta_t = 1$ K.s$^{-1}$, $E_{t1} = 1.3$ eV, and $s_1 = 5\times10^{12}$ s$^{-1}$. The results are shown in Figure 6.8 in which the height of the TL peak is plotted as a function of the dose. A very strong supralinearity is observed, all the way to trap saturation. The dashed line indicates a linear dose-response relationship The degree of supralinearity is much larger than the supralinearity illustrated for $n_1$ in Figure 6.7, indicating that

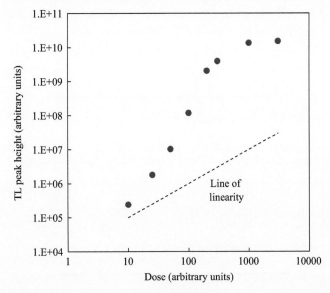

**Figure 6.8** Supralinear dose-response relationship for the 2T1R model of Figure 6.6. The rate equations for trap filling and trap emptying were solved using the parameters shown in the caption to Figure 6.7, along with $E_{t1} = 1.3$ eV, $s_1 = 5\times10^{12}$ s$^{-1}$, and $\beta_t = 1$ K.s$^{-1}$.

competition during the trap emptying phase has a much stronger influence on the final $I_{TL}(D)$ curve (or, by inference, the $I_{OSL}(D)$ curve) than competition during the trap filling phase. It is also clear that $n_1$ cannot be taken as a surrogate for $I_{TL}$ or $I_{OSL}$ for this more complex model. As a result, if the trap 1 centers were also RPL centers, the RPL versus dose-response curve would be entirely different from the TL/OSL dose-response curve.

When the competitors are electron traps as the dose increases so the concentration of empty competitors decreases. (In the 2T1R model, the competitors are empty trap 2 states, of concentration $N_2 - n_2$. As $n_2$ increases so $N_2 - n_2$ decreases.) During trap filling, the empty trap 2 states are competitors to trap 1; during trap emptying they are competitors to the recombination centers, concentration $m$.

---

**Exercise 6.3   Competition during trap emptying (I)**

(a) Write a set of rate equations to describe the flow of charge for the 2T1R model depicted in Figure 6.6.
(b) Choose values for the appropriate parameters and solve the equations to produce the growth of OSL as a function of dose. Assume trap 1 is the optically active trap and that trap 2 is the competitor. To do this calculation you will have to solve the equations during the irradiation phase ($G > 0$, $p = 0$), a short relaxation phase ($G = 0$, $p = 0$), and an optical stimulation phase ($G = 0$, $p > 0$). Here $p$ is the rate of optical excitation from trap 1. The calculation will need to be repeated for different doses, i.e. different values of the irradiation time $t_{irr}$, for the same value of $G$.
(c) Plot the OSL intensity against the delivered dose ($= G t_{irr}$). What is the shape of the dose-OSL response curve? Examine how changing the parameters governing the competition changes the dose-response curve.

---

An entirely different situation is represented in Figure 6.9, which shows a 1T2R model. Here the competitors are the recombination centers, that is, filled hole-traps of concentration $m_2$. During trap filling, these act as competitors to the filling of empty trap 1 states, while during trap emptying they act as competitors to recombination at recombination center 1 (of

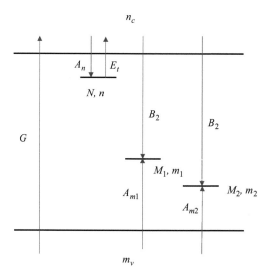

**Figure 6.9**   A 1T2R interactive model.

concentration $m_1$). Also, during trap filling empty hole traps at center 2 (concentration $M_2 - m_2$) are competitors to trapping holes at center 1. If recombination of electrons at recombination center 1 yields TL (or OSL) then complex competition effects during both trap filling and trap emptying can occur, and a difficult-to-predict TL or OSL response to the dose will be obtained.

The rate equations describing filling and emptying for the model of Figure 6.9 are:

$$\frac{dn}{dt} = n_c (N - n) A_n - ns\exp\left\{-\frac{E_t}{kT}\right\}, \tag{6.11a}$$

$$\frac{dm_1}{dt} = m_v (M_1 - m_1) A_{m1} - n_c m_1 B_1, \tag{6.11b}$$

$$\frac{dm_2}{dt} = m_v (M_2 - m_2) A_{m2} - n_c m_2 B_2, \tag{6.11c}$$

$$\frac{dn_c}{dt} = G - n_c (N - n) A_n + ns\exp\left\{-\frac{E_t}{kT}\right\} - n_c m_1 B_1 - n_c m_2 B_2, \tag{6.11d}$$

$$\frac{dm_v}{dt} = G - m_v (M_1 - m_1) A_{m1} - m_v (M_2 - m_2) A_{m2}. \tag{6.11e}$$

During trap filling $T$ is held constant at $T = T_0$ (usually room temperature, ~300 K), and $n = m_1 = m_2 = n_c = m_v = 0$ at time $t = 0$. After irradiation, a period of relaxation where $G = 0$ is followed by a heating period where $G$ remains $= 0$ and $T$ changes according to $T = T_0 + \beta_t t$. Some example results from numerical solutions to these equations are shown in Figure 6.10 for the parameters listed in the figure caption. In this example, the concentration $n_1$ grows slightly sublinearly with dose (Figure 6.10a), but the TL signal due to trap 1 electrons recombining with recombination center 1 is supralinear as a function of

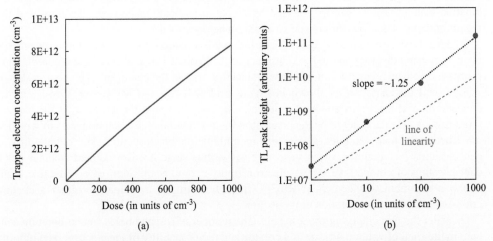

(a)                                                                 (b)

**Figure 6.10** Example results for the 1T2R model of Figure 6.9 in which filled recombination centers are the competitors. (The dose units here are the number of free electron-hole pairs generated by the radiation, per cm³.) Although the trapped electron concentration grows slightly sublinearly with $D$ (figure (a)), the TL peak height grows as a power law (figure (b)), approximately following $D^{1.25}$. The line of linearity is shown for comparison.

dose (Figure 6.10b), over the dose range shown. In this case, the TL signal grows approximately as a power law, $D^{1.25}$. Once again, competition during trap emptying is the main factor in governing whether or not the TL (and similarly the OSL) signal will grow supralinearly with dose.

---

**Exercise 6.4    Competition during trap emptying (II)**

Repeat the calculations of Exercise 6.3 but this time for the 1T2R model of Figure 6.9. What differences do you see between the two cases, and why?

---

It needs to be recognized that just because there is a competing trap and/or a competing recombination center, it does not mean that supralinearity necessarily follows. The critical features of the model are the parameters of the various traps and centers. The discussion in this section merely highlights the fact that competition can produce nonlinear TL or OSL growth with dose, and that competition during trap emptying is more important than competition during irradiation for TL and OSL processes. Of course, the last statement is not true of RPL. Nonlinearity occurs when the competing center is made less effective such that more charge is distributed to the trapping states that ultimately lead to TL or OSL. The competitors can be traps or recombination centers or combinations of these and complex behavior can result with real materials. As a final note, the conditions for supralinearity with competition models are not necessarily consistent with first-order kinetics and quasi-equilibrium.

### 6.2.3   Spatial Effects

The models discussed so far assume an isotropic distribution of traps and recombination centers. Once an electron is released into the conduction band its wavefunction is delocalized and it can recombine with any recombination center or be retrapped into any trap. A question arises if these simple models can be applied to a situation where the distribution of traps and recombination centers is non-isotropic caused by a spatial association between the traps and the recombination sites. How should such a situation be described?

To describe the potential processes occurring in this situation it is useful to move away from the energy band diagram and to consider instead the geometrical aspects of the recombination processes. To begin, consider in this case an electron trap in the center of a sphere of radius $r$, with a shell thickness $dr$, as shown in Figure 6.11 (Mische and McKeever 1989). Let the concentration of electron traps be $N_1$, of which $n_1$ are filled, and for which $\sigma_1$ is the capture cross-section. Similarly, let $M$ be the concentration of potential recombination sites of which $m$ are filled with holes and let $\sigma_m$ be the recombination cross-section. For competitors, a concentration $k$ is assumed, of cross-section $\sigma_k$. At this stage it is not defined whether $k$ is a concentration of competitive empty electron traps, or a concentration of competitive, non-radiative recombination centers (NRRCs) (filled hole traps); $\sigma_k$ could be either a capture cross-section or a recombination cross-section.

Consider the trap emptying process only. The number of trapped holes (recombination centers) in the shell of volume $4\pi r^2 dr$ is $4\pi r^2 m dr$ and the probability of the electron recombining in this shell is therefore $(\sigma_m / 4\pi r^2) \, 4\pi r^2 m dr = \sigma_m m dr$. Define a mean free path $\mu$, where:

$$\mu = \frac{1}{\sigma_m m + \sigma_k k}. \tag{6.12}$$

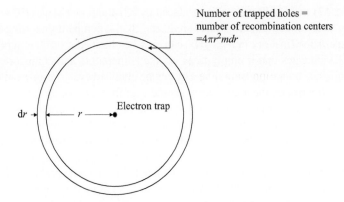

**Figure 6.11**   Schematic notion of an electron trap within a sphere of radius $r$, and shell thickness $dr$. If the concentration of recombination centers is $m$, then there are $4\pi r^2\,mdr$ recombination centers available with which an electron from the trap can recombine if it reaches the shell.

With these definitions the probability that an electron released from a trap at the center of the sphere will travel a distance $r$ and recombine in shell $dr$ is:

$$dP(r) = \sigma_m m dr \exp\left\{-\frac{r}{\mu}\right\}. \tag{6.13}$$

Integrating from $r = 0$ to $r = R$:

$$P(R) = \mu\sigma_m m\left[1 - \exp\left\{-\frac{R}{\mu}\right\}\right]. \tag{6.14}$$

If $R$ is approximately the dimensions of the sample, $d$, then $R \approx d >> \mu$ and $P(d) = \mu\sigma_m m$. For $n_1$ total electron traps the net TL or OSL signal is then proportional to $n_1 P(d) = n_1\mu\sigma_m m$. Several options follow.

(i) If $\sigma_m m >> \sigma_k k$, then $\mu = 1/\sigma_m m$, in which case $P(d) = 1$. This means all electrons will recombine because there is no alternative, and:

$$I_{TL/OSL}(D) \propto n_1(D). \tag{6.15}$$

$I_{TL/OSL}(D)$ is linear if $n_1(D)$ is linear, or is nonlinear if $n_1(D)$ is nonlinear.

(ii) If $\sigma_k k >> \sigma_m m$, then $\mu = 1/\sigma_k k$, in which case $P(d) = \sigma_m m / \sigma_k k$. In this case:

$$I_{TL/OSL}(D) \propto \frac{n_1(D)\sigma_m m(D)}{\sigma_k k(D)}. \tag{6.16}$$

Two possibilities emerge:

(a) If the competitors are recombination centers, then $n_1 = k + m \approx k$ (assuming $\sigma_k \approx \sigma_m$). Thus:

$$I_{TL/OSL}(D) \propto m(D) \tag{6.17}$$

and $I_{TL/OSL}(D)$ is linear if $m(D)$ is linear, or nonlinear if $m(D)$ is nonlinear.

(b) Alternatively, if the competitors are empty electron traps, then let $k = N_k - n_k$ and:

$$I_{TL/OSL}(D) \propto \left(\frac{\sigma_m}{\sigma_k}\right)\left(\frac{n_1(D)m(D)}{N_k - n_k(D)}\right) \tag{6.18}$$

which can be a highly non-linear function from $D = 0$.

Equations (6.17) and (6.18) were also produced by Kristianpoller et al. (1974) using a delocalized energy band model. Equation (6.17) indicates that, if competition during trap emptying dominates, super/supralinearity in $I_{TL/OSL}(D)$ can only be obtained if $m(D)$ is super/supralinear. Equation (6.18) indicates that if empty traps are the competitors and competition during trap emptying dominates, then super/supralinearity can be obtained even if $n_1(D)$ and $m(D)$ are separately linear. Furthermore, the nonlinearity should start from $D = 0$.

It was mentioned earlier that a commonly observed dose-response relationship is the linear-supralinear-sublinear characteristic, an example of which was shown in Figure 6.3. Considering the above, none of the mechanisms for non-linear dose response discussed so far predict this type of characteristic. Although it has been shown how linear responses, sub-linear responses, and super- or supralinear responses can be attained, the linear-supralinear-sublinear dose-response sequence is very particular. Furthermore, an accompanying observation that is often seen with the linear-supralinear-sublinear dose response is that the defects that are thought to be responsible for the TL signal themselves grow linearly with dose over the same dose range. This implies that it is the interaction between these defects during trap emptying, not the growth of the defects themselves during irradiation, that causes the nonlinearity in the dose-response relationship.

Perhaps the archetypical TL dosimetry material that displays these dose-TL response characteristics is LiF:Mg,Ti, known popularly as TLD-100. (For general examples of TLD-100's dose-response relationship see Zimmerman (1971), Portal (1981), Horowitz (1984), McKeever et al. (1995), or many other general texts on TLD dosimetry materials.) If the defects that produce the TL signal grow linearly during irradiation, then competition during irradiation cannot be the cause of the dose-response behavior. It has been shown above that if empty traps are the competitors, and the competition is active during trap emptying (heating), then Equation (6.18) would describe the dose-response curve, and this does not follow the desired linear-supralinear-sublinear curve. Furthermore, the above analysis also shows (Equation (6.17)) that if the competitors during heating were competing recombination centers, then the recombination centers would have to grow in the same manner during irradiation as the TL (i.e. $m(D)$ would have to grow in the same manner as $I_{TL}(D)$, which they do not).

The cause of the linear-supralinear-sublinear dose-response characteristic is explained by including an additional element, namely the spatial association of the traps and recombination centers. Consider an electron trap separated from a recombination center by a distance $s$ which is small enough such that no other defect centers of any type can exist within the volume $4\pi s^3$. Let $K$ be the fraction of recombination centers that are separated from an electron trap by a distance $\leq s$; $1 - K$ is then the fraction that are separated by a distance $> s$ from an electron trap. The situation is depicted schematically in Figure 6.12.

Within the volume $4\pi s^3$ the probability of an electron released from the trap and recombining with the recombination center is $\sigma_m / 4\pi s^2$. Therefore, the probability that the electron will miss the recombination center, travel a distance $R$, and recombine elsewhere is:

$$P(R) = \left(1 - \frac{\sigma_m}{4\pi s^2}\right)\mu\sigma_m m\left[1 - \exp\left\{-\frac{R}{\mu}\right\}\right]. \tag{6.19}$$

The total probability of recombination (either with the nearest-neighbor center or elsewhere in the crystal) with $R \approx d >> \mu$ is:

$$P_K^{tot} = n_1 K\left[\frac{\sigma_m}{4\pi s^2} + \left(1 - \frac{\sigma_m}{4\pi s^2}\right)\mu\sigma_m m\right]. \tag{6.20}$$

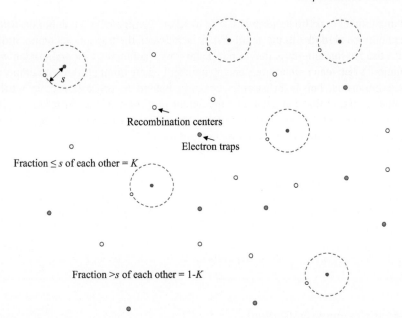

**Figure 6.12** Schematic model of a distribution of trapped electrons (filled circles) and trapped holes/recombination centers (open circles). Some electrons/recombination-center pairs are separated by a distance ≤ $s$, within which distance no other traps or centers of any type exist. The fraction of such traps is $K$ and $(1 - K)$ represents the fraction of all other traps, separated by distances > $s$.

Here, $n_1K$ is the number of traps that are associated in pairs with recombination centers. The first term in square brackets is the probability of recombination within the pair, and the second term is the probability of escaping the pair and recombining elsewhere within the sample.

Now include those electron traps that are not associated with recombination centers, of concentration $(1 - K)n_1$. The total recombination probability is now:

$$P^{tot} = P_K^{tot} + P_{1-K}^{tot} = n_1 K \left[ \frac{\sigma_m}{4\pi s^2} + \left( 1 - \frac{\sigma_m}{4\pi s^2} \right) \mu \sigma_m m \right] + n_1 \left( 1 - K \right) \mu \sigma_m m, \quad (6.21)$$

and

$$I_{TL/OSL} \left( D \right) \propto n_1 \left( D \right) \left[ K \left( \frac{\sigma_m}{4\pi s^2} \right) + \left( 1 - K \frac{\sigma_m}{4\pi s^2} \right) \mu \sigma_m m \left( D \right) \right]. \quad (6.22)$$

With many competitors $(k \gg m)$, $\mu = 1/\sigma_k k$,

$$I_{TL/OSL} \left( D \right) \propto n_1 \left( D \right) \left[ K \left( \frac{\sigma_m}{4\pi s^2} \right) + \left( 1 - K \frac{\sigma_m}{4\pi s^2} \right) \left( \frac{\sigma_m m \left( D \right)}{\sigma_k k \left( D \right)} \right) \right], \quad (6.23)$$

and if the competitors are electron traps $k = N_k - n_k$, then:

$$I_{TL/OSL} \left( D \right) \propto n_1 \left( D \right) \left[ K \left( \frac{\sigma_m}{4\pi s^2} \right) + \left( 1 - K \frac{\sigma_m}{4\pi s^2} \right) \left( \frac{\sigma_m m \left( D \right)}{\sigma_k \left( N_k - n_k \left( D \right) \right)} \right) \right]. \quad (6.24)$$

The first term in square brackets is linear with dose and dominates at low doses. The second term is nonlinear and dominates at high doses.

Stated qualitatively, the linear term is due to local or "geminate" recombination where competitors are not involved due to the small distances between the trap and recombination center. In contrast, the nonlinear term is due to non-local ("non-geminate") recombination where the recombination is between distant traps and centers and where competition with empty electron traps is now important. This is the essential principle behind the processes active in LiF:Mg,Ti, but it should be stated that the details of the actual processes in that material may be more nuanced (Horowitz et al. 2019). These details will be set aside for now until Part II of this book in which the properties of real materials will be discussed.

---

### Exercise 6.5    Spatial Effects

(a) Using Equation 6.24, and chosen values for the various parameters, plot the dose-response curve for a particular set of values.
(b) By changing the sets of values used in the model ($K$, $\sigma_m$, $m$, $N_k$) show how the shape of the dose-response function depends upon these parameters.

---

#### 6.2.4    Sensitivity and Sensitization

Two additional phenomena that are directly linked to the property of supralinearity are sensitivity and sensitization. In the simplest terms the sensitivity $S$ of a material is defined as the response (TL, OSL, or RPL) to a low, "test" dose $D_t$, thus:

$$S = I_{D_t} / D_t, \qquad (6.25)$$

where $I_{D_t}$ is the TL, OSL, or RPL response of the dosimeter to test dose $D_t$.

One of the characteristics of supralinearity is that the response of the material increases as the dose increases, as defined by the supralinearity index in Equation (6.5). It is clear that if the sensitivity $S$ were to be measured anywhere on the linear region of the dose-response curve, the same value for $S$ would be obtained independent of $D_t$, whereas if $D_t$ was chosen on that part of the dose-response curve that is supralinear or sublinear, then the value of $S$ would depend on the test dose used. If the sensitivity measurement is performed before any prior irradiations it may not be known whether the chosen $D_t$ is on the linear part of the dose-response curve or not. This is one of the reasons that the test dose $D_t$ is always kept low.[1]

If the specimen has already been used in prior experiments then the value of $S$ will (or may) depend on the radiation, thermal, and/or optical history of the sample. For RPL, the readout process does not remove the signal (nondestructive) and so sensitivity may then be defined as the incremental increase in the RPL signal following the addition of an incremental dose $\Delta D_t$. In this sense the sensitivity is akin to a derivative, i.e. $S = \Delta I_{RPL} / \Delta D_t$, where $\Delta I_{RPL}$ is the incremental increase in the RPL signal due to the additional dose $\Delta D_t$.

A second reason $D_t$ is kept low has to do with competition. For example, imagine a system in which there exists competing electron traps. At low radiation doses the competing traps are mostly empty and are therefore available to take part in competition. The sensitivity of the dosimetry signal is reduced under such circumstances since the competition is high. However, if the competitors were filled, or partially filled, by prior irradiations, the sensitivity would be increased since the number of competitors is now reduced. This could be detected by

---

[1] There may not always be a linear part of the dose-response curve. Nevertheless, the test dose should be low enough to give a reliable value for $I_{D_t}$ and should be consistently chosen.

comparing the sensitivity of the system before and after a large irradiation by administering a small test dose $D_t$ and measuring the subsequent response. If the large irradiation partially filled the competitors, the sensitivity would increase and this would be detected by an increase in $S$. A complication arises, however, if $D_t$ itself is large. If $D_t$ is comparable to the large dose, it too would alter the number of competitors and although an increase in $S$ may still be detected, the extent of the increase would be false and an incorrect conclusion reached.

The above discussion assumes that the competitors are empty traps. If the competitors are recombination centers the opposite effect, that is, a decrease in sensitivity would be observed after the large irradiation. If the competitors are a mix of traps and centers, either an increase or a decrease could be observed depending upon the specific circumstances.

The general observation of a change in sensitivity after irradiation is termed sensitization (or de-sensitization if the sensitivity is reduced). Sensitization/de-sensitization can also be observed after thermal annealing as competing centers are thermally emptied or their defect structure is modified, or after optical bleaching, especially using UV light, when deep traps and centers can be optically emptied (or filled). In each case, the essential effect is to change the concentration of the competing traps and recombination centers and thereby change the sensitivity.

### 6.2.5   High Dose Effects

#### 6.2.5.1   Loss of Sensitivity

In the discussions of dose-response functions so far, the effect of high doses has been said to induce saturation along with its corresponding sublinearity. Saturation simply implies that all traps are full and no matter how much additional radiation is absorbed no further signal increase is obtained. A frequently observed effect in the dose-response relationship of many, if not most, luminescence dosimeters at high doses is in fact not saturation but a loss of sensitivity such that the signal begins to decrease rather than stay stable as the dose increases. Often this is explained using the unfortunately rather vague term of "radiation damage," without a discussion of what exactly is being damaged and what the damage mechanisms are.

Alteration of the trapping and recombination centers or of the surroundings to those defects by ion displacement or ionization events could in principle occur at any dose, but specific models for such processes and how they cause a decrease in sensitivity at high doses are rarely described to account for the high-dose de-sensitization process. An alternative explanation for a decrease in sensitivity at high doses is available, however. For those luminescence dosimeters that display non-linear dose-response relationships due to competition effects, the sensitivity decrease at high doses can be explained by the very same effects that cause non-linearity at lower doses – that is, competition.

Consider the model shown in Figure 6.13a. During trap emptying, electrons released from the main electron trap (MET) recombine at the radiative recombination center (RRC) to produce TL or OSL. However, the RRC competes with a deep electron trap (DET) and an NRRC. The arrows in the figure are described in the figure caption. Figure 6.13b shows a set of TL peaks resulting from the release of electrons from the METs recombining at the RRCs, as a function of increasing dose. The third figure, Figure 6.13c, shows the resulting dose-response relationship on a log-log plot. In the initial dose region, the dose-response curve displays supra-linear TL growth, with $f(D) \approx 2$, but as the dose increases, $f(D)$ reduces and becomes highly sublinear with a severe decrease in response at high doses. This effect has nothing to do with "radiation damage" and everything to do with competition.

The model of Figure 6.13a contains both empty competing traps (DETs), which decrease with dose, and filled NRRCs, which increase with dose. Competing traps and competing recombination centers act in opposite senses as a function of dose. Initially, both the DETs and the NRRCs

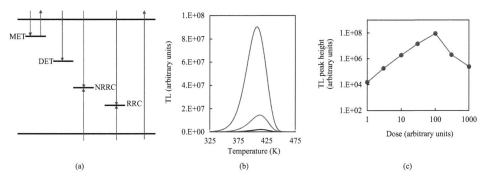

(a)                                     (b)                                     (c)

**Figure 6.13** (a) A 2T2R competition model consisting of an MET and an RRC. Electrons released from MET and recombining at RRC produce either TL or OSL. A DET acts as a competing electron trap and an NRRC acts as a competing recombination center. The arrows indicate possible transitions between energy levels during filling and emptying, including ionization across the energy gap during irradiation, capture of holes by RRC and NRRC, capture of electrons by MET and DET, recombination of electrons with holes at RRC and NRRC, and release of electrons from MET during heating (for TL) or optical stimulation (for OSL). (b) TL resulting from the thermal release of electrons from METs at different doses. (c) The resulting dose-response curve showing a decrease in sensitivity for doses > ~100 dose units. Note, DET and NRRC are considered to be sufficiently deep that electrons and holes trapped at these defects are not released during TL measurement and remain there during subsequent irradiation.

are empty and so the empty electron traps are the only competitors at the start of the irradiation. As a result, at low doses (below ~100 dose units in this particular example) competition by the empty electron traps dominates. As these gradually fill, proportionally more and more electrons per dose are trapped by the METs during trap filling. During trap emptying, more and more released electrons find their way to the RRCs to produce TL, and the net result is that the signal grows supralinearly. As the dose increases further and the effectiveness of the DETs as competitors diminishes, competition from recombination at the filling NRRCs grows and begins to dominate. At high doses competition from the NRRCs is now the dominant competitive process and the sensitivity of the system decreases. Sublinearity and a loss of TL at high doses results.

---

### Exercise 6.6    High Dose

(a) Write the rate equations describing the transitions shown in the model of Figure 6.13a for the irradiation and trap emptying phases.

(b) Assume a set of values for the parameters used in the rate equations. (You choose the values. Your goal is to set the values such that the DET acts as an effective competing trap and the NRRC acts as an effective competing recombination center.)

(c) Using suitable software, solve the rate equations for different doses and evaluate the TL curve at each dose. Assume $\beta_t = 1 \text{ K.s}^{-1}$, $E_t = 1.0 \text{ eV}$, and $s = 10^{12} \text{ s}^{-1}$ and that DET and NRRC are thermally disconnected. (Recall also that $I_{TL} = \left| \dfrac{dm_{RRC}}{dt} \right| = n_c m_{RRC} B_{RRC}$, where $n_c$ = the concentration electrons in the conduction band ($\text{m}^{-3}$), $m_{RRC}$ ($\text{m}^{-3}$) is the concentration of holes at the RRC, and $B_{RRC}$ ($\text{m}^3\text{s}^{-1}$) the recombination probability for electrons with trapped holes at the RRC.)

(d) For doses ranging over several orders of magnitude, do you confirm the dose-response relationship seen in Figure 6.13c? If not, why not?

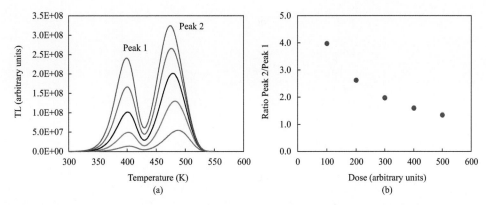

**Figure 6.14**   Changes in TL glow-curve shape as a function of dose. (a) A model based on that of Figure 6.13a, but with two traps (i.e. two METs), leading to two TL peaks, peak 1 and peak 2. The parameters have been chosen such that the two traps fill at different competitive rates as the dose increases, meaning that the ratio of the two traps changes with dose, as shown in part (b).

Although it cannot be claimed that all decreases in the luminescence response at high doses are due to such effects exclusively, it can perhaps be said to be unlikely that any one dosimetry material contains competitors that are of one type only – i.e. either traps or recombination centers. More likely is that a given dosimetry material contains competitors of both types such that competition by both is occurring simultaneously. This "competition between competitors" depends upon their relative concentrations, their relative capture and recombination cross-sections and, of course, the dose. As a result, an effect such as that illustrated in Figure 6.13 is perhaps the norm rather than the exception.

### 6.2.5.2   *TL and OSL Changes in Shape*

Not all TL peaks or individual OSL signals grow at the same rates with dose, nor do they necessarily display the same degree of supralinearity or sublinearity. The result is that the ratio of two TL peaks in a given glow curve, or of two OSL signals in a given OSL curve, may be dose dependent. TL results for some example calculations are shown in Figure 6.14, using a 3T2R model. The model is similar to that shown in Figure 6.13a, except that there are two different METs along with the deep competing electron trap DET, an RRC and an NRRC. Figure 6.14a shows the glow curve with two peaks corresponding to the two METs (using $E_{t1} = 0.9$ eV and $E_{t2} = 1.2$ eV, along with $s_1 = s_2 = 10^{12}$ s$^{-1}$ and $\beta_t = 1$ K.s$^{-1}$). The glow curves are calculated using different doses (arbitrary units) and the ratio of the two peaks is seen to vary with the applied dose as a result of the different competition properties of the different traps and centers (Figures 6.14b).

Similar calculations can be made for OSL. An example is illustrated in Figure 6.15 for the same model as used for Figure 6.14, but with the optical stimulation probabilities for trap 1 and trap 2, namely $p_1 = 8$ s$^{-1}$ and $p_2 = 4$ s$^{-1}$, respectively. The net result is that the overall OSL curve decays faster as the dose increases, not because the decay rates of the individual traps have changed (these are fixed at the above values) but because the balance of the two traps contributing to the net OSL has changed as a function of dose, with trap 1 being more dominant at higher doses.

Changes in the peak-height-ratios of TL peaks in a glow curve or of the overall shape of an OSL decay curve are strong indicators of possible interactive, competition effects in the material under study.

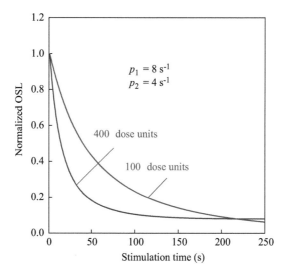

**Figure 6.15**   Changes in CW-OSL decay curve shape as a function of dose. The simulations of OSL use a similar model to that used for Figure 6.14, except that the two METs are emptied optically with the optical detrapping probability of trap 1 being twice as big as that of trap 2. The result is that the overall OSL decay, which is the sum of the OSL decay due to the two separate traps, changes as function of dose, becoming faster as the dose increases.

---

**Exercise 6.7    OSL decay as a function of dose**

(a)  Write rate equations describing OSL using the 3T2R model, with two METs, a DET, an RRC, and an NRRC.
(b)  Using parameter values of your own choosing, solve the equations to show that the ratio of the OSL from the two METs, trap 1 and trap 2, can vary with dose in a similar way to that shown in Figure 6.14 for TL.
(c)  How does the decay of the overall OSL (due to both traps) change as a function of dose (e.g. Figure 6.15)?
(d)  How does the area under the net OSL curve change as a function of dose? How does the initial intensity of the net OSL curve change as a function of dose? Explain your observations.
(e)  Repeat the above calculations using different parameter values in order to understand how these affect the results.

---

### 6.2.6   Charged Particles, Tracks, and Track Interaction

An unstated but assumed condition of the analyses so far is that the radiation is absorbed uniformly throughout the material. The closest one might get to achieving this is through exposure of the dosimeter to a broad beam of energetic photons which passes entirely through the material. Such a case may be achieved, for example, with highly penetrative, high-energy $^{60}$Co or $^{137}$Cs gamma rays. With a concomitant assumption of a uniform distribution of trapping defects within the material, one might envision a situation sketched conceptually in Figure 6.16a wherein are depicted trapped charges (electrons (filled circles) and holes (empty circles)) in a uniformly irradiated dosimetry material. (The diagram does not include spatial association between traps and centers as depicted in Figure 6.12 and discussed in the Section 6.2.3.)

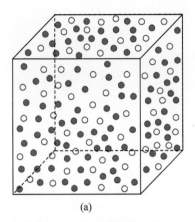 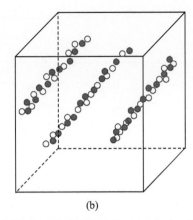

(a) (b)

**Figure 6.16** Schematic visualization of (a) an isotropic distribution of trapped electrons (filled dots) and trapped holes/recombination centers (open dots) in a material irradiated with a highly ionizing, broad photon beam, and (b) a highly non-isotropic distribution of such traps and recombination centers in a material irradiated with a low fluence of energetic, charged particles.

An entirely different situation is sketched in Figure 6.16b, where a notional distribution of trapped charges is shown following irradiation with energetic charged particles. Here, the ionization density is extremely high along the core of the track of the charged particle and drops off rapidly with radial distance away from the track. For example, a 4 MeV alpha particle deposits 90% of its energy within 100 μm of the track center and 98% within 200 μm of the track center leading to a highly non-isotropic distribution of dose within the material.

There are several models to describe the radial distribution of dose in such cases. Some exemplary publications include those by Butts and Katz (1967), Chaterjee and Schaefer (1976), Waligorski et al. (1986), and Geiβ et al. (1998), to which the reader is referred for more specific information. In general, and independent of the details of the models, the radial dose distribution can be said to depend on the particle's energy (in keV), mass (in kg), and charge (in C). An important derived property is the particle's rate of energy loss as the particle penetrates the material, described by its linear energy transfer (LET, in $keV.\mu m^{-1}$). Extremely high doses, as high as $10^6$ Gy or greater, can be expected at the track's core, but the dose drops quickly by orders of magnitude a few nanometers away from the core center, depending on the charged-particle species and its energy.

Figure 6.17 illustrates Monte Carlo calculations of the energy deposition patterns resulting from the passage of three different charged particles in water, showing the high ionization and dose deposition along the core of the track and the spread of energy deposition due to secondary delta electrons away from the trap. The energy deposition density decreases rapidly the further from the track. Note the distance scale, which is approximately the same for each charged particle.

A schematic radial dose distribution (RDD) is shown in Figure 6.18a where the numbers represent approximately what might be expected from a 120.4 $MeV.u^{-1}$ Fe ion in an $Al_2O_3$:C dosimeter. The colored figure below the RDD is a section from a Monte Carlo calculation showing the ionization events from the secondary delta-electron tracks from a 120.4 $MeV.u^{-1}$ Fe ion (see Figure 6.17c), illustrating the pattern of energy deposition. With this consideration it is straight forward to see that a dosimeter irradiated with charged particles yields a highly anisotropic, trapped-charge distribution, and this is without even considering the possible spatial association between traps and recombination centers that may exist within the dosimeter.

A number of important points can now be made.

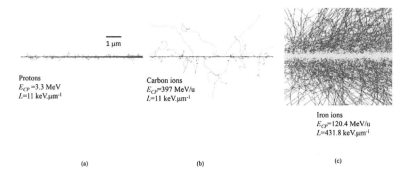

**Figure 6.17** Track structure for three different types of energetic charged particle: (a) 3.3 MeV protons, (b) 397 MeV/u carbon ions (both calculated using the program TOPAS, ver. 3.0.1), and (c) 120.4 MeV/u Fe ions (calculated using GEANT4) . In the first two the LET is the same (11 keV. $\mu m^{-1}$) illustrating that the pattern of energy deposition depends heavily on the energy $E_{CP}$ and charge of the charged particle. A densely ionizing, highly energetic Fe ion ($L = 431.8$ keV.$\mu m^{-1}$) produces a broader and denser pattern of energy deposition. The scale on each figure is approximately the same, as shown. Figures (a) and (b) from Grosshans et al. (2018). Licensed under CC BY 4.0. Figure (c) reproduced from Sawakuchi (2007), with permission of author G. O. Sawakuchi.

### 6.2.6.1 Dose and Fluence Dependence: Low Fluence

Imagine a single track and the net TL or OSL signal resulting from that track. The dose along the track can be very high, but the irradiated volume is very small. The volume irradiated depends upon the energy imparted to the secondary electrons by the primary particles and the subsequent cascade of collision events produced by these secondary electrons as they lose their energy through subsequent ionization events. The secondary electrons (known as delta rays) eventually lose sufficient energy to become trapped at trapping sites within the material; their charge compensating holes also become trapped, and these in turn give the ultimate TL or OSL signals. For a given charged-particle type, of a given charge, the more energetic the particle the broader the track radius and the larger the irradiated volume.

Imagine further that the dose-response characteristic of a dosimeter is linear-supralinear-sublinear. A question arises as to whether the TL/OSL signal (intensity and shape) would be characteristic of the high-dose region of the dose-response curve (the sublinear region), the intermediate-dose region (the supralinear region), or the low-dose (or linear region). This question is illustrated schematically in Figure 6.18b. The figure shows a schematic dose-response curve for a presumed gamma irradiated sample displaying linear-supralinear-sublinear characteristics. The arrows map this curve against the doses expected to be found in the irradiated volume of the charged-particle track. Although at first sight it might be expected that those parts of the track with the highest dose would dominate the TL/OSL signal, two effects mitigate against this. Firstly, at very high doses the sensitivity is often seen to decrease (as was discussed in Section 6.2.5.1). For example, in the schematic illustration shown in Figure 6.18b, a lower TL response is expected for a dose of $\sim 10^7$ Gy than for $\sim 10^5$ Gy. Secondly, the region of high dose irradiation is a very small volume compared to the regions of intermediate and low doses. Which regions will dominate the TL/OSL signal?

Consider two luminescence dosimeters, A and B. (They may be TLDs, OSLDs, or RPLDs.) If the dose deposited in each is the same, but the mass of A (say) is twice that of B, then the luminescence emitted from A will be twice that of B. That is, the luminescence intensity is proportional to the total energy deposited, not the total dose. The luminescence is proportional to dose only if the masses of the dosimeter are the same, or if the signal is normalized to the

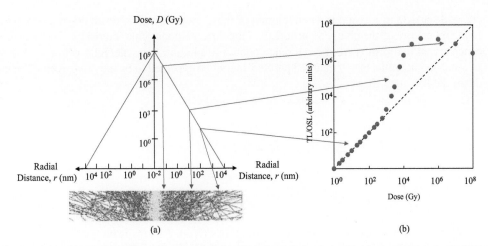

**Figure 6.18** (a) Schematic radial dose distribution (log-log representation) due to an energetic charged particle compared to the ionization pattern from a similar particle (see Figure 6.17c). The numbers represent approximately those that characterize a 120.4 MeV.u$^{-1}$ Fe ion in Al$_2$O$_3$:C (adapted from Sawakuchi et al. 2008b). (b) A schematic dose-response characteristic for a gamma-irradiated luminescence dosimeter with a linear-supralinear-sublinear response, including a loss of sensitivity at high doses. The red arrows show the general mapping of doses in regions surrounding the track core with the dose-response characteristic of the dosimeter. The net TL/OSL response from a single track is a convolution of the dose-related TL/OSL properties from the different regions surrounding the track.

masses of the dosimeters. The energy deposited by the particle within the track $E_{track}$ is determined by the LET, $L$:

$$E_{track} = \int_0^d L.dx \qquad (6.26)$$

where $x$ is the direction of particle travel and $d$ is the sample thickness (assuming the incident particle is perpendicular to the sample surface and travels completely through the sample). The macroscopic dose delivered to the sample is:

$$D_{CP} = L\frac{\Phi}{\rho}, \qquad (6.27)$$

where $\Phi$ is the particle fluence (number of particles per unit area, m$^{-2}$), and $\rho$ is the density of the material (kg.m$^{-3}$).

For two charged particles with the same $L$, but different energies $E_{CP}$ (e.g. the proton and C ions shown in Figures 6.17a and 6.17b) the energy deposited in the track $E_{track}$ is the same and the macroscopic dose $D_{CP}$ is the same. However, the particle with the largest energy $E_{CP}$ will have a wide track radius $R$, as seen in Figure 6.17 due to a larger energy being imparted to the secondary delta electrons. As a result, the irradiated volume (and irradiated mass) for the particle with the larger energy $E_{CP}$ is also larger than for the particle with the smaller energy $E_{CP}$. As a result, the net dose in the track $D_{track}$ is larger for particle with the smaller energy. That is, the microscopic dose $D_{track}$ is different for the two particles despite the fact that the macroscopic $D_{CP}$ is the same.

Since the luminescence properties of the material vary with dose (sensitivity, TL glow-curve shape, OSL decay-curve shape), then the overall luminescence signal following charged-particle irradiation is a convolution of these properties. The result depends upon the LET and energy of the charged particle, and the dose-response characteristics. The luminescence signal that results from a reference photon irradiation (e.g. energetic gamma) of dose $D_{ref}$ and an energetic, heavy charged particle cannot be expected to be the same even if the net doses are the same, $D_{ref} = D_{CP}$.

Gieszczyk et al. (2014) summarized this mathematically:

$$I_{CP} = K \sum_{i=1}^{k} \left( \int_{D_{min}}^{D_{max}} I_{ref}^i(D) h(D) dD \right) \tag{6.28}$$

where $I_{CP} = \sum_{i=1}^{k} I_{CP}^i$ is the net TL or OSL signal due to the charged particle and is the sum of

the TL/OSL signals due to the individual trap types ($I_{CP}^i$). $I_{ref}^i(D)$ is the signal due to the $i^{th}$ trap type (i.e. the $i^{th}$ TL peak or the $i^{th}$ individual OSL signal) following a dose $D$ of reference radiation (usually gamma). The term $h(D)$ is a function describing the frequency of occurrence of a given dose $D$ within a single particle track; it is directly related to the radial dose distribution function within the track. $K$ is a scaling constant. The integration limits are from $D_{min}$, which occurs at the limit of the range of the secondary delta electrons (at $r = R$), and $D_{max}$ which occurs at the track center (at $r = 0$). Assuming multiple trap types contribute to the net TL or OSL signal ($i = 1 \dots k$), Equation 6.28 accounts for the fact that the individual TL or OSL signals vary with dose in different ways, as was illustrated in Figures 6.14 and 6.15.

An efficiency term, normally expressed as a function of the particle's LET, can be defined:

$$\eta_L = \frac{I_{CP} / D_{CP}}{I_{ref} / D_{ref}}, \tag{6.29}$$

where $I_{CP}$ is the luminescence signal due to the charged-particle dose $D_{CP}$, and $I_{ref}$ is the luminescence signal due to the reference dose $D_{ref}$. Dose here is defined as dose to the material. Even if $D_{CP} = D_{ref}$, $\eta_L$ does not necessarily equal 1, and can be > 1 or < 1, depending upon which parts of the radial dose distribution and dose-response curves dominate the luminescence signal (Figure 6.18).

Ideally, one would like to choose a reference radiation that produces the same secondary delta-electron energy spectrum as the charged particle under study. Since this is difficult (as well as impractical since it would change with the particle under study), commonly available and highly ionizing gamma radiation sources, either $^{60}$Co or $^{137}$Cs, are usually used for the reference fields. As to the dose, it is important to select a reference dose that is on the linear part of the dose-response curve so that sensitization effects (see Section 6.2.4) do not occur during reference irradiation.

An additional point to be made is that at low fluence the luminescence is always linear with fluence, and therefore linear with $D_{CP}$ (Equation 6.27). This can be understood by considering that whatever the value of $I_{CP}$ due to one track, then if there are two tracks $I_{CP}$ will double; if there are three tracks $I_{CP}$ will treble, and so on. Therefore, $I_{CP}/D_{CP}$ is a constant at low fluence. The caveat "at low fluence" is necessary because at higher fluence, as tracks get closer to each other such that they overlap or cross, the concept of track interaction has to be considered.

### 6.2.6.2    High Fluence: Track Interaction

Once the particle fluence becomes high enough that electrons from one track have a probability of reaching an adjacent track (track interaction) nonlinear effects can occur as a function of fluence. The nonlinear effects are not simply a result of the tracks interacting but are the result of competition. At this stage in the discussion competing empty electron traps (DETs) and filled NRRCs must be introduced. Consider Figure 6.19 illustrating two conceptual, charged-particle tracks, of radius $R$ and length $l$, where $l \gg R$. The material is assumed to have traps and recombination centers, METs, DETs, RRCs, and NRRCs. The empty DETs and filled NRRCs act as competitors to the TL or OSL process. Within the tracks where the ionization is high, there are high concentrations of filled METs, DETs, RRCs, and NRRCs, whereas in the regions between the traps the opposite holds true, i.e. there are large concentrations of empty METs, DETs, RRCs, and NRRCs. Only empty DETs and filled NRRCs can act as competitors.

An electron released from one of the METs inside a track is unlikely to be captured by empty competing DETs inside the track since the latter are essentially all full in this region (i.e. the intra-track region). However, RRC and NRRC concentrations are large and the probability of producing a TL or OSL photon depends on the relative concentrations of these centers. If an electron can escape the track and enter the inter-track region, it will now be in a region where there are many empty, competing DETs, but few RRCs and NRRCs

Let $P_{intra}$ be the probability that an electron from an MET recombines at an RRC inside the track and produces a TL or OSL photon, and $P_{inter}$ be the probability that an electron from an MET inside one track successfully migrates the inter-track region and recombines with an RRC inside a neighboring track. Then the net probability of producing a photon is:

$$P_{tot} = P_{intra} + P_{inter} < 1. \qquad (6.30)$$

Both $P_{intra}$ and $P_{inter}$ are individually $< 1$. Note that although the probability of capture by an empty DET may be high in the inter-track region, the probability of recombining (at either an RRC or NRRC) is negligible since the concentrations of these are very small in these low-dose regions. (A caveat needs to be added here. This statement assumes that there is no initial concentration of recombination centers in the material. In many dosimetry materials, for

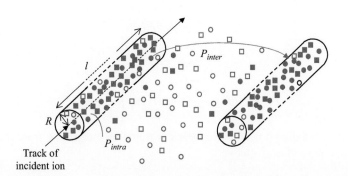

**Figure 6.19** The concept of track interaction. Once the fluence is high enough such that the tracks are close enough, an electron that escapes from one track may be able to migrate and recombine in an adjacent track. The probability that an electron in one track recombines within that track to produce luminescence is $P_{intra}$. The probability that it reaches a nearest-neighbor track and then produces luminescence is $P_{inter}$. The filled circles and squares illustrate filled METs, DETs, RRCs, and NRRCs, while empty circles and squares illustrate empty METs, DETs, RRCs, and NRRCs.

example, $Al_2O_3$:C, this is not true and a large concentration of pre-existing RRCs exist prior to irradiation.)

The relative sizes of $P_{intra}$ and $P_{inter}$ will depend on such parameters as the mean distance between tracks (the track distribution function), the radius of each track (the ionization density within each track), the charge carrier mean-free-paths (different in the intra- and inter-track regions), the concentrations of competing traps and recombination centers, and the temperature (since the mobility will be temperature dependent). Overall, the TL sensitivity will be a function of the charged particles' LET, energy and charge, the material's defect density and distribution, and the temperature–dependencies, which have all been observed experimentally.

A term $P_{intra}^{tot}$ can be defined as the probability that a freed electron from inside the track will either recombine at an RRC within the track or be lost to a competitor (either DET or NRRC) inside the track. Note that $P_{intra}^{tot} \neq P_{intra}$ since the latter is the probability that only recombination at RRCs will occur inside the track. The two terms are only equal if $k_i = 0$. Nevertheless, $k_i = 0$ is an unrealistic assumption and it is always safer to assume that $0 < P_{intra} < 1$ (and $0 < P_{inter} < 1$).

To examine the probability that an electron that escapes a track will reach another track and recombine there, the distribution of adjacent tracks and the probability that a freed electron will reach one of those tracks needs to be considered. This problem has been examined in several publications by Horowitz and colleagues (summarized by Horowitz et al. 2019). These authors considered the probability that an electron will migrate from one track to another to be given by the product of a term $\mu_o$ and a solid angle, geometrical factor $g(r,R)$, where $r$ is the distance between the tracks and $\mu_o$ is the mean free path for the freed electron in the inter-track region (the subscript "$o$" refers to the region outside the tracks). Horowitz and colleagues approximate $g(r,R)$ to $2R/r$. The term $\mu_o$ can be written $\mu_o = 1/(\sigma_k k_o)$ and $k_o = N_{k_o} - n_{k_o}$, where $N_{k_o}$ and $n_{k_o}$ are the total concentrations of DETs and filled DETs, respectively, in this region; $\sigma_k$ is the capture cross-section for the DETs.

If the incident ion fluence is $\Phi$, then the probability that a given track has its $j^{th}$-nearest neighbor track at a distance $r$ may be written $P_j(\Phi,r)$ and for the 1st, 2nd, and 3rd nearest-neighbors it may be written (Lavon et al. 2015):

$$P_1(\Phi,r)\mathrm{d}R = \Phi 4\pi r^2 \exp\left\{-\frac{4\pi r^3}{3}\Phi\right\}\mathrm{d}R, \tag{6.31a}$$

$$P_2(\Phi,r)\mathrm{d}R = \frac{16}{3}\Phi^2 4\pi^2 r^5 \exp\left\{-\frac{4\pi r^3}{3}\Phi\right\}\mathrm{d}R, \tag{6.31b}$$

and

$$P_3(\Phi,r)\mathrm{d}R = \frac{64}{9}\Phi^3 4\pi^3 r^8 \exp\left\{-\frac{4\pi r^3}{3}\Phi\right\}\mathrm{d}R. \tag{6.31c}$$

The probability of finding a neighboring track is found by integrating from $R$ to infinity (the sample dimensions) and is given by:

$$P_o = \int_R^\infty g(r,R)\exp\{-r/\mu_o\}\left[\sum_j P_j(\Phi,r)\mathrm{d}r\right]. \tag{6.32}$$

where the sum is over all nearest-neighbor tracks ($j = 1, 2, 3 \dots$ etc.). Thus:

$$P_{inter} = \left(1 - P_{intra}^{tot}\right)P_o P_{intra} = \left(1 - P_{intra}^{tot}\right)\left[\int_R^\infty g(r,R)\exp\{-r/\mu_o\}\sum_j P_j(\Phi,r)\mathrm{d}r\right]P_{intra}. \tag{6.33}$$

Here, the first term $\left(1 - P_{intra}^{tot}\right)$ represents the probability that the electron escapes the initial track. The middle term $P_o$ is the probability that it successful navigates the inter-track region and reaches the neighboring track, and the final term $P_{intra}$ is the probability that it recombines in the neighboring track to produce a TL or OSL photon. $P_{intra}$ is assumed to be identical for all tracks.

At low $\Phi$, when $r$ is large, then $P_{inter} \ll 1$, $P_{tot} = P_{inter} + P_{intra} \approx P_{intra}$, and $P_{intra} \leq 1$. In this case the TL or OSL output depends only on the properties of the charged particle and the dosimetry material, and scales linearly with $\Phi$. At higher fluences, $r$ decreases and $P_{inter}$ increases, along with a likewise increase in $P_{tot}$. As a result, supralinearity in the TL or OSL signal occurs with increase in fluence.

The supralinearity index of charged-particle irradiation can now be calculated using Equation 6.3, but written here as a function of fluence $\Phi$, namely $f(\Phi)$. (Dose $D$ is proportional to fluence $\Phi$ (Equation 6.27) but the proportionality constant depends on the charged particle used. The supralinearity index is written this way in this text since it is often measured and represented as a function of $\Phi$ in published experimental results. Otherwise, there is no difference between $f(\Phi)$ and $f(D)$.) Recalling that $P_{tot} = P_{inter} + P_{intra}$, then the TL or OSL intensity $I \propto n_i P_{tot}$. Here $n$ is the number of electrons trapped in the METs and only those inside the track (subscript "$i$") are considered since those trapped in the inter-track region are negligible by comparison. Using Equation (6.3), $f(\Phi)$ may be written:

$$f(\Phi) = \frac{\left(n_i P_{tot}\right)_\Phi / \Phi}{\left(n_i P_{tot}\right)_{\Phi_l} / \Phi_l} = \frac{\left(n_i\right)_\Phi \left(P_{intra} + P_{inter}\right) / \Phi}{\left(n_i\right)_{\Phi_l} \left(P_{intra}\right) / \Phi_l} = \frac{\left(n_i\right)_\Phi \left[P_{intra} + \left(1 - P_{intra}^{tot}\right) P_o P_{intra}\right] / \Phi}{\left(n_i\right)_{\Phi_l} P_{intra} / \Phi_l}. \quad (6.34)$$

Since $n_i \propto \Phi$, then:

$$f(\Phi) \propto \frac{P_{intra} + \left(1 - P_{intra}^{tot}\right) P_o P_{intra}}{P_{intra}} = 1 + \left(1 - P_{intra}^{tot}\right) \int_R^\infty g(r,R) \exp\{-r / \mu_o\} \left[\sum_j P_j(\Phi,r) dr\right]. \quad (6.35)$$

The first term (number 1) on the right-hand side of Equation 6.35 indicates linear growth with $\Phi$ and the second term indicate supralinear growth due to interaction between the tracks.

Figure 6.20a shows an example distribution $P_j(\Phi,r)$ for first-nearest-neighbor tracks ($j = 1$; Equation 6.31a) and for several different fluences $\Phi$. As expected, the distribution shifts to the left (smaller $r$) as the fluence increases. Using these values, along with Equation 6.35, the variation of $f(\Phi)$ for first-nearest-neighbor tracks is shown in Figure 6.20b showing linearity ($f(\Phi) = 1$) at low $\Phi$ and supralinearity ($f(\Phi) > 1$) at high $\Phi$.

---

**Exercise 6.8   Track overlap (I)**

Approximating $g(r,R)$ to $2R/r$, estimate $f(\Phi)$ for $\Phi$ varying from $10^{10}$ ions.m$^{-3}$ to $10^{14}$ ions.m$^{-3}$ for second-nearest neighbor tracks ($j = 2$), and for third nearest neighbor tracks ($j = 3$), using Equations 6.31b and 6.31c. Assume $P_{intra} = 0.8$, and numerically calculate the integral in Equation 6.35 from $r = 10^{-8}$ m to $2.5 \times 10^{-5}$ m. Vary the integration step size until you obtain consistent results.

---

Considering a random distribution of charged particles over the incident surface, the mean distance $r_{mean}$ between the centers of two adjacent tracks is:

$$r_{mean} = 2 \sqrt{\frac{1}{\pi \Phi}}. \quad (6.36)$$

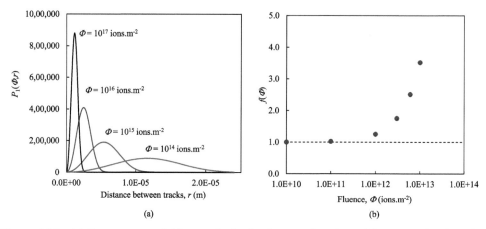

**Figure 6.20** (a) First-nearest-neighbor track distribution as a function of fluence, using Equation 6.31a. (b) The calculated $f(\Phi)$-versus-$\Phi$ curve for first-nearest-neighbor tracks, using Equation 6.35. For purposes of these illustrations, $P_{intra}^{tot}$ was assumed to be 0.8, and $\mu_o$ was assumed to be $10^{-7}$ m. The integral in Equation 6.35 was determined numerically.

The point of track overlap will then occur when the track radius $R = r_{mean}/2$. Thus, the threshold fluence $\Phi_{th}$ for track overlap is:

$$\Phi_{th} = \frac{1}{\pi R^2}.$$ (6.37)

Readers may see a conceptual similarity between the spatial association model described by Mische and McKeever (1989) and discussed in Section 6.2.3, and the track interaction models of Horowitz and colleagues discussed above. In both cases, a released electron may recombine either within the associated pair (Mische and McKeever) or within the track (Horowitz et al.), or may escape the pair or track and recombine elsewhere to produce TL/OSL. There are differences in detail and definitions but the concepts expressed are somewhat similar. These ideas were unified and given a mathematical footing by Horowitz and colleagues (Horowitz and Rosenkrantz 1990; Horowitz et al. 1996a, 1996b; Horowitz 2001) who named the theory the Unified Interaction Model (UNIM). The UNIM is able to predict the dependence of the supralinearity index of both gamma dose $D$ and particle fluence $\Phi$ and will be discussed further in Part II.

---

### Exercise 6.9    Track overlap (II)

On the web page, under Exercises and Notes, Chapter 6, Exercise 6.9, there is a schematic plot of two RDDs (on a log-log plot) at two different fluences, $\Phi_1$ and $\Phi_2$, where $\Phi_2 > \Phi_1$, represented by blue lines for $\Phi_1$ and orange dotted lines for $\Phi_2$. As the fluence increases, the spacing between the tracks decreases such that overlap begins. If $r_{mean}$ is the mean distance between the tracks, then a minimum in the dose ($D_{min}$) occurs when the distance $r = r_{mean}/2$. As the fluence increases, $r_{mean}$ decreases and $D_{min}$ increases.

Also in Exercise 6.9 on the web page is a table of RDDs ($D(r)$ versus $r$) for charged particles of different energies. (The data are approximately the values that might be obtained for energetic protons in water (Waligorski et al. 1986). The RDD has been approximated to $D(r) \propto r^{-2}$ (which is a good approximation for most charged particles).)

(a) Using these data calculate the fluences that would be required for $D_{min}$ to be attained at distances of $r = 2.5 \times 10^2$ nm, $4.1 \times 10^4$ nm, and $6.5 \times 10^4$ nm from a track center.

(b) What are the values of $D_{min}$ at these locations for each of the particles listed in the table (i.e. for particle energies $E_{CP} = 1$ MeV, 10 MeV, 20 MeV, and 50 MeV)?

### 6.2.7 RPL

#### 6.2.7.1 Buildup during Irradiation: A Special Kind of Supralinearity

In the analyses discussed so far regarding competition from either empty traps or filled recombination centers, both during irradiation and during trap emptying, an assumption has been that the concentration of the electron and hole traps (usually denoted as $N$ or $M$) remains fixed during the irradiation and during the TL or OSL readout periods. Only the number of filled traps $n$ or filled recombination centers $m$ vary. A special kind of supralinearity during irradiation arises when the concentration of the center itself ($N$) changes during the irradiation and post-irradiation periods. Examples of this type of behavior were already discussed in Chapter 4 when the buildup of RPL after irradiation was discussed.

As noted in Chapter 4, electronic and ionic transport processes can occur post-irradiation to create the phenomenon of buildup. Examples were discussed for LiF, $Al_2O_3$, and Ag-doped phosphate glasses. In each case, a radiation-induced defect (A) or defects (A and B) reacted post-formation to produce a defect of the type AB. The composite defect AB is the source of the RPL signal. Since such processes can occur following irradiation they also can occur during irradiation and the result is a supralinear growth of the RPL signal with irradiation time (i.e. dose).

Consider an irradiation period $t_{irr}$, divided into $k$ segments, each separated by time intervals $dt$, as shown in Figure 6.21. Consider that the RPL signal due to each dose element (i.e. the dose $dD$ delivered in each time interval $dt$) builds up after the period $dt$ and that the buildup curve has the form $I_{BU}(t)$. Further assume that $I_{BU}(t)$ has the same form for each subsequent dose segment, but is shifted in time by an amount $dt$.

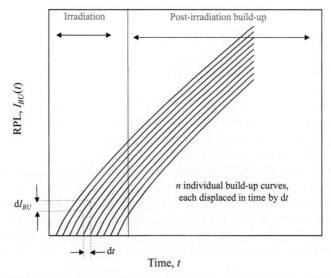

**Figure 6.21** Schematic representation of the buildup of RPL during irradiation divided into elemental units d$I_{BU}$ in time intervals dt.

At any given time $t$ during the irradiation ($t < t_{irr}$) the net intensity at that point ($I_t^{net}$) is $I_t^{net}(t) = \sum_{i=1}^{k_t} I_{BU}^i(t)$, where $k_t$ is the number of buildup segments and $I_{BU}^i(t)$ is the intensity of the $i^{th}$ buildup curve at the time $t$. In other words, $I_t^{net}(t)$ is given by the integral of $I_{BU}(t)$ from $t = 0$ to $t$, namely:

$$I_t^{net}(t) = \frac{1}{t_D} \int_0^t I_{BU}(t)\,\mathrm{d}t. \tag{6.38}$$

As noted in Chapter 4, $I_t^{net}(t)$ may be called the real-time RPL $I_{RT}(t)$ (see Equation 4.13). One assumption, that cannot be tested experimentally, is that the RPL buildup due to each elemental dose d$D$ can be described by the same form $I_{BU}(t)$. Clearly, it is not possible to deliver a dose in an elemental time d$t$, but a dose can be delivered in as short a time period as possible, e.g. 1 s. If the buildup curve obtained in these circumstances passes through zero – i.e. there is negligible signal accumulation (buildup) during the 1 s exposure itself – then it can be considered that the buildup curve is a good approximation to the RPL buildup curve following the elemental irradiation time d$t$, and it is then further assumed that this is the same curve for each elemental irradiation.

In order to evaluate Equation 6.38, an expression for $I_{BU}(t)$ is needed. As discussed in Chapter 4 this may be obtained empirically by fitting the experimentally determined buildup curve, following the short (1 s) irradiation, to obtain the curve shape. An example of a simulated buildup curve obtained by solving the differential Equations 4.5a-f was shown in Figure 4.8b. The curve may be fitted to an empirical expression of the form given in Equation 4.12 and re-written here as:

$$I_{BU}(t) \approx \sum_{j=1}^{k} K_j \left(1 - \exp\left\{-\frac{t}{\tau_j}\right\}\right) \tag{6.39}$$

assuming $I_{BU}(t) = 0$ at $t = 0$. For example, for $k = 2$ and using Equation 6.38 this leads to:

$$I_{RT}(t) = \frac{1}{t_D} \int_0^t I_{BU}(t)\,\mathrm{d}t = \frac{1}{t_D}\left(K_1\tau_1 \exp\{-t/\tau_1\} + K_2\tau_2\exp\{-t/\tau_2\} + (K_1 + K_2)t + C\right), \tag{6.40}$$

where $C$ is the integration constant and $t_D$ is the value of $t_{irr}$ needed to obtain a delivered dose $D$. The parameters $\tau_1$ and $\tau_2$ define the characteristic buildup time for the buildup process; the smaller the values of $\tau_1$ and $\tau_2$, the faster the buildup. Equation 6.40 has the form of an upward curving, i.e. supralinear, growth as a function of irradiation time if the RPL is measured during the irradiation in real time. An example is shown in Figure 6.22 where a buildup curve using Equation 6.39 is shown in Figure 6.22a (with the parameters given in the figure caption). The real-time growth of RPL with irradiation time is shown in Figure 6.22b, using Equation 6.40, where the supralinear growth can be clearly seen.

Note that in a real experiment the RPL intensity is not measured at every infinitesimal interval d$t$, but instead it is measured in a real time interval $\Delta t$. If $\Delta t \gg \tau_j$, then the RPL signal at a given time $t$ will be fully built up by the time the next RPL signal is measured at $t + \Delta t$. Using Equation 6.40 as an example, the exponent terms will be zero under such circumstances and $C = 0$ at $t = 0$ so that $I_t^{net}(t) = (K_1 + K_2)t$. That is, the real-time RPL growth in these circumstances would be linear with irradiation time.

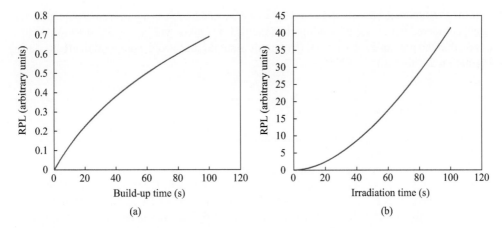

**Figure 6.22**   (a) A simulated RPL buildup curve using Equation 6.39, with $K_1 = 0.2$, $K_2 = 1.5$, $\tau_1 = 25$, $\tau_2 = 250$, $C = -380$. (b) Integral of curve (a) according to Equation 6.40, showing the supralinear growth of RPL during irradiation (real-time RPL).

---

**Exercise 6.10    RPL buildup (I)**

(a) Assume a buildup curve for RPL as given by Equation 6.39 with three components ($k = 3$). Further assume that the time constants are thermally activated according to:

$$\tau_j = \tau_{oj}\exp\{E_j / kT\}.$$

With $\tau_{01} = \tau_{02} = \tau_{03} = 10^{-4}$ s, $K_1 = K_2 = K_3 = 1$ arbitrary unit, $E_1 = 0.3$ eV; $E_2 = 0.33$ eV, and $E_3 = 0.35$ eV, show how the net buildup curve $I_{BU}(t)$ changes with temperature, in twenty degree steps from 300 K to 400 K. Assume a time resolution of $\Delta t = 0.5$ s.

(b) Using the same time resolution calculate the growth of RPL $I_t^{net}(t) = I_{RT}(t)$ during irradiation starting from zero dose. Do the calculation for 300 K and 400 K and for an irradiation period $t_{irr} = 100$ s.

(c) Calculate the RPL buildup curves that would be obtained immediately following the 100 s irradiation, at 300 K and 400 K. (Compare the curve shapes with the data shown in Figure 4.10 obtained for the particular model discussed in Chapter 4.)

(d) If the time required to reach dose $D$ during irradiation is $t_{irr} = t_D$ (dose rate = $D/t_D$), calculate how the real-time RPL growth $I_t^{net}(t) = I_{RT}(t)$ would change if the dose rate was doubled, and if the dose rate was halved, for the same dose $D$.

---

### 6.2.7.2   *Buildup after Irradiation: Linear Response to Dose*

Normally, an RPL measurement is made sometime following the irradiation, not during it. Following the end of the irradiation, beyond $t = t_{irr}$, buildup continues and the RPL signal eventually reaches a steady-state level. Once this has been achieved, and in the absence of strong competition effects during irradiation, the RPL level is a linear function of the absorbed dose. Since the buildup mechanisms are thermally activated (Chapter 4), the buildup process is accelerated at elevated temperature and the time to reach the steady-state level can be reduced by a period of post-irradiation annealing, at a fixed temperature for a fixed time. The final RPL measurements can be then be made after cooling to room temperature.

If the irradiation is very slow (i.e. at very low dose rates, such as natural, environmental dose rates) compared to the rate of buildup, then the RPL response will be a linear function of dose, as noted in the previous sub-section. (Note: this assumes that strong competition effects during irradiation are absent.)

---

**Exercise 6.11    RPL buildup (II)**

After irradiation the RPL continues to build up. However, depending on the dose $D$ and dose rate $dD/dt$ (and therefore the time $t_{irr} = t_D$) a certain amount of buildup will have already occurred during the irradiation period.

(a) On the web page under Exercises and Notes, Chapter 6, Exercise 6.11 you can find the data set for Figures 6.22a and 6.22b, obtained using Equations 6.39 and 6.40, respectively, with the parameters given in the figure caption. How would you use these data to predict what the buildup curve at the end of the irradiation period (100 s) would look like?

(b) Vary the irradiation time and repeat the exercise.

---

# Part II

## Experimental Examples:

## Luminescence Dosimetry Materials

*A fool ... is a man who never tried an experiment in his life.*

– E. Darwin 1792

# 7

# Thermoluminescence

*An experiment in nature .... is capable of different interpretations, according to the preconceptions of the interpreter.*

– W. Jones 1781

## 7.1 Introduction

To demonstrate how the principles introduced in Part I of this book can be used to interpret the properties of luminescence dosimeters, Part II describes several popular dosimetry materials and examines their properties in sufficient detail to illustrate the underlying phenomena. The chapters do not represent reviews of all experimental developments and data related to each material, but rather contain illustrations, through a judicious selection of properties, of how the materials' characteristics can be explained, or partly explained, using the principles presented in Part I.

However, in doing so, the reader should be mindful of W. Jones' 1781 observation and the more recent views of Albert Einstein who stated that: "The scientific theorist is not to be envied, for Nature, or more precisely experiment, is an exorable and not very friendly judge of his work." Einstein went on to note that the only definitive result from an experiment is when it proves a theory wrong; it never can prove a theory right. As scientists we are often tempted to talk about the "laws of physics," or the "laws of nature" when in fact what we mean are "mankind's laws," which we invented to try to explain and describe nature and to predict its properties. When we get it wrong, nature is not at fault; we are. Nevertheless, it is an amazing fact that even with a surfeit of ignorance mankind has been able to use what little knowledge he possesses to develop novel technologies that have proved to be reliable and extremely useful. In our field of luminescence dosimetry perhaps no better example can be given than the use of LiF as a radiation dosimeter. Our lack of detailed knowledge and understanding of the core processes that lead to TL from this material has not prevented the development of LiF-based TL dosimeters, which have served the radiation dosimetry and health physics communities exceptionally well over many decades. Therefore, it is perhaps appropriate that lithium fluoride is the first material examined in Part II.

*A Course in Luminescence Measurements and Analyses for Radiation Dosimetry,* First Edition.
Stephen W.S. McKeever.
© 2022 John Wiley & Sons Ltd. Published 2022 by John Wiley & Sons Ltd.
Companion Website: www.wiley.com/go/mckeever/luminescence-measurements

## 7.2    Lithium Fluoride

Lithium fluoride first emerged as a potential TL dosimetry material in the 1950s with the seminal work of Farrington Daniels and colleagues at the University of Wisconsin, USA (e.g., Schulman et al. 1951). An apocryphal story relates how the group discovered the potential of LiF as a dosimeter after examining a piece of LiF single crystal that was available in the laboratory. To enhance the properties, the group then grew a new, high-quality crystal, only to be baffled by the absence of the same properties. Whether this story is true or not, this initial work prompted the examination of the detailed TL characteristics of LiF and the role of impurities, and led to the realization of how it can be used as a radiation dosimeter based on TL. Sometime later, the material LiF:Mg,Ti was commercialized as a TL dosimeter (TLD) by the Harshaw Chemical Company, Cleveland, USA (now ThermoFisher Scientific) and given the commercial code TLD-100.[1] TLD-100 is still the most widely used TL dosimetry material worldwide and publications concerning the mechanisms of TL production and its dosimetry properties are legion. The key dopants that give the material its desirable properties are Mg and Ti, with an additional role played by OH impurities. Since the advent of TLD-100, research groups around the world have produced their own versions, described in the academic literature and the commercial world under various trade names, along with a plethora of patented claims.

Following the success of LiF:Mg,Ti, other dosimetry versions of lithium fluoride emerged, the most popular of which is LiF triply doped with Mg, Cu, and P, giving rise to LiF:MCP, first introduced by Nakajima and colleagues (Nakajima et al. 1978, 1979) and manufactured by the Solid Dosimetric Detector and Method Laboratory, China as GR-200.[2] As with TLD-100, other research laboratories produced their own versions of LiF:MCP, some of which have been commercialized. The properties of both LiF:Mg,Ti and LiF:MCP are described in the following sections.

### 7.2.1    LiF:Mg,Ti

#### 7.2.1.1    Structure and Defects

LiF has the structure of two, interpenetrating, face-centered-cubic lattices, one for $Li^+$ and one for $F^-$, as illustrated previously in Figure 2.4. Upon the introduction of magnesium (in the form of $MgF_2$) in the crystal growth phase, $Mg^{2+}$ substitutes for $Li^+$ and a $Li^+$-vacancy is introduced to compensate for the additional positive charge. The removal of a $Li^+$ host ion gives the $Li^+$-vacancy ($Li_{vac}^+$) an effective negative charge and coulombic interaction with the additional positive charge of the $Mg^{2+}$ ion leads to the association of the vacancy with the impurity ion to form a $Mg^{2+} / Li_{vac}^+$ dipole along the <110> direction within the lattice (Figure 2.4a). The dipole can be detected via its dielectric properties (using techniques such as thermally stimulated polarization and depolarization currents, TSPC/TSDC, and dielectric loss). Through such examinations, clustering of three dipoles to form $\left[ Mg^{2+} / Li_{vac}^+ \right]^3$ trimers is known to occur, of which there are several possible orientations within the crystal (e.g. Figure 2.4b). As noted in Chapter 2, the ionic radii of $Mg^{2+}$ and $Li^+$ are different, leading to relaxation of the host ions around the defect complexes and a reduction in volume by about 15% in the region of the Mg defect clusters.

---

[1]  TLD-100™ Thermoluminescent Dosimetry Material (thermofisher.com); website accessed December 2020.
[2]  China Thermoluminescence manufacturer, Dosimeter, Dosimetry supplier - Solid Dosimetric Detector & Method Laboratory (DML) (made-in-china.com); website accessed December 2020.

If the system is allowed to equilibrate (on a timescale of days to weeks at room temperature), higher-order clusters and eventually precipitation of the Mg occurs into either stable $MgF_2$ or metastable "Suzuki phase" $6LiFMgF_2$ precipitates. The concentration of Mg is critical to the precipitation process with high Mg concentrations producing precipitated forms of Mg very easily. Since LiF with Mg in precipitated form is of low luminescence efficiency there exists an optimum concentration of Mg for the highest TL efficiency.

The structure of the defects related to $Ti^{3+}$ or $Ti^{4+}$ impurities is less well understood. The $Ti^{3/4+}$ ions also substitute for $Li^+$ but charge compensation appears to involve $OH^-$ ions to form either $Ti^{4+}(O^{2-})$ complexes, or $Ti(OH)_n$ complexes. $OH^-$ ions are also known to cluster with Mg to form $Mg(OH)_m$ complexes (Stoebe and DeWerd 1985; McKeever et al. 1995). $Ti(OH)_n$ centers appear to be the main luminescence emitters in LiF:Mg,Ti whereas $Mg(OH)_m$ defects are non-luminescent. Thus, there is an optimum Ti concentration for maximum TL sensitivity, depending on the Mg and OH concentrations. All practical forms of LiF:Mg,Ti contain a few parts per million (ppm) of OH impurities and optimum concentrations of Mg and Ti appear to be around 180–200 ppm and 10–12 ppm, respectively, for the highest TL sensitivity.

Natural Li comes in isotopic forms $^6Li$ (7.5%) and $^7Li$ (92.5%), and LiF:Mg,Ti also comes in $^6Li$-enriched (95.6% $^6Li$ and 4.4% $^7Li$, for neutron dosimetry) and $^7Li$-enriched (99.93% $^7Li$ and 0.07% $^6Li$) versions. Following the Harshaw product code, these are known as TLD-600 and TLD-700, respectively. Other manufacturers have equivalent product codes.

### 7.2.1.2   TL Glow Curves

The TL from LiF:Mg,Ti is shown in Figure 7.1 where two glow curves are illustrated, one after an anneal at 400 °C for 1 h plus rapid cool, and one after the same heat treatment but followed by an 80 °C anneal for 24 h and a slow cool to room temperature. The dose delivered (~410 mGy, $^{90}Sr/^{90}Y$ beta) is the same for each and a preheat to 90 °C for 30 s was applied before recording the glow curves in order to remove the unstable peak 1. The two glow curves illustrated are the "classic" glow curves for TLD-100 (and TLD-600/700) and have been the subject of much analysis leading to the partial identification of the defects responsible for the peaks. The separation of the glow curve into its individual peaks is the focus of Exercises 7.3 and 7.4 below.

Curve (b) is the form usually adopted for dosimetry. The reason for this is that peaks 1–3 are unstable at room temperature following irradiation whereas the peak known as "peak 5" is both stable and enhanced after the 80 °C/24 h anneal. The behavior of the impurities in the LiF lattice can partially explain the instability of the lower temperature peaks.[3]

As noted above, Mg precipitates if the material is left for long enough at room temperature, during which time the Mg impurities follow the general clustering sequence:

$$\text{Dipoles} \leftrightarrow \text{Trimers} \leftrightarrow \text{Higher-order Clusters} \leftrightarrow \text{Precipiates} \qquad (7.1)$$

Each reaction in this sequence is reversible with reaction rates depending on temperature and Mg concentration. To drive this reaction to the left (i.e. maximize the concentration of dipoles) the sample must undergo high-temperature annealing to dissolve the precipitates and dissociate the clusters. A temperature of 400 °C for 1 h is optimal. To ensure all of the Mg is in its dipolar form, however, the sample must then be very rapidly quenched from the high temperature to room temperature. Failing to do so will allow the reaction to proceed to the right during the cooling process. Samples that have undergone this annealing/quenching treatment yield the

---

[3] The instability of peak 1, the defect for which is unknown, can be explained by thermal detrapping of the trapped electrons in the peak 1 trap at room temperature. It is not considered to be important for dosimetry and is neglected from further discussion.

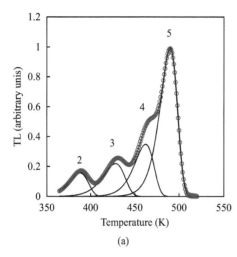
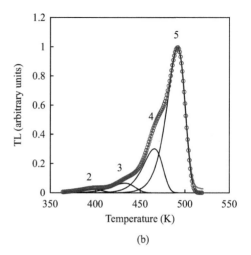

(a)                                                      (b)

**Figure 7.1** (a) Normalized glow curve obtained from LiF:Mg,Ti (TLD-100) after a 400 °C/1 h anneal and a rapid cool to room temperature, followed by beta irradiation to a dose of ~410 mGy. (b) Normalized glow curve obtained after the same dose, but after annealing at 400 °C/1 h plus an 80 °C/24 h anneal and slow cool to room temperature. Both were subjected to a pre-heat at 90 °C for 30 s (to remove the unstable peak 1) and the glow curves were measured at 1 K.s$^{-1}$. (Data kindly provided by S. Sholom.) The peak numbering is the standard numbering used to describe the LiF:Mg,Ti glow curve. Peak 5 is used in dosimetry. The $E_t$ and $s$ values determined from the fits shown are: (a) 1.20 eV, 2.95×10$^{14}$ s$^{-1}$; 1.28 eV, 9.23×10$^{13}$ s$^{-1}$; 1.68 eV, 2.09×10$^{17}$ s$^{-1}$; and 2.11 eV, 5.11×10$^{20}$ s$^{-1}$. (b) 0.96 eV, 1.25×10$^{11}$ s$^{-1}$; 1.23 eV, 1.53×10$^{13}$ s$^{-1}$; 1.71 eV, 2.44×10$^{17}$ s$^{-1}$; and 2.10 eV, 3.63×10$^{20}$ s$^{-1}$, for peaks 2–5, respectively.

glow curve illustrated in Figure 7.1a and do not have ideal TL properties for dosimetry. As reaction 7.1 proceeds to the right with storage time at room temperature, the concentration of $Mg^{2+}/Li_{vac}^{+}$ dipoles decreases and that of trimers increases due to a 3rd-order reaction, namely:

$$3\left[Mg^{2+}/Li_{vac}^{+}\right] \underset{K_2}{\overset{K_1}{\longleftrightarrow}} \left[Mg^{2+}/Li_{vac}^{+}\right]_{trimer}^{3} \tag{7.2}$$

where $K_1$ and $K_2$ are the thermally activated rate constants for the forward and backward reactions, respectively. If the concentration of dipoles is maximized initially, then either storage at room temperature or raising the temperature of the sample will drive the reaction to the right, with rate constant $K_1$. Upon further increase in temperature, however, the reaction reverses, dictated by rate constant $K_2$, as the trimers dissociate back into dipoles.

In Figure 7.2 three data sets are shown for a LiF:Mg,Ti sample after two different annealing treatments. The first treatment (full circles) is 400 °C for 1 h, followed by a quench; the second treatment (open circles) is 400 °C for 1 h plus 80 °C for 24 h followed by a slow cool. In each case, the sample is then pulse annealed to the temperature shown on the horizontal axis ($T_{stop}$). As $T_{stop}$ increases, reaction 7.2 proceeds to the right while upon further increase in temperature the reverse reaction dominates. In Figure 7.2a, the concentration of $Mg^{2+}/Li_{vac}^{+}$ dipoles is shown as measured by TSDC. After the 400 °C/1 h-plus-quench treatment, the concentration of dipoles is initially high but decreases as $T_{stop}$ increases, reaching a minimum around $T_{stop} \approx 150$ °C before increasing back to its original value. After the 80 °C/24 h anneal, however, the concentration of dipoles is already low and only the back reaction in reaction 7.2 is observed as $T_{stop}$ increases.

In Figure 7.2b, the same is seen for the intensity of the optical absorption band at ~380 nm (3.23 eV in Figure 7.3) showing a direct correlation between this band and the $Mg^{2+}/Li_{vac}^{+}$

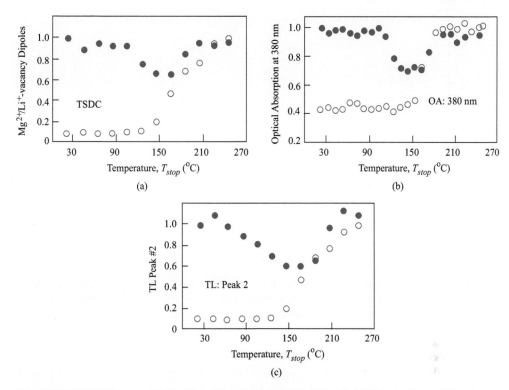

**Figure 7.2** (a) Normalized TSDC intensity (corresponding to the concentration of $Mg^{2+}/Li^+_{vac}$ dipoles), (b) normalized optical absorption at 380 nm, and (c) normalized TL peak 2 intensity, each versus pulse-anneal temperature, $T_{stop}$. For (b) and (c) the samples were irradiated after the pulse anneal to produce the optical absorption or TL signals. Full circles (red) correspond to samples that have been annealed at 400 °C for 1 h then quenched to room temperature. Open circles (blue) are after the same anneal, plus a second anneal at 80 °C for 24 h and a slow cool to room temperature. (Reproduced from Yuan and McKeever (1988) with permission from John Wiley and Sons.)

dipole concentration. (For these data, the sample had to be irradiated after the pulse anneal to induce the absorption bands.) In Figure 7.2c the intensity of TL peak 2 (see Figure 7.1a) is shown; the sample was also irradiated after the pulse anneal to induce this signal. The same general trend is observed for TL peak 2 as was observed for TSDC and 380 nm optical absorption, namely a decrease followed by an increase, but the reaction rates are clearly not the same; the TL decreases sooner than the TSDC and optical absorption signals and the increase is slower at the higher temperatures.

This behavior is explained (Townsend et al. 1983; McKeever 1984) by invoking an association between the $Ti(OH)_n$ complexes and the $Mg^{2+}/Li^+_{vac}$ dipoles, thus:

$$TiOH_n + Mg^{2+}/Li^+_{vac} \overset{K_3}{\underset{K_4}{\leftrightarrow}} \left[ TiOH_n / Mg^{2+}/Li^+_{vac} \right]_{complex} \tag{7.3}$$

The thermally activated forward and reverse reaction constants are now $K_3$ and $K_4$ and thus the decrease and increase with increasing pulse-anneal temperature exhibits greater complexity due to both reactions 7.2 and 7.3 occurring simultaneously (Yuan and McKeever 1988).

The above is an example of how additional experimental techniques (in this case TSDC and optical absorption) are required to infer the identity of defects responsible for specific TL peaks and

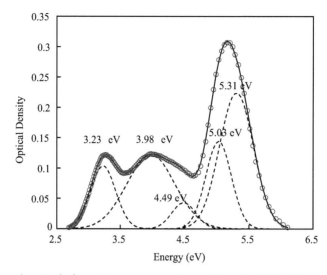

**Figure 7.3** Typical optical absorption spectrum from LiF:Mg,Ti (TLD-100) measured at liquid nitrogen temperature and separated into Gaussian components. The sample was annealed at 400 °C for 1 h and quenched to room temperature, then irradiated with 367 Gy of [60]Co gamma irradiation at room temperature.

features. In this case, the TL peak in question is peak 2 from the LiF glow curve. In general, dielectric properties, conductivity, optical absorption, photoluminescence, and electron spin resonance are among the experimental techniques that can be used to assist in the identification of the defects responsible for a given TL peak, bearing in mind that any one TL peak is the result of at least two defect types – a trap and a recombination center. Distinguishing between them and identifying the role of a specific defect in the TL (or, in general, OSL or RPL) process is not straightforward.

As a result of the reactions described above, peak 2 is not used in TL dosimetry since it is unstable at room temperature. However, by annealing the material at 80 °C for 24 h, reaction 7.2 proceeds to the right and although precipitation can still occur (reaction 7.1), the precipitation rate is very slow at room temperature such that the concentration of $\left[ Mg^{2+} / Li^{+}_{vac} \right]^{3}$ trimers is essentially stable on the timescales required for reliable dosimetry. TL peak 5 is maximized with respect to the other TL peaks by this treatment. Figure 7.4 illustrates that TL peak 5 is associated with the optical absorption band at ~310 nm (3.98 eV in Figure 7.3) in that they both anneal over the same temperature range after irradiation. In the same way that peak 2 is suggested to be caused by a $\left[ TiOH_n / Mg^{2+} / Li^{+}_{vac} \right]_{complex}$, it is suggested that peak 5 is related to a complex of $\left[ Mg^{2+} / Li^{+}_{vac} \right]^{3}$ trimers and $Ti(OH)_n$ defects.

Not only does the shape of the TL glow curve in this material vary with annealing temperatures/times, but it also depends on the rate of cooling after the annealing because of the clustering effects that occur during the cooling period. Even quenching rates as fast as 500 K.min$^{-1}$ are not fast enough to prevent some clustering of dipoles into trimers. Furthermore, in addition to the normal kinetic effects expected at different heating rates (Chapter 3) clustering effects may also occur during heating in a TL measurement (Taylor and Lilley 1982a, 1982b, 1982c; Bos et al. 1992).

Identifying the defects responsible for the TL peaks and the recombination centers is only part of the challenge in understanding the TL from this, or any, material. Even after six or seven decades of research, the detailed mechanisms governing TL in LiF:Mg,Ti remain elusive. Sagastibelza and Alvarez Rivas (1981) proposed that TL in all alkali halides is initiated by the creation of free excitons (an unstable, coupled, electron-hole pair) during irradiation. The excitons

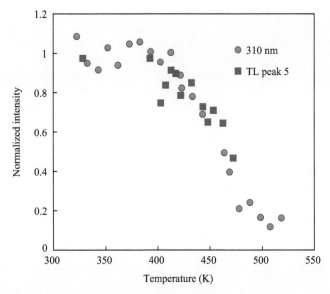

**Figure 7.4** Thermal stability of the 310 nm optical absorption band and TL peak 5 as a function of preheat temperature following irradiation. (Reproduced from McKeever (1984) with permission from AIP Publishing.)

relax to form *F-H* pairs in the lattice. Rogalev et al. (1990) proposed that the high ionization potential of $Mg^{2+}$ causes the *F*-center to release its electron in the vicinity of the Mg-related defects to form $Mg^+$. Thus, the trimer complexes that cause peak 5 have their $Mg^{2+}$ reduced to $Mg^+$. The *H*-centers then form $F_3^+$-centers. During heating, the $Mg^+$ ions release an electron, re-forming $Mg^{2+}$ and initiating electron-hole recombination near $Ti(OH)_n$ centers to produce TL. The details of this recombination process are still unclear. Alternatively, the *H*-centers (interstitial fluorine atoms) localized near the Mg-defects have been suggested to recombine with the *F*-centers followed by energy transfer to Ti via unknown mechanisms (Sagastibelza and Alvarez Rivas 1981; Taylor and Lilley 1982b; Delgado and Delgado 1984).

Whatever the precise mechanism producing TL in LiF:Mg,Ti, the close association of the trap ($\left[Mg^{2+} / Li_{vac}^+\right]^3$ trimers) with the recombination center ($Ti(OH)_n$ complexes) is accepted by most observers. It is proposed that this association gives this material many of its particular TL properties, including its emission spectra, kinetics, and its dependence on dose and the ionization density of the absorbed radiation. The close association between the trap and the recombination center gives the potential for a localized or semi-localized transition model, as discussed earlier in Chapter 3, Section 3.8.2. The consequences of such a model on the TL response to dose were discussed in Chapter 6, Section 6.2.3. In addition to those general discussions, the following sections will discuss the properties specific to LiF:Mg,Ti.

**Exercise 7.1  Clustering**

One of the early questions regarding clustering of $\left[Mg^{2+} / Li_{vac}^+\right]$ dipoles was whether or not they first formed stable $\left[Mg^{2+} / Li_{vac}^+\right]^2$ pairs (dimers) and then formed $\left[Mg^{2+} / Li_{vac}^+\right]^3$ trimers by the addition of a third dipole, or whether the stable cluster was the trimer formed by the reaction of three dipoles. In other words, is the loss of dipoles characterized by a second-order reaction or a third-order reaction? Experiments with dielectric loss and TSDC/TSPC, which monitor the dipole concentration directly, indicate that the reaction is third order and that stable dimers are not formed. If the optical absorption band at 380 nm and TL peak 2 in LiF:Mg,Ti are related to dipoles, as indicated in the above text, they too should decay following a third-order reaction during ageing. However, since TL peak 2 is also associated with $Ti(OH)_n$ defects, its decay kinetics could be quite complex (see Figure 7.2, for example). For the 380 nm absorption band, however, one way to support the identity of it being caused by dipoles is to plot the intensity of the absorption band against ageing time, $t_{age}$ and to determine the decay kinetics. If the kinetics are found to be the same as those for dipoles (as measured by TSDC or dielectric loss) then the association of the 380 nm absorption band with dipoles is supported.

Consider the following experiment: A TLD-100 sample is annealed at 400 °C for 1 hour, quenched to room temperature, and then left to age for different amounts of time, before being irradiated and its optical absorption spectrum recorded. The height of the absorption band at 380 nm is then measured at different ageing times, $t_{age}$. On the web folder under Exercises and Notes, Chapter 7, Exercise 7.1, can be found an Excel spreadsheet with some experimental measurements of the 380 nm absorption band at room temperature as a function of $t_{age}$. Using these data, demonstrate that the decay of the absorption band is consistent with a third-order reaction.

### 7.2.1.3   TL Emission Spectra

A TL emission spectrum for a sample of TLD-100 after annealing at 400 °C/1 h and rapid cooling is shown in Figure 7.5. Such spectra for LiF:Mg,Ti have been recorded and analyzed over several decades since the first such measurement by Fairchild et al. (1978a). An example is the analysis by Townsend et al. (1983) who used data similar to that illustrated in Figure 7.5 to show that the main peak of the emission shifts, from longer wavelength to shorter wavelength, depending on TL peak. For example, Townsend et al. (1983) report that peak 2 emits with a maximum at 460 nm, peak 3 at 435 nm, and peaks 5 and 6 at 420 nm. The TL from Ti-only doped specimens displays just two TL peaks, both of which emit at 410 nm. In all cases the emission is suggested to be from Ti (in the form of $Ti(OH)_n$ complexes). For the Mg-doped

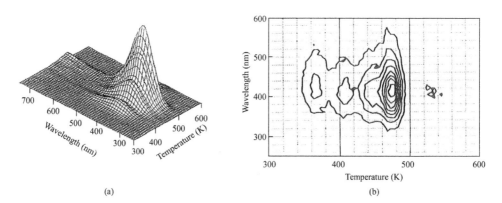

(a)                                    (b)

**Figure 7.5**   (a) Isometric plot of the TL emission spectrum from LiF:Mg,Ti (TLD-100) as a function of glow-curve temperature, following a 400 °C/1 h anneal, rapid cool to room temperature, irradiation with 5 Gy $^{90}$Sr/$^{90}$Y beta and heated at 3 K.s$^{-1}$. (Reproduced from Piters et al. (1993) with permission from AIP Publishing.) (b) Contour plot of the TL emission spectrum after the same annealing and irradiation, but at a heating rate of 0.1 K.s$^{-1}$. (Reproduced from Piters and Bos (1993a) with permission from Oxford University Press.)

materials, the presence of Mg is suggested to perturb the emission away from 410 nm to longer wavelengths, depending on the form of the Mg defect. Thus, the TL emission from peak 2, due to $Mg^{2+}/Li_{vac}^{+}$ dipoles, peaks at 460 nm, whereas the emission from peak 5, due to $\left[Mg^{2+}/Li_{vac}^{+}\right]^{3}$ trimers, has a maximum at 420 nm. (See Fairchild et al. 1978a; Townsend et al. 1983; Delgado and Delgado 1984; McKeever 1984; Piters et al. 1993; Piters and Bos 1993a.)

Although the wavelengths identified above represent the peak of the emission spectrum, the emission spectrum for each TL peak is made up of more than one individual emission bands. Analysis of the glow-curve emission spectrum shows that each peak is in fact made up of the same set of emission bands, but that these bands vary in relative intensity and that the apparent shift in the maximum of the emission is due to varying ratios between the peaks. This observation was still explained on the basis of a close association between the Mg-related traps and the Ti-related recombination centers (Biderman et al. 2002).

---

**Exercise 7.2    Optical Absorption Bands**

On the web site, under Exercises and Notes, Chapter 7, Exercise 7.2, you will find an Excel file with optical absorption data for two different LiF samples (single crystals) – one with 100 ppm Mg, 0.55 mm thick, and irradiated with 2 kGy of $^{90}Sr/^{90}Y$ beta particles; and one with 450 ppm Mg, 1.20 mm thick, and irradiated with 0.5 kGy of $^{90}Sr/^{90}Y$ beta particles. The files give the optical densities (absorbance) before and after irradiation. For each sample:

(a) Subtract the background (zero dose) spectra and plot the absorption coefficient against photon energy.
(b) Fit each spectrum to the sum of a series of Gaussian absorption bands. Compare your fits to those shown in Figure 7.3. How many individual absorption bands do you think are present?

---

### 7.2.1.4    TL Glow-Curve Analysis

The glow curve from LiF:Mg,Ti is one of the most analyzed glow curves in the literature. Multiple different analyses have been attempted and published and, overall, a consensus has been reached on the kinetics and the values of the $E_t$ and $s$ parameters that describe the glow-curve shape. Many of the methods described in Chapter 5 (partial-peak, whole-peak, peak-shape, peak-position, curve-fitting, and isothermal-decay methods) have been attempted and results published. Although initially there was some debate about the order of kinetics the general agreement now is that the kinetics for all the major TL peaks in LiF:Mg,Ti are first-order, and this will be assumed from this point forward. (See McKeever (1985) for a summary of the pre-1985 literature in which the kinetic order for TL from LiF:Mg,Ti was debated.)

Less-well established is the number of peaks in the TL glow curve. This may seem strange given the fact that the glow curve from this material is so well studied and is used as a reference glow curve to test various analytical routines, especially peak fitting. (See, for example, the papers on the GLOCANIN project already referred to in Chapter 5 (Bos et al. 1993, 1994).) Nevertheless, uncertainty arises because of the profound changes to the glow-curve shape that occur with different pre-irradiation annealing conditions, increasing dose, and increasing ionization density of the incident radiation. Overlap of several of the peaks, especially around the main peak used in dosimetry (peak 5), makes resolution of the various peaks difficult. The resulting complexity of the overall TL curve shape does not allow complacency when stipulating how many TL peaks exist in the glow curve. Experiments such as $T_m$-$T_{stop}$ as well as the application of multiple different analysis methods become essential.

With low-dose, low-ionization-density radiation (for example, energetic gamma irradiation) the classic four-peak structure already shown in Figure 7.1a is obtained. The obtained $E_t$ and $s$ values are given in the Figure caption. Similar examples are found in the GLOCANIN data set.

---

**Exercise 7.3    GLOCANIN Analysis I**

Exercise 5.9 used two simulated glow curves, one of which (Refglow002) approximated the LiF:Mg,Ti glow curve for low doses after a 400 °C/hr anneal and rapid quench to room temperature. The GLOCANIN exercise also provided several experimental glow curves to the participants of which two (Refglow003 and Refglow004) were experimental curves obtained after similar treatment of TLD-100. The data for these reference curves can be found in an Excel file on the accompanying web site (Exercises and Notes, Chapter 7, Exercise 7.3).

(a) Open the above-mentioned files and plot the two glow curves and the background against temperature. Repeat with the background subtracted.

(b) Using curve-fitting analysis, as described in Chapter 5, Section 5.2.5, fit the glow curve to a series of first-order Randall-Wilkins peaks. Use a different number of peaks for the fits; start with 4 peaks, then 5, then 6. (To start, use the answers you obtained when fitting the synthetic glow curve in Exercise 5.9, Refglow002, as starting values for the parameters $E_t$ and s for the 4 TL peaks.) Compare the FOM values obtained with each fit. Which gives the "best fit" and are there any caveats to this conclusion?

(c) Compare your answers with the other published results of the GLOCANIN analysis, given in Bos et al. (1993, 1994) and Kitis et al. (1998).

(d) Also fit the data of Figure 7.1a, also given in the web folder under Exercises and Notes, Chapter 7, Exercise 7.3. Compare your results with those from Refglow003 and Refglow004.

---

Figure 7.1b illustrates the changes that occur to the LiF:Mg,Ti glow curve following additional annealing at 80 °C/24 h. Additionally, the GLOCANIN project published two reference curves (Refglow007 and Refglow008) that resulted from storage at 45 °C for 7 days following the 400 °C/1 h anneal. In each case, the additional treatment after the 400 °C/1 h anneal resulted in similar changes to the glow curve.

---

**Exercise 7.4    GLOCANIN Analysis II**

The data for TL curves for Figure 7.1b and Refglow007 and Refglow008 can be found in Excel files on the accompanying web site (Exercises and Notes, Chapter 7, Exercise 7.4).

(a) Open the above-mentioned files and plot the glow curves after background subtraction. Compare the shapes.

(b) Using curve-fitting analysis, as in Exercise 7.3, fit the glow curve to a series of first-order, Randall-Wilkins peaks. Once again, try using a different number of peaks for the fits. (Use the "best fit" answers you obtained when fitting the Refglow003 and Refglow004 files as starting values for the parameters $E_t$ and $s$ for the TL peaks.) Again, compare the FOM values obtained with each fit. Which gives the "best fit" and, again, are there any caveats to this conclusion?

(c) Also fit the data of Figure 7.1b, also given in the web folder under Exercises and Notes, Chapter 7, Exercise 7.4. Compare your results with those from Refglow007 and Refglow008.

A particular observation regarding the values of $E_t$ and $s$ for peaks 2–5 in LiF:Mg.Ti is the dependence of the values on the cooling rate and heating rate used in the experiments (Bos et al. 1992). Since both of these affect the degree of clustering, it is concluded that the degree of clustering changes $E_t$ and $s$ for these peaks. This may partially explain the range of $E_t$ and $s$ values reported in the literature since not all research groups use exactly the same cooling rates following pre-irradiation annealing, nor do they all use the same heating rates during TL measurement.

### 7.2.1.5   Changes to the Glow-Curve Shape with Dose and Ionization Density

Changes to the glow-curve shape also occur when subjecting LiF:Mg,Ti to high doses of radiation and to radiation with high ionization density (e.g. energetic charged particles). Figure 7.6 shows typical glow curves following high gamma dose and charged-particle (alpha-particle) irradiation, separated into individual peaks. The gamma dose used in Figure 7.6a was high (5 kGy) and the high-temperature (> 200 °C) TL peak structure, typical of this material, is clearly revealed. Although the overall dose delivered to the alpha-irradiated sample (Figure 7.6b) was much lower, the high-temperature peaks are still revealed indicating that the TL signal in this case is characteristic of the high-dose regions of the alpha particle tracks (Chapter 6, Section 6.2.6).

Figure 7.6 also illustrates the different TL peaks in the material as claimed by two research groups (Horowitz et al. 2002a, 2002b; Bilski et al. 2007). Of special interest is the apparent appearance of new, additional peaks in the glow curve following alpha irradiation. In particular, three peaks, termed peaks 5a, 5, and 5b, can be observed in the region of peak 5 after alpha irradiation. Special importance is placed by Horowitz et al. (2002a, 2002b) on peak 5a, which is said to only appear prominently in high-ionization-density irradiations (such as charged-particle irradiation).

Figure 7.7 is an illustration of the model proposed by Horowitz and colleagues for the production of the TL peaks in the region of peak 5. To understand Figure 7.7, consider first Figure 6.12 where a schematic distribution is illustrated of localized electron-hole pairs, separated by recombination centers and competing traps. The electron and hole traps in the electron-hole

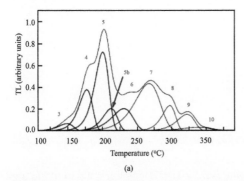 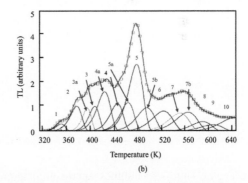

**Figure 7.6**   Glow curves from TLD-100 separated into its component peaks using curve fitting. (a) The sample was annealed at 400 °C for 1 hr followed by 100 °C for 2 h; the dose was 5 kGy gamma irradiation and $\beta_t = 2$ °C.s$^{-1}$. (Reproduced from Bilski et al. (2007) with permission from Elsevier.) (b) The sample was annealed at 400 °C for 1 hr followed by a slow cool to room temperature, then irradiated with 5 MeV alpha particles (fluence $5\times10^9$ cm$^{-2}$) from an [241]Am source and heated at $\beta_t = 1.3$ K.s$^{-1}$. (Reproduced from Horowitz et al. (1999) with permission from Oxford University Press.) Note: peak 5b in Figure (a) is changed here from the original nomenclature used by Bilksi et al. (2007), who labelled it peak 5a. The relabeling used here is to make the nomenclature consistent in the two illustrated glow curves, and with the text.

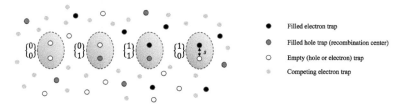

**Figure 7.7** Conceptual notion of trap/recombination center pairs as used by Horowitz and colleagues (Horowitz et al. 2002a, 2002b) to describe the TL response to dose in the region of peak 5.

pairs are considered to be close enough that no other trap or center can exist within the volume $4\pi s^3$. Horowitz and colleagues applied this notion to the traps and recombination centers in an effort to explain TL peaks 5 and 5a in LiF:Mg,Ti . In Figure 7.7, the ellipse signifies the trap/ recombination-center pair wherein the hole trap (recombination center, viz., $Ti(OH)_n$) is located next to or near the electron trap (viz. $\left[Mg^{2+} / Li_{vac}^+\right]^3$ trimers). Different charge states are proposed to exist for the overall complex, namely, one trapped electron and one trapped hole; one trapped electron only, one trapped hole only, and no trapped electron or hole. Using a similar nomenclature to that adopted by Mandowski (2008), these four states can be represented by:

$$\begin{bmatrix} \bar{n} \\ \bar{m} \end{bmatrix} = \begin{bmatrix} 1 \\ 1 \end{bmatrix}, \begin{bmatrix} 1 \\ 0 \end{bmatrix}, \begin{bmatrix} 0 \\ 1 \end{bmatrix}, \text{ or } \begin{bmatrix} 0 \\ 0 \end{bmatrix} \qquad (7.4)$$

where $\bar{n}$ and $\bar{m}$ are integers representing the occupancy of the electron and holes traps within the trap-center complex. (Compared with the notation of Mandowski (2008) described in Chapter 3, an excited state of the electron is not included in Equation 7.4 or Figure 7.7.)

Horowitz et al. (2002a, 2002b) propose that peak 5a occurs when the freed electron recombines with the local hole in "geminate" recombination, whereas peak 5 occurs when the electron escapes the local complex and recombines with recombination centers in the bulk of the crystal outside the complex ("non-geminate" recombination). These concepts may be represented by the reactions:

$$\begin{bmatrix} 1 \\ 1 \end{bmatrix} \rightarrow \begin{bmatrix} 0 \\ 0 \end{bmatrix} \quad \rightarrow \text{ TL peak 5a,} \qquad (7.5)$$

and

$$\begin{bmatrix} 1 \\ 1 \end{bmatrix} \rightarrow \begin{bmatrix} 0 \\ 1 \end{bmatrix} + e^-, \qquad (7.6a)$$

$$e^- + LC \rightarrow TL\, peak\, 5, \qquad (7.6b)$$

where LC signifies a luminescent recombination site (viz., a $Ti(OH)_n$ center not part of the trap/ center pair).

TL peak 4 is said to be due to the release of a hole from a trap/center pair without an electron, i.e.:

$$\begin{bmatrix} 0 \\ 1 \end{bmatrix} \quad \rightarrow \begin{bmatrix} 0 \\ 0 \end{bmatrix} + h^+, \qquad (7.7a)$$

$$h^+ + \left(unknown\, recombination\, center\right) \rightarrow TL\, peak\, 4. \qquad (7.7b)$$

Peak 5b is said to be caused by the release of an electron from an electron-only occupied trap/recombination center pair (Y.S. Horowitz, personal communication) thus:

$$\begin{Bmatrix} 1 \\ 0 \end{Bmatrix} \rightarrow \begin{Bmatrix} 0 \\ 0 \end{Bmatrix} + e^-, \tag{7.8a}$$

$$e^- + LC \rightarrow TL \text{ peak 5b}. \tag{7.8b}$$

Recalling Equation 6.24 from Chapter 6:

$$I_{TL/OSL}(D) \propto n_1(D)\left[ K\left(\frac{\sigma_m}{4\pi s^2}\right) + \left(1 - K\frac{\sigma_m}{4\pi s^2}\right)\left(\frac{\sigma_m m(D)}{\sigma_k(N_k - n_k(D))}\right) \right], \tag{6.24}$$

the first term in the square brackets corresponds to "geminate recombination" (reaction 7.5) and the second term to "non-geminate" recombination (reaction 7.6). The suggestion of Horowitz and colleagues is that the linear term in the growth of the glow curve in the region of the main dosimetry peak with increasing dose is actually due to the growth of peak 5a, while the non-linear part is due to peak 5. Thus, peak 5a should be the dominant peak at low gamma doses, rather than peak 5. However, this is not seen experimentally. At low gamma doses, only peak 5 can be clearly seen in the glow curve (e.g. Figure 7.1 and GLOCANIN Refglow003 and 004). Furthermore, if peak 5 is due to reaction 7.6, described by the second term in Equation 6.24, the growth of peak 5 would be expected to be supralinear from zero dose. This too is not seen.

An additional possible reaction is:

$$\begin{Bmatrix} 1 \\ 1 \end{Bmatrix} \rightarrow \begin{Bmatrix} 0 \\ 1 \end{Bmatrix} + e^-, \tag{7.9a}$$

$$e^- + \begin{Bmatrix} 1 \\ 1 \end{Bmatrix} \rightarrow \begin{Bmatrix} 1 \\ 0 \end{Bmatrix}. \tag{7.9b}$$

Here, an electron released from one filled trap/center pair recombines with a hole in an adjacent trap/center pair. Since this reaction is initiated by the same electron release as reaction (7.6a), then presumably this reaction is also a contributor to peak 5 (following the scheme of Horowitz and colleagues). It may, however, emit at a slightly different wavelength due to the different atomic configurations of the recombination center in reaction 7.6b compared to that in reaction 7.9b. One might also expect:

$$\begin{Bmatrix} 1 \\ 1 \end{Bmatrix} \rightarrow \begin{Bmatrix} 0 \\ 1 \end{Bmatrix} + e^-, \tag{7.10a}$$

$$e^- + \begin{Bmatrix} 0 \\ 1 \end{Bmatrix} \rightarrow \begin{Bmatrix} 0 \\ 0 \end{Bmatrix}, \tag{7.10b}$$

where the recombination event in reaction 7.10b occurs at a different $\begin{Bmatrix} 0 \\ 1 \end{Bmatrix}$ complex than the one on the right-hand-side of reaction 7.10a.

Finally, it must be assumed that the following reactions are also possible:

$$\begin{bmatrix} 1 \\ 0 \end{bmatrix} \rightarrow \begin{bmatrix} 0 \\ 0 \end{bmatrix} + e^{-}, \tag{7.11a}$$

$$e^{-} + \begin{bmatrix} 1 \\ 1 \end{bmatrix} \rightarrow \begin{bmatrix} 1 \\ 0 \end{bmatrix}, \tag{7.11b}$$

or

$$e^{-} + \begin{bmatrix} 0 \\ 1 \end{bmatrix} \rightarrow \begin{bmatrix} 0 \\ 0 \end{bmatrix}. \tag{7.11c}$$

Since these are initiated by the same electron release event as reaction 7.8a, they would presumably also contribute to TL peak 5b (à la Horowitz and colleagues), but since the recombination events 7.8b, 7.11b, and 7.10c involve hole traps in slightly different atomic configurations in the lattice, one might expect slightly different emission bands for each. Multiple recombination events triggered by single electron release events is consistent with the emission spectra observations described earlier.

Also possible but not included in the above description is transport of the electron from the trap to a recombination site via an intermediate state, such as a common excited state or band tails states, or even via thermally assisted tunneling for the case of geminate recombination. These may be feasible given the defect-rich nature of the trap/center complexes involved. In this case, the notation of Mandowski (2008) should be used, namely:

$$\begin{bmatrix} \overline{n_e} \\ \overline{n_g} \\ \overline{m} \end{bmatrix} = \left\{ \begin{matrix} 0 \\ 1 \\ 1 \end{matrix} \right\}, \left\{ \begin{matrix} 1 \\ 0 \\ 1 \end{matrix} \right\}, \left\{ \begin{matrix} 0 \\ 0 \\ 1 \end{matrix} \right\}, \left\{ \begin{matrix} 0 \\ 1 \\ 0 \end{matrix} \right\}, \left\{ \begin{matrix} 1 \\ 0 \\ 0 \end{matrix} \right\} \text{ or } \left\{ \begin{matrix} 0 \\ 0 \\ 0 \end{matrix} \right\} \tag{7.12}$$

where $\overline{n_e}$ and $\overline{n_g}$ are integers representing the electron in either the excited state or the ground state, and configuration $\left\{ \begin{matrix} 1 \\ 1 \\ 1 \end{matrix} \right\}$ is not allowed. Use of these configurations opens an entirely new set of reactions and increases the potential complexity further.

In the Unified Interaction Model (UNIM), using the concepts shown in Figure 7.7 and outlined above, Horowitz and colleagues (e.g. Nail et al. 2002) used an equation for TL, thus:

$$I_{TL}(D) = K \frac{\sigma_m}{4\pi s^2} n_1 + n_1 \left( 1 - \frac{K\sigma_m}{4\pi s^2} \right) \left[ \sum_{i=1}^{3} \int_{0}^{d} g(R_i) \exp\left\{ -\frac{R_i}{\mu} \right\} P_i(m, R_i) dR_i \right] \tag{7.13}$$

where the original equation by Horowitz and colleagues has been re-written here using the parameters previously defined in this book, specifically: $\sigma_m$ – the recombination cross-section, $s$ – the distance between the trap and the recombination center in the trap/center pair (Figure 6.12), $K$ – the fraction of traps, of concentration $n_1$, that are within a trap-center pair (Figure 6.12), $d$ – the crystal dimension, $R_i$ – the distance between neighboring trap/center pairs with $i$ either 1, 2, or 3 for the first 3 nearest-neighbors. The distribution functions $g(R_i)$ and $P_i(m,R_i)$ are the three-dimensional solid angle factor over which geminate recombination can occur, and the $i^{th}$-nearest-neighbor probability distribution function, respectively. This equation for the TL intensity includes geminate recombination (reaction 7.5) and pair-to-pair recombination (reactions 7.10 or 7.11) only and differs from Equation 6.24, which includes geminate (reaction 7.5) and non-geminate recombination (reaction 7.6) only. Neither Equation 6.24 nor 7.13 include all the possibilities noted above and thus predictions of the growth of TL with dose based on these equations are necessarily limited. Clearly, the possible recombination events in LiF:Mg,Ti involving $\left[\mathrm{Mg}^{2+} / \mathrm{Li}_{\mathrm{vac}}^{+}\right]^{3}$ trimers as traps and $\mathrm{Ti(OH)}_n$ defects as recombination sites are myriad and unravelling this complexity remains a major challenge.

---

**Exercise 7.5   TL glow curves and Emission Spectra from LiF:Mg,Ti**

Considering all the potential recombination pathways noted above for LiF:Mg,Ti involving the states $\begin{bmatrix}1\\1\end{bmatrix}, \begin{bmatrix}1\\0\end{bmatrix}, \begin{bmatrix}0\\1\end{bmatrix}$ and $\begin{bmatrix}0\\0\end{bmatrix}$, plus those traps and recombination centers that are not included in these states, i.e. that are not part of a localized trap-center pair (see Figure 7.7) what do you conclude, in principle, about the possible number of TL peaks and their emission spectra from this material? Think about the different environments in which the traps are located and how this might affect the trap depths and frequency factors. Think about the different environments in which the different recombination sites are located and how this might affect the emission spectra. Examine published papers on TL peak numbers, positions, and emission spectra analysis in LiF:Mg,Ti, especially in the region of peak 5. Are they qualitatively consistent with these ideas?

Now consider allowing for excited states, namely: $\begin{bmatrix}0\\1\\1\end{bmatrix}, \begin{bmatrix}1\\0\\1\end{bmatrix}, \begin{bmatrix}0\\0\\1\end{bmatrix}, \begin{bmatrix}0\\1\\0\end{bmatrix}, \begin{bmatrix}1\\0\\0\end{bmatrix}$ and $\begin{bmatrix}0\\0\\0\end{bmatrix}$. What additional complexity might these add? How would you proceed to determine if concepts such as excited-state tunneling, or semi-localized transitions were occurring in the LiF:Mg,Ti system? Are there published experimental data that support or oppose such ideas – e.g. isothermal decay of peak 5?

---

Because of the complexity noted above, a distribution of trapping states rather than discrete, signal-value trap depths and frequency factors may be more appropriate to describe the TL in this material (Y.S. Horowitz, personal communication). Figure 7.8 shows an analysis of the data from Figure 7.1a using Gaussian distributions for the $E_t$ and $s$ values in accordance with the discussion in Chapter 5, Sections 5.2.5.1 and 5.2.5.4. For a generalized distribution of $E_t$ and $s$ (i.e. $g(E_t,s)$), and using first-order kinetics, the TL glow-curve may be written:

$$I_{TL}(T) = \int\limits_{E_{t1}}^{E_{t2}} \int\limits_{s_1}^{s_2} g(E_t,s) I_{RW}(T) \mathrm{d}E_t \mathrm{d}s \tag{7.14}$$

where $I_{RW}(T)$ is the Randall-Wilkins form for a first-order TL peak. $E_{t1}$ and $E_{t2}$, and $s_1$ and $s_2$, are the limits of the $E_t$ and $s$ distributions, respectively. This may be approximated to:

$$I_{TL}(T) = \sum\limits_{E_{t1}}^{E_{t2}} \sum\limits_{s_1}^{s_2} g(E_t,s) I_{RW}(T) \tag{7.15}$$

For a Gaussian distribution in $E_t$:

$$g(E_t) = A_{E_t} \exp\left\{-B_{E_t}(E_{tm} - E_t)^2\right\} \tag{7.16}$$

where $E_{tm}$ is the mean of the $E_t$ distribution and $A_{E_t}$ and $B_{E_t}$ are constants. Similarly, for a Gaussian distribution in $s$:

$$g(s) = A_s \exp\left\{-B_s\left(s_m - s\right)^2\right\}$$ (7.17)

where $s_m$ is the mean of the $s$ distribution and $A_s$ and $B_s$ are constants. Thus, the two-dimensional distribution $g(E_t, s)$ is:

$$g(E_t, s) = g(E_t)g(s) = A\exp\left\{-B_{E_t}\left(E_{tm} - E_t\right)^2\right\}\exp\left\{-B_s\left(s_m - s\right)^2\right\}$$ (7.18)

where $A$ is the amplitude of the $g(E_t, s)$ distribution. If the resolution in the $E_t$-distribution is such that there are $n$ $E_t$-values, and the resolution in the $s$-distribution is such that there are $m$ $s$-values, then the experimental glow curve may be described by the sum of $n$ x $m$ individual glow curves, and $E_{tm}$, $s_m$, $A$, $B_{E_t}$ and $B_s$ may be used as free parameters to obtain a best fit to the glow curve. The result of fitting the glow curve in Figure 7.1a to Equation 7.15, using Equation 7.18, is shown in Figure 7.8. Figure 7.8a shows the best fit to the glow curve (FOM = 1.49%), while Figures 7.8b and 7.8c show the resulting $g(E_t, s)$ distribution. More details concerning the fitting procedure for Figure 7.8 are given on the web site under Exercises and Notes, Chapter 7, Figure 7.8.

There are some limitations with this procedure, including the (a priori) assumption of four Gaussian distributions. Nevertheless, the narrowness of the resulting distributions in $E_t$ and $s$ space suggests that the glow curve is, in fact, a sum of four glow peaks described by single-values for $E_t$ and $s$ for each peak, as was already shown in Figure 7.1a.

This result could have been expected for a number of reasons. Firstly, the shape of the glow peaks are asymmetrical and described perfectly by a Randall-Wilkins curve. Furthermore, especially in the case of peak 5, the peak is quite narrow. For a distribution of $E_t$ and $s$ values a broader, non-Randall-Wilkins TL peak would be expected (see Chapter 5). Secondly, experimental $T_m$-$T_{stop}$ data indicate clear steps in the $T_m$-$T_{stop}$ curve, as illustrated in Figure 7.9. Such clear steps are not expected for TL curves caused by a distribution of trapping centers. Nevertheless, Figures 7.6a and 7.6b show examples of significant changes to the glow curve for high gamma doses and for charged-particle irradiation. Attempts to fit these glow curves using $E_t$ and $s$ distributions may yield a different result.

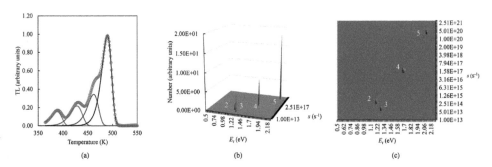

**Figure 7.8**   (a) Fit to the glow curve of Figure 7.1a assuming four Gaussian distribution functions for $E_t$ and $s$. (b) and (c) $E_t$ and $s$ distributions obtained from the fit, shown as an isometric plot (b) and a contour plot (c). The mean values $E_{tm}$ and $s_m$ for each of the glow peaks are: 1.20 eV, $3.59\times10^{14}$ s$^{-1}$; 1.28 eV, $9.43\times10^{13}$ s$^{-1}$; 1.68 eV, $1.95\times10^{17}$ s$^{-1}$; and 2.10 eV, $4.48\times10^{20}$ s$^{-1}$. More details describing the fitting procedure can be found on the web site.

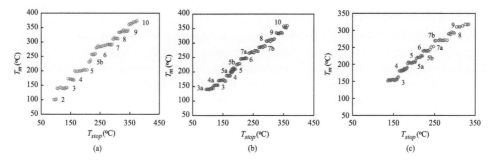

**Figure 7.9** $T_m$-$T_{stop}$ curves for TLD-100. The samples were annealed at 400 °C for 1 hour and heated after irradiation at 1.3 °C.s$^{-1}$. The irradiation sources were: (a) 50 Gy $^{60}$Co gamma; (b) 5×10$^9$ cm$^{-2}$ 5 MeV alpha, and (c) 3.4×10$^{10}$ cm$^{-2}$ 3 MeV $^4$He ions. The numbers refer to the identified individual peaks. (Refer to Figure 7.6). (Adapted from Horowitz et al. (1999).)

The obtained $E_{tm}$ and $s_m$ values given in the figure caption to Figure 7.8 should be compared to the $E_t$ and $s$ values obtained in Figure 7.1b. For example, if the half-life ($\tau_{1/2}$) of TL peak 2 at $T = 300$ K is calculated using the values of $E_{tm}$ and $s_m$, and $E_t$ and $s$, along with the expression $\tau_{1/2} = \ln(2)/\exp\{-E_t/kT\}s$, values of 2.7 days and 3.9 days are obtained, respectively. However, isothermal decay measurements, for example by Taylor and Lilley (1978) and by Bos and Piters (1993), yield measured values for $\tau_{1/2}$ at 300 K of 0.86 days and 0.78 days, respectively, with correspondingly smaller apparent values of $E_t$ and $s$. How can such clear discrepancies be reconciled? The answer lies in the defect clustering reactions previously mentioned in Sections 7.2.1.2 and 7.2.1.4. It was noted that Bos et al. (1992) had observed that the calculated $E_t$ and $s$ values for the various peaks were dependent on the cooling rate before irradiation and the heating rate during the TL measurement. Emission spectra changes were also observed depending upon the heat treatment (Section 7.2.1.3). Both observations were interpreted as being caused by variations in the extent of clustering among the Mg defects and the Ti defects, although the precise mechanisms are elusive. Clustering can also be expected to occur during isothermal annealing, dependent upon the isothermal storage conditions.

Piters and Bos (1993b) simulated the TL curves that would be obtained in a hypothetical system if the defect that was the source of a given TL peak was allowed to interact with another defect in an interaction of the type $A^* + B \rightarrow C^*$ where defect $A^*$ (defect $A$ with a trapped electron) is allowed to react with defect $B$ to form defect $C^*$. TL results from reaction $A^* \rightarrow A + e^-$, but both of the above reactions result in the loss of $A^*$. With this model, Piters and Bos show that a first-order, TL peak can be formed and very-well fitted to the Randall-Wilkins equation to yield an effective activation energy $E_{eff}$ and frequency factor $s_{eff}$ that will change with heating rate in a manner dependent upon the relative values of the trap depth and interaction enthalpy. Both of the above reactions also take place during isothermal annealing, but one dominates over the other at short storage times, and vice-versa at long storage times. The net result is that a slower fading rate is obtained if the fading rate is calculated from $E_{eff}$ and $s_{eff}$ for the TL peak than if it is measured from isothermal decay. This is exactly what is found for LiF TLD-100.

It is very difficult to predict exactly what $E_{eff}$ and $s_{eff}$ values will be determined from peak-fitting of TL glow peaks from LiF:Mg,Ti for given circumstances (impurity contents, annealing times and temperatures, cooling rates, heating rates), and what fading rates will be measured during isothermal annealing. However, information is contained in these measurements, and in measurements of emission spectra, concerning defect clustering in this system – information that is important for the overall dosimetric properties of the material. The properties can be said to be deterministic but, unfortunately, not yet wholly predictive.

### 7.2.1.6   Competition

Whatever the details of the TL emission process, it appears that competition is the defining mechanism in dictating the shape of the dose-response function in this material. Localized (geminate) recombination (reaction 7.5) cannot be the sole mechanism of TL production since this predicts straightforward linear growth. Some type of non-localized (non-geminate) recombination, such as those described in reactions 7.6 and/or 7.8–7.11, must be active in order to produce non-linearity. The non-linearity is a result of competition caused by deep electron traps that reside in the region outside the trap-center pairs and which remain empty after the initial irradiation. In this sense, whether or not the TL as a function of dose is described by Equation 6.24 or 7.13 (for example), or some combination of both, is immaterial. Both predict supralinearity at high doses because of the competition.

It should also be remembered that non-radiative luminescence centers can also act as competitors. These are filled hole traps and also exist in the region outside the trap-center pairs. As discussed in Chapter 6, these may become particularly active at high doses when there are many of them and fewer empty, competing, electron traps, such that the dose-response curve not only becomes sublinear but the response even decreases at high dose.

Considering the competing, empty electron traps, it is important to remember that sublinear filling of these traps is not required for supralinearity to be observed. Recalling Equation 6.24, the nonlinear part contains the term:

$$n_1(D) \left( \frac{\sigma_m m(D)}{\sigma_k (N_k - n_k(D))} \right)$$

where $N_k$-$n_k$ is the concentration of empty, competing traps. Even if $N_k \gg n_k$, there still remains the non-linear term $n_1(D)m(D)$. Similarly, with Equation 7.13, the nonlinear term resides in the expression for $P_i(m,R_i)$ (Mahajni and Horowitz 1997), no matter the extent to which the competitors are filling.

Although the detailed structure of the competing traps is unknown, several characteristics can be listed:

- Since Peak 5 is an electron trap, the competitors must be electron traps.
- The concentration of filled competitors will increase over the same temperature range that peak 5 appears, i.e. over the same temperature range that the traps responsible for peak 5 release their electrons.
- The competing traps must be thermally disconnected over the temperature range of peak 5, i.e. they must be thermally disconnected deep traps, TDDTs, and electrons will not be released from the TDDTs in the temperature range of peak 5.
- Since there is no requirement that the competitors fill during irradiation, the competing traps can either grow weakly or strongly during irradiation. However, since there is only weak competition during irradiation for peak 5, then the competitors can be expected to grow only weakly during irradiation.

A search for suitable competitors that fulfill these characteristics has revealed that the defect responsible for optical absorption at 5.5 eV is a likely source of the competing traps (McKeever 1990). Figure 7.10a shows the change in intensity of this band compared with the intensities of several other prominent absorption bands in this material as a function of post-irradiation, pulse-anneal temperature ($T_{stop}$) while Figure 7.10b shows similar data, but as a function of the temperature during TL readout. The data in Figure 7.10a are taken at –196 °C (at which temperature the "5.5 eV" band appears at 5.7 eV) while the data in Figure 7.10b are taken at the

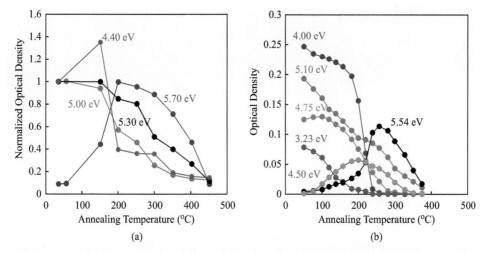

**Figure 7.10**   Changes in various optical absorption bands from irradiated LiF:Mg,Ti as a function of temperature. (a) As a function of the pulse-anneal temperature ($T_{stop}$) after irradiation. In this experiment, the absorption spectrum was recorded at liquid nitrogen temperature (–196 °C), heated to $T_{stop}$, cooled again to –196 °C and the absorption spectrum re-measured. The spectrum was then fitted to a series of Gaussian functions to separate the individual absorption bands. At –196 °C, the "5.5 eV" band appears at 5.7 eV. (Adapted from Landreth and McKeever (1985).) (b) As a function of temperature during continuous heating of the sample after irradiation. Absorption spectra were repeatedly taken during heating and separated into Gaussian bands. In addition to the 5.5 eV band (here denoted 5.54 eV) there is a band at 4.5 eV. (Adapted from Bos and De Haas (1998).)

temperature indicated. As previously shown (in Figure 7.4) TL peak 5 is related to the band at 4.00 eV (Figure 7.10b). (Note that the band at 4.40 eV in Figure 7.10a is likely a combination of the 4.00 eV and 4.50 eV bands shown in Figure 7.10b.) During annealing of the 4.00 eV band, the 5.5 eV band is seen to grow strongly, before decaying at temperatures > 250 °C. This has all the characteristics expected for a competing trap to the TL peak 5 traps. The 5.5 eV band is related to Mg impurities and does not appear if Mg is absent (McKeever 1984, 1990). It has been proposed to be a Mg atom on an anion site with a cation vacancy (Radzhabov and Nepomnyachik 1981; Chernov et al. 2001).

The 5.5 eV defects may not be the only competing traps in this material. A particular observation is that increasing the OH content reduces the sensitivity and removes supralinearity. Example data are shown in Figure 7.11. It is proposed (Stoebe and DeWerd 1985) that adding Mg enables the formation of non-luminescent $Mg(OH)_m$ centers, which act as non-luminescent competitors to the luminescent $Ti(OH)_n$ centers and to the peak 5 electron traps. Thus, as the OH content increases, so the sensitivity decreases. The loss of supralinearity shown in Figure 7.11 can be inferred from Equation 6.23:

$$I_{TL/OSL}(D) \propto n_1(D) \left[ K \left( \frac{\sigma_m}{4\pi s^2} \right) + \left( 1 - K \frac{\sigma_m}{4\pi s^2} \right) \left( \frac{\sigma_m m(D)}{\sigma_k k(D)} \right) \right] \tag{6.23}$$

where $k(D)$ is the concentration of competing centers, which can be either empty competing traps, or filled competing hole traps (non-radiative recombination centers). If the dominant competitors are non-luminescent centers, then $k(D)$ increases with dose. In the case of peak 5

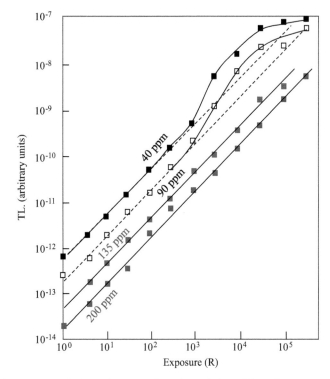

**Figure 7.11**   TL response of TL peak 5 as a function of dose (given here as exposure in R) for samples of LiF:Mg,Ti with different OH contents, as indicated. (Reproduced from Stoebe and DeWerd (1985) with permission from AIP Publishing.)

TL in LiF:Mg,Ti, it is presumed that $n_1(D)$ and $m(D)$ are the concentrations of filled peak 5 traps and filled $Ti(OH)_n$ luminescent center concentrations, respectively. If $k(D)$ is the concentration of non-luminescent, filled $Mg(OH)_m$ centers, and if $n_1(D)$, $m(D)$, and $k(D)$ each fill linearly with dose, then Equation 6.23 predicts that the TL signal will also be linear with dose, as observed in Figure 7.11 for high OH concentrations. At lower OH concentrations, however, competition is dominated by empty electron traps and $k(D) = N_k - n_k(D)$. Equation 6.24 now applies so that at higher exposures (> several 100 R), the response becomes nonlinear.

### 7.2.1.7   *Photon Dose-Response Characteristics*

The general linear-supralinear response of peak 5 to increasing dose, as observed in Figure 7.11 for lower OH contents, is a characteristic of LiF:Mg,Ti TLD-100. Indeed, many peaks in this material display similar characteristics, as is observed in Figure 7.12 for peaks 4–9 (using the numbering system illustrated in Figure 7.6a). Note that the onset of supralinearity occurs at different doses for each peak. In general, the higher the temperature of the TL peak, the lower the dose at which supralinearity occurs and the larger the maximum extent of the supralinearity. This effect is also shown in Figure 7.13, which displays the supralinearity function, $f(D)$ for peaks 4, 5, and 5b. Similar data can be found for the other peaks in Massillon et al. (2006).

By using the UNIM and Equation 7.13, Horowitz and colleagues have simulated $f(D)$ curves under a variety of conditions (summarized in Horowitz et al. 2019). These authors assumed the growth of trapped electrons and holes to follow the simple saturating exponential function given in Equation 6.2. With this simplified assumption, and extracting $\alpha$ values from the growth of optical absorption bands with dose, $f(D)$ curve shapes similar to those seen in Figure 7.13

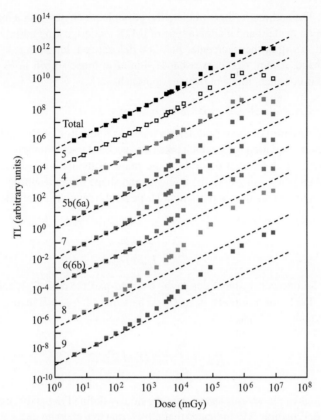

**Figure 7.12**   TL as a function of dose for the total signal, and for peaks 4–9. The dashed line corresponds to the linear response. Individual data sets have been normalized for clarity of the display. (Reproduced from Masillon et al. (2006) with permission from IOP Publishing.) The peak numbers are as shown in Figure 7.6a. Note that Masillon et al. (2006) called peak 5b "peak 6a," and peak 6 "peak 6b" (shown in parentheses on the figure). "Total" refers to the net area under the entire glow curve.

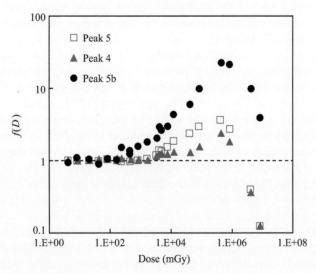

**Figure 7.13**   Plots of $f(D)$ for peaks 4, 5, and 5b using the dose-response curves of Figure 7.12. (Adapted from Masillon et al. (2006).)

can be obtained. The authors further simulate the possible processes with a multiple-trap/multiple-recombination-center band model (a type of IMTS model), taking into account both localized recombination within the trap/center pair and delocalized, long-range recombination via the conduction band. Again, $f(D)$ curve shapes similar to those shown in Figure 7.13 can be predicted with this complex, multi-parameter approach.

### 7.2.1.8    *Charged-Particle Dose-Response Characteristics*

In Chapter 6, the variation of the supralinearity factor as a function of charged-particle fluence was discussed, using Equation 6.35, which takes into account track interaction and competition effects. Also discussed in Chapter 6 was how the shape of the radial dose distribution (RDD) profile is critical in determining both the intensity and shape of the resulting TL glow curve.

With respect to the intensity of a particular TL peak, the relative TL efficiency was defined in Equation 6.29 as:

$$\eta_L = \frac{I_{CP} / D_{CP}}{I_{ref} / D_{ref}} \tag{6.29}$$

where $I_{CP}$ is the luminescence signal due to the charged-particle dose $D_{CP}$, and $I_{ref}$ is the luminescence signal due to the reference dose $D_{ref}$. The value of $I_{CP}$ in Equation 6.29, for the $i^{th}$ glow peak, is given by:

$$I_{CP}^i = \int_{D_{min}}^{D_{max}} I_{ref}^i (D) h(D) dD \tag{7.19}$$

where the meanings of the various terms were previously defined in the discussion of Equation 6.28 in Chapter 6. Equation 7.19 indicates that the TL due to a charged particle is a convolution of the RDD for that particle (defined by $h(D)$) and the dose-response function for the reference radiation for that peak, $I_{ref}^i (D)$. This ultimately means that the ratio of the relative efficiencies of two different peaks, say $i = 1$ and $i = 2$, is dependent on the LET of the charged particle, $L$, for a given particle species.

For the $i^{th}$ peak, even if $D_{CP} = D_{ref}$, $\eta_L$ does not necessarily equal 1, and can be $> 1$ or $< 1$, depending upon which parts of the RDD dominate the luminescence signal and where this is represented on the dose-response function (i.e. the linear region, the supralinear region, or the sublinear region; refer to Figure 6.18).

These effects are quite prominent on LiF:Mg,Ti. Figure 7.14a shows the glow curves from TLD-700 after irradiation with 50 mGy of either $^{60}$Co gamma or $^{20}$Ne charged particles ($L = 31.6$ keV/$\mu$m). It is clear that in the region of peak 5, $\eta_L < 1$ whereas for the region of the high-temperature TL emission (HTE), $\eta_L > 1$. The ratio peak 5/HTE is a function of $L$.

The exact value of $\eta_L$ for a specific region of the glow curve depends on the energy, charge and LET of the incident particle. For the same value of $L$, the dose from a charged particle with greater charge $Q$ is deposited in larger volumes around the track. Depending upon the shape of the RDD, if the TL/OSL signal is dominated by that region of the RDD that corresponds to the linear region of the dose-response curve, then $\eta_L = 1$. On the other hand, if the supralinear region dominates, then $\eta_L > 1$, and if the sublinear region dominates, then $\eta_L < 1$.

Figure 7.14b illustrates example data for the efficiency of peak 5 versus $L$ for 3 different particles ($^4$He, $^{12}$C, and $^{29}$Si) of varying energies. The data of Figure 7.14 are illustrative but not definitive, however. That is, the actual values of $\eta_L$ for the different TL regions are also critically dependent on the details of the TL readout protocols. In particular, the efficiencies are

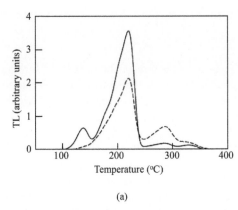
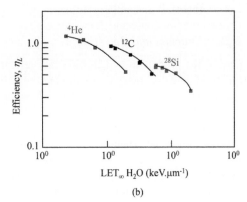

(a)                                        (b)

**Figure 7.14** (a) TL glow curves from TLD-700 following 50 mGy dose of either $^{60}$Co photons (full line) or $^{20}$Ne charged particles (dashed line). (b) Peak 5 efficiency $\eta_L$ as a function of the LET of the charged particle for $^4$He, $^{12}$C, and $^{29}$Si ions. (Reproduced from Berger and Hajek (2008) with permission from Elsevier.)

found to vary with pre-irradiation annealing conditions, cooling rates, reference irradiation (e.g. $^{60}$Co, $^{137}$Cs, x-rays), the individual sample used, the TL heating rate, and the method of peak intensity evaluation (overall height of the glow curve at a given temperature, peak height following individual peak deconvolution, area under the curve between set temperature limits, etc.). This leads to a non-universality in the reported data for $\eta_L(L)$, which is important due to the increasing use of TLDs (and OSLDs and RPLDs) in radiobiology, ion therapy in medicine, and space dosimetry for astronauts (Berger and Hajek 2008).

Despite the experimental differences and the resulting variations in the efficiency values measured, theoretical models for track structures are able to predict qualitatively the relationship between of the TL efficiency and LET for most particle types. Quantitative agreement with experiments, however, is limited by the afore-mentioned experimental issues and by uncertainties in the developed models (e.g. Berger and Hajek 2008; Horowitz et al. 2011, 2012).

---

**Exercise 7.6   TL peak 5 efficiency for different charged particles**

Examine Table 3 in the paper by Berger and Hajek (2008).

(a) Plot the efficiency of TL peak 5 $\eta_L$ (written as $\eta_{HCP}$ by Berger and Hajek) versus LET for the particles listed in Table 3 and for the two labs mentioned in the paper (ATI in Austria and DLR in Germany). Include error bars in the plots.

(b) Examine the dose-response curve and $f(D)$ curves for peak 5 given in Figures 7.12 and 7.13. Using these data and the plots of the $\eta_L$ data in part (a), what do you conclude about the following:

(c) Variations in $\eta_L$ from lab-to-lab.

(d) Variations in $\eta_L$ as a function of particle charge $Q$.

(e) Variations in $\eta_L$ as a function of LET $L$.

(f) Which part of the dose-response curve dominates the $\eta_L$ behavior, for different particles and for different LET, $L$?

### 7.2.2   LiF:MCP

#### 7.2.2.1   *Structure and Defects*

Many of the properties of LiF:MCP are reflective of the structure of the host (LiF) but the detailed TL and dosimetry properties are critically dependent on the structure of defects caused predominantly by the presence of the dopants Mg, Cu, and P. The concentration of Mg in LiF:Mg,Ti is less than 0.02 M% whereas in LiF:MCP the optimal concentration is an order of magnitude higher, at around 0.2%. At such large Mg concentrations, precipitation of secondary Mg phases might be expected. x-ray Diffraction (XRD) spectra, however, show that the dominant secondary phases in LiF:MCP are related to P. Phosphorous is present at high levels, 1–5 M% depending on the growth details. Zha et al. (1993) note that at such high P levels, unwanted secondary phases may be present. The exact phases present seem to depend on the growth details. Figure 7.15 shows an XRD spectrum for LiF:MCP, MCP-N (Poland). Apart from the dominant lines due to LiF, lines due to various polymorphs of lithium phosphate ($Li_3PO_4$) can also be observed (Ayu et al. 2016). Sun et al. (1994) observed $Li_4P_2O_7$ phases in material grown by them (China), along with an unidentified phase, possibly a mixed-metal pyrophosphate, $Li_6Cu(P_2O_7)_2$. Phases due to precipitation of Mg, i.e. $MgF_2$ and $6LiFMgF_2$, have not been reported in LiF:MCP, unlike in LiF:Mg,Ti (Bradbury and Lilley 1977). The role played by the secondary phosphate phases in TL from LiF:MCP, if any, is unclear.

Several papers have discussed the optimum concentrations of Mg, Cu, and P for maximum sensitivity. The optimum concentrations of Mg and P are noted in the previous paragraph, but the optimum Cu content depends on the oxygen content and valency, namely 0.002 M% for $Cu^{2+}$ and 0.004 M% for $Cu^+$ (e.g., Shoushan 1988; Horowitz and Horowitz 1990; McKeever et al. 1993a, 1993b). Apart from maximizing the sensitivity, the optimum dopant concentrations are also found to reduce the high-temperature TL. One of the negative features of LiF:MCP is the loss of sensitivity when heated during TL readout to over ~240 °C. To prevent sensitivity

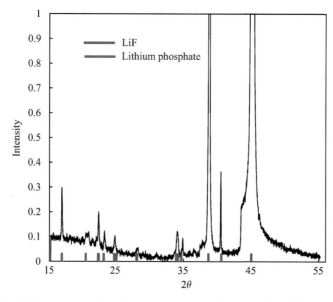

**Figure 7.15**   An XRD spectrum for LiF:MCP (MCP-N, obtained from the Institute of Nuclear Physics in Krakow, Poland). (Data kindly provided by K. Remy.)

loss, the readout temperature is limited and the high-temperature TL is not zeroed during normal TL measurement. Often, therefore, the sample is held at 240 °C for a few seconds as a compromise between minimizing sensitivity loss and maximizing zeroing of the signal. Growth conditions that reduce the high-temperature signal are preferred.

Similar to LiF:Mg,Ti, the main TL features and trapping structures of LiF:MCP are understood to be related to Mg, probably $\left[ Mg^{2+} / Li_{vac}^{+} \right]^{3}$ trimers. Phosphorous is presumed to act as a luminescent recombination site (McKeever 1991; Sun et al. 1994; Meijvogel et al. 1995; Bilski et al. 1996), although the detailed ionic structure(s) of the site(s) and the luminescence process(es) are unknown. The role of Cu is less clear, but is suggested to form complexes with O and P. The role of oxygen appears important due to oxidation/reduction reactions with P following irradiation (Sun et al. 1994).

### 7.2.2.2   TL Glow Curves

A typical glow curve from LiF:MCP at low doses is shown in Figure 7.16. All the TL peaks illustrated are believed to be due to electron traps (McKeever et al. 1995) and the similarity between this glow curve and that of LiF:Mg,Ti (Figure 7.1) is clear, with a dominant, narrow peak around 210 °C (~480 K; $\beta_t = 1.0$ K.s$^{-1}$). The individual peaks shown in the figure arise from fits to the glow curve, described in Section 7.2.2.4 below. For this figure the sample has been preheated to 90 °C for 10 s during which time the low temperature peaks were removed.

The main peak, denoted peak 4, is used for dosimetry and its counterpart in LiF:Mg,Ti is peak 5. As noted above, the higher temperature TL from LiF:MCP, beyond the main peak, is problematic in that the reusability of the material is adversely affected if this is not removed (corresponding traps emptied) during TL readout. However, heating much beyond peak 4 (>~240 °C) leads to an irreversible change to the sensitivity. Therefore, a figure of merit (FOM) for a good LiF:MCP dosimeter includes not only the sensitivity of peak 4 (the higher the

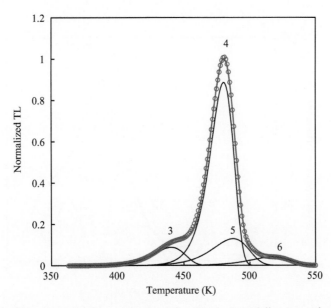

**Figure 7.16**   TL glow curves for the same material as Figure 7.15 following a dose of ~410 mGy, preheating to 90 °C for 10 s, and heating at $\beta_t = 1$ K.s$^{-1}$. (Data kindly provided by S. Sholom.) The glow curve is fitted to four, first-order TL peaks.

better), but also the intensity of the high-temperature TL beyond peak 4 (the lower the better). Following Horowitz and Horowitz (1990) a relative FOM (i.e. RFOM, relative to LiF:Mg,Ti) may be defined as:

$$\text{RFOM} = \frac{\text{Sensitivity of peak 4 in LiF:MCP / Sensitivity of Peak 5 in LiF:Mg,Ti}}{\text{Intensity high-temperature TL / Intensity of peak 4 in LiF:MCP}} \quad (7.20)$$

where the "high-temperature" TL is that beyond peak 4 and is approximately the intensity of the peak shown as peak 6 in Figure 7.16. The numerator (ratio of sensitivities of the dosimetry signals) can be as high as 15–20, whereas the RFOM can be as high as 2400 (McKeever et al. 1995).

The optical absorption spectrum for LiF:MCP is dose dependent in terms of the intensities of the various bands. A typical spectrum showing all the major bands is shown in Figure 7.17. The main features are the $F$-band, and bands due to $F$-center complexes, particularly $F_2$- and $F_3$-centers in different charge states. These are important in the production of RPL from this material and are discussed in more detail in Chapter 9. Also observed are various other bands, considered (but not confirmed) to be related to Mg.

### *7.2.2.3 TL Emission Spectra*

A typical emission spectrum from this material is shown in Figure 7.18. The spectrum is characterized by a broad, asymmetric emission peaking at ~370 nm, with a low-energy (long-wavelength) tail, as described in several published studies (e.g. McKeever 1991; Meijvogel et al. 1995; Mandowska et al. 2010). Emission spectra deconvolution reveals a complex array of overlapping emission bands, the relative intensities of which vary with TL temperature, dopant levels, dose, and thermal treatments. An example is shown in Figure 7.19 following a large gamma dose of 20 kGy. The number of bands increases as the dose increases. The fit shown is the minimum number of bands (Gaussian) required for an acceptable fit.

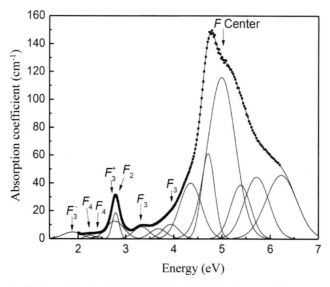

**Figure 7.17** Optical absorption for LiF:MCP (MCP-N) following 200 kGy gamma irradiation at room temperature. (Reproduced from Remy et al. (2017) with permission from Elsevier.)

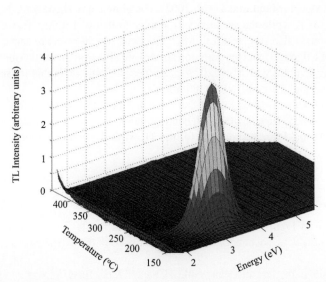

**Figure 7.18**    An isometric plot of TL-versus-temperature-versus-energy for LiF:MCP (MCP-N) after a dose of 1 kGy gamma and heating at 1 K.s$^{-1}$. (Data kindly provided by S. Sholom.)

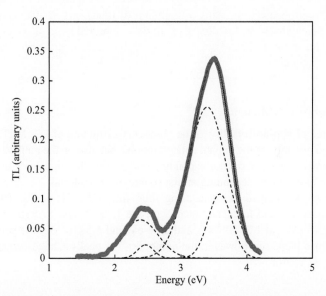

**Figure 7.19**    Example TL emission spectrum for LiF:MCP (MCP-N) following a dose of 20 kGy gamma. The spectrum was obtained by recording an isometric spectrum (e.g. Figure 7.18) and summing along with wavelength axis over all temperature values, then converting from wavelength to energy and correcting by $1/\lambda$.$^2$ The central wavelengths of the fitted bands are 345 nm, 365 nm, 503 nm, and 520 nm. (Data kindly provided by S. Sholom.)

A particular feature of LiF:MCP, and LiF:Mg,Ti, is that the emission spectrum shape is strongly dependent on the Mg concentration, despite the hypothesis that the luminescence sites in each are related to P and Ti, respectively. Interesting experimental observations emerge when Ti replaces P in LiF:MCP (producing LiF:MCT) and P replaces Ti in Li:Mg,Ti (producing LiF:Mg,P) (Mandowska et al. 2002). The glow-curve shape for LiF:MCT is similar to that for LiF:MCP, while the sensitivity is similar to that of LiF:Mg,Ti. Curiously, the TL emission spectrum is also similar to LiF:MCP. Likewise, the glow-curve shape of LiF:Mg,P is similar to that of LiF:Mg,Ti and the emission spectra are also similar. Furthermore, the emission spectrum for LiF samples with Mg <0.02% peaks near 400 nm while that for samples with Mg concentrations around 2% peak below 400 nm.

In order to reconcile the observations that the emission intensity in LiF:MCP is related to P but that the emission wavelength is dependent upon Mg, it is proposed that trap/center complexes of Mg and P are responsible for the TL, with the concentrations of each affecting the wavelength and the intensity of emission.

---

**Exercise 7.7    Analysis of emission spectra for LiF:MCP**

The fit shown in Figure 7.19 uses just four emission bands. The number of bands reported elsewhere varies depending upon the conditions of the experiment and, especially, the dose.

On the web site under Exercises and Notes, Chapter 7, Exercise 7.7, can be found the raw data shown in Figure 7.19, as a function of wavelength.

(a) Plot the data versus wavelength and compare to Figure 7.19.
(b) Convert the spectrum to an energy axis, correcting for the conversion from fixed wavelength bandwidth to fixed energy bandwidth. Compare again to Figure 7.19.
(c) Fit the resulting spectrum using a minimum of four Gaussian bands. Compare the peak wavelengths of the fitted bands with those reported in the literature.
(d) Increase the number of bands, and compare again with the published literature. What do you conclude?

---

### 7.2.2.4    TL Glow-Curve Analysis

There is a degree of non-universality in the glow-curve structure of LiF:MCP. The shape and intensity depend not only on the dopant concentrations, but also on the method of material production and growth, and there is a corresponding non-universality in the positions of the different peaks in the glow curve, depending on the source of the material.

Figure 7.16 shows a fit to the data using four, first-order TL peaks. The values of the activation energies ($E_t$) and frequency factors ($s$) are 1.39 eV and $5.6 \times 10^{14}$ s$^{-1}$, 2.24 eV and $3.4 \times 10^{22}$ s$^{-1}$, 1.48 eV and $1.3 \times 10^{14}$ s$^{-1}$, and 1.54 eV and $6.5 \times 10^{13}$ s$^{-1}$ for peaks 3–6, respectively. As the dose increases, however, the number of the peaks changes dramatically, as discussed in the next section.

---

**Exercise 7.8    LiF:MCP glow-curve analysis**

In the web file under Exercises and Notes, Chapter 7, Exercise 7.8, you will find the data for the glow curve shown in Figure 7.16.

Fit the glow curve to a summation of first-order peaks. Vary the number of peaks and obtain a best fit in each case. Compare your results with those in the literature.

What do you conclude?

### 7.2.2.5 Changes to the Glow-Curve Shape with Dose and Ionization Density

LiF:MCP is well known for the remarkable changes to the glow-curve shape and complexity that occur as the dose increases. The changes are also dopant-dependent in that varying the amount of Mg, P, or Cu yields different glow-curve shapes as a function of dose (e.g. Bilski 2002). For standard or "normal" LiF:MCP, e.g. MCP-N, the glow-curve shape is independent of dose over the dose range used for personal dosimetry, and remains reasonably invariant up to ~1 kGy. At higher doses, however, significant changes begin to appear on the high-temperature side of peak 4. While peak 4 saturates, the higher-temperature TL grows; the whole glow-curve shape evolves with dose.

Figure 7.20 illustrates the changes to the LiF:MCP glow curve as a function of gamma dose from 1 kGy to 300 kGy, while Figure 7.21 illustrates the complex changes to the emission spectrum, illustrated after two large gamma doses (30 kGy and 200 kGy). New emission bands at longer wavelength (lower energy) emerge as the dose increases, as well as the new TL peaks. (Compare with Figure 7.16 for lower doses; see also Bilksi et al. 2008; Mandowska et al. 2010; Gieszczyk et al. 2013.)

The highest-temperature peak is usually called "peak B" and it appears at high gamma doses, or after charged-particle or neutron irradiation (Obryk et al. 2010). One of the properties of this peak is that at very high doses it decreases and shifts to higher temperatures (Figure 7.20). The emission from peak B is in the region of 3.0 eV to 3.5 eV (Figure 7.21). This wavelength region also features significant optical absorption bands, primarily due to $F_3$-centers (Figure 7.17). All the absorption bands show strong annealing in the temperature range of Peak B, and vary strongly with dose (Remy et al. 2017). Therefore, following the discussion in Chapter 5 (Section 5.2.7.4) regarding self-absorption, the possibility of self-absorption affecting the shape, size, and position of peak B must be considered.

It may also be noted that the absorption coefficient after large doses is extremely high. Figure 7.22 shows an approximate, semi-transparent, overlay of the isometric TL emission

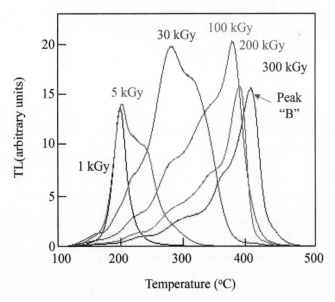

**Figure 7.20** Changes to the glow-curve structure as a function of dose for very high gamma doses; $\beta_t = 0.5$ °C.s$^{-1}$. (Data kindly provided by K. Remy.)

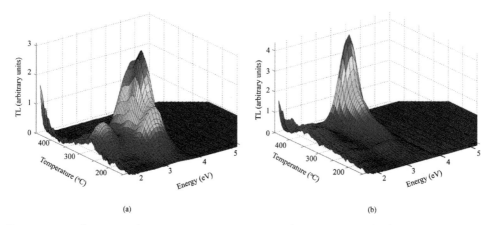

**Figure 7.21**    Changes to the TL emission spectrum as a function of very high gamma doses; $\beta_t = 1.0\ °C.s^{-1}$. (a) 30 kGy; (b) 200 kGy. (Data kindly provided by S. Sholom.)

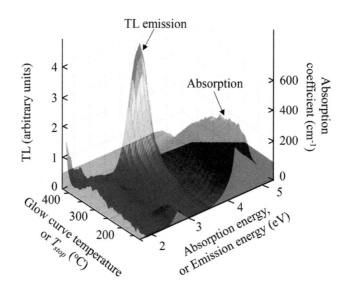

**Figure 7.22**    A semi-transparent overlay of the isometric TL emission spectrum (from Figure 7.21b) and an isometric plot of the optical absorption as a function of post-irradiation (200 kGy gamma), pulse-anneal temperature ($T_{stop}$; all absorption spectra taken at room temperature). (Data kindly provided by K. Remy.)

spectrum from LiF:MCP (MCP-N) (data from Figure 7.21) with an isometric plot of the variation of the absorption spectrum as a function of post-irradiation pulse anneal temperature ($T_{stop}$) from the same material, each following a dose of 200 kGy gamma. It is observed that the TL emits (i.e. is measured) at those temperatures and at those wavelengths (the TL emission wavelengths and optical absorption wavelengths) where the absorption is weakest. By comparison, there is little or no TL in regions where the absorption is highest, again suggesting that self-absorption effects may be very strong in this material.

**Exercise 7.9    TL and self-absorption**

(a) Using a simulated, first-order, TL glow peak (you choose the $E_t$ and $s$ values) plot the peak shape and position for an arbitrary dose. Consider that the TL peak increases with dose according to:

$$I_{TL} = I_{TLmax}\left(1 - \exp\left\{-\alpha_{TL}D\right\}\right) \tag{1}$$

where $I_{TL}$ is the TL peak height at dose $D$, $I_{TLmax}$ is the maximum TL peak height at saturation, and $\alpha_{TL}$ is a constant. Show how the peak is expected to grow with dose.

(b) Now consider that the wavelengths with which the TL emits overlap a strong optical absorption (OA) band in the same wavelength region. Consider that the absorption band decays over the same temperature range as the TL peak appears. Choose a function to describe the decay of the OA band over this temperature range. What is the effect of absorption on the TL emission on the shape of the TL peak?

(c) Similarly consider that the optical absorption centers grow as:

$$I_{OA} = I_{OAmax}\left(1 - \exp\left\{-\alpha_{OA}D\right\}\right) \tag{2}$$

where $I_{OA}$ is the absorption band height at dose $D$, $I_{OAmax}$ is the maximum OA band height at saturation, and $\alpha_{OA}$ is a constant. Also assume $\alpha_{TL} \neq \alpha_{OA}$ – i.e. the TL and OA peaks do not grow at the same rate.

  Using Equations (1) and (2), and your chosen decay function for the decay of the OA band with increasing temperature, show how the shape and position of the TL peak varies with dose.

(d) Repeat the above calculations but vary the $E_t$ and $s$ values for the TL peak and the values of $\alpha_{TL}$ and $\alpha_{OA}$. What do you conclude with respect to what is observed for LiF:MCP TL at high doses?

---

*7.2.2.6    Photon Dose-Response Characteristics*

Unlike LiF:Mg,Ti, the photon dose-response characteristics of LiF:MCP are linear-sublinear. An example of the dose-response function $f(D)$ is shown in Figure 7.23 for several variants of LiF:Mg,Cu,P (with different Mg concentrations) compared to LiF:Mg,Ti. Each MCP variant shows linearity ($f(D) \approx 1$) for $D < {\sim}10$ Gy, becoming sublinear ($f(D) < 1$) for $D > 10$ Gy. This is in contrast to the linear-supralinear-sublinear behavior of LiF:Mg,Ti (Bilski 2002).

  Since supralinearity in LiF:Mg,Ti is understood to be caused by the competition effect of unfilled, deep electron traps, the observation of linearity only (before sublinearity) in LiF:MCP is assumed to indicate that competitors are absent or not prominent in this material. However, post-irradiation, step-annealing measurements of optical absorption in LiF:MCP indicate that an absorption band at ~5.7 eV grows during the production of the main TL peak 4. This is very like the similar experimental observations in LiF:Mg,Ti, in which material the 5.5 eV defect is hypothesized to be the electron competitor (see Section 7.2.1.6). Peak 4 in LiF:MCP also coincides with the rapid annealing of the $F_2/F_3^+$ bands, while in samples with no Mg (i.e. Cu and P only) no increase in the 5.7 eV band is observed. This has led to a suggestion that the 5.7 eV band is a Mg/$F$-center complex (Remy et al. 2017). If the 5.7 eV band in LiF:MCP is similar to the 5.5 eV band in LiF:Mg,Ti, then it is not clear why it would act as a competitor in LiF:Mg,Ti, but not in LiF:MCP.

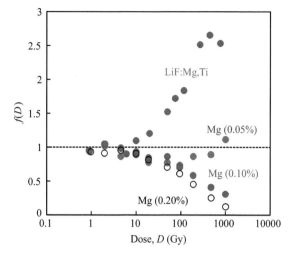

**Figure 7.23**  The dose-response characteristics of LiF:Mg,Ti compared to that for several variants of LiF:MCP, expressed as the dose-response function $f(D)$. The reference dose for the evaluation of $f(D)$ was 1 Gy in all cases. (Reproduced from Bilski (2002) with permission from Oxford University Press.)

The TL mechanism for peak 4 appears to involve Mg and P along with $F$- and $F_n^m$-centers in as-yet unidentified defect complexes. As with LiF:Mg,Ti, the notion of localized transitions within or between trap-center/recombination-center pairs must also be considered as a possibility for LiF:MCP, as must the notion of $F$- and $H$-center recombination processes as already suggested for LiF by Sagastibleza and Alvarez-Rivas (1981) and others (Rogalev et al. 1990; Mysovsky et al. 1995; Chernov et al. 1998). However, apart from these generalized statements it is clear that the TL mechanism in this complex material is far from understood.

### 7.2.2.7  Charged-Particle Dose-Response Characteristics

The linear-sublinear dose-response characteristics of this material have a direct effect on its response to charged particles. While the concept of trap/recombination-center pairs can be suggested for this material as well as for LiF:Mg,Ti, there has been no detailed UNIM-type model proposed to account for the ionization density dependence of the TL signal. Using the concepts already discussed in this book, however, it is possible to make some general, qualitative statements about what can be expected.

The shape and intensity of the TL glow curve due to charged particles will be the net result of a convolution of the RDD and the TL dose-response relationship $f(D)$ for a reference radiation (Equation 7.19). Using the expression for $\eta_L$ (Equation 6.29), the TL efficiency for a particular charged particle of LET $L$ will depend on whether or not the TL signal is dominated by that region of the RDD that corresponds to the linear region of the dose-response curve, for which $\eta_L = 1$, or to the sublinear region, for which $\eta_L < 1$. An example dependence of the efficiency $\eta_L$ for LiF:MCP, compared to LiF:Mg,Ti, for several types of charged particle is shown in Figure 7.24. As expected, the efficiency is never > 1 for LiF:MCP in contrast to that for LiF:Mg,Ti. Is it possible to say more than this? The next section examines this question.

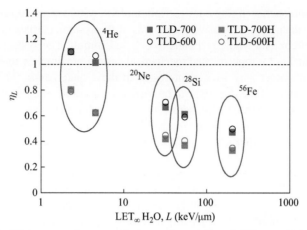

**Figure 7.24**  TL Efficiency $\eta_L$ for several charged particles of different LET, $L$, and for different TLDs. (The LET is expressed as unrestricted LET in water.) The TLD materials are LiF:Mg,Ti TLD-600 and TLD-700, along with their counterparts for LiF:MCP, denoted here as TLD-600H and TLD-700H (where "H" stands for high sensitivity). (Adapted from Berger and Hajek (2008).)

### 7.2.3  Approximately Right; Precisely Wrong

This chapter on TL, using LiF as a representative TLD material, has provided examples of some of the principles discussed in Part I regarding thermoluminescence. In addition, it has also provided an opportunity to discuss a basic question in science and engineering, namely whether or not one should accept being approximately right or strive for greater precision with the risk being ultimately, and precisely, wrong. It is clear that the specific TL processes active in the main forms of LiF dosimetry materials – LiF:Mg,Ti and LiF:MCP – do not need to be pinned down with precision before the empirical properties of the material can be used in dosimetry. These LiF compounds, although exhibiting highly complex TL properties, are perhaps the best TLD materials available commercially today, and yet we do not know with confidence how the TL process works in either of them.

Many efforts have been expended over the years to elucidate the TL properties, with a degree of progress being achieved through using an array of scientific tools, including TL, conductivity, optical absorption, photoluminescence, RPL, EPR, exoelectron emission, and others. This has led some to be confident enough to try to develop models at the microscopic scale and to use such models to explain the detailed dose-response properties – for example the UNIM model in LiF:Mg,Ti. The general difficulty with approaches of this kind is that significant detail regarding the traps, recombination centers and charge transitions between them needs to be included in the model and the response of these elements to changes in dose and ionization density also have to be known. The further one tries to achieve a precise description of the processes, the more in-depth knowledge about those processes is required – knowledge that is actually very sparse and much of which relies upon interpretation and informed guesses. The danger is that a model may be developed that is based on estimation on top of hypothesis on top of conjecture. Of course, the estimations, hypotheses, and conjectures may, in fact, all be correct; they may all be good representations of the truth. Nevertheless, the obvious danger is that striving for a precise model description of a complex system may lead along erroneous paths so that, ultimately, we may be precise, but wrong.

The alternative approach is to stop our description of the system at an empirical level. Rather than describing the detailed workings of the innards of the system, perhaps we should simply look at how the system as a whole responds to external stimuli. The system then becomes a single component and not a sophisticated mechanical system of complex, interacting parts. In this regard, let's not forget why we are interested in these materials in the first place. It is because we wish to know if the material can be used as an accurate, reliable dosimetry system and if we can predict its response under thermal (or optical) stimulation to radiation of different types and doses. We want to ask, and answer, questions such as: can we use this dosimeter at (say) high, energetic-photon doses? Can we use it for low doses for (say) environmental dosimetry? Can it be used in a clinical setting for dosimetry of (say) energetic protons? Can it be used for estimation of doses in a space environment where it is exposed to a mixture of highly energetic charged particles? What is the best way to calibrate the system? Finally, what can we expect for the accuracy and reliability of the measurement? To answer these and other similar questions, do we really need to know what defect causes peak X in a dosimeter, and how? In all cases, the answer to the last question is probably "no". All we need to know is how the signal (the TL peak) behaves with respect to absorbed radiation.

In contrast, if we are more interested in the classical and quantum mechanical processes in which defects in solids give rise to luminescence or other properties, then an entirely different suite of questions needs to be asked, such as: what is the structure of the defect that causes emission at such-and-such a wavelength? What are the energy levels involved? Why is a defect that is stable over certain temperature ranges unstable over others? Why does the system emit at a certain wavelength when dopant X is present, but at a different wavelength when it is not? The list of possible questions is lengthy. Detailed knowledge of the internal mechanics of the system is necessary in these cases, to varying degrees.

For dosimetry, however, we do not have to probe so deeply. A good example of this is provided by the work of Parisi and colleagues (Parisi et al. 2018, 2019a, 2020; Parisi 2020), based on earlier work by Olko and others (Olko et al. 2002). These authors used a Monte Carlo code to describe the radial specific energy distribution $d(z)$ in LiF detectors due to the penetration of charged particles of different types and different LETs. They then modelled the charged-particle efficiency relative to $^{60}$Co gamma rays using the expression:

$$\eta_L = \frac{\left[\int_0^{+\infty} d(z)\, r(z)\, dz\right]_{CP}}{\left[\int_0^{+\infty} d(z)\, r(z)\, dz\right]_{60_{Co}}} \tag{7.21}$$

where $d(z)$ and $r(z)$ represent the probability density distribution of the specific energy and the specific energy response function, respectively.

Specific energy $z$ is a microdosimetric quantity and is the energy imparted $\varepsilon$ to matter of mass $m$ ($z = \varepsilon/m$, in units of Gy; Rossi and Zaider 1996). The distribution function $d(z)$ is calculated using suitable software (Parisi et al. use PHITS, the Particle and Heavy Ion Transport code System (Sato et al. 2015)), while $r(z)$ is approximated by the dose-response function $f(D)$. The latter is experimentally measured (for example Figure 7.23, for both LiF:Mg,Ti and LiF:Mg,Cu,P) and the data fitted to a suitable mathematical function to describe the experimentally measured $f(D)$. The approximation $r(z) \approx f(D)$ is valid since the microdosimetric quantity $z$ may be approximated by $D$ if the energy is distributed uniformly over the site volume such that the average value of specific energy $z$ is approximately the absorbed dose $D$. This is a good approximation for sparsely ionizing radiation, such as $^{60}$Co gamma, which is the radiation used to establish $f(D)$. In general, the approximation seems good for energies > 1MeV

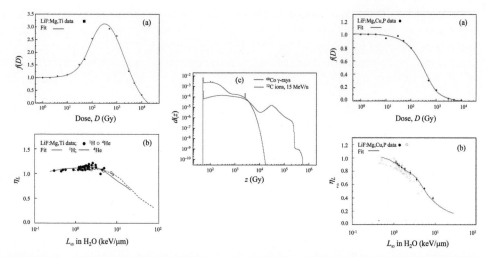

**Figure 7.25**    (a) Supralinearity index (dose-response function) $f(D)$ for LiF:Mg,Ti (left) and LiF:MCP (right), using $^{60}$Co gamma. (b) Calculated and measured relative efficiencies as a function of unrestricted LET in water for $^{1}$H particles in LiF:Mg,Ti, and LiF:MCP. (c) Calculated $d(z)$ for $^{60}$Co photons and $^{12}$C ions. (Adapted from Parisi et al. (2018, 2019a, 2020).)

(Parisi 2020). The site size, i.e. the size necessary for the approximation $r(z) \approx f(D)$ to hold, is the only free parameter in the calculation and is varied in order to provide a fit to the calculated value of $\eta_L$ using Equation 7.21 and the measured value using Equation 6.29.

Example results from this approach are shown in Figure 7.25 for LiF:Mg,Ti and LiF:MCP. In part (a) of the figure are shown the measured $f(D)$ curves as a function of gamma dose. In part (b) are the measured $\eta_L$ data for $^{1}$H ions for LiF:Mg,Ti and LiF:MCP compared with the calculated values using Equation 7.21. The approach is able to calculate the relative charged-particle efficiency for a wide range of particle types using a site size of 40 nm.

Similar calculations by Parisi and colleagues (Parisi et al. 2019b) show that the same procedure can also be used to model the photon energy response of the two materials, using Equation 7.21 but with photon irradiation, not charged-particle irradiation. Predictions of the relative photon energy response in the range 12 keV to 1250 keV are shown to match the experimentally measured photon responses for both LiF:Mg,Ti and LiF:MCP, again using a site size of 40 nm.

In all of the calculations for $d(z)$, irrespective of the particle type, a best fit to the experimental data is found using a site size of 40 nm. This is the only free parameter used to obtain the fits. One might question such a large site size (40 nm) when the interatomic spacing in LiF is only 0.4 nm. However, one should recall the discussion in Chapter 2 which indicated that so-called point defects are not points at all and the actual size, that is the changes in potential and in the positions of the surrounding ions, extends over many lattice spacings. In the LiF-based dosimetry materials the site that produces the TL may be a large complex of a Mg-related defect and a Ti-related defect (in LiF:Mg,Ti) or a P-related defect (in LiF:MCP). Considering this, a site size of 40 nm is not difficult to contemplate. Apart from this inference, the microdosimetric model of Parisi and colleagues does not require precise knowledge of the traps and recombination centers, or the charge transitions that occur between them. Instead, the model is an excellent example of an approximation ($r(z) \approx f(D)$) yielding accurate results. The model is able to describe and predict how the efficiency of a TLD material will vary with the type of radiation, and it achieves this using one calculation, one free parameter, and only one set of easily measured empirical data.

# 8

# Optically Stimulated Luminescence

*The loveliest theories are being overthrown by these damned experiments....*

– J. von Liebeg 1834

## 8.1 Introduction

Following the discussion of example TL materials in Chapter 7, Chapter 8 covers the archetypical OSL material, $Al_2O_3$:C. Aluminum oxide doped with carbon arrived on the scene of luminescence dosimetry as a proposed TL dosimeter with the advent of high-sensitivity $Al_2O_3$:C grown in a highly reducing atmosphere using the Stepanov crystal growth technique (Akselrod and Kortov 1990; Akselrod et al. 1993). Its sensitivity compared to TLD-100 was stated to be about 50 times higher and its non-tissue-equivalence was deemed to be a problem only at low photon energies; this could be corrected through accurate calibration. As a TL material, however, it suffers from two disadvantages, namely thermal quenching of the TL signal and sensitivity to visible light. The former means that the sensitivity is partly lost since the material has to be heated to produce TL, and increasing the heating rate leads to a further decrease in sensitivity as the peak shifts to higher temperatures. The latter means that the material has to be protected from exposure to ambient light following irradiation. The material was subsequently commercialized by Harshaw (called TLD-500) and by Victoreen in the USA, but it did not challenge the position of TLD-100 in the TLD field.

The breakthrough for $Al_2O_3$:C came when the optical sensitivity of the TL was examined, whereupon it was clear that this sensitivity could be used to advantage by using the material not as a TL material, but as an OSL material. During optical stimulation it was observed that the material emitted strong luminescence and, because the OSL could be measured at room temperature, thermal quenching was not an issue. The result was that the OSL emission was even more sensitive than the TL emission (Markey et al. 1995; Bøtter-Jensen and McKeever 1996; McKeever et al. 1996; Akselrod and McKeever 1999; McKeever and Akselrod 1999).

CW-OSL and POSL are the two OSL measurement methods commonly used for dosimetry using $Al_2O_3$:C. The POSL method was adopted by Landauer Inc. (USA) using $Al_2O_3$:C films (grains deposited on and sandwiched between polycarbonate strips) in the dosimetry system

*A Course in Luminescence Measurements and Analyses for Radiation Dosimetry,* First Edition.
Stephen W.S. McKeever.
© 2022 John Wiley & Sons Ltd. Published 2022 by John Wiley & Sons Ltd.
Companion Website: www.wiley.com/go/mckeever/luminescence-measurements

commercially marketed as Luxel[®].[1] This technology increased in popularity to eventually cover ~25% of the world's personal dosimetry market in the 2010–2020 decade. Subsequent OSL technology includes InLight[®].[2]

## 8.2 Aluminum Oxide

### 8.2.1 Al₂O₃:C

#### 8.2.1.1 *Structure and Defects*

The lattice structure of $Al_2O_3$ is hexagonal close-packed (hcp) with the aluminum ions occupying 2 out of 3 octahedral interstices in the hcp lattice (Figure 8.1). The $O^{2-}$ ions occupy sites of $C_2$ symmetry and are arrayed in equilateral triangles, one above and one below the plane of the $Al^{3+}$ ions. The $Al^{3+}$ ions occupy sites of distorted octahedral ($O_h$) symmetry. Nearest-neighbor and next-nearest-neighbor Al-O bond lengths are 0.186 nm and 0.197 nm, respectively.

An important feature of oxygen-deficient $Al_2O_3$, and one which is particularly relevant to its luminescence properties, is the creation of oxygen-vacancy centers which have either trapped 2 electrons as charge compensators to form $F$-centers, or 1 electron to form $F^+$-centers, and simple aggregates thereof ($F_2$, $F_2^+$, etc.; e.g. Evans 1993). These prominent defects can be observed in optical absorption and photoluminescence spectra in as-grown material, before irradiation, as shown in Figures 8.2a and 8.2b. The peak of the $F$-center absorption is at ~205 nm corresponding to the $1S$-$1P$ transition, and the emission peaks at ~420 nm due to the

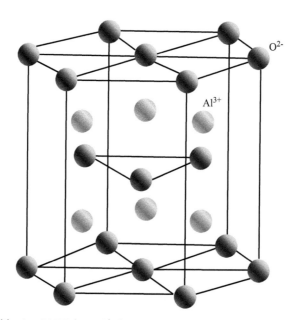

**Figure 8.1** Crystal lattice (HCP) for α-$Al_2O_3$.

---

[1] Luxel®+ – Dosimetry Badges I LANDAUER; accessed January 2021.
[2] InLight Model 2 – Dosimetry Badges I LANDAUER; accessed January 2021.

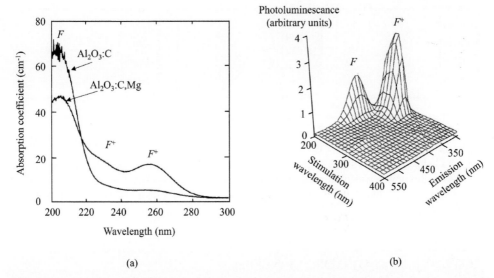

**Figure 8.2**   (a) Optical absorption spectrum of C-doped and C,Mg-doped $Al_2O_3$ illustrating the *F*- and $F^+$-center absorption bands in unirradiated material. The Mg-doping gives rise to stronger $F^+$-absorption. (b) Isometric photoluminescence excitation and emission spectra of unirradiated $Al_2O_3$ showing the emission and absorption for the *F*- and $F^+$-centers. (Reproduced from McKeever et al. (1999) with permission from Oxford University Press.)

$3P$-$1S$ transition; the $F^+$-center absorption peaks are between ~230 nm and ~260 nm, corresponding to $1A$-$2A$ and $1A$-$1B$ transitions, with emission in the UV with a maximum at ~325 nm ($2P$-$1S$).

Carbon is freely introduced into $Al_2O_3$ and dissolves in substitutional or interstitial positions. The role of carbon depends on whether it occupies Al-vacancy sites, O-vacancy sites, or interstitial sites. Density functional theory (DFT) calculations by Zhu et al. (2014) for C-doped $\alpha$-$Al_2O_3$ indicate that in oxygen-deficient $Al_2O_3$, carbon tends to occupy oxygen substitutional sites ($C_O$) in preference to Al substitutional ($C_{Al}$) or interstitial ($C_i$) sites and can take on charge states of +1, 0 or –1, depending on the position of the Fermi level. The DFT calculations also indicate that, relative to the top of the valence band, C mostly induces electron-occupied gap states at various levels in the lower half of the band gap below the equilibrium Fermi level (hole traps), for $C_O$, $C_{Al}$, and $C_i$. However, $C_{Al}$ and $C_i$ also induce several unoccupied levels (electron traps) above the Fermi level. Thus, there exist mostly deep hole traps with some deep electron traps due to C doping. Additionally, with $C^{4-}$ substituting for $O^{2-}$, oxygen vacancy centers are required for charge compensation, resulting in the formation of *F*- and $F^+$-centers.

### 8.2.1.2   *OSL Curves*

Examinations of changes in the TL and OSL sensitivities with the level of C doping indicate that ~5000 ppm C is the optimum amount for maximum sensitivity, with no change in the position of the main TL peak (near 465 K) as the C concentration increases, but with an observed faster OSL decay (Yang et al. 2008). Typical CW-OSL, LM-OSL, and POSL curves for $\alpha$-$Al_2O_3$:C are shown in Figure 8.3. The OSL kinetics are first-order but the CW-OSL decay rate is more than one simple exponential, indicating that it is caused by the sum of at least two and possibly more processes. Similarly, the LM-OSL and POSL curves cannot be described by one first-order process.

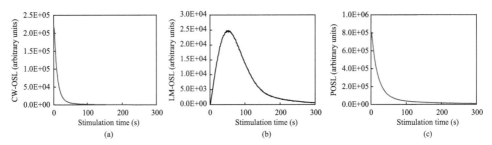

**Figure 8.3**   (a) CW-OSL, (b) LM-OSL, and (c) POSL from Al$_2$O$_3$:C following $^{90}$Sr/$^{90}$Y beta irradiation with 0.8 Gy for (a) and (b) and 3 Gy for (c). For CW-OSL the laser power was 30 mW.cm$^{-2}$, while for LM-OSL the power was increased with a linear ramp to this power over 300 s. For POSL the pulse frequency was 4 kHz and the mean power was 2.65 mW. (Original data kindly provided by S. Sholom.)

---

### Exercise 8.1   Al$_2$O$_3$:C OSL Curve Shapes

The data from Figure 8.3 are given on the web site under Exercises and Notes, Chapter 8, Exercise 8.1.
(a) Plot the CW-OSL, LM-OSL, and POSL data.
(b) Fit each data set to the sum of 1, 2, and 3 first-order OSL curves. Which gives the best fit?
(c) What do you conclude about the number of processes likely to be contributing to the OSL in this material at the stimulation wavelength used to obtain these data? In a real experiment, what additional data would you need to confirm the number of processes occurring in the production of OSL?

---

#### *8.2.1.3   Emission and Excitation Spectra*

Some additional information concerning the number of processes (traps) contributing to the OSL in Al$_2$O$_3$:C can be obtained from an examination of the excitation spectrum necessary for OSL production. One of the most informative ways to do this is to simultaneously record the excitation and the emission spectra and to present the data as an isometric plot, as shown in Figure 8.4a. Comparing Figure 8.4a with Figure 8.2b the original photoluminescence excitation bands for the $F$- and $F^+$-bands are observed, but in addition there is a radiation-induced stimulation band extending from ~240 nm to beyond 340 nm (the limit of the stimulation wavelengths used in this particular experiment). Such a broad excitation band may result from several traps producing OSL, with emission at 420 nm, and from a continuous array of states in the conduction band to which the electrons are excited. This is also suggested in Figure 8.4b where the excitation spectrum for OSL emission at 420 nm is compared with the excitation spectrum for photoconductivity, demonstrating that both are the same. Several inferences can be drawn from these observations. Firstly, there is a suggestion that several traps contribute to the OSL; secondly, the OSL process involves delocalized transitions that involve the generation of free charge carriers giving rise to photoconductivity; and thirdly, the main emission is at 420 nm, suggesting a recombination event of the type $e^- + F^+ \rightarrow F^* \rightarrow F + h\nu_{420nm}$, where a free electron recombines with an $F^+$-center, producing an excited $F$-center ($F^*$) which relaxes with the production of a photon at 420 nm. Also observed is emission peaking around 510 nm and excited around 300 nm, identified as $F_2$-centers induced by the irradiation (Evans et al. 1994). The energy levels associated with the $F$-, $F^+$-, and $F_2$-centers are shown in Figure 8.5.

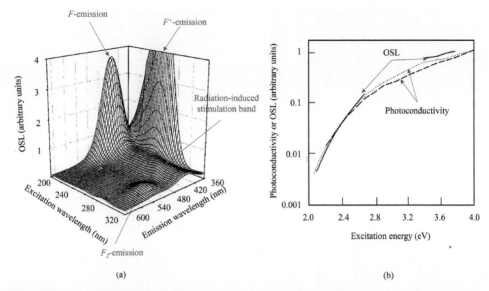

**Figure 8.4** (a) OSL simulation and emission spectra for Mg-doped $Al_2O_3$:C after a dose of 300 Gy. The radiation induces a stimulation band from ~240 nm to > 340 nm with peak emission corresponding to the *F*-center (420 nm), $F^+$-center (320 nm), and with some $F_2$-center emission (at ~500 nm). (b) A comparison of the photoconductivity excitation spectrum and OSL (at 420 nm) excitation spectrum following a beta dose of 300 Gy. (Reproduced from Whitley and McKeever (2000) with permission from AIP Publishing.)

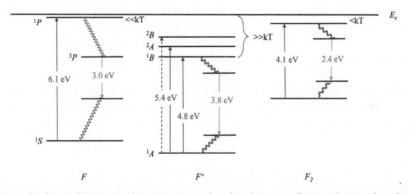

**Figure 8.5** Flat band diagram indicating energy levels relative to the conduction band edge ($E_c$) and the main absorption and emission transitions for *F*-, $F^+$-, and $F_2$-centers in $Al_2O_3$. Blue arrows indicate absorption transitions; green arrows indicate radiative transitions, and red lines indicate non-radiative (phonon) transitions. Energy levels and energies are taken from Evans et al. (1994).

The above conclusions are supported by correlations between OSL and TL, and between TL and optical absorption. Correlations between TL and OSL in $\alpha$-$Al_2O_3$:C are well established. The main TL peak in this material is complex. Its position shifts in ways that are not predicted from conventional TL kinetics but that can be explained by invoking multiple, overlapping peaks, not all of which are sensitive to light. Nevertheless, a subset of the traps responsible for the TL peaks have been shown to be the source of the OSL (e.g. Walker et al. 1996; Akselrod and Akselrod 2002; Dallas et al. 2008; Biswas et al. 2013).

Correlations between the TL peaks and optical absorption bands (*F*- and $F^+$-centers) demonstrate that the main TL peak (and, by inference, the OSL signals) are due to electrons traps.

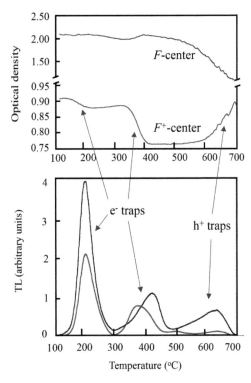

**Figure 8.6**   Top: Variation of the *F*- and *F*⁺-center absorption bands in irradiated Al₂O₃:C as a function of temperature during heating after irradiation. Bottom: TL glow curves for two different Al₂O₃:C samples following irradiation. The positions of the main TL peaks are compared with the annealing steps for the *F*- and *F*⁺-bands, enabling the identification of electron and hole detrapping. (Adapted from Polf (2000).)

This is demonstrated in the data of Figure 8.6 where the loss of $F^+$-centers and the gain of $F$-centers is interpreted as the release of electrons from traps and, conversely, the loss of $F$-centers and the gain of $F^+$-centers is interpreted as the release of holes from traps, following reactions of the type:

$$e^- + F^+ \rightarrow F \ \left(F - \text{band increases and } F^+ - \text{band decreases}\right) \tag{8.1}$$

and

$$h^+ + F \rightarrow F^+ \left(F - \text{band decreases and } F^+ - \text{band increases}\right). \tag{8.2}$$

Thus, TL peaks around 200 °C (the main TL dosimetry peak and the source of OSL) is interpreted as being due to electron traps, as are the TL signals around 400 °C, while very deep traps, with TL peaks around 600 °C to 700 °C are due to hole traps.

Ionization of $F$-centers by UV (205 nm) absorption, inducing the opposite reaction to Equation 8.1 (namely $F \rightarrow F^+ + e^-$), leads to the growth of the TL peaks near 400 °C, consistent with these being due to deep electron traps (e.g. Nikiforov et al. 2001; Kortov et al. 2002, 2006). This process ($F$ to $F^+$ conversion) also occurs during irradiation (with gamma, beta, etc.)

via hole trapping (Equation 8.2) as does the opposite ($F^+$ to $F$ conversion) via electron trapping (Equation 8.1). The key to the overall balance of these reactions during irradiation is the concentration of the deep electron and hole traps, the presence of which is inferred in Figure 8.6 and by the experiments of the Kortov and Nikiforov group.

Additionally, although OSL does not suffer from thermal quenching if performed at room temperature (but not so if performed at elevated temperature), TL certainly is affected by this phenomenon (e.g. Akselrod et al. 1998; Nikiforov et al. 2001). This, in turn, leads to a heating rate dependence of TL whereby the TL efficiency is reduced at higher heating rates. Quenching can be caused by the thermal ionization of $F$-centers. The ionized electrons may then be captured by deep traps and thus the degree of quenching can be dependent upon the availability of empty deep traps. As a result, the degree of filling of deep traps is found to affect the degree of thermal quenching. However, the effect of deep traps is found to be equivocal in experiments, with some researchers reporting increased thermal quenching if the deep traps are filled (Akselrod et al. 1998) and some reporting decreased thermal quenching if the deep traps are filled (Kortov et al. 1999). Fortunately, no such complications exist with OSL if measured at or near room temperature.

---

### Exercise 8.2   *F*- and *F$^+$*-bands

The data of Figure 8.6 indicate that the $F$-band and $F^+$-band act in opposite directions as the temperature is increased following irradiation. This is understood to be a consequence of Reactions 8.1 and 8.2. The traps responsible for the main TL peak(s) near 200 °C are also thought to be responsible for the OSL signal from this material. Figure 8.4 shows that the OSL emits at 420 nm, which is characteristics of the emission from $F$-centers, inferring that the OSL traps are electron traps.

The following observations are also true:

(a) The concentration of both $F$-centers and $F^+$-centers is large even before irradiation (e.g. Figure 8.2). What does this imply about the order of kinetics of the OSL (and TL) production from this material?

(b) In some $Al_2O_3$ samples the concentration of $F$-centers increases after irradiation; in other samples it decreases after irradiation. What might govern whether the $F$-center concentration will increase or decrease as a result of the irradiation? What will the $F^+$-center concentration do?

(c) According to Reactions 8.1 and 8.2, for every gain/loss of an $F$-center there must be a loss/gain of an $F^+$-center. However, the loss/gain of optical density for $F$-centers in Figure 8.6 is not matched by the gain/loss of optical density for $F^+$-centers over the same temperature ranges. Why not?

---

One inference of the above observations might be that the charges released during OSL and during production of the main TL peak are only electrons. However, the high temperature side of the main TL peak has been associated with emission at ~325 nm, especially in samples containing Mg. This emission has been identified as being due to the relaxation of excited $F^+$-centers (Figure 8.5), caused by Reaction 8.2, above. The inference from this, therefore, is that the main TL peak might be a mix of electron traps and hole traps. Studies of POSL in $Al_2O_3$:C, where the emission spectrum can be monitored either during the stimulation pulse, or between the stimulation pulses, shows that both 420 nm and 325 nm (i.e. both $F$- and $F^+$- emissions)

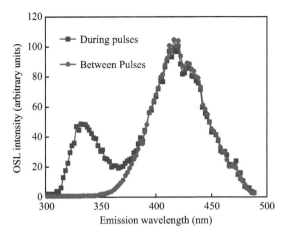

**Figure 8.7** POSL emission spectrum from $Al_2O_3$:C following 200 Gy of $^{60}$Co gamma irradiation at room temperature measured either during the POSL pulses or between pulses. Each data point represents a separate measurement at the shown fixed wavelength and the standard deviation of each data point, estimated from repeated measurements, was less than 5%. The shoulder at ~430 nm is thought to be an artefact of the spectral response correction algorithm. (Reproduced from Yukihara and McKeever (2006) with permission from AIP Publishing.)

appear during the stimulation pulses, but only 420 nm emission is observed between the pulses (Figure 8.7). The $F^+$-centers have a short luminescence lifetime (<~7 ns) and this luminescence signal appears only while the stimulation light is on, whereas the $F$-centers have a long luminescence lifetime (~35 ms) and their luminescence decay is still observed following the end of each stimulation pulse. The area under the $F$-band is about 3–4 times that under the $F^+$-band. Both $F^+$-and $F$-luminescence appear in CW-OSL and LM-OSL. These observations lead to the suggestion that there are both electron and hole detrapping and recombination processes occurring during TL and OSL in $Al_2O_3$:C, with the electron processes dominating.

Figure 8.4b shows the variation in the CW-OSL emission at 420 nm (and photoconductivity, PC) as a function of stimulation energy for constant illumination flux and represents the variation in the photoionization cross-section as a function of stimulation energy. An analysis of LM-OSL/LM-PC in $Al_2O_3$:C reveals a contribution from three main traps at room temperature with photoionization cross-sections ranging from a small value of $3.3\times10^{-20}$ cm$^2$ to a large value of $1.5\times10^{-18}$ cm$^2$ using 526 nm stimulation, depending on the sample and using Equation 2.17 for the cross-section (Whitley and McKeever 2001). The latter analysis made no a-priori assumptions as to the number or distribution of trapping centers giving rise to the LM-OSL and LM-PC signals and analyzed the data using an equation of the type given in Equation 5.58 for LM-OSL/LM-PC and a fixed stimulation wavelength (526 nm in this case). The results from Whitley and McKeever (2001) are shown in Figure 8.8 for three different $Al_2O_3$:C samples. Similarly, using the $t_{max}$-$t_{stop}$ analysis for LM-OSL curves (Section 5.3.4.1), in addition to analysis of CW-OSL curves, Soni and Mishra (2016) found two main traps contributing to the OSL for the sample studied by them, with photoionization cross-sections of $1.51\times10^{-18}$ cm$^2$ and $5.02\times10^{-19}$ cm$^2$ at a stimulation wavelength of 470 nm.

The above analyses of the photoionization cross-sections at room temperature do not include the possibility of a temperature dependence. From thermally activated OSL (TA-OSL) experiments, the main OSL signal (OSL originating from the same traps that cause the main TL peak) can be shown to be slightly thermally activated, with an activation energy of 0.03 eV (Mishra et al. 2011). This was interpreted as an available energy level 0.03 eV above the

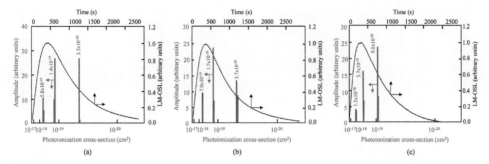

**Figure 8.8** LM-OSL curves for three different $Al_2O_3$:C samples (a, b, and c) showing the results of deconvolution and the obtained photoionization cross-sections that give rise to the OSL signals. (Reproduced from Whitley and McKeever (2001) with permission from AIP Publishing.)

ground state and from which optical excitation to the conduction band occurred. At still higher temperatures, thermal assistance from deeper states becomes possible. TA-OSL from very deep traps has been observed with activation energies of 0.268 eV and 0.468 eV. The latter are said to correspond to ionization from states with photoionization cross-sections of $5.82\times10^{-20}$ $cm^2$ and $3.7\times10^{-22}$ $cm^2$, respectively, at a stimulation wavelength of 470 nm (Soni et al. 2012).

### Exercise 8.3 CW-OSL and LM-OSL

(a) Using the values for the photoionization cross-sections at a stimulation wavelength of 470 nm as found by Soni and Mishra (2016), namely $1.51\times10^{-18}$ $cm^2$ and $5.02\times 10^{-19}$ $cm^2$, reconstruct the CW-OSL and LM-OSL curves obtained by them. To do so you will need refer to the published paper by these authors to extract the experimental parameters (stimulation power for CW-OSL, ramp rate for LM-OSL, etc.). Compare your reconstructed results with the data published in their paper. Are the cross-section values $1.51\times10^{-18}$ $cm^2$ and $5.02\times10^{-19}$ $cm^2$ consistent with the results obtained?

(b) Simulate the $t_{max}$-$t_{stop}$ experiment (called $T_{max}$-$T_{stop}$ in their paper) and compare with the published results.

Chruścińska (2016) used Equation 2.18 with VE-OSL (Section 5.3.10) to account for phonon coupling and to determine the optical trap depths, rather than the photoionization cross-sections responsible for the OSL. The benefit of this approach is that the optical trap depth is independent of stimulation wavelength whereas the cross-section is wavelength dependent. The disadvantage is that VE-OSL is a difficult method to apply in practice. Chruścińska concluded that the optical trap depth $E_o$ of the main dosimetry traps in $\alpha$-$Al_2O_3$ is represented by a distribution of states between ~2.0 eV and ~3.2 eV, as illustrated in Figure 8.9a. The optical trap depth distribution obtained by Chruścińska from VE-OSL is compared in Figure 8.9b with similar distributions obtained by Whitley and McKeever (2000) from photoconductivity data. The latter authors likewise concluded that the optical trap depths for the main dosimetry traps are distributed and also determined distributions for the optical traps depths of the deeper traps. The shapes of the distributions depend upon the dose, with a narrower distribution for the main dosimetry traps being characteristic of lower doses and a wider distribution appearing at higher doses.

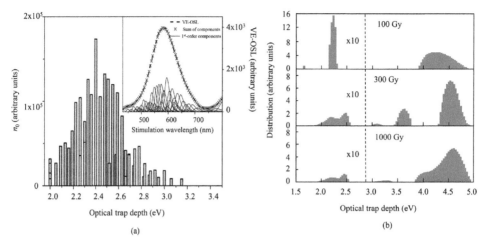

**Figure 8.9**   (a) A VE-OSL scan over the wavelength range from 800 nm to 430 nm at the rate of 0.2 nm.s$^{-1}$ and for a dose of 1.6 Gy. The inset shows the VE-OSL curve and its fit to a sum of first-order curves. The parameter $n_0$ represents the concentration of traps at each value of the optical trap depth. (Reproduced from Chruścińska (2016) with permission from Elsevier.) (b) Distributions of optical traps depths in α-Al$_2$O$_3$:C obtained from photoconductivity data at different doses showing changes to the optical trap depth distributions as the dose increases. (Reproduced from Whitley and McKeever (2000) with permission from AIP Publishing.)

The above results are qualitatively consistent with an analysis of thermally stimulated conductivity (TSC) by Agersnap Larsen et al. (1999) that indicated a dose-dependent distribution of thermal activation energies $E_t$ for the main dosimetry traps. However, a quantitative comparison between the $E_t$ distribution determined by these authors and the $E_o$ distributions of Figure 8.9 is ill-advised since Agersnap Larsen et al. assumed a fixed value for the frequency factor $s$ (namely, $s = 10^{13}$ s$^{-1}$). Whitley et al. (2002) also analyzed TSC but allowed for distributions in both $E_t$ and $s$ and used Equation 3.73 to simultaneously analyze a series of TSC curves obtained at several different heating rates with the process described in Chapter 3, Section 3.5. The results were already shown in Figure 3.23 and they illustrate a series of multiple, narrowly spaced (in $E_t$) traps, each with their own discrete values of $E_t$ and $s$, contributing to TSC over the same temperature range as the main TL peak. Comparing the results of Whitley et al. (2002) and Agersnap Larsen et al. (1999) shows $E_t$ values spread over the range ~1.3 eV to ~1.5 eV in both cases with $s$ values ranging from ~$10^{12}$ s$^{-1}$ to $10^{14}$ s$^{-1}$.

Taking all the experimental results together (from various types of OSL experiment, plus TL, TSC, and photoconductivity), it may be speculated that the main dosimetry traps in α-Al$_2$O$_3$:C are a set of closely spaced centers, both in energy and physically. As the concentration increases (i.e. as the dose increases) spatial overlap of the centers causes them to become distributed in optical and thermal trap depth (and, by inference, in photoionization cross-section and frequency factor). The details remain elusive, but vary among samples of different origin, with both TL and OSL analyses being strongly sample dependent; different sources of Al$_2$O$_3$:C produce different results. Nevertheless, most experimenters agree that the kinetics of both TL and OSL are first-order. (The reader should beware of published analyses of the main TL peak without correction for thermal quenching and/or peak overlap.) The conclusion of first-order kinetics is consistent with the fact that there is a large concentration of pre-existing recombination centers ($F$- and $F^+$-centers) in the material even before irradiation.

## Exercise 8.4    Thermal and Optical Trap Depths in $\alpha$-Al$_2$O$_3$:C

A useful and informative exercise is to review the literature for publications on attempts to analyze OSL and TL curves from $\alpha$-Al$_2$O$_3$:C. Look for:

(a) Different methods of analysis and experimental approaches.

(b) Different equations used in the analyses.

(c) Different assumptions used in the analyses.

(d) Different parameters values obtained.

(e) Different conclusions obtained.

What do you conclude?

### 8.2.1.4  *Temperature Dependence*

As noted, OSL does not suffer from thermal quenching if performed at or near room temperature. A popular type of OSL measurement, however, is thermally assisted OSL (TA-OSL) in which the sample is heated during OSL measurement in order to determine which traps play important roles in OSL production. When performed on Al$_2$O$_3$:C, thermal quenching has to be considered. An example of a TA-OSL curve, compared to a TL curve obtained under similar conditions, is shown in Figure 8.20 (denoted "TA-OSL Expt." in the figure). The OSL output increases steadily as the temperature increases to ~400 K (at a heating rate of 2 K.s$^{-1}$) beyond which a sharp decrease is observed. The increase is interpreted as the increased instability of shallow traps allowing more of the released electrons to recombine to produce OSL. The reason for the decrease is twofold. One is the thermal emptying of the main traps, the primary source of the OSL signal (and the corresponding main TL signal), and the second is thermal quenching of the 420 nm emission from *F*-centers.

Using a simple model of 2 shallow traps, 1 main trap, 1 deep trap, and 1 recombination center (a 4T1R model), including thermal quenching, Markey et al. (1996) simulated a TA-OSL curve and produced the results also shown in Figure 8.10 ("TA-OSL Sim."). This straightforward model was able to replicate the main features observed in the experiment, using the parameters listed in the figure caption.

### 8.2.1.5  *Photon Dose-Response Characteristics*

With the above information concerning the components involved in the production of OSL from Al$_2$O$_3$:C, what can be expected from the behavior of the OSL signal as a function of absorbed dose? It has been established (from OSL, TL, and TSC) that there are multiple, different electron traps, at least some of which act as sources for OSL. Also established (from emission spectra) is that there are some hole traps. There are also deep electron traps and deep hole traps present in the material (see Figure 8.6). The observation of photoconductivity during optical stimulation (and TSC during heating) show that detrapping processes involve charge transport via the delocalized bands. Thus, there are multiple opportunities for competitive trapping and recombination events to occur during the production of OSL and the consequent likelihood of supralinear and/or sublinear growth of the OSL signal as a function of dose. There has been no direct evidence to date of a spatial association between traps and recombination centers in the form of trap/center pairs in this material. (This is not to say that it does not occur and in fact may be inferred from the discussion in 8.2.1.3. With such large defect densities it is even likely. However, there are no obvious experimental TL or OSL properties – e.g. tunneling – that can be definitively attributed to spatial association of traps and centers in Al$_2$O$_3$:C.) Since it can be expected

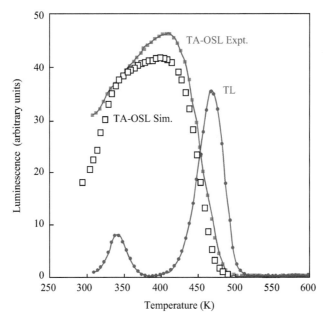

**Figure 8.10**   Experimental TA-OSL (TA-OSL Expt.) and TL from irradiated (1.5 Gy) Al$_2$O$_3$:C measured during heating at 2 K.s$^{-1}$, with a detection window centered on the *F*-center emission (~420 nm). These data are compared with simulated results (TA-OSL Sim.) for a 4T1R model with the following parameters: $i$ = 1–4 traps, with concentrations ($N_i$), capture probabilities ($A_i$), trap depths ($E_{ti}$) and frequency factors ($s_i$) of $N_1 = 10^{12}$ cm$^{-3}$, $N_2 = 5\times10^{10}$ cm$^{-3}$, $N_3 = N_4 = 3\times10^{11}$ cm$^{-3}$; $A_1 = A_2 = A_3 = A_4 = 10^{-10}$ cm$^{-3}$.s$^{-1}$; $E_{t1}$ = 0.77 eV, $E_{t2}$ = 0.85 eV, $E_{t3}$ = 1.21 eV, $E_{t4}$ = 3.0 eV; $s_1 = 5\times10^{13}$ s$^{-1}$, $s_2 = 10^{12}$ s$^{-1}$, $s_3 = 8\times10^{11}$ s$^{-1}$, $s_4 = 10^{12}$ s$^{-1}$. The recombination center concentration was $M_1 = 5\times10^{12}$ cm$^{-3}$ with a recombination probability of $A_m = 10^{-9}$ cm$^3$.s$^{-1}$. The optical transition rates from levels 3 and 4 were $10^{-2}$ s$^{-1}$ and $10^{-3}$ s$^{-1}$, respectively. Thermal quenching was given by $1/(1+10^{17}\exp\{-1.55/kT\})$. (Adapted from Markey et al. (1996).)

that each component of the OSL signal will grow with dose at different rates it is likely that the shape of the OSL curve (CW-OSL or LM-OSL) will change as the dose increases.

These characteristics can be observed experimentally in Al$_2$O$_3$:C, to greater or lesser extents depending on the particular sample. Specifically, the rate of decay of the CW-OSL curve is seen to increase as the dose increases (Yukihara et al. 2004; Biswas et al. 2009) as demonstrated in Figure 8.11 for Luxel$^\circledR$. This behavior is exactly in accordance with that predicted in Chapter 6, Section 6.2.5.2 and illustrated in Figure 6.15, and is due to the differential growth with dose of the traps contributing to the OSL signal. Since the rate of decay changes as the dose increases, the dose-response function depends upon whether the OSL intensity is measured as the initial intensity, or as the total area under the curve, as indicated in Figure 8.12 for Luxel$^\circledR$ and TLD-500.

In Figure 8.12, the OSL intensity for TLD-500 is observed to decrease at high dose, in a manner predicted in Chapter 6, Section 6.2.5 and Figure 6.13c, for those materials with a high concentration of non-radiative hole traps (or recombination at hole traps with radiative emission outside the detection window).

Competition from electron traps can be seen to cause a slight degree of supralinearity in the dose-response curve, especially when using the initial OSL intensity. This difference in the degree of supralinearity between the initial OSL intensity and the OSL area is perhaps due to

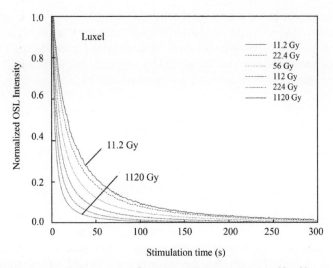

**Figure 8.11** CW-OSL curves from Luxel® Al$_2$O$_3$:C beta irradiation ($^{90}$Sr:$^{90}$Y) to different doses showing the increase in the decay rate with increasing dose. (Reproduced from Yukihara et al. (2004) with permission from Elsevier.)

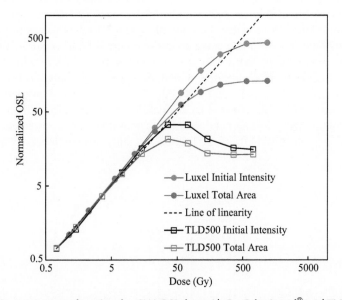

**Figure 8.12** Dose-response function for CW-OSL from Al$_2$O$_3$:C for Luxel® and TLD-500 chips as a function of beta irradiation ($^{90}$Sr:$^{90}$Y). The OSL is measured either as the initial intensity or as the total area under the curve. (Reproduced from Yukihara et al. (2004) with permission from Elsevier.)

there being slightly more competitors available when the initial traps empty compared to when the later traps empty. However, in either case, supralinearity is not strong and this material displays linear growth of TL and OSL over a large dose range, from the μGy range to ~10 Gy, a dynamic range of approximately 7 orders of magnitude. At first sight this may be surprising since both TL and OSL are accompanied by their conductivity counterparts (TSC and photoconductivity) indicating delocalized transitions and, therefore, competition effects may be expected. However, one has to remember that large concentrations of recombination sites (both *F*- and *F*⁺-centers) exist in this material even before irradiation (Figure 8.2) such that the probability of charge being caught by competitive, non-radiative centers during TL or OSL is outweighed by the probability of the charge undergoing radiative recombination processes. In other words, the relative concentration of competitors, and the subsequent probability of competition, are low.

### 8.2.1.6   *Charged-Particle Dose-Response Characteristics*

The characteristics of the CW-OSL curves as a function of dose also manifest themselves in the observed responses to energetic charged particles. As was discussed in the previous chapter with regard to TL from LiF, and using the principles outlined in Chapter 6, Section 6.2.6, the net OSL from Al₂O₃:C is a convolution of the radial dose distribution around the charged-particle track and the dose-response function, summarized by Gieszczyk et al. (2014) in Equation 6.28. The relationship between the changes in shape of the CW-OSL decay curve from Al₂O₃:C and ionization density is well documented (Yukihara and McKeever 2006; Jain et al. 2007) and, as might be expected, the shape of the CW-OSL decay curve varies with the LET of the charged particle (Yasuda and Kobayashi 2001; Yasuda et al. 2002). Figure 8.13a shows typical results, while Figure 8.13b demonstrates how the CW-OSL efficiency from Al₂O₃:C Luxel® dosimeters varies over a wide range of LET values for several charged particles. Here, the relative OSL efficiency ($\eta_L$) is calculated using Equation 6.29, and the OSL signal is defined as the OSL area under the CW-OSL curve (Sawakuchi et al. 2008a).

To be noted in Figure 8.13 is that the value of $\eta_L$ is > 1 for low values of $L_\infty^{water}$ due to the supralinearity of the dose-response curve shown in Figure 8.12. This can be more clearly seen in terms of the supralinearity index $f(D)$, as shown in Figure 8.14 for beta irradiation. Two $f(D)$ curves are shown, one for OSL detection of both *F*- and *F*⁺-emissions, and one for detection of the *F*-emission only. It has been noted that both of these emissions are observed during OSL, although *F*⁺-emission is characterized by a much faster luminescence lifetime (<7 ns) compared to the *F*-emission (~35 ms). As a result, both emissions can be observed during the stimulation pulse (e.g. in CW-OSL or LM-OSL) but only *F*-emission is observed after the

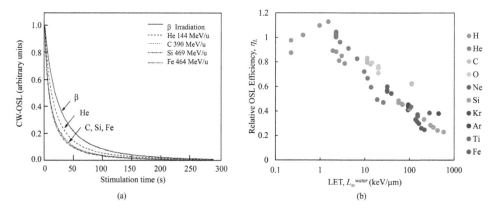

(a)   (b)

**Figure 8.13**   (a) Example CW-OSL curves for different charged particles, as indicated, each obtained for the same total dose (100 mGy). (From Yukihara et al. (2006).) (b) CW-OSL versus LET ($L_\infty^{water}$) for Al₂O₃:C (Luxel®) for a wide range of charged particles. The figure was plotted using data extracted from Table I of Sawakuchi et al. (2008a). Both *F*- and *F*⁺-emission was detected and the OSL was defined as the area under the CW-OSL curve integrated over 300 s, less the background.

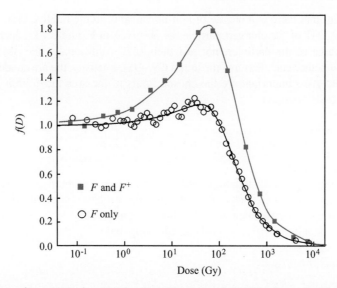

**Figure 8.14** Supralinearity index *f(D)* for CW-OSL from $Al_2O_3$:C measured with both the *F*- and *F*⁺-emissions detected, and with only the *F*-emissions detected. (Adapted from Sawakuchi et al. (2008a).)

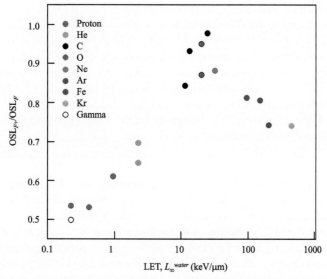

**Figure 8.15** Ratio of the *F*⁺- and *F*-components of the POSL spectrum from $Al_3O_3$:C Luxel® detectors using data from Table 2 from Yukihara et al. (2006); see also Figure 8.7.

stimulation pulse (e.g. in POSL). On the basis of Figure 8.14 it is clear that CW-OSL and POSL will therefore have different dose-response characteristics and, as a result, they will also have different responses to charged particles. This latter property can be seen in Figure 8.15 where the ratio of the two emissions is plotted as a function of LET for several charged particle types, showing the relative emission to be non-constant as the LET varies.

It is evident from the above discussion of the changes in the OSL curves shapes (CW-OSL, LM-OSL and POSL) with dose, charged particles and emission wavelength, that there is broad similarity between the OSL properties of $Al_2O_3$:C OSLDs and the TL properties of LiF TLDs.

It is also evident that there is a degree of non-universality in the relative OSL efficiency as a function of the LET of the charged particle, as observed in Figure 8.13, depending upon the energy and charge of the particles, not just their LET. Additionally, the observed behavior depends upon whether one measures the initial CW-OSL intensity, the area under the CW-OSL curve, or POSL. Also important are the chosen limits of the area integration, and the wavelengths of the OSL detection.

---

**Exercise 8.5    OSL and TL response versus LET**

Refer to the publication:
"Relative optically stimulated luminescence and thermoluminescence efficiencies of $Al_2O_3$:C dosimeters to heavy charged particles with energies relevant to space and radiotherapy dosimetry" by Sawakuchi et al. (2008) *J. Appl. Phys.* **104**, 124903.
Examine the data in Table 1, and plot the following data:
(a) Relative OSL efficiency versus LET ($L_\infty^{water}$) for:
   (1) Luxel®, filter set #1
   (2) Luxel®, filter set #2
   (3) Single crystal, filter set #1
(b) Relative TL efficiency versus LET for:
   (1) $Al_2O_3$:C, single crystal, Corning 5-58
   (2) LiF:Mg,Ti, TLD-100, Schott BG-59
(c) From the shapes of the plots that you obtain what can you say about the following?
   (1) Which of the above combinations (of readout method (OSL or TL), material ($Al_2O_3$:C or LiF:Mg,Ti), and filter set (#1, #2, 5-58 or BG-59)) is most likely to display strong supralinearity in its gamma dose-response curve? Strong sublinearity?
   (2) Is there a non-universality on the LET-dependence of the OSL efficiency from $Al_2O_3$:C? Under what circumstances could it be used as an "LET meter"?
   (3) For either $Al_2O_3$:C OSL or LiF TLD-100 TL, can you speculate on how you would approach separating the gamma dose from a charged-particle dose in a mixed gamma/charged particle field?

---

Despite these potential limitations, for a constrained set of conditions, much useful information can be attained concerning the incident charged-particle radiation from the measured OSL signal. Examples have been demonstrated by Sawakuchi and colleagues who have used the changes in the OSL properties of $Al_2O_3$:C as a function of particle LET to demonstrate the application of OSL to calibrate the LET of therapeutic proton beams for use in cancer treatment (e.g. Sawakuchi et al. 2010, 2014; Granville et al. 2014). The principle is illustrated in Figure 8.16 and example results are shown in Figure 8.17. CW-OSL curves are illustrated schematically in Figure 8.16a, with the corresponding stimulation and gate detection schemes indicated above and the CW-OSL shown below. Since the CW-OSL is measured during the stimulation pulse both $F$- and $F^+$-emissions are measured. Similar curves, stimulation and gate schemes are shown for POSL in Figures 8.15b and 8.15c. Two detection channels are used in the POSL; one (Channel $A$, of length $t_A$) is for measurement of OSL during the stimulation pulse, during which time both $F$- and $F^+$-emission are detected, and the second (Channel $B$, of length $t_B$) is measured after the stimulation pulse during which time only $F$-emission is detected. This is repeated for many pulses, depending on the pulse frequency, for a total stimulation time similar to that of the CW-OSL measurement.

In schemes (a) and (b), the curves shapes can be quantified by defining:

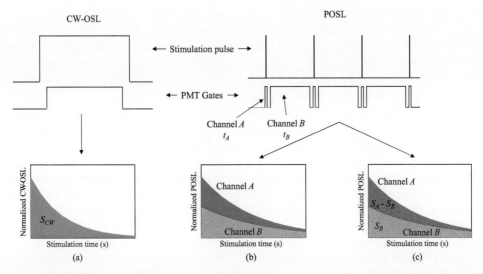

**Figure 8.16** Different methodologies for using OSL to measure the LET of charged-particle beams. Method (a) uses CW-OSL, while methods (b) and (c) use POSL; (a) and (b) use numerical methods to define the curve shape, while (c) uses a numerical method to define the $F^+$-to-$F$ emission ratio. The details are described in the text.

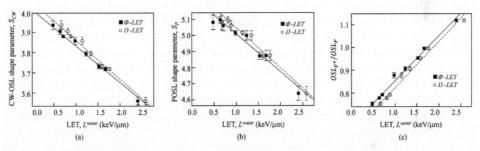

**Figure 8.17** Results using the methods outlined in Figure 8.16 for determining the LET of therapeutic proton beams: (a) CW-OSL curve shape, (b) POSL curve shape, and (c) the POSL $F^+/F$ emission ratio, each shown as a function of either fluence-averaged LET ($\Phi$-LET) or dose-averaged LET ($D$-LET). (Reproduced from Granville et al. (2014) with permission from IOP Publishing.)

$$S_{CW} = \frac{\sum_{t=1}^{t=10} I_{CW-OSL}\left(t\right)}{I_{CW-OSL}\left(t=1\right)} \tag{8.3}$$

for CW-OSL (curve (a)), and:

$$S_P = \frac{\sum_{t=1}^{t=10} I_{POSL-A}\left(t\right)}{I_{POSL-A}\left(t=1\right)} \tag{8.4}$$

where $I_{POSL-A}$, is the POSL signal from Channel $A$ (curves b and c). In both Equations 8.3 and 8.4, $t$ is the total stimulation time.

Alternatively, the ratio of the $F^+/F$ emission intensities can be quantified (Yukihara and McKeever 2006) using:

$$\frac{OSL_{F^+}}{OSL_F} = \frac{\sum_{t=1}^{t=10} I_{POSL-A}(t) - \left(\frac{t_A}{t_B}\right)\sum_{t=1}^{t=10} I_{POSL-B}(t)}{\sum_{t=1}^{t=10} I_{POSL-B}(t)} = \frac{S_A - \left(\frac{t_A}{t_B}\right)S_B}{S_B} \qquad (8.5)$$

where $I_{POSL-B}$ is the POSL signal from Channel $B$.

Using these definitions, Granville et al. (2014) calibrated different therapeutic proton beams with $Al_2O_3$:C OSLDs. Example data are shown in Figure 8.17 where the LET is calculated as either fluence $\Phi$ averaged or dose $D$ averaged ($\Phi$-LET or $D$-LET, respectively).

For a single particle type these are promising methods for LET determination in charged-particle fields. For more complex fields, such that experienced by astronauts in space, however, OSL methods can still be useful, despite the degree of non-universality of the efficiency observed in Figure 8.13b. The reason for this lies in the energy spectrum of charged particles in the space environment, ranging from 10 MeV to 1000 MeV, for which energies the low-LET part of each particle's spectrum dominates. In other words, it is the "envelope" of the low-LET part of a plot such as that shown in Figure 8.13b that is of importance. Example data showing the CW-OSL response of $Al_2O_3$:C to only high-energy, charged particles is shown in Figure 8.18 where a single function is obtained, independent of the particle type. The experimental data are compared to a theoretical convolution of the radial dose distribution (using the Butts-Katz model) and the dose-response function for CW-OSL, as explained by Sawakuchi (2007).

Combining the information so far leads to a consideration of what can be expected for the OSL from $Al_2O_3$:C response as a function of charged-particle fluence. The result will depend upon the LET and energy of the charged particle, the mode of OSL measurements (CW-OSL, POSL), the chosen OSL signal (initial peak height, area under the curve), and the OSL wavelength ($F$, $F^+$, or both). Figure 8.19 shows several example results. It is important to recall that as long as the particle tracks do not overlap or interact, the response must be linear as a function of particle fluence. This is because whatever response is measured for one track will simply double for two tracks, triple for three tracks, and so on. This is observed in the data of Figure 8.19. What happens when the tracks overlap depends very much on the particle type and

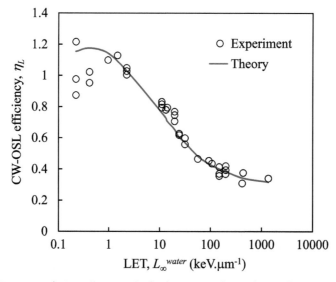

**Figure 8.18**  CW-OSL relative efficiency for high-energy, charged particles versus LET, compared with calculated values obtained from the radial dose distribution and the dose-response function. (Adapted from Sawakuchi (2007).)

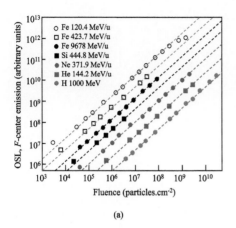
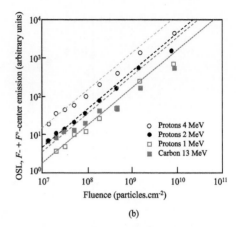

(a)                       (b)

**Figure 8.19** (a) Fluence-response characteristics for high-energy, charged particles, using CW-OSL and monitoring only the $F$-center emission. The dashed lines indicate the lines of linearity $(f(\Phi) = 1)$ for each particle showing that the responses are either linear or sublinear at high fluence, due to track overlap. (Reproduced from Sawakuchi et al. (2008b) with permission from Elsevier.) (b) Fluence-response characteristics of $Al_2O_3$:C for low-energy, charged particles, using CW-OSL but monitoring both $F$- and $F^+$ emission. Mostly sublinearity $f(\Phi) < 1$ is observed, especially at high fluences. (Adapted from Gaza et al. (2006).)

the dose-response curve. Track overlap occurs when the outer regions of the tracks interact and this will occur sooner (at lower fluences and at greater inter-track distances) for those particles with higher energies compared to those particles with lower energies.

As shown in Figures 8.12 and 8.14, varying levels of supralinearity (usually slight) and sub-linearity (usually strong) can be observed in the dose-response functions depending upon the sample type and the measurement method. Taken all together, experimental fluence-response curves show linearity at low fluences for all particle types, samples, and OSL methods, but with varying degrees of supralinearity (slight) and/or sublinearity (stronger) as the fluence increases. All these features can be observed in the data of Figures 8.19a and 8.19b.

### 8.2.2 A Final Observation

In case the reader may be feeling that the story outlined in this chapter regarding the OSL emission from $Al_2O_3$:C is entirely resolved it should be stated that there are still fundamental questions and puzzles that need to be settled and answers that need to be teased out. In Figure 8.7 and throughout this chapter the emission in the UV region was described as $F^+$-emission since it coincides with the photoluminescence emission from $F^+$-centers, illustrated in Figure 8.2. The lifetime of the UV emission signal is faster than the LED pulse width used in the measurement of the OSL data of Figure 8.7, which is also consistent with the short lifetime expected for $F^+$-centers. To explain the concurrence of $F^+$- and $F$-emission during OSL (and TL), an overlap of processes was invoked, but exactly how much overlap can really be expected? For example, how likely is it that the $E_t/s$ values for the electron trap(s) that yield $F$-emission and the corresponding values for the hole trap(s) that yield $F^+$-emission be such that the temperature dependencies for the detrapping processes from the two types of trap would be the same? Examine the data shown in Figure 8.20. Here an experiment similar to that in Figure 8.7 has been conducted, using a small gamma dose of ~20 mGy and a pre-heat to the temperature indicated before measuring the OSL. As with Figure 8.7, the OSL was measured either during the stimulation pulse, when both the UV and the $F$-center emission was monitored, or after the

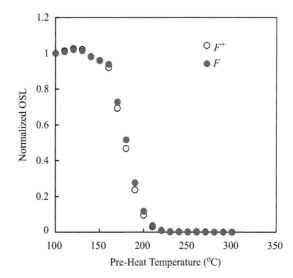

**Figure 8.20** Comparison of the dependence on pre-heat temperature of the $F^+$- and $F$-emissions during and after the POSL stimulation pulse in $Al_2O_3$:C. (Original data kindly provided by E. Yukihara.)

pulse when only the $F$-emission was measured, allowing separation of the two signals. Each signal is seen to follow exactly the same temperature dependence, raising the question – how can they be caused by two different trap types – one electron and one hole? A coincidence? Perhaps, but unlikely. Or perhaps the UV emission is not $F^+$-emission after all, but is due to an unknown, radiation-induced, UV-emitting center activated by recombination with an electron? Exactly what is(are) the main trap(s) causing the OSL and TL signal(s)? Are they complexes involving $F$- and/or $F^+$-centers, and possibly C-related hole centers, with multiple recombination pathways? There are many unanswered questions and so much is yet to be learned about this complicated material such that students and researchers should be kept busy for a considerable time to come. The story of the most-studied (and successful) OSL material, $Al_2O_3$, remains far from complete. As we strive to apply theories, old or new, to explain the OSL properties of this or any other material we are always coming up against the candid observation of J. von Liebeg, whose 1834 quotation opened this chapter.

# 9

# Radiophotoluminescence

*If a man will begin with certainties, he shall end in doubts; but if he will be content to begin with doubts, he shall end in certainties.*

F. Bacon 1605

## 9.1 Introduction

When interpreting RPL during signal readout, consideration of charge transfer processes between two defect types (i.e. traps and recombination centers) is not necessary. As a result, one may be tempted to think that the design and understanding of RPL dosimeters should be easier than that of TL or OSL dosimeters. This would be a mistake. When it is recalled that there are few, if any, TLD or OSLD materials for which either the traps or the recombination sites are definitively identified, and that in most such materials spatial association between traps and recombination centers is probably the norm rather than the exception, then it should not be imagined that identifying the key defects in RPL materials should be any easier than it is for TL or OSL materials.

Although the phenomenon of RPL is quite common, to date only a handful of materials have been demonstrated to have utility in dosimetry applications. In some of these, identification of the RPL defect(s) is still the subject of debate, while in others the identification is more settled. In this chapter, three materials that typify RPLDs are discussed. Phosphate glass is used particularly in personal and environmental dosimetry. The same material, along with $Al_2O_3$:C,Mg and LiF, is also used in charged-particle detection in which application they are known as Fluorescent Nuclear Track Detectors (FNTDs). However, $Al_2O_3$:C,Mg is the original dosimetry material in this latter application and is still the FNTD leader. The following sections describe how each material works in these applications and illustrate the RPL processes involved.

## 9.2 Phosphate Glasses

### 9.2.1 Ag-doped Phosphate Glass

#### 9.2.1.1 *Formulation, Growth, and RPL Centers*

The original Ag-doped phosphate glass ($AlNa_3O_8P_2$:Ag) was formulated by the Toshiba Corporation, Japan, and is known commercially as FD-7. As described by Perry (1987), FD-7

---

*A Course in Luminescence Measurements and Analyses for Radiation Dosimetry,* First Edition.
Stephen W.S. McKeever.
© 2022 John Wiley & Sons Ltd. Published 2022 by John Wiley & Sons Ltd.
Companion Website: www.wiley.com/go/mckeever/luminescence-measurements

was produced after trial-and-error changes to the elemental composition of various glasses (termed FD-1 to FD-6) by varying the percentages of Li, Na, P, O, Al, Ag, and B, before finally settling on FD-7. The latter consists of 31.55% P, 51.16% O, 6.12% Al, 11.00% Na, and 0.17% Ag, by weight; Li, Mg, and B were not used in the final FD-7 formulation (Yokota and Nakajima 1965). The O:P ratio in this composition makes this a polyphosphate glass (Fan et al. 2011) and results in an effective atomic number of 12.04 and a density of 2.61 g.cm$^{-3}$. The Ag content can be varied, but 0.17% is the standard dopant level in FD-7. The basic molecular unit is the $PO_4$ phosphate group (Figure 9.1, shown here with three $Na^+$ ions and one $Al^{3+}$ ion) in which P-tetrahedra form via strong covalent bonds with oxygen. The tetrahedra are classified according to the number of bridging oxygens per tetrahedron, and the classification defines the number of tetrahedral linkages. Depending on the number of linkages, Na and Al readily form links to the $PO_4$ tetrahedra (Brow 1993, 2000; Brow et al. 1993).

Synthesis is via melt-quenching using orthophosphates $Na_3PO_4$ and $AlPO_4$ along with oxides $SiO_2$ and $Ag_2O$ as starting materials. The reactants are melted at ~1200 °C for several hours, followed by rapid cooling and a secondary anneal at 500 °C for 1 hour. After cooling, the glass may be cut and polished into the desired shapes and sizes for dosimetry purposes. Commercial FD-7 glass dosimeters are available from Chiyoda Technol Corporation as rods or plates under a variety of commercial code names.[1] Depending on the application, the weakness of the P-O-P bonds gives the material poor surface degradation properties in the presence of moisture, and various recipes are used to strengthen the material by the addition of $Al_2O_3$, $Fe_2O_3$, $Pb_2O_3$, and $SiO_2$ during glass synthesis, giving the material greater weathering resistance (see Iwao et al. 2021, for an example). These improvements in mechanical and chemical stability induce only minor changes to the general RPL properties and dosimetry characteristics.

Although radiophotoluminescence (RPL) from silver-doped phosphate glasses is a well-known and reliable radiation dosimetry technique (Piesch et al. 1986, 1990; Perry 1987; Yamamoto 2011) and has been in use as a dosimetry method for decades (Schulman et al. 1951), the details of the RPL mechanism are still topics of continuing research. Nevertheless, the consensus is that the process begins with oxidation/reduction of $Ag^+$ ions by interaction with radiation-induced free holes and electrons to form $Ag^{2+}$ and $Ag^0$ via the reactions $Ag^+$ +

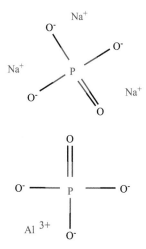

**Figure 9.1** $AlNa_3O_8P_2$ alkali aluminophosphate glass. (E.g. https://pubchem.ncbi.nlm.nih.gov/compound/Levair#section=Structures; accessed April, 2021.)

---

[1] English (c-technol.co.jp); accessed March 27, 2021.

$h^+ \rightarrow Ag^{2+}$ and $Ag^+ + e^- \rightarrow Ag^0$, respectively (for example, Kurobori et al. 2010; Miyamoto et al. 2010a, 2010b, 2011, 2014; Yamamoto 2011). Furthermore, it is understood that clustering of the Ag species occurs, forming centers of the type $Ag_m^{n+}$ where n and m are integers and m = n + 1 (Yokota and Imagawa 1967; Dmitryuk et al. 1989, 1996). For example, $Ag^+$ reacts with $Ag^0$ to form a $Ag_2^+$ dimer (or pair).

Another well-known and important (for RPL) defect species in irradiated Ag-doped phosphate glass is the so-called phosphorous oxygen hole center (POHC, or $PO_4^{2-}$ center) formed by the trapping of a hole at a $PO_4^{3-}$ site, i.e. $PO_4^{3-} + h^+ \leftrightarrow PO_4^{2-}$. The latter reaction is shown as reversible since the trapped hole is unstable at room temperature and thermal stimulation can re-form the $PO_4^{3-}$ complex. The freed hole may then be trapped by the other $Ag^+$ ions and be stabilized by forming further $Ag^{2+}$ ions.

The above centers may be identified in electron spin resonance (EPR) spectra, an example of which is shown in Figure 9.2. There is strong overlap between the various EPR signals from the phosphate and Ag centers. The $Ag^0$ signals transform into $Ag_2^+$ pairs at room temperature via the association of $Ag^0$ atoms with $Ag^+$ ions. An example of a corresponding optical absorption curve is shown in Figure 9.3 following irradiation.

There are two schools of thought that have emerged regarding the mechanism of RPL production in this material. In both concepts the key to understanding the RPL mechanism is to also understand the phenomenon of 'buildup' wherein the RPL signal increases with time following the end of the radiation. One mechanism and model for RPL in Ag-doped phosphate glass was already discussed in Chapter 4 but it is worthwhile comparing the two main models that have been proposed in the literature for RPL in this material at this point in the discussion. Figure 9.4 illustrates two flat-band energy diagrams and compares the two models side by side.

Figure 9.4a shows the model of Miyamoto, Kurobori, and colleagues in which $Ag^+$ ions trap either holes or electrons creating $Ag^{2+}$ and $Ag^0$ centers. In this model, as described by Yamamoto et al. (2020), blue RPL (at ~460 nm) is emitted from $Ag^0$, while the main orange RPL signal is due to $Ag^{2+}$. The latter is suggested to build up more slowly than the blue emission due to the slower diffusion of holes to $Ag^+$ sites compared to the diffusion of electrons. The slowness is exacerbated by the trapping and detrapping of holes at unstable POHC sites (not shown in Figure 9.4a). The second model, as described in Chapter 4 and advocated initially by Dmitryuk

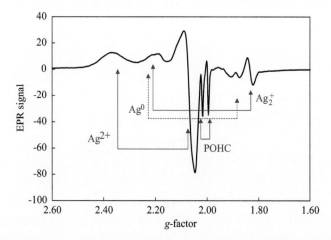

**Figure 9.2**  EPR spectrum at room temperature from Ag-doped FD-7 glass (viz. a GD-305 glass rod from Chiyoda Technol Corporation) following 1.5 kGy beta irradiation. (Original data kindly provided by S. Sholom.)

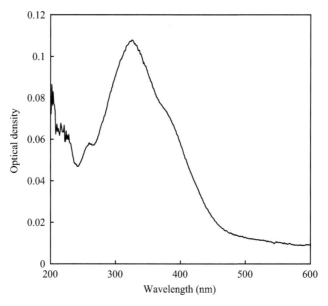

**Figure 9.3**   Optical absorption spectrum for irradiated Ag-doped phosphate glass (54 Gy beta at room temperature). (Original data kindly kindly provided by S. Sholom.)

and colleagues, and later by McKeever, Sholom, and colleagues, is that the main orange RPL signal at ~630 nm is from $Ag_2^+$ pairs, formed by the reaction between $Ag^0$ and $Ag^+$ ions. The RPL signal builds up due to the diffusion of $Ag^+$ ions, which is a slower process than the diffusion of electronic species (electrons or holes). The $Ag^+$ diffusion model was proposed from the observed growth of the $Ag_2^+$ EPR and OA signals and the decay of the $Ag^0$ EPR and OA signals during RPL buildup (McKeever et al. 2019, 2020a).

*9.2.1.2   Emission and Excitation Spectra: RPL Decay Curves and Signal Measurement*

When excited with a constant stimulation source (CW-RPL) a strong, zero-dose background fluorescence signal (photoluminescence, PL), peaking around ~460 nm with a short

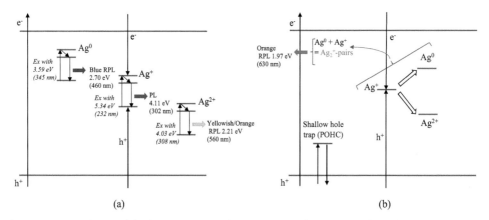

**Figure 9.4**   (a) The model of Miyamoto et al. (2010a, 2010b, 2011) and Kurobori et al. (2010), and (b) the model of Dmitryuk et al. (1989) and McKeever et al. (2019) for RPL from Ag-doped alkali phosphate glass.

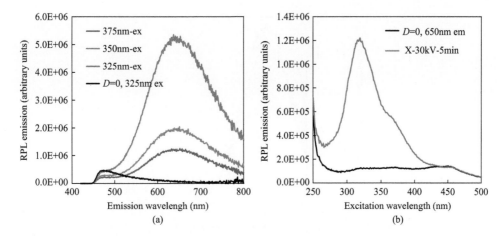

**Figure 9.5** (a) Emission and (b) excitation spectra for Ag-doped phosphate glass. In (a) the zero-dose spectrum, excited at 325 nm, shows a peak at ~460 nm with a long tail extending to longer wavelengths. (The spectrum is truncated at short wavelengths due to the use of a band-pass filter to eliminate the excitation source.) After irradiation (five minutes of 30 kV x-rays) a strong emission peaking at ~630 nm is observed. The same emission is also observed but at weaker intensity for 350 nm and 375 nm excitation. In (b) the excitation spectra for zero-dose and for five minutes of x-rays are illustrated. The excitation spectrum after irradiation is to be compared with the absorption spectrum of Figure 9.3. (Original data kindly provided by S. Sholom.)

luminescence lifetime, is emitted concurrently with the radiation-induced RPL signal. The latter peaks at ~630 nm and with a slightly longer lifetime. Emission and excitation spectra for unirradiated and irradiated Ag-doped phosphate glass are shown in Figure 9.5. The zero-dose, background fluorescence spectrum can be observed in Figure 9.5a ($D = 0$ Gy, excited at 325 nm) with a long tail extending into the long-wavelength region. This is compared with the RPL emission spectrum observed peaking at ~630 nm after irradiation with five minutes of 30 kV x-rays and also excited at 325 nm. The other two curves in Figure 9.5a show the spectra after excitation at 350 nm and 375 nm, respectively. All spectra were measured under CW excitation.

The corresponding excitation spectra can be seen in Figure 9.5b for emission at ~650 nm, for zero dose and after five minutes of 30 kV x-rays. The excitation spectra can be compared to the absorption spectrum shown in Figure 9.3.

---

**Exercise 9.1   Absorption and Excitation Spectra**

Compare Figures 9.3 and 9.5b (orange curve). The first shows the total absorption after irradiation whereas the latter shows the absorption of light that induces orange RPL emission (near 650 nm). On the web folder under Exercises and Notes, Chapter 9, Exercise 9.1, can be found an Excel spreadsheet with normalized versions of the two spectra for ease of comparison. What are the similarities? What are the differences? Can you propose reasonable explanations for the differences and for the similarities?

---

Although the peak of the emission is around 630 nm, the emission peak itself is broad and seems to be composed of several individual components. Figure 9.6 shows one example fit of

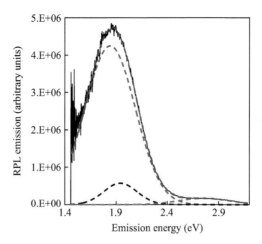

**Figure 9.6**    Emission spectrum (325 nm excitation) from Ag-doped phosphate glass deconvolved into three Gaussian bands. (Original data kindly provided by S. Sholom.)

the emission into a series of Gaussians, assuming that each of the components can be described by Gaussian-shaped bands. Such deconvolutions always carry with them a degree of uncertainty and a wider series of experiments – at different doses, excitation wavelengths, and temperatures, for example – would be advised before confidence could be placed on such fits. Nevertheless, one suggested inference of multiple contributions to the ~630 nm emission band is that the band is caused by luminescence from both $Ag_2^+$-pairs and $Ag^{2+}$ ions, in some way a combination of both of the models presented in the previous section.

---

**Exercise 9.2    RPL Buildup**

In Figure 9.6, the emission spectrum was fitted to three Gaussian curves. Specifically, in the region of the orange emission, two Gaussians were used. The data for Figure 9.6 were taken two hours after irradiation (5 Gy beta) of a phosphate glass sample. On the web folder under Exercises and Notes, Chapter 9, Exercise 9.2, can be found an Excel spreadsheet showing the change in the emission spectrum as a function of time after the irradiation, from time = 0 (i.e. as soon as possible after the irradiation) to a time of two hours after irradiation. Fit the data to a series of Gaussian emission bands and plot how each band changes as a function of time after irradiation. How many bands do you think best describe the emission spectrum? How do they vary with time? How did you deal with the background (for dose, $D = 0$)?

---

Since CW-RPL signals are composed of both the fast-decaying fluorescence background (zero-dose) signal as well as the radiation-induced RPL signal, even when measured in the long-wavelength region, an improvement in the differentiation between the two can be made by using pulsed stimulation, which is now the normal mode used in modern RPL measurements (P-RPL). A typical P-RPL decay recorded after the end of the stimulation pulse is observed in Figure 9.7, for dose of ~0.5 Gy, normalized to the value 50 ns after end of the stimulation pulse. The zero-dose fluorescence signal is typified by a fast component (~0.3 μs) and a weaker, longer-lived component. The data in Figure 9.7 has had this component subtracted and is therefore the RPL signal only. Fitting the radiation-induced RPL signals reveals three main components, plus a weak, long-lived component, with lifetimes listed in the figure caption. Similar, but slightly different RPL decay times have been reported in the published literature (Schneckenburger et al. 1981; Perry 1987 (and references therein); Kurobori et al. 2010; Tanaka et al. 2016, and others). The reported values depend on a variety of factors, including Ag content, alkali type (e.g. Na, Li, etc.), and dose.[2]

The RPL signal is usually defined over a particular time interval, minus the background signal over an equivalent time interval. Thus, for the first time an RPL dosimeter is used, the signal may be defined, for example:

$$S_D = S_{t_1-t_2}^{D} - S_{t_1-t_2}^{D=0} \tag{9.1}$$

where $t_1$ and $t_2$ define a particular time interval after the stimulation pulse, and $S^D$ and $S^{D=0}$ signify the RPL curves after dose $D$ and after zero dose, respectively. The given dose $D$ can then be determined from:

$$D = \frac{S_D}{f} \tag{9.2}$$

where $f$ is the RPL increment per unit dose and is a constant obtained from calibration (assuming a linear dose-response function).

However, it is to be recalled that RPL is not a destructive measurement. That is, the signal is not removed by measurement. If the dosimeter is not annealed after use (e.g. 400 °C for 1 hour) to remove any pre-existing signal and the dosimeter is simply used a second time, or a third time, etc., there will always be a pre-existing signal before each subsequent use. This signal is

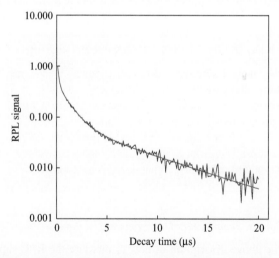

**Figure 9.7** A normalized RPL decay curve from Ag-doped phosphate glass after a beta dose of ~0.5 Gy. The curve has had the zero-dose photoluminescence and background signal subtracted and the first data point is taken 50 ns after the end of the stimulating laser pulse (375 nm). The time resolution is 100 ns. The data are fitted (red line) to a sum of four exponential components with decay constants of 0.13 μs, 1.15 μs, and 6.5 μs, plus a weak, slow component of ~27.0 μs. (Original data kindly provided by S. Sholom.)

---

[2] Other factors affecting the measurement of the time constants include purely experimental considerations such as the stimulation pulse shape and the time resolution of the measurements. Much care has to be taken to ensure experimental reliability and veracity.

generally termed the "pre-exposure signal," $S_{pre}$, and corresponds to a "pre-dose" of $S_{pre}/f$. Even if the sample is annealed after each use, the background signal ($S^{D=0}$) will always be present. As described by Perry (1987), any pre-existing signal may be accounted for in the following manner. If $S_{post}$ is the post-irradiation signal then Equation 9.1 may be replaced more generally by:

$$fD = S_{post} - \left(S_{pre} + S^{D=0}\right) \tag{9.3}$$

where it is understood, without explicitly stating, that each $S$ value has to be defined between a time interval $t_1$ to $t_2$. $S_{post}$ includes the radiation-induced RPL signal due to the dose of interest, $fD$, plus all previous radiation-induced RPL signals, $S_{pre}$, plus the original zero-dose, fluorescence signal, $S^{D=0}$. Rearranging:

$$D = \frac{S_{post} - \left(S_{pre} + S^{D=0}\right)}{f} = \frac{\Delta S}{f} \tag{9.4}$$

Here, $\left(S_{pre} + S^{D=0}\right)/f$ represents the full pre-dose. If the dosimeter has been used multiple times, it may be that $S_{pre} >> S^{(D=0)}$. In practice, the pre-dose signal has to be determined before each irradiation and subtracted from the post-irradiation signal, and therefore it is the incremental change in signal that is proportional to dose, $\Delta S = fD$, assuming a linear response to dose. The accumulation of the pre-dose signal with re-use can place a limitation on the minimum detectable dose for subsequent later uses and the ability to measure low, environmental doses may require annealing of the dosimeter to remove all accumulated pre-dose signals.

In a real instrument, $S^{D=0}$ may consist of not just the fluorescence from the dosimeter, but also stray fluorescence from the optical components from the instrument itself.

---

### Exercise 9.3    RPL Decay Curves and Signal Processing

On the web folder under Exercises and Notes, Chapter 9, Exercise 9.3, can be found an Excel spreadsheet showing the several RPL decay curves from Ag-doped phosphate glass for very low doses. Also included is an average pre-dose signal $S_{pre}$. Assume each measurement is for a separate sample but that the average pre-dose signal is the same for each sample. Using Equations 9.1–9.4, and knowing the doses applied as given in the spreadsheet, choose values for $t_1$ and $t_2$ and estimate the value for $f$. Select different values for $t_1$ and $t_2$ and re-calculate $f$. What do you conclude?

---

### 9.2.1.3    Buildup Curves: Temperature Dependence; UV Reversal

Not mentioned in the above description of how to measure the RPL signal is the previously noted buildup phenomenon, that is the growth of the RPL signal with time following the end of the irradiation period. Figure 9.8a shows the changes in the optical absorption spectrum with elapsed time following x-irradiation (~1 Gy) showing the buildup of the absorption in the 250 nm to 350 nm region, and a decrease in the 375 nm to 500 nm region. The corresponding changes to the emission spectrum (in this case for a 3 Gy beta dose and for an excitation wavelength of 375 nm) are shown in Figure 9.8b. It is very clear that the RPL buildup signal is in the orange wavelength (emission) region, peaking at ~630 nm, and covers a broad excitation range, from ~250 nm to at least 400 nm. However, superimposed on this broad RPL absorption/excitation spectrum is a feature from ~375 nm to 500 nm, which does not build up after irradiation

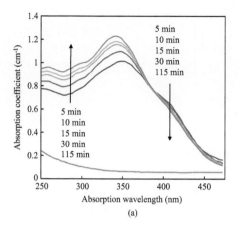 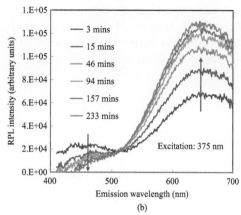

**Figure 9.8** (a) Changes to optical absorption spectra of Ag-doped phosphate glass following 30 kV x-irradiation (1 Gy) showing regions of absorption increase and regions of absorption decrease. (Reproduced from Miyamoto et al. (2011) with permission from Elsevier.) (b) Corresponding changes to the RPL emission spectrum following 3 Gy beta irradiation. (Reproduced from McKeever et al. (2019) with permission from Elsevier.).

but in fact behaves in the opposite sense, and emits in the blue region of the spectrum, near ~460 nm. That is, there appears to be two opposing effects occurring during buildup in the optical absorption range from ~375 nm to 500 nm – one is a buildup of absorption due to buildup of the orange-emitting RPL centers, and one is a decrease in absorption due to a decrease of the blue-emitting centers. The net effect is a slight decrease in absorption in this wavelength range (Figure 9.8a). If the ~460 nm emission feature is due to $Ag^0$ centers, as proposed by Miyamoto and others (Miyamoto et al. 2010a, 2010b, 2011, 2014), then this is the expected behavior for the model in which $Ag_2^+$ centers are the main defects causing RPL emission at 630 nm since, during buildup, $Ag^0$ centers would be expected to decrease while $Ag_2^+$ centers increase due to the reaction $Ag^0 + Ag^+ \rightarrow Ag_2^+$.

Figure 9.9a summarizes the changes to the orange and blue RPL emission signals due the elapsed times following irradiation, while Figure 9.9b shows the correlation between the buildup of the optical absorption bands in the 250 nm to 350 nm region and the buildup of the orange RPL signal.

The buildup phenomenon is highly temperature dependent. Figure 9.10 shows a series of buildup curves obtained at different temperatures. The RPL signal at 630 nm in this material shows significant thermal quenching (Miyamoto et al. 2014; McKeever et al. 2020a; see also Section 3.7.1) and the data in Figure 9.10 are shown after correction for quenching using an equation of the form of Equation 3.89, with $W = 0.289$ eV and $C = 6860$ (McKeever et al. 2020a).

Buildup also occurs during irradiation with the result that the growth of RPL with irradiation time during irradiation is non-linear, as previously discussed in Section 6.2.7. Since the rate of buildup increases with temperature, the curvature of the RPL growth during irradiation is also temperature-dependent, to the extent that the curvature can be removed and the RPL versus irradiation time can be made linear if the irradiation temperature is sufficiently high (Yamamoto et al. 2020). The buildup process is clearly thermally activated. Analysis of the data of Figure 9.10 revealed two thermally activated processes with activation energies of 0.17 eV and 0.45 eV (McKeever et al. 2020a).[3]

---

[3] A suggested interpretation of these activation energies is that 0.17 eV corresponds to the diffusion of $Ag^+$ ions in the formation of $Ag_2^+$-pairs, while 0.45 eV is the energy required to release trapped holes from POHC sites to form $Ag^{2+}$ ions.

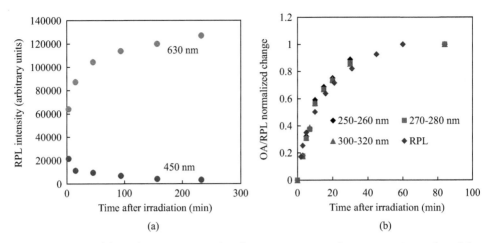

**Figure 9.9** (a) Buildup of RPL emission bands at ~630 nm and ~450 nm (reproduced from McKeever et al. (2019) with permission from Elsevier), and (b) comparison of the RPL buildup at ~630 nm with the changes in optical absorption from several different regions of the absorption spectrum, following irradiation. (Original data kindly provided by S. Sholom.)

In general, whether the growth during irradiation is linear or nonlinear depends upon the balance between the rate of irradiation and the rate of buildup. Low dose rates and/or fast buildup tend to produce linear growth during irradiation, while faster dose rates and/or slower buildup tend to produce non-linear (supralinear) growth (Section 6.2.7). However, in the latter case, if the dosimeter is left for a sufficient time after the irradiation has ceased so that the buildup process is allowed to be completed, the RPL signal will eventually reach its full, "mature" level. This process can be accelerated by heating the sample post-irradiation to a

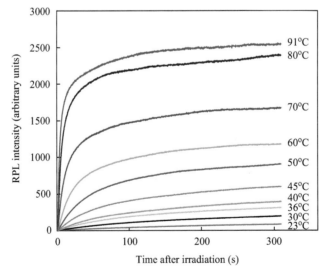

**Figure 9.10** RPL buildup curves for Ag-doped phosphate glass following 500 mGy of 10 MV LINAC irradiation at the temperature indicated and after correction for thermal quenching using Equation 3.89 with $W = 0.289$ eV and $C = 6860$; see text. (Original data kindly provided by S. Sholom.)

sufficiently high temperature, e.g. 70 °C, and holding it there for, say, thirty minutes to 1 hour, after which period the RPL will have reached maturity and be ready for readout. Such post-irradiation annealing is sometimes called "curing". For routine environmental dosimetry where the dose rates are very low, or personal dosimetry where not only are the dose rates expected to be quite low but readout of the RPL signal is usually several days after the exposure period, post-irradiation curing can be dispensed with and accurate dosimetry can still be performed. The 70 °C anneal is only required for higher dose rates and/or rapid readouts.

---

### Exercise 9.4    RPL measurement during irradiation

On the web folder under Exercises and Notes, Chapter 9, Exercise 9.4, can be found a figure showing the output of a Cherenkov detector (in blue) placed in the beam of a LINAC x-ray unit, alongside a silver-doped phosphate RPL detector at the end of an optical fiber. The RPL detector is being continually pulsed with a laser beam (375 nm) and luminescence is detected after each laser pulse and integrated between pulses. The integral is continually monitored and plotted (in black) as a function of the time under the LINAC beam.

Explain the shape of the RPL curve observed in the above data.

---

An important property of Ag-doped phosphate glasses is the dependence of the buildup properties on composition. Variations in the type and concentration of the alkali used in the base glass, and in the concentration of the Ag dopant can have profound changes on the buildup properties. For a fixed base composition, using the base composition of FD-7 (31.55% P, 51.16% O, 6.12% Al, 11.00% Na) for example, the rate of buildup varies dramatically with the Ag dopant content, as illustrated in Figure 9.11. In particular, the greater the Ag concentration, the faster the buildup. For glasses with Ag concentrations above ~1 mol % higher-order molecular species $Ag_m^{n+}$, with $m = n + 1$ and with $m \geq 2$, form readily via reactions of the type $Ag^+ + Ag_2^+ \rightarrow Ag_3^{2+}$, etc. Such molecular species emit at shorter wavelengths, around 470 nm with

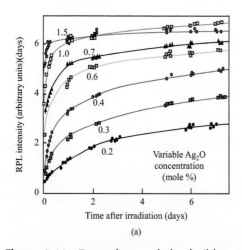
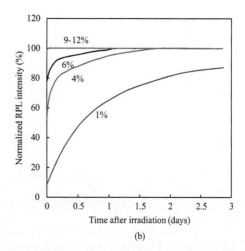

**Figure 9.11**  Dependence of the buildup of RPL from Ag-doped phosphate glass on Ag concentration. (a) For Ag concentrations of 0.2 to 1.5%. (Adapted from Andreeva et al. (1985).) (b) For Ag concentrations from 1% to 12%. (Adapted from Yokota and Imagawa 1967). The buildup rate is faster the higher the concentration of silver.

excitation wavelengths around 370 nm (Hsu et al. 2010). Thus, fast buildup may be associated with rapid formation of $Ag_2^+$ and higher-order molecular silver clusters.

A further remarkable property of buildup is that the process can be reversed by UV bleaching. The effect is shown in Figure 9.12a after different irradiation doses. The Ag-doped phosphate glasses have been irradiated and annealed at 70 °C to ensure full buildup of the RPL signal after irradiation. Then the samples have been exposed to 254 nm light for the times indicated, showing a clear decrease of the RPL signal. The reduced RPL can be partially regained by re-annealing at 70 °C, but not to 100% of its pre-UV value, as illustrated in Figure 9.12b, which shows the corresponding changes to the RPL emission spectrum. The EPR signals due to $Ag_2^+$ centers follow the same behavior as the orange RPL signal; the $Ag^{2+}$ EPR signals, however, are unchanged. The blue 460 nm emission and the $Ag^0$ EPR signals behave in the opposite manner to the $Ag_2^+$ and RPL signals (McKeever et al. 2019).

### 9.2.1.4 Photon Dose-Response Characteristics

The RPL decay curve in Figure 9.7 illustrates the shape of the curve at a reasonably low dose (~0.5 Gy). If the shape of the decay curve is invariant with dose the dose-response relationship can be expected to be linear when the signal is defined in a consistent fashion, such as described in Section 9.2.1.2. Example dose-response functions are illustrated in Figure 9.13, from very low doses in Figure 9.13a (up to ~100 µGy), and up to ~1 Gy in Figure 9.13b. Linearity up to higher doses (several 10s of Gy) is usual.

In fact, some changes to the decay rate of the RPL after the stimulation pulse are observed as the dose increases, but not for all components (Perry 1987). Example data are shown in Figure 9.14 where it can be observed that the largest changes are to the longer-lived components of the RPL decay such that if the RPL signal is defined at short times after the pulse, the observed effect on the measured RPL signal is not enough to affect the dose-response characteristics.

At very high doses, however, other experimental complications set in. Since RPL requires stimulation by light in the UV part of the spectrum, it suffers from absorption of the stimulation

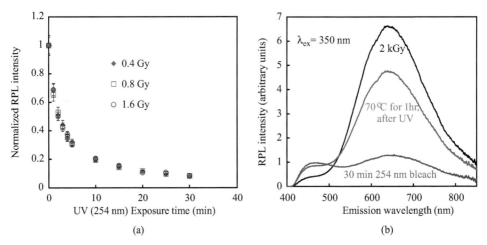

(a)                                    (b)

**Figure 9.12** (a) UV reversal of the buildup effect during illumination with 245 nm light following complete buildup, after three different beta doses. (b) RPL emission spectra after 2 kGy irradiation showing the spectrum after full buildup and after UV reversal (30 minutes of 254 nm bleaching), and after partial recovery of the buildup (70 °C/1 hour anneal). (Reproduced from McKeever et al. (2019) with permission from Elsevier.)

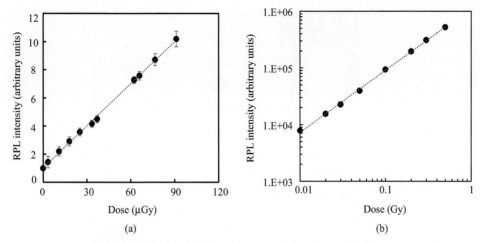

**Figure 9.13**   Dose-response functions at low dose. (a) Zero to 100 μGy, and (b) 0.01 Gy to 1 Gy. (Original data kindly provided by S. Sholom.)

light after high doses. Figure 9.15 shows several optical absorption curves for doses ranging from zero to 300 kGy (beta) for Ag-doped phosphate glass. It is clear that stimulation in the UV part of the spectrum, anywhere in the range 300–400 nm, will result in significant absorption of the stimulating light such that as the dose increases the penetration of the absorbing light will decrease and the stimulated volume will likewise decrease. The noise in the data in Figure 9.15 for 300 kGy corresponds to essentially zero penetration of transmitted light and complete absorption of the incident light at the surface of the glass sample. The effect of a reduced penetration of the stimulation light can be seen in Figure 9.16 showing calculations, using the Beer-Lambert Law,[4] of the fractional loss of stimulation light intensity as the optical absorption

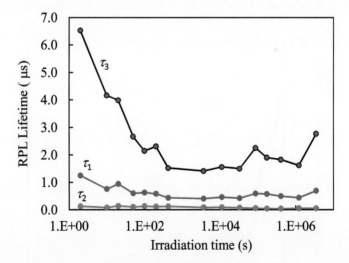

**Figure 9.14**   Changes to the orange RPL decay constants as a function of dose. The irradiation source was a $^{90}Sr:^{90}Y$ source delivering approximately 1 kGy.h$^{-1}$. (Analysis by author from original data provided by S. Sholom.)

---

[4] Beer-Lambert Law: If $I_o$ is the intensity of incident light and $I_x$ is the intensity of light at depth $x$ inside a sample, and $\alpha$ is the absorption coefficient, then $I_x = I_o \exp\{-\alpha x\}$.

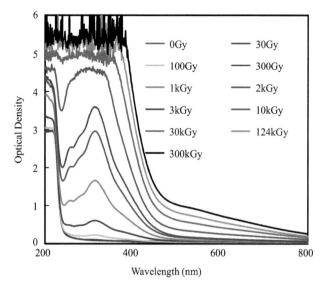

**Figure 9.15**   Optical absorption from Ag-doped phosphate glass after doses ranging from 0 Gy to 300 kGy beta. (Reproduced from Sholom and McKeever (2020) with permission from Elsevier.)

coefficient increases for a 1 mm thick sample for different doses for a specimen of Ag-doped phosphate glass with optical absorption spectra as shown in Figure 9.15, estimated at 325 nm. For a plane parallel beam, only ~10% of the sample is stimulated after a dose of ~10 kGy. This problem has long been recognized (see Perry (1987) and references therein).

Figure 9.17 shows how the RPL intensity changes as a function of dose, for high doses, for both the orange emission (at ~630 nm) and the blue emission (at ~470 nm). Two sets of curves are shown for each emission – one shows the raw data taking no account of the absorption of

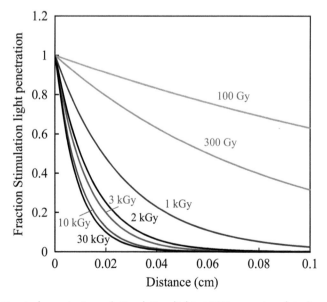

**Figure 9.16**   Estimated penetration of stimulation light at 325 nm using data from Figure 9.15 for a 1 mm thick sample, as a function of dose.

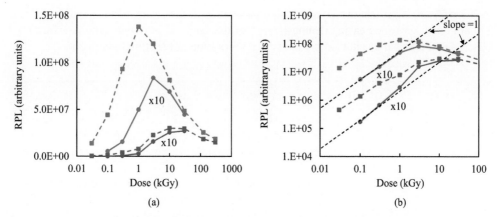

**Figure 9.17** Dose-response functions at high dose (a) linear–linear, and (b) log–log. Data are shown both for orange emission (~630 nm; orange dots and lines) and blue emission (~470 nm; blue dots and lines), taken from the peak of the emission spectra minus the zero-dose value. Two data sets are illustrated; one shows the data without correction for absorption of the stimulation light as the dose increases (dashed lines; see Figures 9.15 and 9.16) and the other shows the same data after an approximate correction for the stimulation light penetration (full lines). The dose at which the orange emission reaches its maximum shifts slightly to higher doses after correction. (Original data kindly provided by S. Sholom.)

the stimulation light (dashed lines) and one shows the data after approximate correction allowing for absorption of the stimulation light and correcting for the reduction in the stimulated volume (full lines). For the orange emission, the peak of the emission before strong sublinearity becomes evident shifts slightly after correction from ~1 kGy to ~2kGy. There is little change for the blue emission after correction. Also observed is that the ~470 nm blue emission increases strongly at about the same dose that the orange emission decreases. After correction, the growth of the ~630 nm emission grows linearly (see dashed 1:1 line on the log–log plot of Figure 9.17b) up to ~1 kGy after which sublinearity occurs, whereas the growth of the 470 nm emission grows supralinearly. The interplay between the two emissions appears to be correlated. The corresponding RPL emission spectral changes are shown in Figure 9.18.

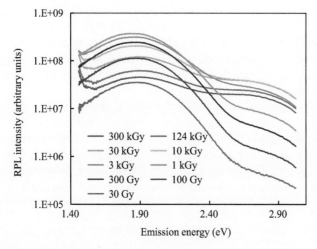

**Figure 9.18** Changes to the emission spectrum of Ag-doped phosphate glass as a function of dose. (Original data kindly provided by S. Sholom.)

These observations that the 470 nm RPL emission is still increasing beyond 1 kGy while the 630 nm emission is decreasing demonstrates that the decrease in orange emission for $D > 1$ kGy is a real effect and is not simply a result of absorption of the stimulation light.

---

### Exercise 9.5    Emission spectra at high doses

On the web folder under Exercises and Notes, Chapter 9, Exercise 9.5, can be found experimental data for Figure 9.18, namely the emission spectra as a function of dose for high doses. Separate each spectrum into Gaussians. Compare your results with the results of Exercise 9.2. Do you get the same number of bands at high doses as you did at the lower doses? Follow the separated emission bands as a function of dose. What do you conclude?

---

The ~470 nm emission observed at high doses in Figures 9.17 and 9.18 is not the same as the original ~450 nm-to-460 nm emission observed at low doses in Figure 9.8b. In the latter case, there is a strong argument that the 450 nm-to-460 nm signal is related to $Ag^0$ (Miyamoto et al. 2010a, 2010b) and it clearly decreases as a function of time after irradiation, during buildup. In contrast, the ~470 nm emission observed in Figure 9.18 is a strong signal that grows with dose and does not decay after irradiation. Following the observations of Dmitryuk et al. (1996), it is possible that this latter radiation-induced emission is caused by a higher-order molecular species of silver, such as $Ag_3^{2+}$, previously identified to emit at ~470 nm in this material (Hsu et al. 2010). The growth of the ~470 nm emission band would then be at the expense of the ~630 nm emission band according to the reaction $Ag^+ + Ag_2^+ \rightarrow Ag_3^{2+}$. Furthermore, the luminescence lifetime of this blue RPL signal is much faster than the orange RPL signal and is of the order of a few ns; this is expected for a molecular species such as $Ag_3^{2+}$ (Hsu et al. 2010).

#### 9.2.1.5    *Charged-Particle Dose-Response Characteristics*

As has been noted in previous chapters, irradiation with charged particles results in a highly spatially inhomogeneous dose distribution in dosimetry materials. Although the overall dose may be small, the local dose in the region of the charged-particle track is high and the net luminescence signal that results is a convolution of the radial dose distribution around the charged-particle track and the photon dose-response function. For RPL in glasses, these considerations were applied initially by Lommler et al. (1992) using essentially the same approaches as applied later by others to TL materials and OSL materials with the resulting prediction of a decrease in the relative efficiency of RPL for charged-particle irradiation with respect to gamma irradiation for increases in LET.

Even though the overall dose imparted to the RPL dosimeter may be low, the characteristic properties of the RPL signal following charged-particle irradiation are similar to those observed following high doses of low-LET irradiation. An example is shown in Figure 9.19 where the emission spectrum from an x-irradiated sample of Ag-doped phosphate glass is compared to that from the same material following alpha-particle irradiation. Whereas the ratio of the blue-to-orange emission following x-irradiation is low, it has increased substantially after α-irradiation, reminiscent of the situation observed at high doses of low-LET irradiation

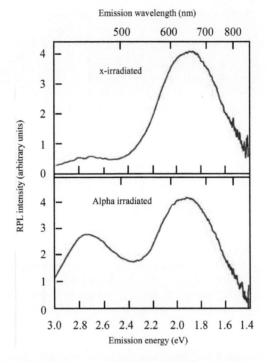

**Figure 9.19**   RPL emission spectra from Ag-doped phosphate glass (370 nm excitation) following x-irradiation (30 kV at 20 mA) and α-irradiation (5.48 MeV alpha particles, 4MBq $^{241}$Am source). Spectra measured at 100 K. (Adapted from Miyamoto et al. (2014).)

(Figure 9.18). The blue-to-orange ratio suggests that the RPL is being emitted from a high-dose region of the sample, despite a low, overall, absorbed dose.

Although charged-particle fluence-response curves for RPL from Ag-doped phosphate glasses are not readily available in the literature, several predictions can be made using the observations already noted in the previous sections. At low fluences, a linear fluence/dose-response function can be expected. Whatever is observed for one track is simply multiplied by the number of tracks as the fluence increases, until the tracks begin to overlap. At higher fluences when track overlap occurs, more of the sample volume will be characterized by high-dose regions. Since the RPL efficiency decreases at high doses (Figure 9.17) a drop in efficiency can be expected as the fluence increases due to track overlap, and a change in the ratio of blue-to-orange RPL emission will occur. This will occur especially for particles with the highest energy (and widest track radius) and highest LET (and highest intra-track doses).

An example of a dose-response function for protons is illustrated in Figure 9.20a, showing a similar shape to that for beta, as in Figure 9.17a (and also gamma, as demonstrated by Fuerstner et al. 2005). The dose at which the RPL reaches a maximum for protons occurs at several hundred Gy, compared to ~1kGy for beta and gamma. As a result of this dose-response function, the relative RPL efficiency for protons decreases with increased LET, as illustrated in Figure 9.20b. For heavier charged particles possessing higher LET values and higher energies, the decrease in the relative RPL efficiency as a function of LET is similar (and similar to that seen

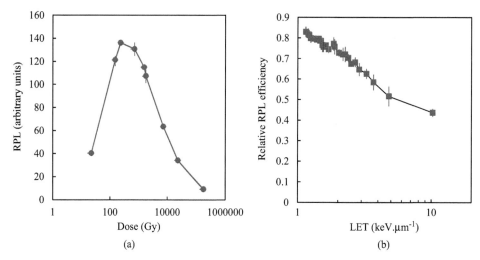

**Figure 9.20** (a) Orange RPL dose-response function for protons for Ag-doped phosphate glass. No correction has been made for absorption of the stimulation light at higher proton doses. (Adapted from Fuerstner et al. (2005).) (b) Relative orange RPL efficiency versus LET for protons. (Reproduced from Chang et al. (2017) with permission from IOP Publishing.)

for both TL and OSL). Figure 9.21 shows an example. Note that there does not appear to be a region where the relative efficiency is greater than 1, consistent with the observation that there is no supralinear region in the low-LET dose-response function (Figure 9.17). The situation for blue RPL is predicted to be different but, to date, no RPL versus LET data for blue RPL is available in the literature.

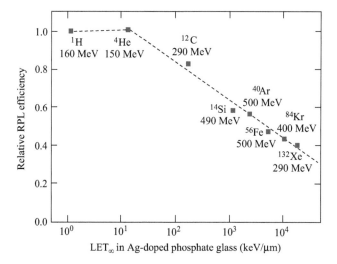

**Figure 9.21** Relative RPL efficiency of 630 nm emission (excitation 337 nm) from Ag-doped phosphate glass as a function of LET of various charged particles. (Reproduced from Kurobori et al. (2017) with permission from Elsevier.)

## 9.2.2   Final Remarks Concerning RPL from Ag-doped Phosphate Glass

RPL from Ag-doped phosphate glass is another example of how a property (RPL) of a material can be used with accuracy and confidence in an application (radiation dosimetry) without having full knowledge of the defects and physical mechanisms that give rise to that property. A complete understanding of the RPL phenomenon in phosphate glass requires much further research, particularly on the role of $Ag_m^{n+}$ molecular clusters (and possibly Ag nanoparticles). Further research may focus on the potential of using blue RPL emission in dosimetry, especially under charged-particle irradiation. There remains much to be done.

## 9.3   Fluorescent Nuclear Track Detectors

### 9.3.1   Al$_2$O$_3$:C, Mg

#### 9.3.1.1   Introduction

Fluorescent Nuclear Track Detectors (FNTDs) are RPL materials in which the RPL centers are used to detect and track the passage of ionizing charged particles through the material. The ionization produced within the track of the particle creates the RPL centers, which emit luminescence upon subsequent stimulation with an external source of light. Since the ionization is localized around the path of the particle, the pattern of the emitted RPL is also spatially localized around the charged-particle path and in this way the track of the particle can be observed. The keys to enabling this technology are sophisticated methods to scan and stimulate the RPL emission at different locations and depths within the material with high-enough resolution to enable the dimensions of the track to be revealed. With sufficient resolution and suitable methods for intensity discrimination, information about the energy and LET of the charged particle can be extracted.

The first material to be developed for this application was crystalline Al$_2$O$_3$ doubly doped with C and Mg, by Akselrod and colleagues in the mid-to-late 2000s (Akselrod and Akselrod 2006; Akselrod et al. 2006a, 2006b; Sykora et al. 2008a; Sykora and Akselrod 2010a; Akselrod and Sykora 2011). The same group also developed the same material for use in neutron fields by using recoil protons from polyethylene converters or alpha particles from the (n,α) reaction from $^6$Li converters (e.g. $^6$LiF) (Sykora et al. 2007, 2008b, 2009; Sykora and Akselrod 2010b, 2010c; Akselrod et al. 2014a). Although publications on other FNTD materials have since appeared in the literature (described in the following sections), Al$_2$O$_3$:C,Mg remains the standard for FNTD applications. Most recent developments have been in equipment design and improved mathematical algorithms for signal analysis, and in applications of the techniques in radiobiology (Akselrod et al. 2014b, 2020; Kouwenberg et al. 2016, 2018; Sawakuchi et al. 2016; Akselrod and Kouwenberg 2018).

#### 9.3.1.2   RPL in Al$_2$O$_3$:C,Mg

As previously described in Chapter 8, the structure of crystalline α-Al$_2$O$_3$ is hexagonal close-packed with the aluminum ions occupying 2 out of 3 octahedral interstices in the hcp lattice (see Figure 8.1). The O$^{2-}$ ions are arrayed in equilateral triangles, one above and one below the plane of the Al$^{3+}$ ions. Fundamental defects in Al$_2$O$_3$ are two-electron, oxygen vacancies (*F*-centers), one-electron, oxygen vacancies (*F*$^+$-centers), and aggregates of these ($F_2$, $F_2^+$, etc.).

As described in Section 8.2.1.1, carbon can occupy oxygen sites, aluminum sites, or interstitial sites, with oxygen-substitutional sites preferred. When $C^{4-}$ substitutes for $O^{2-}$, oxygen vacancy centers are required for charge compensation, resulting in the formation of additional $F$- and $F^+$-centers.

When magnesium is introduced into $Al_2O_3:C$ during crystal growth $Mg^{2+}$ ions substitute for $Al^{3+}$ ions and promote the creation of more $F$- and $F^+$-centers. These appear clearly in optical absorption spectra, as already shown in Figure 8.2a. At longer wavelengths (>300 nm), additional absorption features appear at ~435 nm (Figure 9.22). Optical anisotropy measurements, in which the light absorption and photoluminescence emission from the 435 nm centers are monitored as a function of the polarization of the incident and emitted light, with respect to the crystal's c-axis channel, reveal that the structure of the center giving rise to the 435 nm absorption band is an aggregate of two $Mg^{2+}$ ions and two $F^+$-centers, as shown in Figure 9.23 where the alignment with respect to the c-axis is found to be ~38°-to-39° (Sanyal and Akselrod 2005).

When stimulated at 435 nm the $F_2^{2+}(2Mg)$-center emits at 520 nm. Bleaching at 248 nm (5.0 eV) photochromically converts the 435 nm center into new absorption bands, with main absorption peaks at 335 nm and 620 nm (Ramírez et al. 2005). Excitation into these bands produces emission at 750 nm, while bleaching at 335 nm converts these new centers back to the $F_2^{2+}(2Mg)$-centers. The anisotropy measurements of Sanyal and Akselrod (2005) support the interpretation that the $F_2^{2+}(2Mg)$-center and the new center into which the $F_2^{2+}(2Mg)$ defect photochromically converts are both aligned in the same direction with respect to the c-axis of the crystal, and these authors suggest that the new center is a $F_2^{2+}(2Mg)$-center that has gained an electron, namely a $F_2^{+}(2Mg)$-center, thus:

$$F_2^{2+}(2Mg) + e^- \rightarrow F_2^{+}(2Mg) \tag{9.5}$$

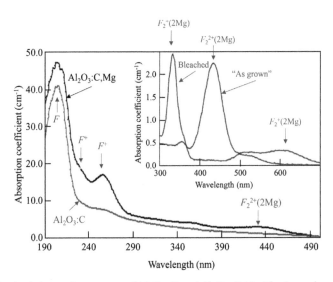

**Figure 9.22**  Optical absorption spectra of $Al_2O_3:C$ and $Al_2O_3:C.Mg$. The inset shows an expansion of the long-wavelength region for $Al_2O_3:C.Mg$ for an "as received" sample, and after photochromic conversion of $F_2^{2+}(2Mg)$-centers into $F_2^{+}(2Mg)$-centers. In this case, photochromic conversion was via 2-photon absorption using a high-intensity, pulsed laser at 435 nm. Photoconversion can also be achieved by excitation at 248 nm (5.0 eV). (Reproduced from Akselrod et al. (2003) with permission of Springer Nature.)

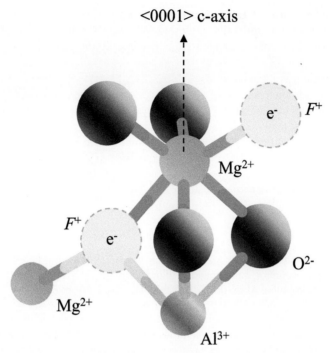

**Figure 9.23**   Structure of the $F_2^{2+}$(2Mg)-center as determined by Sanyal and Akselrod (2005).

The absorption bands associated with the $F_2^+$(2Mg)-centers are noted in the inset of the absorption spectrum of Figure 9.22.

Reaction (9.5) can also occur during irradiation. Ionization creates free electrons which become trapped at $F_2^{2+}$(2Mg)-centers creating $F_2^+$(2Mg) centers. Since the latter are photoluminescent and do not exist in "as grown" Al$_2$O$_3$:C.Mg samples, they are RPL centers. They may be detected by stimulating an irradiated sample into the $F_2^+$(2Mg) absorption bands and detecting luminescence at 750 nm. The excitation and emission bands before and after irradiation are shown in Figures 9.24a and 9.24b, respectively; the preferred excitation band for FNTD operation is the long-wavelength band and for practical purposes a laser at 635 nm is normally used in FNTD instruments, with RPL detection at 750 nm.

### 9.3.1.3   FNTD Imaging of Charged-Particle Tracks

Figure 9.25 displays a typical FNTD image of alpha particle tracks in Al$_2$O$_3$:C,Mg. The image was taken using a confocal microscope, stimulating the sample at 635 nm, and imaging the luminescence at 750 nm. Technical details describing how a typical FNTD image is obtained can be found in the original papers, but can also be found on the web site accompanying this book under Chapter 9, Exercises and Notes, Figure 9.25. As explained in the notes on the web site, the size and intensity of each spot, for a given particle type, is dependent upon the angle of incidence of the charged particle and the focal spot size of the laser beam (Bartz et al. 2014). Improvements in optics and image processing are such that even tracks due to delta rays from photon irradiation can be imaged.

By adjusting the depth at which the laser is focused inside the FNTD crystal (the z-axis) and scanning the laser two-dimensionally at each depth (an x-y scan) a volumetric scan of the crystal can be performed so that a pseudo-3D image of the passage of a charged-particle track through the FNTD detector can be obtained. An example is shown in Figure 9.26, which shows

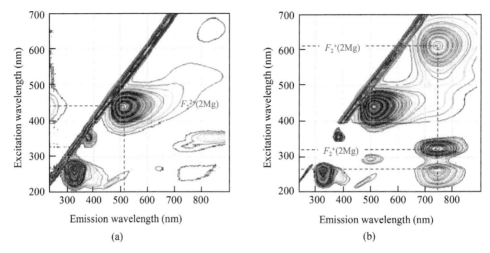

**Figure 9.24**  Excitation-emission spectra for (a) "as grown" and (b) irradiated (100 Gy, 40 kVp x-rays) showing the excitation and emission bands for the $F_2^{2+}(2Mg)$- and $F_2^+(2Mg)$-centers. (Reproduced from Sykora and Akselrod (2010c) with permission from Elsevier.)

a 3-dimensional construction of the passage of a proton through the crystal, entering at an angle $\theta$ with respect to the vertical. The image is composed of eight superimposed images, obtained with incremental depths of 5 μm, and aligned so that the centroid of the luminescence spot (shown in negative contrast) indicates the straight-line path of the proton. An animation of Figure 9.26 can be found on the web site, under Exercises and Notes, Chapter 9, Figure 9.26 Animation, showing the passage of the proton through the FNTD crystal.

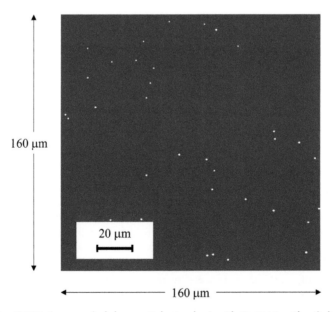

**Figure 9.25**  An FNTD image of alpha particle tracks in $Al_2O_3$:C,Mg. The field of view is as indicated, and the scan time was 81 ms with 10 pixels per micron. The high resolution inherent in this image also allowed the authors to detect photon-induced (x-ray) delta electron tracks. (Reproduced from Akselrod et al. (2020) with permission from Elsevier.)

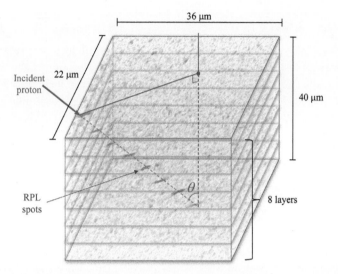

**Figure 9.26** "3-D" projection of the passage of a proton through an Al$_2$O$_3$:C,Mg FNTD crystal, entering at an angle $\theta$. An animation of this figure can be found on the accompanying web site. (Reproduced from Akselrod et al. (2006a) with permission from Elsevier.)

One of the properties of FNTDs is the ability to perform spectroscopy and determine the LET of the high-energy particles detected. Using suitable imaging processing and signal averaging techniques, track size and brightness analysis can be applied to each FNTD image. After correction for the track incidence angle, and issues such as the microscope's field-of-view and spherical aberration, distributions of RPL intensity are found to be dependent upon the LET of different charged particles, suggesting the potential use of FNTDs as LET spectrometers. Example luminescence intensity distributions for a number of different charged particles are shown in Figure 9.27.

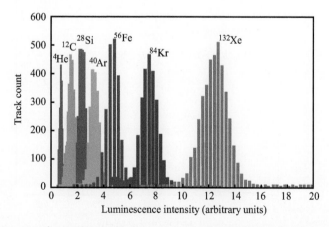

**Figure 9.27** Mean track amplitude histograms for various ions and energies in Al$_2$O$_3$:C:Mg FNTDs. The ions are: 149 MeV/u $^4$He, 135 MeV/u $^{12}$C, 444 MeV/u $^{28}$Si, 500 MeV/u $^{40}$Ar, 500 MeV/u $^{56}$Fe, 400 MeV/u $^{84}$Kr, and 187.5 MeV/u $^{132}$Xe. (Reproduced from Sykora et al. (2008a) with permission from Elsevier.)

### Exercise 9.6    FNTD: luminescence amplitude histograms

A plot such as that shown in Figure 9.27 requires several image processing corrections in order to arrive at the final amplitude histograms. A description of the necessary corrections is given in Bartz et al. (2014). They include FNTD material sensitivity variations, angle dependence, and energy deposition (range of depth) variations. Review this publication to gain an understanding of the basic corrections needed to arrive at the final histograms.

In general, the method works well for a single particle type where the luminescence-versus-LET relationship is linear (e.g. protons), but it becomes more complex when trying to determine the LET of different particle types and at different fluences. The relationship between luminescence intensity and LET is shown in Figure 9.28 for several heavy ions, in bare beams and behind wedge absorbers. The relationship is highly non-linear and demonstrates that even at high LET the RPL has not saturated. Different measurement schemes have been suggested (e.g. spatial-frequency analysis, or track-width analysis) but each is found to have difficulties.

#### 9.3.1.4    FNTD for Neutron Detection

Neutrons, being non-ionizing, do not themselves produce RPL centers, but can produce ionizing recoil protons from hydrogen-containing converters (e.g. polyethylene, PE), or alpha particles via the (n,α) reaction with, say, $^6$Li or $^{10}$B compounds. The number of proton or α-particle tracks can then be related to neutron dose. Figure 9.29 displays a typical FNTD image of recoil proton tracks in $Al_2O_3$:C,Mg acquired after irradiation with 60 mSv of AmBe fast neutrons and using a high-density PE converter to generate the recoil protons.

An animated illustration of proton recoil tracks passing through an $Al_2O_3$:C,Mg FNTD detector can be found on the accompanying web site under Exercise and Notes, Chapter 9, Figure 9.29 Animation.

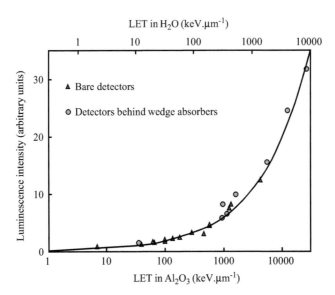

**Figure 9.28**  Luminescence amplitude for track images in $Al_2O_3$:C.Mg FNTD detectors as a function of charged-particle LET. (Reproduced from Sykora et al. (2008a) with permission from Elsevier.)

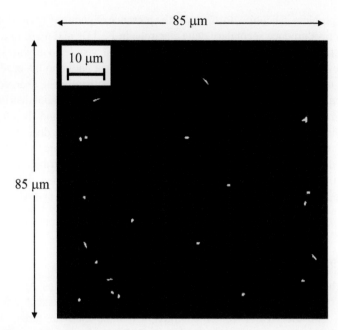

**Figure 9.29** $Al_2O_3$:C,Mg FNTD image of recoil proton tracks from fast neutrons with a high-density PE converter. An animation of this figure can be found on the accompanying web site. (Reproduced from Sykora et al. (2008b) with permission from Elsevier.)

A neutron dose-response calibration curve is illustrated in Figure 9.30 where the recoil proton track density is calibrated against the fast neutron dose. Using curves such as this, neutron dose $D$ can be determined from:

$$D = \frac{N - B}{S} \tag{9.6}$$

where $N$ is the track count (in $mm^{-2}$) after neutron irradiation, $B$ is the background track count before irradiation, and $S$ is the sensitivity (in $mm^{-2}.mSv^{-1}$).

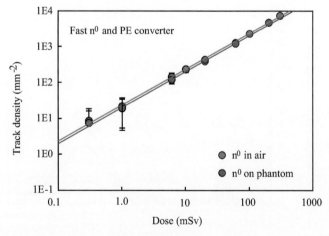

**Figure 9.30** Track density versus neutron dose for fast neutrons, using a PE converter and an $Al_2O_3$:C,Mg FNTD. Measurements were taken in air and against a phantom. (Reproduced from Sykora et al. (2008b) with permission from Elsevier.)

A persistent problem in neutron dosimetry is the presence of gamma rays in almost all neutron fields. Gamma rays produce secondary delta electrons in FNTDs and these in turn give rise to secondary electron tracks, producing a background signal visible in high-resolution FNTD imaging. However, although there is a non-uniform, statistical nature to the production of delta rays, even in a uniform gamma field, neutrons (or, more specifically, the charged particles produced by proton or α-particle converters) leave an entirely different spatial distribution of tracks within the FNTD detector. As a result, the neutron-induced tracks, and therefore the neutron dose, can be distinguished from the delta-ray tracks, i.e. the gamma dose, by the spatial-frequency distribution of the track images. By applying spatial-frequency analysis of the track images acquired by the FNTD, the very different spatial-frequency distributions of the two types of charged particle (delta-rays and protons/alphas) can be distinguished and separated. In this way, the neutron dose can be separated from the gamma dose (Sykora and Akselrod 2010a).

### 9.3.2   LiF

#### *9.3.2.1   RPL in LiF*

RPL from LiF has been known and used in dosimetry for many decades. Alkali halides in general have been the archetypical materials for the study of color centers and indeed it was from such materials that the term *F*-center emerged, from the German farbe, meaning color. Following irradiation, color centers readily form in LiF, based upon *F*-centers, namely $F^-$ ion vacancies occupied by a single electron. Strong *F*-center absorption appears at 5 eV (~247 nm; Figure 9.31). Also observed as the dose increases is the growth of clusters of *F*-centers, such as $F_2$-, $F_3$-, $F_3^+$-, $F_4$, etc. All can be observed in the absorption spectrum at high-enough doses, and the most prominent is the absorption peak at ~2.78 eV (~446 nm) consisting of an overlap of $F_2$- and $F_3^+$-centers.

Luminescence from radiation-induced $F_2$- and $F_3^+$-centers (i.e. RPL) has been used in dosimetry (Levita et al. 1976; Miller and Endres 1990; Kovacs et al. 2000), radiation imaging (Baldacchini et al. 2005; Montereali et al. 2010), and lasers (Mirov and Basiev 1995; Mirov and Dergachev 1997).

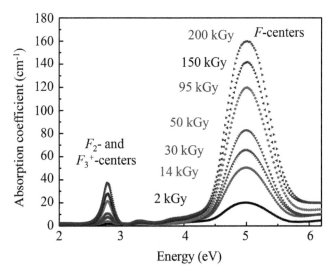

**Figure 9.31**   Optical absorption in beta-irradiated, pure LiF. (Original data kindly provided by K. Remy.)

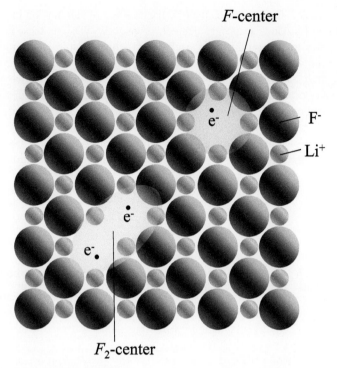

**Figure 9.32**   *F*- and *F₂*-centers in LiF.

### 9.3.2.2   FNTD

Bilksi and colleagues examined the application of RPL from LiF in track detection, with attention being on pure LiF as well as conventional TLD materials LiF:Mg,Ti and LiF:MCP as potential FNTDs. The defects of interest are $F_2$- and $F_3^+$-centers (Figure 9.32), which form under ionizing radiation via the formation of *F*-centers and anion vacancies, $V_a^+$, thus:

$$F + V_a^+ \rightarrow F_2^+ \tag{9.7}$$

$$F_2^+ + e^- \rightarrow F_2 \tag{9.8}$$

and

$$F_2^+ + F \rightarrow F_3^+ \tag{9.9}$$

Similar processes occur during continued irradiation to form $F_3$, $F_4$, and $F_4^+$, etc., with the optimum concentrations for $F_2$- and $F_3^+$-centers depending upon the irradiation conditions and impurity levels (Mirov and Dergachev 1997). When stimulated with light around 446 nm, both centers luminesce (RPL) with a broad emission peaking at ~670 nm ($F_2$-centers) and ~525 nm ($F_3^+$ centers) (Figure 9.33). The emission is very bright for both beta and alpha particles and its use as an FNTD has been demonstrated for protons, alpha particles, heavy ions, and neutrons (Bilski and Marczewska 2017; Bilksi et al. 2017, 2018, 2019). Typical FNTD images obtained using RPL from $F_2$-centers in irradiated LiF single crystals are shown in Figure 9.34 after irradiation with $^{56}$Fe ions. An advantage of using LiF as an FNTD for neutrons is that if LiF enriched in $^6$Li is used the (n,α) reaction can be exploited to detect the resulting α-tracks without the need for a neutron converter.

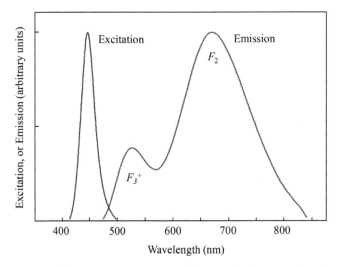

**Figure 9.33**   Excitation and emission spectra (RPL) of $F_2$– and $F_3^+$-centers from beta-irradiated LiF. (Reproduced from Bilski and Marczewska (2017) with permission from Elsevier.)

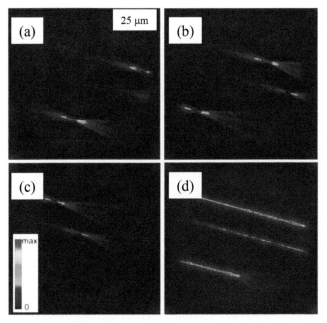

**Figure 9.34**   FNTD images of $^{56}$Fe (145 MeV/u) ion tracks in LiF using the 670 nm luminescence from $F_2$-centers, registered at depths of: (a) 10 μm; (b) 20 μm; and (c) 35 μm. The image in (d) is a maximum intensity projection of the whole stack of images (35 images, from 10 μm to 45 μm, in 1 μm steps). Details of the optics and image processing necessary to obtain the image can be found in the original paper. (Reproduced from Bilski et al. (2019) with permission from Elsevier.)

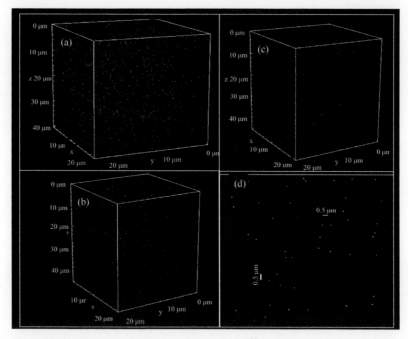

**Figure 9.35** Three-dimensional FNTD images of (a) [137]Cs delta rays tracks, (b) 160 MeV/u protons, and (c) 500 MeV/u [56]Fe ions in Ag-doped phosphate glass. (d) A 2-D representation of [56]Fe-ion tracks in the same material. (Reproduced from Kurobori et al. (2017) with permission from Elsevier.)

### 9.3.3 Alkali Phosphate Glass

*9.3.3.1 FNTD*

RPL in alkali phosphate glass has been extensively discussed already in this chapter. The principles of using the RPL centers in FNTD technology are the same as with $Al_2O_3$:C.Mg and LiF. Three-dimensional RPL track images after irradiations with photons and heavy charged particles are shown in Figure 9.35 for: (a) 1 Gy of [137]Cs gamma photons, (b) 5 Gy of protons and (c) 5 Gy of [56]Fe ions. Also shown (d) is a two-dimensional image of [56]Fe-ion tracks. The tracks shown in Figure 9.35a illustrate those detected using blue RPL emission (in blue) and orange RPL (in red). The experimental details describing how each image was obtained are described by Kurobori et al. (2017).

As with $Al_2O_3$:C,Mg FNTDs, Ag-doped phosphate glasses can also be used as LET discriminators, as demonstrated by Kodaira et al. (2020) and illustrated in Figure 9.36. A reported disadvantage of FNTD measurements in this material, not seen with conventional RPL measurements, is an apparent fading effect with repeated FNTD scans. The wavelength used by Kodaira et al. (2020) to excite the RPL signal in the FNTD measurements was 405 nm. The use of a focused beam increases the laser power density such that two-photon excitation may induce optical dissociation of the RPL centers with each scan, as observed in femtosecond laser writing of optical memory devices using the similar materials (Bellec et al. 2009, 2010).

A potential advantage of Ag-phosphate glass FNTDs is an apparent linear dependence of the luminescence track intensity on LET, as indicated in Figure 9.37. However, when compared to Figure 9.28 it can be seen that the range of LET over which the measurements have been reported is much narrower for Ag-doped phosphate glass than for $Al_2O_3$:C,Mg.

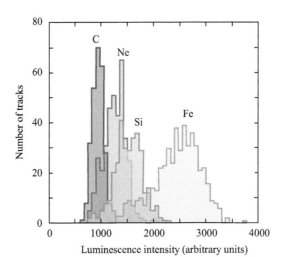

**Figure 9.36** Luminescence intensity distributions in Ag-doped phosphate glass FNTDs for 384 MeV/u C ions, 372 MeV/u Ne ions, 446 MeV/u Si ions, and 423 MeV/u Fe ions. (Reproduced from Kodaira et al. (2020) with permission from Elsevier.)

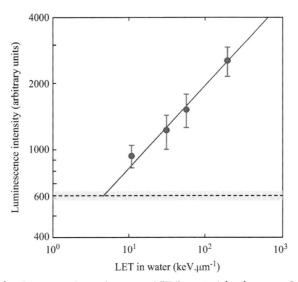

**Figure 9.37** Track luminescence intensity versus LET (in water) for the same C, Ne, Si, and Fe ions as in Figure 9.36. The dashed horizontal line is the mean background level and the grey shading represents one standard deviation of the background. (Reproduced from Kodaira et al. (2020) with permission from Elsevier.)

---

**Exercise 9.7   FNTDs: Comparisons**

Examine the literature describing the three materials offered as FNTDs in this chapter ($Al_2O_3$:C,Mg, LiF, and Ag-doped phosphate glass). From a comparison of the RPL properties of the three materials (dose-response characteristics, stimulation wavelengths, emissions wavelengths, luminescence lifetimes, etc.) can conclusions be drawn about the potential of the three materials for FNTDs in future applications in heavy charged-particle and/or neutron dosimetry? Justify your conclusions.

# 10

# Some Examples of More Complex TL, OSL, and RPL Phenomena

## *The Aluminosilicates*

*The aim of science is to seek the simplest explanation of complex facts. ... The guiding motto in life should be, Seek simplicity and distrust it.*

A.N. Whitehead 1920

## 10.1 Introduction

In Part I of this book, fundamental phenomena governing the production of TL, OSL, and RPL in luminescence dosimetry materials were described so that an experimenter may use these ideas as part of their "tool kit" to be delved into when examining the properties of real materials. Part II has included discussions of some real (and popular) TL, OSL, and/or RPL materials in which several of the fundamentals introduced in Part I have been employed to assist in explaining the phenomena observed. Nevertheless, it has been demonstrated how the detailed mechanisms remain elusive even with the most popular and engineered materials, namely LiF, $Al_2O_3$, and phosphate glasses. Importantly, these vagaries have not prevented these and other materials from being employed to wonderful effect in the world of radiation dosimetry.

To complete Part II the discussion now turns to a few examples of what may be considered as more-complex materials – in particular, natural minerals and engineered glasses. The fields here are wide and deep. Luminescence studies of natural minerals have enjoyed prominence in the scientific literature for many decades and their use in radiation dosimetry, especially in geological and archaeological dating, have filled textbooks and journals alike. Similarly, especially in recent years, the study of luminescence phenomena in engineered glasses has grown with the advent of optical communications, lasers, and optical data storage, in addition to dosimetry. Faced with such a wide selection of potential substances, this chapter has, per force, reverted to severe "cherry picking". No apology is made for this. Bearing in mind that the subject of the book is luminescence for use in radiation dosimetry, two related aluminosilicate materials have

*A Course in Luminescence Measurements and Analyses for Radiation Dosimetry*, First Edition.
Stephen W.S. McKeever.
© 2022 John Wiley & Sons Ltd. Published 2022 by John Wiley & Sons Ltd.
Companion Website: www.wiley.com/go/mckeever/luminescence-measurements

been selected that are important is this regard – namely, the natural aluminosilicate mineral, feldspar (used in luminescence dating), and engineered aluminosilicate glasses (from smartphones, used in retrospective luminescence dosimetry). Even here not all properties of these materials can be discussed but, instead, individual phenomena are isolated and examined to see how what has been discussed in Part I can be used to good effect, even with these more complicated systems.

## 10.2 Feldspar

### 10.2.1 Structure and Defects

Feldspar is a naturally occurring aluminosilicate mineral consisting of potassium, sodium, and calcium cations surrounded by a negatively charged network of tetrahedra of silicon, aluminum, and oxygen atoms. They form a family of solid solution compounds with end-members $KAlSi_3O_8$ (orthoclase, Or), $NaAlSi_3O_8$ (albite, Ab), and $CaAl_2Si_2O_8$ (anorthite, An), as shown in the compositional phase diagram in Figure 10.1. Only limited solid solution occurs, and a large region of immiscibility is present in the phase diagram. Frequent impurities that substitute for the structural alkali ions include Li, Rb, Cs, Mg, Sr, Ba, Fe, Mn, Y, Ln, and other rare earths. Thus, there are many opportunities for traps and efficient luminescence emitters to be present within feldspar minerals.

The structure (Figure 10.2) is based on tetrahedra of silicon or aluminum ions surrounded by four oxygen ions forming long, twisted ("crankshaft") chains of tetrahedra throughout the crystal. Charge balance for the $(SiO_4)^{4-}$ or $(AlO_4)^{5-}$ tetrahedra is provided by alkali ions $Na^+$, $K^+$, or $Ca^{2+}$ in the open interstices, with the framework being sufficiently elastic to adjust to the different sizes of the cations. The structures of the potassium feldspars may be monoclinic (e.g. orthoclase, $KAlSi_3O_8$, and sanidine, $K,NaAlSi_3O_8$), or triclinic (e.g. microcline, also $KAlSi_3O_8$, and anorthoclase, $Na,KAlSi_3O_8$), depending on the details of the growth conditions (particularly temperature and cooling conditions). The growth conditions also dictate whether the

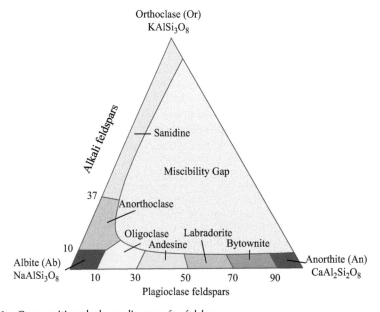

**Figure 10.1** Compositional phase diagram for feldspar.

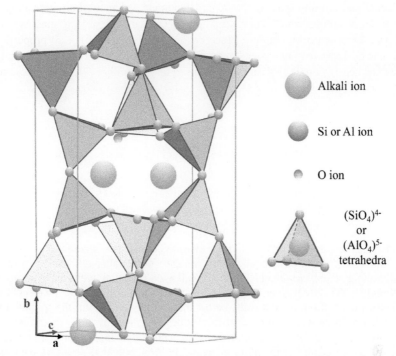

Alkali ion

Si or Al ion

O ion

$(SiO_4)^{4-}$
or
$(AlO_4)^{5-}$
tetrahedra

**Figure 10.2** Structure of feldspar based on $(SiO_4)^{4-}$ and $(AlO_4)^{5-}$ tetrahedra with alkali ions ($Na^+$, $K^+$, or $Ca^{2+}$) in interstitial positions.

distribution of silicon and aluminum is random (disordered) as in sanidine, or has a regular (ordered) distribution as in microcline, or is only partially ordered as in orthoclase. All plagioclase feldspars are more ordered than the alkali feldspars.

On the web page associated with this book, under Exercises and Notes, Chapter 10, Figure 10.2, Notes for Figure 10.2, you will find the link to a web page with an interactive crystal structure for feldspar allowing you to rotate the structure, examine the bond angles, add or remove the tetrahedra, etc., in order to give a clearer picture of the complex structure of this material.

### 10.2.2 Energy Levels and Density of States

The complex structure of feldspars, with mixtures of alkali species, lattice elasticity, different impurities of varying sizes, as well as ordered and disordered silicon and aluminum distributions, leads to strain-induced variations in bond lengths and bond angles, and microdensity fluctuations. This in turn leads to significant (several tenths of an electron-volt) fluctuations in the band gap width, in the manner previously discussed in Chapter 3 (Section 3.7.3.4). As a result, electrons can be localized in potential wells at the bottom of the conduction band edge in a manner resembling trapping with a localized wavefunction, or be localized at higher energies with a more delocalized wavefunction, as depicted in Figure 10.3. The conduction band edge is depicted as a continuously variable edge of local potential wells. At one such deep well, the wavefunction $\Psi_1(r)$ (where $r$ is distance) is confined around the potential well. At another location, the potential well is shallower and the wavefunction $\Psi_2(r)$ is more spread out and less localized. The states are derived from the conduction band, and overlap; electric field effects are strong (Redfield 1963). The density $N(E)$ of such band-tail states is described by the so-called Urbach tail:

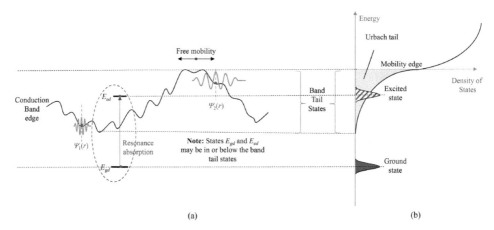

**Figure 10.3** Conceptual representation of random potential fluctuations, due to varying bond lengths and angles, impurity disorder, and microdensity fluctuations, leading to band-tail states in disordered materials. Electrons localized in deeper potential wells have more-localized wavefunctions (e.g. $\Psi_1(r)$), while electrons in shallower potential wells have more-extended wavefunctions (e.g. $\Psi_2(r)$). At high enough energies, above the mobility edge, the electrons are free to move and conduct. The potential wells below the mobility edge form the so-called band-tail states and give rise to the exponential density of states described by Urbach, and known as the Urbach tail. In feldspars, a defect level exists with an excited state ($E_{ed}$), which can be detected by OSL (IRSL) when stimulated with infra-red light. The OSL signal is induced with resonance absorption with a stimulation energy near ~1.4 eV, corresponding to a resonant transition from the defect ground state $E_{gd}$ to the excited state $E_{ed}$. In principle $E_{ed}$ (and even $E_{gd}$) can lie within the Urbach tail.

$$N(E) \propto \exp\left\{\frac{E}{E_U}\right\} \tag{10.1}$$

where $E_U$ is the characteristic width of the Urbach tail. The exponential Urbach density of states extends into the gap such that the band gap edge is no longer easily defined and free electrons have to be excited higher into the conduction band, above the mobility edge, before free mobility can occur. The mobility edge is the demarcation level between localized and delocalized states. It is the short-range, intrinsic disorder of the lattice that produces the mobility edge and the band-tails states rather than the coulombic fields due to point defects (Mott and Davis 2012).

Defect levels, deeper within the band gap are highly localized. In the depiction in Figure 10.3, the main defect giving rise to infra-red-stimulated luminescence (IRSL) is assumed to possess a ground state at energy level $E_{gd}$ and an excited state at $E_{ed}$. The latter is shown in the Figure to reside within the band-tail states, although in general this is not necessarily the case. Even the ground state $E_{gd}$ can reside within the band-tail states, depending on the value of $E_U$. A trapped electron in the ground state of the defect can be excited to the excited state by absorption of a photon, from where it can be thermally excited higher into the band, by absorption of thermal energy. Alternatively, with sufficient optical energy, the electron can be excited directly from the ground state higher into the band. Either process can lead to trap emptying. Once in the excited state, other processes are also available to the electron, as are detailed in the following sections.

### 10.2.3　Emission Spectra

Infra-red, red, green, violet-blue, and UV emission bands can be observed in most feldspars, depending upon the source of excitation and the irradiation conditions. Ultra-violet (3.0 eV to 4.0 eV) and violet-blue (2.4 eV to 3.4 eV) emissions are believed to be associated with deep recombination centers of, as yet, unidentified origin (Prasad et al. 2016). Green emission (2.0 eV to 2.4 eV) is often associated with $Mn^{2+}$ (Krbetschek et al. 1997, and references therein) or may be due to a larger defect cluster (Prasad et al. 2016). The red emission at 1.65 eV is established as being due to the spin-forbidden $^4T_1 \rightarrow {}^6A_1$ energy level transition in $Fe^{3+}$ (Poolton et al. 1996). All of the above emissions can be observed in IRSL/OSL and TL, while RPL emission following infra-red excitation is observed at 1.30 eV and 1.41 eV, from an as-yet-unidentified defect.

### 10.2.4　OSL Phenomena

#### 10.2.4.1　Band Diagram

OSL (and TL) phenomena in feldspars can best be described using a modified version of Figure 10.3, as shown in Figure 10.4. Here the same concepts are now represented on a traditional flat-band diagram where the band-tail states are now represented as non-continuous, flat lines, at different energy levels (in dark grey). The localized defect level (trap) is depicted as a flat line (in blue) with both its ground state $E_{gd}$ (full line) and its excited state $E_{ed}$ (dashed line) shown. Also shown is a recombination level (hole trap) in green. For simplicity, only one species of defect state and one species of recombination center are illustrated.

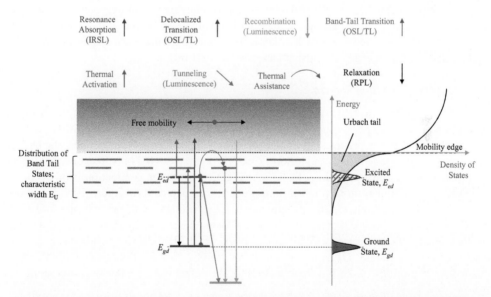

**Figure 10.4**　Translation of Figure 10.3 into a flat-band diagram. Here the band-tail states are represented by flat lines, with longer lines representing shallower, more-delocalized states, and shorter lines representing deeper, more-localized states. In reality the states are distributed, not discrete. Possible transitions are indicated with colored lines, explained in the text and in the figure legend. The density-of-states diagram is included for comparison and reference to Figure 10.3. (Note, not all possible transitions are included.)

A weakness of the flat-band diagram is that states are represented as flat lines suggesting discrete states whereas Equation 10.1 describes the band-tail states as an exponential distribution.[1] The conduction band edge is in fact the mobility edge and above this energy the electrons are free to move. Below this energy, within the energy region represented by the band-tail states, electron transport can occur via hopping from one band-tail state to another, indicated by the red curved arrow.[2] The higher the energy of the band-tail state, the more delocalized the state, represented by longer flat lines in this figure.

A number of possible electron transitions are indicated. In dark red is a resonant transition from the ground state to the excited state (at infra-red wavelengths). From the excited state recombination with the hole in the recombination center can occur via one of two mechanisms – either by thermally assisted activation via a band-tail state (curved red arrow) followed by recombination (shorter green arrow), or by tunneling from the excited state (orange arrow). In either case, luminescence is emitted and since both pathways were initially stimulated by absorption of infra-red light, the luminescence emitted is infra-red-stimulated luminescence, or IRSL. Note that relaxation from the excited state to the ground state is also possible (black downward arrow). Also note that if the temperature is high enough, thermal activation out of excited state is possible higher into the conduction band (red arrow) where the electron possesses enough energy to undergo free mobility and ultimate recombination (longer green arrow).

The purple arrow indicates that the trapped electron may also be excited directly from the ground state to the higher energies of the conduction band by either heating or stimulating with sufficiently energetic visible light. Once free to move through the crystal, recombination can then occur (longer green arrow) with the emission of luminescence. If the luminescence is stimulated by heat, the luminescence is TL; if stimulated by visible light the emission is OSL. Similarly, a thermal or optical transition directly into the band-tail states may be possible (blue arrow), also producing TL or OSL. An additional potential recombination route not illustrated in Figure 10.4 to prevent cluttering the diagram is tunneling from one of the band-tail states. The physics of this process would be identical to that of excited-state tunneling. With this background, OSL properties of feldspar can now be described.

### 10.2.4.2   OSL Excitation Spectra

The OSL properties of feldspars are dominated by emission stimulated by the resonant absorption of infra-red photons near 1.4 eV. From the first observations of this IRSL signal by Hütt et al. (1988) it was proposed that infra-red (IR) photons alone were not energetic enough to ionize the trap (unidentified) and absorption of IR photons served only to excite the trapped electrons to an excited state from which thermal activation to the conduction band occurred, followed by recombination to produce luminescence, as depicted in Figure 10.4. Excitation spectroscopy for many other feldspars indicates that the resonant peak varies from ~1.41 eV to ~1.47 eV (see, for example, Poolton et al. 2002a). Using the simple hydrogen atom as a model for the defect with an electron effective mass of $m_e^* = (0.79 \pm 0.02)\, m_e$ (where $m_e$ is the free electron mass) for $NaAlSi_3O_8$ with a relative permittivity $\varepsilon_r = 2.33$, Poolton et al. (2002a) calculated that the energy of the 1s – 2p transition for the defect is $1.48 \pm 0.04$ eV, very similar to

---

[1]  One view of Equation 10.1 is that it approximates an exponential array (a "ladder") of closely spaced, discrete levels rather than a continuous distribution.

[2]  Diffusion analysis by Morthekai et al. (2012) indicates a hopping range of only 1.1 nm to 1.8 nm. An alternative motion may be percolation rather than hopping/diffusion. A characteristic of percolation is that the decay of luminescence follows a stretched exponential (Jiang and Lin 1990). Pagonis et al. (2012) have demonstrated that Pulsed-IRSL decay can be described by a sum of exponential and stretched exponential terms, suggesting that motion via the band-tail states may be more complex that simple hopping.

the measured resonance energies for the IRSL emission. They further calculated the extent of the radial wavefunctions $\Psi_{1s}(r)$ and $\Psi_{2p}(r)$ for the ground 1s and the excited 2p states:

$$\Psi_{1s}(r) = C_1 \exp\left\{-\frac{r}{a}\right\}, \tag{10.2}$$

$$\Psi_{2p}(r) = C_2 \left(\frac{r}{a}\right) \exp\left\{-\frac{0.5r}{a}\right\}, \tag{10.3}$$

where $r$ is radial distance, $C_1$ and $C_2$ are constants, and $a$ is the effective Bohr radius. The calculation showed that the extents of the wavefunctions are such that tunneling could occur even from the ground state if the recombination centers were located within one unit-cell's distance of the trapping center, even without IR stimulation. (For Na-feldspar the unit cell dimensions $a$, $b$, and $c$ are 0.81 nm 1.28 nm and 0.72 nm, respectively; Figure 10.2). This is believed to be the source of anomalous fading of the IRSL signal from this material, even when the material is kept in the dark, post-irradiation (Visocekas 2002). With IR stimulation, tunneling from the excited state can extend even out to several unit-cell's distances, meaning that some of the observed IRSL emission must be due to excited-state tunneling. Tunneling is an athermal process and therefore it should not require thermal assistance. However, a thermal dependence of the IRSL emission is nevertheless observed, as illustrated in Figure 10.5 for

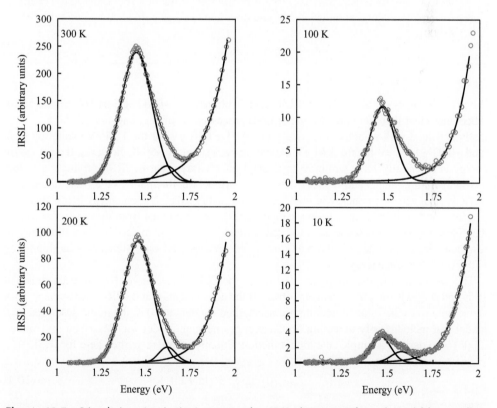

**Figure 10.5** Stimulation (excitation) spectra for IRSL from samples of Na-feldspar (albite) irradiated at 300 K but measured at the temperatures indicated. The experimental data have been fitted to Gaussian excitation bands superimposed on an exponentially rising continuum. (Original data kindly provided by N. Poolton; see also Poolton et al. (2009).)

sodium feldspar, NaAlSi$_3$O$_8$ (albite). Shown in this figure are excitation spectra for the production of OSL emission between 300 nm and 380 nm following irradiation with x-rays at room temperature. Each spectrum is measured at a different temperature, ranging from 300 K to 10 K, over the stimulation range 1.00 eV to 1.95 eV (from Poolton et al. 2009). The main features of the spectra are resonance peaks and a rising continuum at higher energies. In the case of albite, the main peak is at 1.44 eV, with a weaker feature at ~1.61 eV. (Poolton et al. 2009 also measured similar spectra for potassium feldspar KAlSi$_3$O$_8$ (orthoclase), and observed a main peak at 1.46 eV with overlapping smaller features at ~1.56 eV and ~1.26 eV.) It is noticeable that the emissions from both the rising continuum and the resonance peaks decrease in intensity as the temperature decreases. (In the case of orthoclase the resonance peaks disappear completely.)

---

**Exercise 10.1    Excitation Spectra (I)**

On the web site accompanying this book may be found data for the excitation spectra at 300 K and 10 K for IRSL from albite and orthoclase. (Data from Poolton et al. (2009) and kindly provided by N.R.J. Poolton.) The Excel file can be found under Exercises and Notes, Chapter 10, Exercise 10.1.
(a) Plot the four spectra and compare shapes.
(b) Fit the data assuming Gaussian-shaped bands for the resonance peaks, and an expression of the form of Equation 10.7 for the continuum. (Equation 10.7 is discussed later.)
(c) Compare the fitted parameters that you obtain with those published by Poolton et al. (2009).

---

*10.2.4.3    OSL Curve Description*

When the intensity of the resonance IRSL peak is plotted as $\ln(I_{IRSL})$ against $1/T$ (Figure 10.6), it becomes clear that there are two major OSL processes occurring: one, above ~50 K is thermally activated with an activation energy of ~0.04 ± 0.01 eV for the 1.44 eV excitation peak (and ~0.03 ± 0.01 eV for the 1.61 eV excitation peak). Below ~50 K, however, the activation energy is essentially zero, within experimental uncertainty. That is, IRSL production in this low temperature range is athermal. Likewise for orthoclase, the main 1.46 eV excitation peak has a thermal activation energy of ~0.11 ± 0.01 eV with zero activation energy below this temperature (and zero IRSL at very low temperatures). The inference from these observations is that there are two processes leading to IRSL emission from these materials when stimulated at resonance energies: one is a thermally activated process, and one is excited-state-tunneling with zero activation energy.

This can be seen further in Figure 10.7 where the IRSL decay curves for albite are shown at 10 K and at 300 K for stimulation at 1.46 eV. If the above thesis that the IRSL emission at 10 K is due only to excited state tunneling is correct, then the data at 10 K should be described by a model that includes only tunneling. In Chapter 5, one such model was described by Huntley (2006) who derived a simple analytical function, Equation 5.59, re-written here for IRSL:

$$I_{IRSL}(t) = I_{IRSL}(0)\exp\left\{-\rho'\left(\ln[P_1 t]\right)^3\right\}, \tag{10.4}$$

with $\rho'$ being a dimensionless number representing the density distribution of recombination centers (see Chapter 5) and $P_1$ (in s$^{-1}$) being the rate constant for tunneling. Figure 10.7a shows a fit to this simple equation, suggesting that indeed the data can be described by a

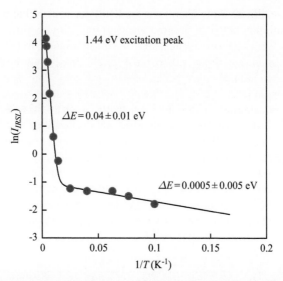

**Figure 10.6**   Temperature dependence of the IRSL for stimulation at 1.44 eV for Na-feldspar (albite). Fits to the data indicate a thermally activated process for the 1.44 eV excitation band with an activation energy of ~0.04 eV. Below ~50 K the IRSL process is essentially athermal. (Original data kindly provided by N. Poolton; see also Poolton et al. (2009).)

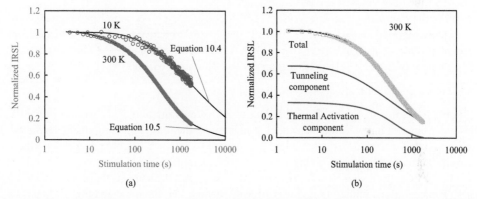

**Figure 10.7**   (a) IRSL decay curves at 10 K and 300 K, following irradiation at 300 K. The 10 K data can be fitted to Equation 10.4, indicating that tunneling only is responsible for the IRSL signal at this temperature. At 300 K the data must include a thermal component and are fitted to Equation 10.5, which includes a first-order thermal term plus a tunneling term. In (b) the 300 K fit is resolved into the thermal and tunneling components. (Original data kindly provided by N. Poolton; see also Poolton et al. (2009).)

straightforward tunneling expression. It should be said that this equation itself does not prove that the tunneling is from the excited state since it could apply to the ground state just as easily. However, without the 1.46 eV IR stimulation there is almost no OSL at this wavelength, indicating that the signal being monitored is from the excited state.

At 300 K, however, it is not possible to achieve a satisfactory fit to the IRSL decay using Equation 10.4 alone. At this higher temperature a thermal component is necessary, the evidence for which is clear from Figure 10.6. Therefore, the IRSL decay at 300 K is fit to an equation of the form:

$$I_{IRSL}(t) = C_1\exp\{-t/\tau\} + C_2\exp\{-\rho'\left(\ln[P_2t]\right)^3\}, \tag{10.5}$$

where the tunneling rate constant at 300 K ($P_2$) is not necessarily the same as $P_1$ at 10 K since there may now be additional thermal excitation into the excited state in addition to the optical excitation from the IR photons. The term $\rho'$ is the same in both cases, however. The thermal component is described by the constant $\tau$, which is dependent upon the thermal activation energy and the temperature. Note that this term assumes first-order kinetics for the thermal component. This is chosen in this case since it is assumed that the delocalized band is not involved in the recombination process; rather the thermal transition is assumed to take place only via the band-tail states and is therefore a localized transition only, resulting in first-order kinetics, as described in Chapter 3, Section 3.8.2.1.

---

**Exercise 10.2    IRSL Decay Curves**

On the web site accompanying this book may be found data for the IRSL decay curves for albite at 300 K and 10 K. (Data from Poolton et al. (2009) and kindly provided by N.R.J. Poolton.) The Excel file can be found under Exercises and Notes, Chapter 10, Exercise 10.2.

(a) Plot the decay curves and compare shapes.
(b) Fit the data using Equations 10.4 and 10.5. Compare your fitted parameters with those reported in this book, and with the parameters obtained by Poolton et al. (2009).
(c) Poolton et al. used a different expression for the decay curve at 300 K. Justify (or not) the use of Equation 10.5 rather than the equation used by Poolton et al.

---

Figure 10.7a illustrates the fit of the data at 300 K to Equation 10.5, while Figure 10.7b shows the same fit but with the thermal component and the tunneling component separated. It is observed that in this case, the tunneling component still dominates, but nevertheless, there is a strong thermal component, giving rise to the temperature dependence observed in Figure 10.6. In the fit to the data at 300 K, the value of $\rho'$ was kept the same as that found in the fit of the 10 K data, but the tunneling rate was allowed to vary. The fits indicated that $P_2 = 0.60$ s$^{-1}$, while $P_1 = 0.11$ s$^{-1}$.

The other main feature of the data in Figure 10.5 is the rising continuum as the excitation energy increases. As has been previously discussed in Chapters 3 and 5, it is normally expected that when scanning the excitation wavelength to determine those wavelengths required to produce OSL an excitation spectrum will be obtained that reflects the photoionization cross-section for the defect being emptied. The form of the photoionization cross-section in an ordered, crystalline structure is expected to show a sharp onset as a function of stimulation energy at an energy value corresponding to the optical trap depth for the defect, followed by a shape determined primarily by the density-of-states function in the delocalized band. Various shapes for these functions were described in Chapter 2 (Section 2.2.1.2) and illustrated in Figure 2.9. However, for a disordered material such as a glass, or (in this case) a natural mineral with a high degree of disorder to its crystal structure, there is a concentration of band-tail states described by the Urbach tail and the band edge is no longer well defined. As a result, as the excitation energy increases, initial excitation out of the trap is into the band-tail states and not into the higher energies of the conduction band and is described by a photoionization cross-section that is governed by the density of states for the band-tail states. Thus:

$$\sigma_p(E) \propto \exp\left\{\frac{E - E_o}{\Delta E}\right\}. \tag{10.6}$$

Here, $E$ is the excitation energy, $E_o$ is the optical trap depth and $\Delta E$ is the range of band-tail states accessible by the electrons. Note that this is not necessarily the full width of the Urbach tail since, in general, the ground state may not be below the band-tail states, but within it. As drawn in Figure 10.4, however, the defect ground state is below the lowest band-tail state, but this may not be true in general. Therefore, if the OSL transition is governed by stimulation into the band-tail states, and the photoionization cross-section is described by Equation 10.6, it follows that the OSL intensity will also follow an exponential dependence on the excitation energy:

$$I_{OSL}(E) = C\exp\left\{\frac{E - E_o}{\Delta E}\right\}, \tag{10.7}$$

where $C$ is a proportionality constant. Using Equation 10.7, fits to the continuum for the spectra in Figure 10.5, at 10 K, 100 K, and 200 K reveal $\Delta E = 0.10\,\text{eV}$, and 0.11 eV at 300 K. All the fits consistently give $E_o \sim 2.11$ eV $\pm 0.01$eV for the optical trap depth.

The data of Figures 10.5–10.7 were each obtained by irradiating at 300 K but measuring the OSL/IRSL at the temperatures indicated. However, if the irradiation is performed at 10 K, not only can the defect level be filled, but the band-tail states can also be filled. That is, each of the potential wells in Figure 10.3 can be populated with electrons. Clearly, this is only possible at very low temperatures but for feldspar, where the depth of the potential wells can be tenths of an electron-volt, 10 K is low enough to achieve this. This was demonstrated by Poolton et al. (2009) who irradiated the same albite samples as were used to obtain the data of Figures 10.5–10.7 with x-rays at 10 K and then scanned the excitation wavelengths, also at 10 K, over the same infra-red range (1.0 eV to 1.9 eV). Instead of a resonance peak superimposed on a rising continuum (Figure 10.5), only a rising OSL continuum was obtained, which followed a pure exponential as a function of stimulation energy. The signal originated from the emptying of the band-tail states themselves, and was orders of magnitude higher than that shown in Figure 10.5 at 10 K. Fitting the spectrum to Equation 10.8:

$$I_{OSL}(E) = C\exp\left\{\frac{E}{E_U}\right\}, \tag{10.8}$$

gives a value for $E_U$ of 0.32 eV, assuming the density of states in the band-tail states is given by the Urbach rule. Similar results were obtained for orthoclase, while Li and Li (2013) estimate the characteristic width of the band-tails states to be between 0.3 eV and 0.4 eV in K-feldspar from the thermal stability of the IRSL signal.

The maximum stimulation energy in Figure 10.5 is 1.95 eV with the energy range corresponding to the infra-red range. At higher energies and shorter stimulation wavelengths, OSL (rather than IRSL) can be observed as electrons become excited above the band-tail states and higher into the conduction band where free mobility becomes possible. Figure 10.8 shows an excitation spectrum for the same albite sample irradiated at 300 K and stimulated at 10 K but over a wider energy range, from 1.70 eV to 2.8 eV. A number of observations can be made. Firstly, fitting Equation 10.7 to the data from 1.7 eV to 2.25 eV yields $\Delta E = 0.17$eV and $E_o \approx 2.11$ eV. The $\Delta E$ value is larger than the 0.1 eV value found earlier from the data of Poolton et al. (2009) over the

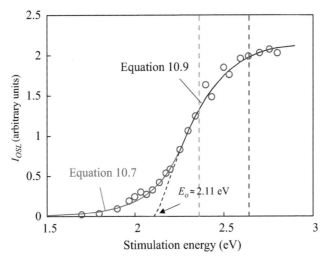

**Figure 10.8** Stimulation (excitation) spectra for OSL from samples of Na-feldspar (albite) irradiated at 300 K but measured at 10 K. The data are recorded over a wider energy range than in Figure 10.5 and show that as the stimulation energy increases the shape of the $I_{OSL}(E)$ function changes from one that follows the density of states in the Urbach tail (Equation 10.7), to one that follows the density of states in the conduction band (Equation 10.9). The dashed line is the extrapolation of Equation (10.9) showing the intersection with the energy axis at $E_o = 2.11$ eV. Infra-red stimulation will promote trapped electrons from the ground state of the main trap into the band-tail states, whereas green and blue stimulation (vertical dashed lines) will promote the electrons into the conduction band, above the mobility edge. (Original data kindly provided by N. Poolton; see also Kars et al. (2013).)

more-narrow energy range, while the value of $E_o = 2.11$ eV is the same and is to be compared with the thermal trap depth $E_t = 1.92$ eV estimated for the same center by Li and Li (2013).

Secondly, at higher stimulation energies, the data depart markedly from the form of Equation 10.7 and begin to flatten as the energy increases. The interpretation here is that the freed electrons have now reached the free mobility range of conduction band energies and the photoionization cross-section now follows the density of states in the conduction band. A fit of this portion of the data, from 2.25 eV to 2.8 eV, to Equation 2.17 from Chapter 2, re-written here as:

$$\sigma_p(E) \propto \frac{(E - E_o)^{3/2}}{E\left[E + E_o\left(m_o / m_e^* - 1\right)\right]^2}, \tag{10.9}$$

also gives a value of $E_o \approx 2.11$ eV (see Figure 10.8). This value is taken to be a good approximation to the optical trap depth considering that the band edge is ill-defined in this material. In general, estimations of $E_o$ in other feldspars range from ~2.0 eV to ~2.5 eV.

---

## Exercise 10.3   Excitation Spectra (II)

On the web site accompanying this book may be found data for the IRSL decay curves for albite at 10 K (data from Poolton et al. (2009) and kindly provided by N.R.J. Poolton) and for microcline at 7 K (data extracted digitally from Riedesel et al. 2019). The Excel file can be found under Exercises and Notes, Chapter 10, Exercise 10.3.

(a) Plot the data on suitable intensity axes and compare shapes.

(b) Fit the data using Equations 10.7 and 10.9. Over what ranges of energies do you apply each equation? Compare your fitted parameters with the parameters mentioned in this book, plus those published in Poolton et al. (2009) and Riedesel et al. (2019).

(c) What conclusions do you reach?

OSL signals may be observed when stimulated with green light (525 nm/2.36 eV) or blue light (470 nm/2.64 eV). Examination of Figure 10.8 shows that green light will excite electrons from the defect into the bottom edge of the conduction band states, whereas excitation by blue light will excite the electrons higher into the conduction band (vertical dashed lines in Figure 10.8). Referring to Figure 10.4, these transitions correspond to the purple arrow. Recombination can then occur directly from the conduction band (long green arrow). It may also be noted that green light and blue light may stimulate electrons out of deeper traps in addition to the main, IR-sensitive trap. If this is the case, especially with green light, some of the excited electrons may only reach the band-tail states and not be excited directly into the conduction band. In this case, green-light-stimulated OSL may be characterized by a mixture of localized and delocalized transitions.

Excitation into the conduction band, and therefore justification for the use of Equation 10.9 (or similar equations; Chapter 2) at higher energies, is shown by the observation of photoconductivity when irradiated feldspar is stimulated with sufficiently energetic light. Short (2005) exposed irradiated orthoclase to green light (515 nm/2.41 eV) and observed photoconductivity, but there was no such signal when the sample was stimulated with IR light (850 nm/1.45 eV). One can assume that if the sample was stimulated with even more energetic light (e.g. blue at 470 nm/2.64 eV) photoconductivity would also be observed. Interestingly, pre-heating removes the green-light-stimulated photoconductivity and Short (2005) suggested that the preheat emptied shallow traps from where excitation high into the conduction band occurred, but left deep traps from where excitation only into the band-tail states occurred. One can speculate that blue-light-stimulated photoconductivity would be less affected by the preheat.

Figure 10.9 illustrates a CW-OSL decay curve following blue stimulation from potassium feldspar grains (separated from an aeolian sand; Gong et al. 2012) following irradiation at room temperature. The OSL is detected in the 290 nm to 370 nm range. The decay curve is fitted with three exponential curves, assuming first-order kinetics for each component. The three components are called the "fast," "medium," and "slow" by the authors. By comparing how the blue-stimulated OSL changes if the sample is stimulated with infra-red light before stimulation with blue light (in an experiment called "post-IR blue OSL," or pIR-BLSL), and conversely how the IRSL signal changes after first stimulating with blue light (i.e. "post-blue IRSL," or pBL-IRSL) the authors conclude that the fast and medium components are associated with the same trap that gives rise to the IRSL signal, and that the slow component comes from a deeper, separate trap. How there are two components (fast and medium) associated with one trap is not explained, but recall that the identification of two components rests on the assumption of simple first-order kinetics. It was discussed in Chapter 3 how CW-OSL curves can have several complexities of shape due to interactive kinetics yet can still be fitted with a summation of exponentials, leading to misleading conclusions. Simple analyses should always be treated carefully.

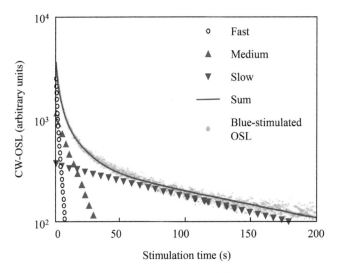

**Figure 10.9**   CW-OSL decay curve at room temperature following room temperature irradiation and a 280 °C/10s pre-heat. The curve is fitted to three first-order exponentials, termed "fast," "medium," and "slow". (Reproduced from Gong et al. (2012) with permission from Elsevier.)

### 10.2.5   TL Phenomena

#### 10.2.5.1   Glow-Curve Description

The possible recombination pathways discussed thus far for the various OSL phenomena – ground-state tunneling, excited-state tunneling, band-tail-state tunneling, band-tail-state hopping, conduction-band transport – have profound consequences on the possible shape and subsequent analysis of TL glow curves from feldspars. Since the interest in this text is on dosimetry properties, the discussion is limited to TL glow curves above room temperature. Example glow curves, from Sfampa et al. (2015), are shown in Figure 10.10 for potassium feldspar (microcline). The general form of the glow curve is a peak around 100 °C, which is particularly prominent in this case, followed by a less-structured, broad set of peaks from ~100 °C to beyond 300 °C. TL glow curves from other feldspar samples show similar structures, with the 100 °C peak less prominent in some, and the broad peak less structured in some. The overall features are similar from feldspar to feldspar, however.

Figure 10.10a illustrates a set of TL curves recorded at different times after stimulating a sample with blue light (470 nm) at room temperature, after irradiation but before TL measurement. The experiment is repeated multiple times on a freshly irradiated sample each time and for different blue-light stimulation times from 0 s to 2000 s. Figure 10.10b shows a similar set of data, but this time for stimulation with infra-red light (875 nm). From the analysis and discussion above on OSL, it is known that IR light stimulates trapped electrons to the excited state of the main IRSL defect, from where both thermal excitation via the band-tail states and excited-state tunneling lead to recombination and IRSL emission, while blue-light stimulation excites trapped electrons into the conduction band from where recombination leads to OSL emission (frequently termed BLSL). Thus, the decay of the TL signals in Figure 10.10 occurs through a variety of mechanisms, depending upon the light source. Independent of whether the stimulation is via blue light or infra-red light, there will also be a strong thermal component, especially to the decay of the 100 °C TL peak due to the instability of the trapped electrons giving rise to it during the stimulation period at room temperature.

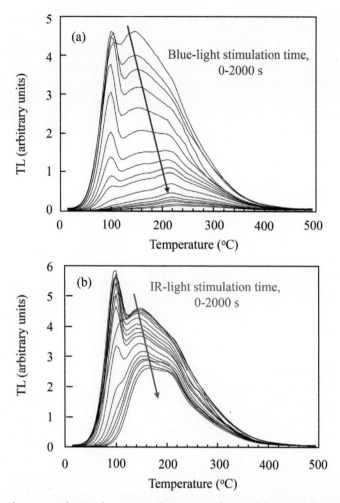

**Figure 10.10** Changes in the TL glow curves from microcline (K-feldspar) following irradiation at room temperature and stimulation with (a) blue light (470 ± 20 nm), or (b) infra-red light (875 ± 40 nm), for different stimulation times (over the range from 0 s to 2000 s) before recording the remaining TL (heating at 2 °C.s$^{-1}$). (Reproduced from Sfampa et al. (2015) with permission from Elsevier.)

For blue-light stimulation, the dominant process for loss of trapped electrons, and therefore loss of TL, is by stimulation into the conduction band. Since all parts of the glow curve decay during blue-light stimulation, several traps, from shallow to deep, must be emptied by the energetic blue light and the traps empty primarily via delocalized transitions. For IR stimulation, however, the light is not energetic enough to ionize the traps by direct stimulation to the conduction band and trap emptying occurs via localized transitions, through thermally assisted band-tail hopping and/or excited-state tunneling. (Tunneling from the band-tail states may also occur.) It is not surprising, therefore, that both the extent of the glow curve decay (which in itself could simply be a function of different stimulation power) and, more importantly, the shape of the glow curve differs depending upon the stimulation wavelength.

---

**Exercise 10.4    TL Lost versus OSL or IRSL**

Figures 10.10a and 10.10b show the decay of TL as a function of stimulation time with either blue light (470 nm) or IR light (875 nm). During stimulation BLSL or IRSL is emitted. The data can be replotted as TL lost as a function of stimulation time, as well as cumulative BLSL or cumulative IRSL as functions of stimulation time. On the web site accompanying this book may be found such data for the TL glow curves shown in Figure 10.10 (and their accompanying BLSL and IRSL decay curves). Note that the TL lost data are for blue-light stimulation only. (The data are digitized from Sfampa et al. (2015), wherein full experimental details can be found.) The Excel file can be found under Exercises and Notes, Chapter 10, Exercise 10.4.

(a) Plot the data on appropriate intensity axes and compare shapes. Is TL lost proportional to BLSL gained? If not, why not?

(b) Sfampa et al. analyze these data using the assumption of tunneling from the main defect (i.e. the defect responsible for the IRSL resonance signal). Is it appropriate to analyze all three data sets using this one model? If yes, why so? If not, why not?

(c) If it is not appropriate, what alternative models/processes should be considered?

---

### 10.2.5.2    TL Analysis

As shown in Figures 10.7 and 10.9, the shape and analysis of the CW-OSL decay curves for IRSL and blue-stimulated OSL are quite different, with blue-stimulated OSL being described by a simple superposition of exponentials while the IRSL decay curve shape is a combination of thermal and tunneling terms. The TL mechanism must also be a complex mixture of localized (tunneling/band-tail hopping) and delocalized (thermal excitation to the conduction band) processes. Multiple traps must participate in the TL process. Depending upon the trap/recombination-center identities and distributions, and the glow curve temperature, either localized recombination or delocalized recombination will dominate the TL signal. Furthermore, given the complex nature of the density of states and the lack of a clear conduction band edge (Figure 10.3), a distribution of thermal activation energies for the release of the trapped charge is possible, not to say probable.

Several attempts have been made to evaluate the thermal activation energies ($E_t$) in various feldspar samples. (See, for example, Pagonis et al. (2014); Pagonis (2019) for references to the early literature.) The $E_t$ values obtained range generally from ~1.6 eV to ~2.0 eV. As discussed previously (Section 5.2.1), a useful method for detecting distributions of activation energies is to apply the $T_m$-$T_{stop}$ method, coupled with application of the Initial Rise Method (IRM) to estimate $E_t$. An example is shown in Figures 10.11 and 10.12 for a predominantly andesine sample. For clarity, Figure 10.11a shows the resulting TL curves for $T_{stop}$ values ranging from 50 °C to 150 °C, while Figure 10.11b shows the glow curves for $T_{stop}$ from 160 °C to 400 °C, in 10-degree intervals. The results of IRM analyses on each glow curve are shown in Figure 10.12. The $E_t$-$T_{stop}$ data indicate the presence of a shallow trap (in this particular sample) with a trap depth of 0.38 ± 0.01 eV for $T_{stop} <$ ~100 °C. As $T_{stop}$ increases beyond 100 °C $E_t$ increases continuously up to ~260 °C, above which $T_{stop}$ value $E_t$ stabilizes at 1.48 ± 0.07 eV. Similar measurements on other feldspars reveal that the above general pattern is common to all, with some differences in the $E_t$ values obtained.

To interpret Figure 10.12 it is useful to divide the data into three regions:

$T_{stop} < 100\,°C$: The simplest interpretation of the fixed $E_t$ value below 100 °C is to propose a shallow, discrete trap at ~0.38 eV corresponding to the low-temperature glow peak below 100 °C

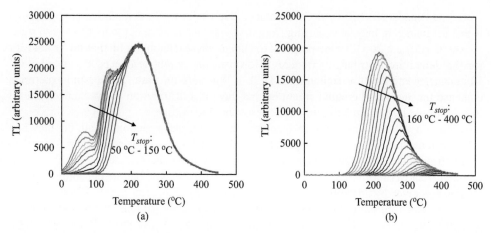

**Figure 10.11** TL from an andesine sample irradiated at room temperature (21.3 Gy) and heated at 2 °C.s⁻¹ with the emission detected at 395 nm ± 50 nm for multiple values of $T_{stop}$. $T_{stop}$ is varied from (a) 50 °C to 150 °C, and (b) 160 °C to 400 °C. (Original data kindly provided by V. Pagonis; see also Pagonis et al. (2014).)

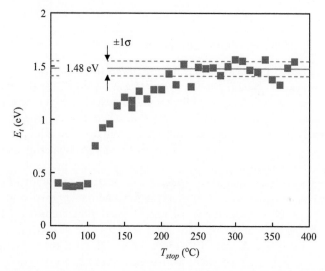

**Figure 10.12** Variation of $E_t$ with $T_{stop}$ for the series of glow curves shown in Figure 10.11. The mean value for $E_t$ of $T_{stop}$ > 260 °C is 1.48 eV ± 0.07 eV, as indicated in the figure. (Adapted from Pagonis et al. (2014).)

in Figure 10.11. This proposal assumes that this part of the glow curve is produced purely through thermal stimulation.

$T_{stop}$ > 260 °C: To explain the constant $E_t$ value for $T_{stop}$ > ~260 °C, extending over a broad temperature range up to the end of the glow curve around 400 °C, the excited-state tunneling model described in Section 3.8.1.3 can be applied. Here it is considered that: (i) electrons are thermally stimulated from the ground state to the excited state; (ii) the traps and recombination sites are randomly distributed; and (iii) tunneling from the excited state of the traps to the nearest-neighbor recombination centers occurs. It has been demonstrated that this model is able to reasonably describe the experimental TL glow curve shape after pre-heating to $T_{stop}$ > ~260 °C

(Kitis and Pagonis 2013). The model assumes a fixed value for $E_t$. A modification to this model is that the tunneling may be occurring from the band-tail states rather than the excited state.

*100 °C < $T_{stop}$ < 260 °C:* One proposal to explain these data might be that the ground states for the defect are distributed in energy. However, one might also expect a distribution in IR-resonance energies for excitation of IRSL if this were the case. There is no experimental evidence for this. An alternative might be that there is a distribution in the excited states, but this too would lead to a distribution in IR-resonance energies.

A third consideration may not be related to trap energy distributions at all, but may be the result of tunneling. (The author is grateful to M. Jain for this suggestion.) Consider that TL results from thermally assisted tunneling from the excited state of a defect, with a tunneling constant $\alpha$ and a tunneling distance $r$. In Chapter 3 it was shown that the tunneling probability (in s$^{-1}$), was given by an expression $P(r) = P_0 \exp\{-\alpha r\}$ (Equation 3.102). If $n$ is the number of trapped electrons ready to undergo tunneling, it can be shown that the recombination rate, proportional to the TL emission intensity, is given by the first-order expression: $I_{TL}(T) \propto dn/dt \approx nP_0\exp\{-(E/kT) - \alpha r\}$, where $E$ is the activation energy from the ground state to the excited state. (See Jain et al. (2015) and Exercise 10.5.) This means that the TL is governed by an effective activation energy $E_{eff} = E + kT\alpha r$. As $T_{stop}$ increases, $r$ increases and so too does the measured activation energy $E_{eff}$. Similar discussions and analyses are presented by Polymeris et al. (2017).

---

**Exercise 10.5  Excited-State Tunneling**

Consider a tunneling model of a trap with a ground state and an excited state, and a nearby recombination center separated by a distance $r$. Consider a total concentration of trapped electrons $n$ of which $n_g$ are in the ground state and $n_e$ are in the excited state, and a total concentration of recombination centers $m$, where $n = m$. Ground-state electrons may be thermally excited into the excited state, and may relax with probability $B$ (s$^{-1}$). While in the excited state they may also tunnel to the recombination center with probability $P(r) = P_0\exp\{-\alpha r\}$.

(a) Write a series of rate equations to describe the thermal excitation of electrons from the ground state to the excited state, relaxation, and recombination via tunneling.

(b) Assume $r$ is such that the tunneling rate is small compared to the relaxation rate $B$, and that quasi-equilibrium holds such that d$n_e$/d$t$ = 0. Assume $E$ and $s$ are the activation energy and frequency factor, respectively, for thermal excitation from the ground state to the excited state. Derive an expression for the rate of recombination due to tunneling (proportional to the TL emission) and show that it is given by: $I_{TL}(T) \propto dn/dt \approx nP_0\exp\{-(E/kT) - \alpha r\}$.

(c) Justify why the previous expression for $I_{TL}(T)$ does not contain either $B$ or $s$.

---

A fourth consideration might be that the data reflect not a distribution of the excited states or ground states, but a distribution of the band-tail states and that the primary tunneling mechanism occurs from the band-tail states. With this consideration such low values for $E_t$ at the beginning of this $T_{stop}$ range are difficult to understand if taken at face value. It is at this point that the discussion of the limitations of the IRM analysis that were discussed in Chapter 5 (Section 5.2.1.2) must be recalled, especially the difficulties encountered with strongly overlapping TL signals. Experimental criteria in this type of analysis include: (a) the resolution of $T_{stop}$ values (no more than 1 to 5 degrees is recommended; the data of Figure 10.12 were obtained using 10 degrees); (b) correct selection of the $T_{min}$ value used to determine the slope of the initial rise plot, bearing in mind the signal-to-noise ratio, with too low a value tending to bias the results toward low $E_t$ values; (c) selection of the $T_c$ values (less than ~5% of the $T_m$ value). Unfortunately there is no information on the $T_{min}$ and $T_c$ values used in the analysis of $E_t$. (See Chapter 5 for the definitions of $T_{min}$ and $T_c$.) The possibility exists, therefore, that the $E_t$ values shown in Figure 10.12 in the mid-$T_{stop}$ range are to some extent artefacts of the experiment, as discussed in Chapter 5, and not to be taken literally. This is not to say that a distribution in $E_t$ does not exist, only that the data in this $T_{stop}$ range must be interpreted with care.

---

**Exercise 10.6   IRM analysis of TL**

Figure 10.11 shows a set of TL glow curves for andesine feldspar following different values of $T_{stop}$, from 50 °C to 400 °C. On the web site accompanying this book may be found an Excel file containing the original data used for this figure (Sheet #1). Also in the same Excel file (Sheet #2) may be found the original data for several additional TL curves for the same material. These latter data were obtained after irradiation and subsequent stimulation with IR light (870 nm ± 40 nm) for 100 s at 50 °C, before heating to the $T_{stop}$ values indicated and recording the full glow curve. (All data are from Pagonis et al. (2014), and have been kindly provided by V. Pagonis.) The Excel file can be found under Exercises and Notes, Chapter 10, Exercise 10.6.

(a) From the data in the Excel files plot all the TL glow curves; compare the glow curves for $T_{stop}$ = 50 °C with and without IR stimulation. What are the immediate inferences that you can make regarding the effect of the IR stimulation on the glow curve?

(b) Apply the Initial Rise Method of analysis to the leading edge of all the glow curves in the file, both with and without the IR stimulation. In doing so apply the principles described in Chapter 5 when applying this method, particularly concerning the determination of the values of $T_{min}$ and $T_c$ (see Section 5.2.1.2). Evaluate $E_t$ from the leading edge (the initial rise) of each glow curve.

(c) Plot $E_t$ versus $T_{stop}$ for the data with and without IR stimulation. What do you conclude?

(d) Compare your results with those of Pagonis et al. (2014), Figure 3. Discuss differences (if any).

---

## 10.2.6   RPL Phenomena

### 10.2.6.1   RPL Emission and Excitation Spectra

In the excitation diagram of Figure 10.4, a relaxation transition from the excited state to the ground state (black arrow) is indicated following the resonant excitation from the ground state to the excited state (dark red arrow). This relaxation transition is radiative (Prasad et al. 2017; Kook et al. 2018; Kumar et al. 2018). If excited at 1.40 eV (855 nm), an emission peaking near 1.30 eV (955 nm) is observed in irradiated in K-feldspar. Similarly, if excited at 1.49 eV (830 nm) a second emission peaking at ~1.41 eV (880 nm) is observed. Both emissions originate from the same principle defect and therefore appear to represent either different excited states of the same center or slight perturbations of the same excited state but in two different environments (Kumar et al. 2018, 2020). Since these emissions are not observed in unirradiated material they represent RPL.[3]

---

[3] In the above-referenced publications RPL from feldspar has been given the name "infrared photoluminescence," or IRPL. In order to be consistent with all other discussions within this book, however, the conventional name of RPL is used here.

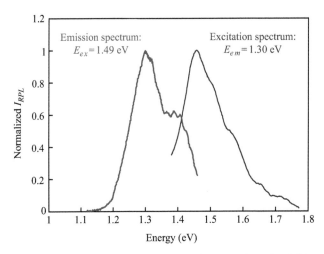

**Figure 10.13**    Excitation and emission spectra for RPL from irradiated K-feldspar. (Adapted from Prasad et al. (2017) and Kumar et al. (2020).)

Figure 10.13 shows example excitation and emission spectra for the RPL emission from K-feldspar. The excitation spectrum shown in this figure ends near 1.8 eV but further excitation peaks appear at higher energies. The exact position of the excitation peaks (and the emission peaks) depends on the feldspar type. The highest-energy excitation bands (~3.2 eV to 3.5 eV), and the lowest-energy excitation peak at 1.49 eV are believed to correspond to excited states above and below the conduction band edge (mobility edge), respectively. The central excitation band (~2.0 eV to 2.5 eV) is thought to arise from excitation from $E_o$ to the conduction band edge (Kumar et al. 2020). The RPL lifetime of both the 1.30 eV emission and the 1.41 eV emission have been measured at ~20 μs.

### 10.2.6.2  RPL Temperature Dependence

When continually stimulated at room temperature with photons of energy either 1.40 eV or 1.49 eV, the RPL intensity initially decays and then stabilizes after sufficient time, depending upon the stimulation power (Figure 10.14). During the decay of the RPL signal, IRSL is emitted and ~50% of the RPL signal is lost. That is, ~50% of the trapped charge is excited out of the trap in the processes described above for IRSL emission, while ~50% remains unaffected. The remaining 50% is interpreted as being due to recombination centers that are located at distances beyond a reasonable tunneling distance during the course of the experiment. Poolton et al. (2002a) calculated that excited-state tunneling can extend out to several unit-cell distances and in fact estimated that the probability of tunneling from the trap's excited state to the recombination center decreases with distance $r$ between the trap and the center, and drops from its maximum (around $r \approx 0.6$ nm) to about 0.03% of its maximum value at $r \approx 2.8$ nm. Considering the dimensions of the unit cell, tunneling is therefore unlikely at distances much longer than this.

As the temperature increases, however, hopping/diffusion of the electrons among the band-tail states may occur leading to additional recombination and further reduction in the RPL signal. An observation by Prasad et al. (2017) is that there is little-to-no loss of RPL signal at 7 K, or at temperatures even as high as 77 K. Tunneling is an athermal process, so why is there no loss of RPL

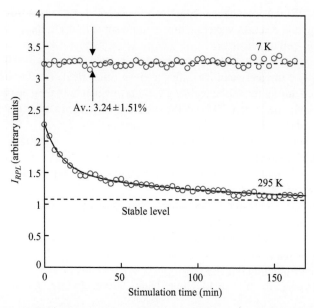

**Figure 10.14**  Decay of the RPL signal from K-feldspar (emission integrated between 1.10 eV and 1.33 eV) at 7 K and 295 K excited by 1.40 eV light (1.5 mW.cm$^{-2}$). The data at 295 K are fit to two exponential curves plus a constant representing a stable background at long stimulation times. (Adapted from Prasad et al. (2017).)

at low temperatures, and how can IRSL still be observed at temperatures as low as 10 K (Figures 10.5–10.8)? Examination of Figure 10.5 shows that the IRSL signal at 300 K is approximately two orders of magnitude larger than the same signal at 10 K (from the areas under the IRSL peaks at 1.47 eV). Figure 10.14 indicates that the RPL signal at 295 K decays by ~50% before stabilizing. Therefore, one might expect that the RPL loss would be no more than (50/100)% = 0.5% at 10 K or less. This is less than the noise for the measurement in Figure 10.14 at 7 K and therefore such a small loss of RPL would not be seen experimentally at these temperatures.[4]

### 10.2.7  What Can Be Concluded?

As a summary of the potential electron-hole transport and recombination processes in irradiated feldspars the following outline statements can be made:

- Infra-red absorption induces a resonant transition from a defect ground state to an excited state, at energies around or a little above 1.4 eV. The ground state is at an optical trap depth ~2.0 eV to ~2.5 eV below the mobility edge. The thermal trap depth for the same defect level is ~1.92 eV.
- Without excitation (optical or thermal), electron tunneling from the ground state to the recombination center can occur over approximately one unit-cell's dimension, inducing loss (fading) of the signal even if stored in the dark and at low temperature. This may give rise to the well-known "anomalous fading" of TL and OSL and "after-glow" from this material.

---

[4]Note, Figure 10.5 is for Na-feldspar at 300 K and 10 K while Figure 10.14 is for K-feldspar at 295 K and 7 K; therefore direct comparisons are difficult, but approximate comparisons remain valid.

- Once excited by IR light, localized recombination occurs via tunneling from the excited state (at which energy the electron wavefunction is more spread-out) to the recombination center over several unit-cell dimensions.
- At room temperature, excited-state tunneling dominates between near-neighbor, trap-center (donor–acceptor) pairs and stabilizes after a given time. IRSL (in the visible wavelength range) is emitted during tunneling and decays to zero when tunneling ceases.
- During this stimulation period, RPL (in the infra-red wavelength range) is also emitted, stabilizing to a constant value as the IRSL signal decays to zero.
- At higher temperatures, thermally induced diffusion/hopping (or perhaps percolation) may occur from the excited state through the band-tail states so that additional recombination between traps and centers at farther distances becomes important. Further decay of the RPL signal occurs; more IRSL occurs.
- Recombination at these greater distances may be by direct recombination from the band-tail states, or by tunneling from the band-tail states; this issue is not yet clear.
- More than one IR-sensitive defect exists in most feldspars, each one giving rise to similar processes to those described above.
- More-energetic stimulation (e.g. by blue light) can excite the ground-state electrons directly to higher energies in the conduction band allowing free mobility and ultimate recombination.
- A loss of energy of these energetic electrons may (or may not) also occur ("thermalization") reducing the free electron energies to the band-tail energies.
- Thermal excitation (e.g. during TL) at higher temperatures may also excite electrons into higher energies in the conduction band.
- There may also be deeper, localized traps also contributing to OSL and TL.

Given these multiple possibilities, several of the models for OSL and TL in particular that are discussed in the literature are often somewhat mixed. Several are limited in scope in that they tend to emphasize, say, localized recombination but do not include delocalized recombination, or vice-versa. Additionally, the available discussions are often constructed around the assumption of only one primary defect, i.e. that which also gives rise to the IRSL signal. Excitation spectra for IRSL clearly indicate more than one excitation resonance, which may in turn suggest more than one such defect, while blue-stimulated OSL indicates one or more traps that become fully ionized (delocalized transitions) when stimulated with sufficiently energetic photons. Together, such data lead to a suggestion that TL may also result from several defects, and attempts to analyze TL glow curves assuming just one main trap with one dominant mechanism should be considered as no more than a first approximation. The words of Mayank Jain are appropriate here: "A model is a tool to test our hypothesis, and not a proof of reality."

Despite its long history, the topic of OSL and TL (especially) from feldspars remains ripe for further examination.

## 10.3   Aluminosilicate Glass

Aluminosilicate glasses have long been industrially manufactured and studied because of their use in the ceramics and optics industries and, more recently, in the mobile electronic device industry. Low-alkali content aluminosilicate glass is used as display glass in personal electronic devices while higher-alkali content, ion-exchanged material is used as protective glass. For example, Gorilla Glass® is a type of aluminosilicate glass manufactured by Corning Inc. (USA) and produced for its surface strength and scratch resistance. It is chemically strengthened using an ion exchange process in which the glass is submerged in molten potassium salt at 400 °C, during which process sodium ions are replaced by potassium ions. Several generations of Gorilla Glass (GG) are available on different mobile phones and other mobile electronic devices.[5] A variety of mobile device manufacturers use GG; other device manufacturers use similar aluminosilicate glasses. The material is included in this book because of its use as a potential TL and/or OSL radiation dosimetry material for retrospective, emergency dosimetry applications in which the glass from personal electronic devices is proposed as a surrogate for personal radiation dosimeters such as TLDs or OSLDs for the general public in the event of a widespread radiological accident or incident (ICRU 2019; McKeever and Sholom 2019; McKeever et al. 2020b).

### 10.3.1   Structure and Composition

Aluminosilicate glasses contain 57–60% silicon dioxide ($SiO_2$) and 16–20% aluminum dioxide ($Al_2O_3$), along with small amounts of 5–7% lime (CaO), 6–12% magnesium oxide (MgO), boron trioxide ($B_2O_3$), and other cations.[6] An example composition distribution is illustrated in Figure 10.15. $SiO_2$ is the main glass-forming, network-forming component while the alkali and alkaline earth elements are network modifiers, interrupting network polymerization. $Al_2O_3$ acts as an intermediate component and can act as both a network-forming and a network-modifying element. The Al ions can be 4-fold or 5-fold coordinated depending upon the Al:alkali ratio. In 4-fold coordination, charge balance for $(AlO_4)^{5-}$ tetrahedra is provided by $Na^+$, $K^+$, or $Ca^{2+}$, but may also be provided by the formation of so-called oxygen triclusters where oxygen bridges three tetrahedra instead of two. The triclusters are assumed to be bonded to two Si and one Al in order to avoid Al-O-Al bonds (the "Al avoidance" principle). A possible aluminosilicate structure, without alkali or other cations shown, is illustrated in Figure 10.16.

As with all glasses, aluminosilicate glasses lack long-range order and are characterized by variations in bond angles, bond lengths, local density fluctuations, and disorder. As a result, Figures 2.5b and 10.3 are representative of the energy band diagrams and densities of states for these materials and much that has been learned from the description of feldspar OSL and TL can be applied to an understanding of the OSL and TL properties of these materials also. The

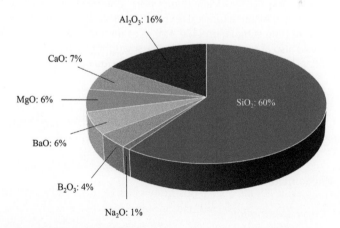

**Figure 10.15**   Example composition distribution for aluminosilicate glass.

---

[5] https://www.corning.com/gorillaglass/worldwide/en.html (accessed August, 2021).
[6] https://matmatch.com/learn/material/aluminosilicate-glass (accessed June, 2021).

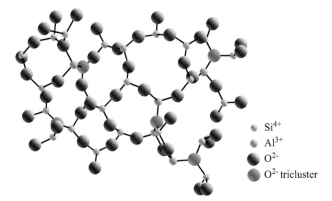

**Figure 10.16** Example possible aluminosilicate glass structure (without alkali or alkaline ions) showing variable bond angles and lengths, with oxygen triclusters in which oxygens are bound to two Si ions and an Al ion. (Adapted from Ando et al. (2018).)

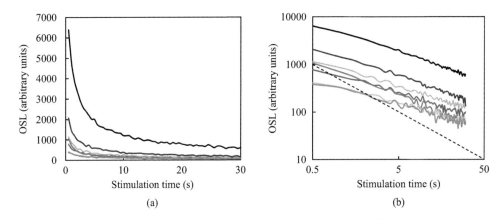

**Figure 10.17** A collection of OSL decay curves from aluminosilicate glasses from a variety of smartphones from different manufacturers shown on (a) linear–linear, and (b) log–log scales. Each CW-OSL curve is from blue stimulation (470 nm) at room temperature 30 s after beta ($^{90}$Sr:$^{90}$Y) irradiation, which was also at room temperature. The smartphones include iPhones X, XR, XS, and Galaxy S10 and Note 10. (Original data kindly provided by S. Sholom.)

published literature, however, is much sparser and relevant data have been obtained almost exclusively from glasses used in mobile phone displays and protective glass.

### 10.3.2   OSL Phenomena

#### *10.3.2.1   OSL Curve Description*

Figure 10.17 represents a typical set of CW-OSL curves obtained from a variety of aluminosilicate glasses (GG protective glass) for a collection of smartphones from different manufacturers. All follow stimulation with blue light at room temperature after irradiation and the curves are shown on a linear–linear plot (Figure 10.17a) and on a log–log plot (Figure 10.17b). The latter plot shows an almost-linear shape indicating an approximately power-law decay of the form $t^{-k}$.

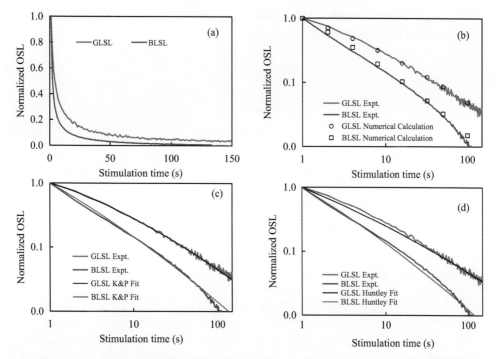

**Figure 10.18** (a) Normalized BLSL and GLSL CW-OSL curves from an aluminosilicate glass (GG) from an iPhone XR. (b) Comparison of BLSL and GLSL data with numerical calculations using $n_0 = \rho_0 = 10^{22}$ m$^{-3}$, $P_o = 10^{12}$ s$^{-1}$, $\alpha_e = 3.52 \times 10^8$ m$^{-1}$, $B = 4 \times 10^7$ s$^{-1}$, $\sigma_p \Phi_{green} = 9.56$ s$^{-1}$, $\sigma_p \Phi_{blue} = 152.0$ s$^{-1}$. (c) Fits of the BLSL and GLSL data ("K&P Fit") to the analytical expression of Kitis and Pagonis (2013), with the same $n_0$ and $\sigma_p \Phi$ values as for Figure 10.18b, plus $\rho'_{green} = 9.58 \times 10^{-4}$ and $\rho'_{blue} = 9.61 \times 10^{-4}$. (d) Fits of the BLSL and GLSL data ("Huntley Fit") to the analytical expression of Huntley (2006), with $\rho'_{green} = 3.35 \times 10^{-4}$ and $\rho'_{blue} = 4.06 \times 10^{-4}$, and $P_{o(green)} = 1.4 \times 10^{11}$ s$^{-1}$ and $P_{o(blue)} = 1.4 \times 10^{10}$ s$^{-1}$. (Original data kindly provided by S. Sholom.)

In these particular cases $k$ is approximately 0.50 to 0.62, depending on the sample. A dashed line with a value of $k = 1.0$ is shown for comparison.

In Chapter 3 it was shown how a time dependence of the form $t^{-k}$ may be indicative of tunneling, in this case optically assisted tunneling from an excited state. As an indication that the CW-OSL from such materials can be described by tunneling, Figure 10.18a illustrates blue-light-stimulated CW-OSL curves (BLSL) and green-light-stimulated CW-OSL curves (GLSL) from an aluminosilicate GG sample. In Figure 10.18b, the same experimental data are shown on a log–log plot and compared to a numerical calculation of OSL due to optically assisted tunneling. To do so, an equation of the form of Equation 3.119 was used, except that the thermally assisted term $s \exp\{-\Delta E / kT\}$ was replaced by an optically assisted term $\sigma_p \Phi$, thus:

$$I_{CW-OSL}(t) = \int_0^\infty \frac{\delta}{\delta t} \left[ 4\pi n_0^2 r^2 \exp\left\{ -\frac{4}{3}\pi n_0 r^3 \right\} \exp\left\{ -\left(\frac{\sigma_p \Phi}{B}\right) P_0 \exp\{-\alpha_e r\} t \right\} \right] dr. \quad (10.10)$$

All terms are defined in Chapter 3. In the calculation the integral over all values of $r$ at different times $t$ was determined numerically and the values of the various parameters used in the calculation are given in the figure caption.

In Section 5.3.9 it was noted that Kitis and Pagonis (2013) derived an analytical approximation to Equation 10.10, namely:

$$I_{CW-OSL}(t) = 3n_0\rho'F(t)^2\left(\frac{1}{t}\right)\exp\left\{-\rho'F(t)^3\right\} \tag{10.11a}$$

where

$$F(t) = \ln\left(1 + z\sigma_p\Phi t\right), \tag{10.11b}$$

with all parameters previously defined. Figure 10.18c shows fits of Equations 10.11a and 10.11b to the BLSL and GLSL data.

Huntley's approximation (Huntley 2006) was also mentioned in Section 5.3.9, namely:

$$I_{CW-OSL}(t) = I_{OSL}(0)\exp\left\{-\rho'\left(\ln\left[P_0t\right]\right)^3\right\}, \tag{10.12}$$

and Figure 10.18d shows a best fit of this equation to the experimental data.

---

### Exercise 10.7    OSL curve shapes

On the web site accompanying this book can be found an Excel file under Exercises and Notes, Chapter 10, Exercise 10.7, giving 470 nm- stimulated OSL (BLSL) decay curves for aluminosilicate glass from four different smartphone glasses, as indicated. (Original data courtesy S. Sholom.) Determine if these OSL decay curves are consistent with the concept of optically assisted tunneling by fitting the data to Equations 10.11a, 10.11b, and 10.12.

---

#### 10.3.2.2    OSL Excitation Spectrum

Excitation spectra for the production of OSL from aluminosilicate glasses are scarce. One such spectrum, over a very limited energy range, is shown in Figure 10.19 for back protective GG from two smartphones (iPhones 8 and 11). A simple, exponential-like increase in OSL is observed as the excitation energy is increased in a manner similar to the excitation spectra continuum for feldspar shown in Figure 10.5, or the low-$E$ part of Figure 10.8. For the feldspar case the data were explained by invoking the optical excitation of electrons from traps into the band-tail states rather than the higher-energy states of the conduction band and the exponential increase is a reflection of the exponential density of states in the Urbach tail (Figure 10.4). In a similar manner the few data in Figure 10.19 can be fitted to Equation 10.7. Although the fits are somewhat uncertain due to the sparsity of data, the values of $E_o$ (optical trap depth) of 1.91 eV and 2.02 eV are obtained for the iPh11 and iPh8 glasses, respectively, along with characteristic widths for the Urbach tail of $\Delta E = 0.30$ eV and 0.26 eV, respectively.

The emission spectrum for OSL is not yet available in the literature but is presumably similar to that for TL, as discussed below.

The conclusion of these analyses is that the CW-OSL, both blue stimulated and green stimulated, is well described by an optically excited tunneling model through excitation into the band-tail states. As a result, tunneling can be expected to have a strong effect on the fading of OSL from these materials. Figure 10.20 shows two BLSL curves from samples of GG, one taken two minutes after irradiation and one three days after irradiation, for storage at room temperature in the dark. Fading can be due to a combination of thermal and athermal effects

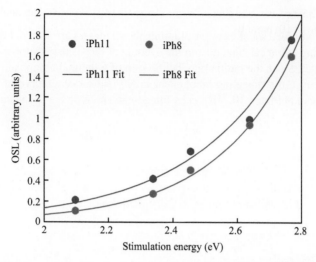

**Figure 10.19**    Excitation spectra for aluminosilicate glass from two smartphones (iPhones 8 and 11) over the energy range from ~2.1 to ~2.8 eV showing an apparently exponential rise in OSL intensity with increasing stimulation energy. Approximate fits of the two data sets to Equation 10.7 yield values for the optical trap depths $E_o$ of 2.04 eV and 1.91 eV for iPhones 8 and 11, respectively, and the Urbach tail characteristic widths $\Delta E$ of 0.25 eV and 0.30 eV, respectively. (Original data kindly provided by S. Sholom.)

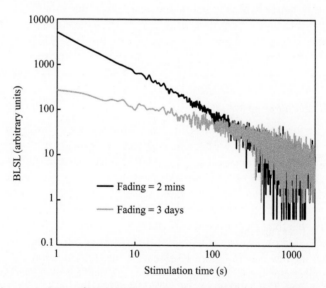

**Figure 10.20**    A comparison of two BLSL curves from GG from an iPhone XR stimulated at 470 nm for different times after irradiation (two minutes, and three days). The samples were kept in the dark at room temperature between irradiation and optical stimulation. (Original data kindly provided by S. Sholom.)

with the athermal component being due to tunneling. Since the samples were stored in the dark at room temperature after irradiation, the tunneling is not optically stimulated. A thermally stimulated, excited-state tunneling component cannot be ruled out, but since the temperature is low (room temperature) ground-state tunneling, as is the case with feldspar, is perhaps more likely.

### 10.3.2.3    OSL Fading

Typical fading curves for a set of example aluminosilicate glasses are seen in Figure 10.21a, with a "universal" fading curve calculated from the average of these curves shown in Figure 10.21b (on a log–log scale). Since the fading may be the sum of a thermal component and a tunneling component, and since the tunneling component may yield a decay curve of the form $t^{-k}$, the average fading curve in Figure 10.21b can be fitted to an approximate decay curve of the form:

$$I_{OSL}(t) = A\exp\left\{-\frac{t}{\tau}\right\} + Bt^{-k} \tag{10.13}$$

where $\tau$ is the temperature-dependent time constant for the thermally activated component, and $A$ and $B$ are constants. A fit of this expression to the data is also shown in Figure 10.21b, with the thermal and tunneling components illustrated separately.

---

### Exercise 10.8    OSL fading I

On the web site accompanying this book can be found an Excel file under Exercises and Notes, Chapter 10, Exercise 10.8, giving data for the fading (in the dark at room temperature) of 470 nm-stimulated OSL (BLSL) from aluminosilicate glass. (Original data courtesy S. Sholom.) The data are an average from seven different smartphone glasses and the standard deviations of the averages are indicated.

(a) Determine if the fading is due to thermal fading or tunneling, or both.

(b) If both, which component dominates, and over which time range?

---

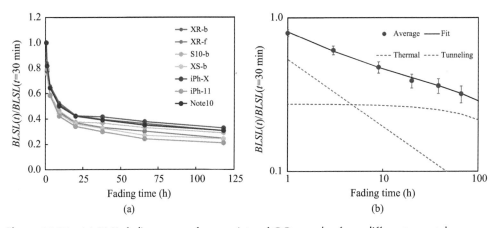

**Figure 10.21**    (a) BLSL fading curves for a variety of GG samples from different smartphones, as indicated. (Note "b" and "f" refer to protective glass taken from the back or the front of the phone, respectively.) (b) A "universal" fading curve calculated from the average for the data from part (a) and fitted to Equation 10.13, with the thermal and tunneling components shown separately. (Original data kindly provided by S. Sholom.)

---

### Exercise 10.9    OSL fading II

Figure 10.20 illustrates two blue-stimulated OSL curves (BLSL) from aluminosilicate glass, one taken two minutes after irradiation and one three days after irradiation (stored at room temperature in the dark). On the web site accompanying this book similar data (in an Excel file) for a different sample can be found under Exercises and Notes, Chapter 10, Exercise 10.9. (Original data courtesy S. Sholom.)

(a) Using the data obtained after fading of two minutes show that the OSL can be described by optically stimulated tunneling.

(b) Apply a similar analysis to the data obtained after three days fading. Can it also be described by tunneling only? If not, why not? (Note, assume the value of $\sigma_p\Phi$ for part (b) is the same as that used in part (a).)

(c) Try adding an ionization component to the OSL emission in the analysis of the OSL curve after three days fading. What mathematical form would this ionization component take? (Assume simple first-order kinetics.) Do you obtain a better fit? Why?

(d) How reasonable is it to assume that $\sigma_p\Phi$ is the same in part (b) as it was in part (a)?

### 10.3.2.4   *Potential Uses in Radiation Dosimetry*

The aim of TL and OSL studies of mobile phone glass is to determine the utility of these materials to act as retrospective, emergency dosimeters in situations where large numbers of the general population may have been potentially exposed to radiation during an accidental, or intentional, release of radiation. Examples include large-scale nuclear power plant accidents and potential terrorist events, possibly involving radiation exposure devices, radiation release devices ("dirty bombs"), or improvised nuclear devices. In any of the above scenarios large numbers of the general public may become exposed to harmful levels of radiation and require medical treatment (ICRU 2019). Since it is unlikely that members of the public would carry with them conventional radiation dosimeters (TLDs, OSLDs, etc.) to indicate the dose of radiation they received, surrogates for these devices have been sought among the common items that people generally carry with them during their daily routine. Since one such popular item is a smartphone, the display or protective glasses used in these devices have been the subject of study for these applications. The above-described OSL properties suggest that tunneling is a major characteristic of OSL from these materials and, as a result, fading is an important issue. This in turn limits the effectiveness of using OSL from smartphone protective glass as a dosimetry method to the first ~100 h after the initial irradiation. Within this time period, however, a measurable signal can still be obtained and a correction for fading can perhaps be made using a "universal" fading curve, such as that shown in Figure 10.21.

The other property desired of a useful dosimeter is a linear dose-response function. Figure 10.22 shows the dose-response functions of a selection of glasses from three smartphones, as indicated. In Figure 10.21 fading of the OSL signal, described by Equation 10.13, was found to be significant in these materials. It should be realized that fading will also be occurring during the irradiation period. Depending upon the dose rate the fading may be strong (especially at longer irradiation times and higher doses) and, as a result, it must be accounted for during the irradiation period $t_{irr}$ in order to obtain an accurate dose-response curve shape. If the delay between the end of the irradiation and the OSL measurement is $t_{del}$, then an effective-delay time $t_{eff}$, can be defined where:

$$t_{eff} = t_{irr}/2 + t_{del} \tag{10.14}$$

By varying $t_{del}$ such that $t_{eff}$ is kept the same for all irradiation times $t_{irr}$ (doses), the effects of fading during irradiation can be eliminated along with the resultant distortions to the

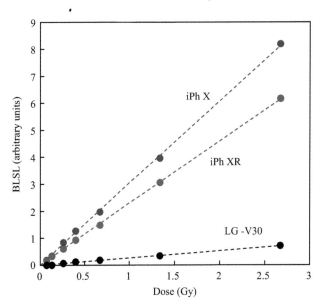

**Figure 10.22**  Dose-response functions for aluminosilicate glass from three different smartphones. In each case fading during irradiation has been corrected by keeping the effective delay time $t_{eff}$ constant for all doses ($t_{eff}$ is defined in Equation 10.14) by varying the delay time $t_{del}$ between the end of the irradiation and the start of the OSL measurement. After correction for fading during irradiation, the dose-response functions are linear. (Original data kindly provided by S. Sholom.)

shape of the dose-response curve. When this is done, the linear dose-response functions of Figure 10.22 are obtained.[7]

Clearly, fading is an important issue when using OSL from aluminosilicate glass in dosimetry and correction for fading of the OSL signal is required. Corrections based on expressions such as that given in Equation 10.13, or other empirical fading expressions, are possible if a reliable universal fading curve can be obtained.

### 10.3.3   TL Phenomena

#### 10.3.3.1   Glow-Curve Description

The optical excitation spectrum of Figure 10.19 was interpreted as being caused by an exponential increase in the photoionization cross-section of the optically active trap due to the exponential density of states in the Urbach tail, as expected for glass materials. However, the optical trap depth itself $E_o$ was assumed to be a fixed, discrete value in the approximate analysis of the data. Furthermore, only one trap was assumed to contribute to the OSL signal. In general, both of these assumptions are questionable. At a fixed wavelength of stimulation only one trap may dominate the signal, but the idea of a discrete optical trap depth in a glass material is less comfortable.

With regard to TL, one can expect that more than one trap will contribute to the TL glow curve over the wide temperature range normally employed in glow curve measurements.

---

[7] This correction procedure should be used for all materials that exhibit significant fading and where $t_{irr}$ is comparable to $t_{del}$ in order to determine the true shape of the dose-response function.

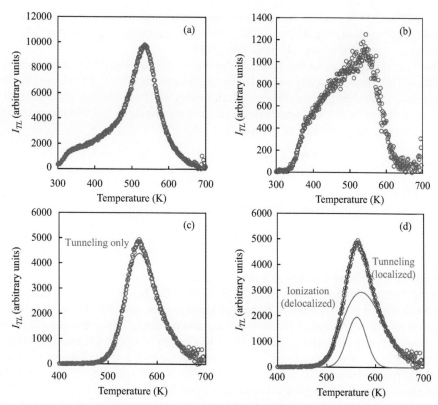

**Figure 10.23**   (a) Typical TL glow curve from an aluminosilicate glass (protective glass from an iPhone 11) following beta irradiation at room temperature. (b) TL lost due to OSL (i.e. TL before BLSL minus TL after BLSL). (c) TL after $T_{stop}$ = 543 K and best fit using Equations 10.15a, 10.15b assuming thermally assisted tunneling, with $E_t$ = 1.38 eV and $s$ = 2.2x10$^{13}$ s$^{-1}$. (d) Same data as part (c) but best fit using Equations 10.15a, 10.15b plus a Gaussian, assuming thermally assisted tunneling (localized recombination) with $E_t$ = 1.33 eV and $s$ = 2.1x10$^{12}$ s$^{-1}$, and ionization (delocalized recombination) from a distribution of states. All data at 1 K.s$^{-1}$. (Original data kindly provided by S. Sholom.)

Additionally, one can expect that each individual glow peak will be characterized by distributions in $E_t$ and in $s$. For systems which display tunneling, one might expect that thermally assisted tunneling will contribute to the TL emission in addition to delocalized recombination via thermal excitation high into the delocalized bands. The net result will be a complex glow curve, not easily interpreted, and its detailed nature may be impossible to unravel.

Figure 10.23a illustrates a typical TL glow curve for an aluminosilicate protective glass from a smartphone (iPhone11). It is clear that there are several overlapping, detrapping processes occurring and unpacking the individual mechanisms presents a significant challenge. The most prominent TL feature appears around 540 K. Figure 10.23b shows the TL that is lost during OSL measurement (blue stimulation, as in Figure 10.18). Some of the TL lost will be due to thermal instability especially at low glow-curve temperatures, but it appears that at least the main TL feature near 540 K is associated with the production of OSL. To isolate this TL feature, Figure 10.23c shows the TL curve after heating an irradiated sample to a temperature

$T_{stop}$ = 543 K (270 °C) before recording the remaining glow curve. Since optically assisted tunneling appears to explain the main OSL signal, it is reasonable to ask if thermally assisted tunneling can explain the main TL signal near 540 K.

It has been discussed above that Kitis and Pagonis (2013) developed an analytical expression to describe OSL due to optically assisted tunneling (Equations 10.11a, 10.11b) and this analytical expression was shown to describe the OSL curve very well from this material (Figure 10.18c). These authors also developed an equivalent analytical expression to describe thermally assisted tunneling during the production of TL, namely:

$$I_{TL}(t) = \frac{3n_0 \exp\left\{-\rho' F(t)^3 s\right\}\left(E_t^2 - 6k^2 T^2\right) z \rho' F(t)^2 \beta}{E_t k s T^2 z - 2k^2 s T^3 z + \exp\left\{E_t / kT\right\} E_t \beta}, \quad (10.15a)$$

where all terms have been previously defined but where $F(t)$ is re-defined in the case of TL as:

$$F(t) = \ln\left[1 + \frac{z s k T^2}{\beta E_t} \exp\left\{-\frac{E_t}{kT}\right\}\left(1 - \frac{2kT}{E_t}\right)\right]. \quad (10.15b)$$

Note that the TL is described here as a function of time rather than a function of temperature.

Figure 10.23c also shows a best fit of the glow curve for $T_{stop}$ = 543 K to Equations 10.15a and 10.15b. The fit is not perfect. At these high temperatures, however, it is likely that thermal ionization processes (delocalized transitions) are also occurring. Since this is a glass material we can expect $E_t$ and $s$ to each be described by a distribution of values. In Chapter 3 (Section 3.5 and Figure 3.21b) it was shown that a Gaussian distribution of $E_t$ and $s$ will produce a Gaussian-shaped TL peak. Therefore, assuming a trap with Gaussian distributions for $E_t$ and $s$, Figure 10.23d shows a best fit of the glow curve for $T_{stop}$ = 543 K to Equations 10.15a and 10.15b, plus a Gaussian-shaped TL peak to account for delocalized ionization processes which may also be occurring. The fit is much improved.

None of this definitively establishes that the glow curve from aluminosilicate glasses is a result of thermal ionization processes (delocalized transitions) plus thermally activated tunneling processes (localized transitions) but it does indicate that the data are consistent with such models and that these processes cannot be ruled out. The notion of optically and thermally assisted tunneling processes occurring during OSL and TL is self-consistent at least.

---

### Exercise 10.10    TL and thermally assisted tunneling

Figures 10.23c and 10.23d show a TL glow curve from aluminosilicate glass after a $T_{stop}$ = 543 K (270 °C) fitted assuming either tunneling only (c) or tunneling plus thermal ionization (d). On the web site accompanying this book can be found an Excel file containing the data for this glow curve, along with similar data for three additional $T_{stop}$ values (at 280 °C, 290 °C, and 300 °C). The file can be found under Exercises and Notes, Chapter 10, Exercise 10.10. (Original data courtesy S. Sholom.)

(a) Fit each of these four TL curves to Equations 10.15a and 10.15b assuming tunneling only (similar to Figure 10.23c). Observe how the calculated $E_t$ and $s$ values and how the $\rho'$ value change with $T_{stop}$. How would you interpret these changes?

(b) Repeat the fitting but now add a thermal ionization component (i.e. delocalized recombination) by assuming that this component produces a Gaussian-shaped TL peak as a function of glow-curve temperature (similar to Figure 10.23d). Observe how the calculated $E_t$ and $s$ values and how the $\rho'$ value now change with $T_{stop}$. How do you interpret these changes?

(c) What are your overall conclusions?

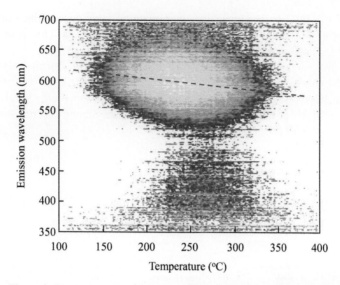

**Figure 10.24** TL emission spectrum from an aluminosilicate protective glass (iPhone XR) following 20 kGy beta at room temperature. The dashed red line highlights the shift in the center of the emission as a function of glow curve temperature. (Original data kindly provided by S. Sholom.)

### 10.3.3.2   TL Emission Spectrum

The emission spectrum for TL from these glasses is shown in Figure 10.24. The emission is broad and featureless, and centered in the red around 600 nm. It is seen to blue-shift steadily as the glow-curve temperature increases (see also Kim et al. 2019, "category B" glass). Such a smooth progression may merely reflect changes in lattice expansion; emission wavelength distortions due to defect clustering (due to trap-recombination center localization effects) are difficult to discern with such a broad emission band. Also observed weakly in Figure 10.24 (and in Kim et al. 2019) is emission around 400 nm, over approximately the same temperature range. It is likely that all recombination processes in this material, independent of the recombination pathway (localized or delocalized) use the same recombination center(s).

### 10.3.3.3   TL Analysis

While it may be feasible to fit a glow curve with a mixture of delocalized and localized analytical expressions (e.g. Randall-Wilkins equations plus tunneling equations), it is very difficult to do so in practice with a glow curve for which the number of separate detrapping processes is unknown, and when it is not known which of these processes are localized and which are delocalized. In addition, in cases when the trap depths and frequency factors are likely to be distributed in value, as with glasses, the net effect is a practical barrier which is very difficult to overcome.

To try to estimate the number of separate processes occurring in the TL glow curve a $T_m$-$T_{stop}$ analysis may be performed, an example of which is shown in Figure 10.25. There are clearly, at least, three main processes, but following the principles outlined in Chapter 3 it is also clear that the traps are distributed in energy and/or frequency factor since the $T_m$ values shift steadily with increase in $T_{stop}$.

As a first approach to analyzing this glow curve McKeever and Sholom (2021) adopted the principles described in Chapter 5 for analyzing glow curves formed from distributions of $E_t$ and

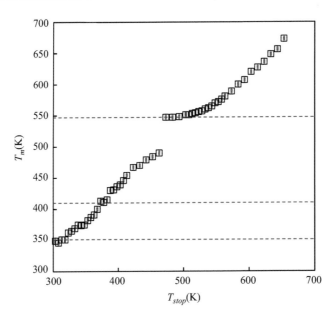

**Figure 10.25**    $T_m$-$T_{stop}$ data for TL from an aluminosilicate glass (iPhone11; $\beta_t = 2\,\text{K.s}^{-1}$). The red dashed lines show the approximate positions of potential individual TL peaks. These represent only the minimum number of peaks in the glow curve. (Adapted from McKeever and Sholom (2021).)

$s$ values. They first fitted the glow curve to a sum of Gaussian-shaped TL peaks and found that four Gaussian expressions adequately described the complete glow curve.[8] Next they estimated the mean energies for the corresponding trap depths and frequency factors, $E_{tm}$ and $s_m$, by monitoring how the Gaussian TL peaks changed position (i.e. temperature) with heating rate and by using Hoogenstraaten's analysis (Chapter 5) to estimate $E_{tm}$ and $s_m$. They then assumed four Gaussian distributions of $E_t$ and $s$ centered around the four $E_{tm}$ and $s_m$ values and constructed a total of $k$ first-order glow peaks, assuming that the net glow curve is made up of the sum of these glow peaks according to the general expression given in Chapter 5, Equation 5.22, and re-written here as a summation, namely:

$$I_{TL}\left(T\right) = \sum_{E_{t1}}^{E_{t2}} \sum_{\log(s_1)}^{\log(s_2)} g\left(E_t, \log\left(s\right)\right) I_{RW}\left(T\right). \tag{10.16}$$

$I_{RW}\left(T\right)$ is the Randall-Wilkins expression given in Chapter 3 (Equation 3.17) but here the distribution $g\left(E_t, \log\left(s\right)\right)$ replaces the term $n_0$. Also note that McKeever and Sholom described the distribution in $s$ in terms of a distribution in $\log(s)$.

The distribution $g\left(E_t, \log\left(s\right)\right)$ is given by:

$$g\left(E_t, \log\left(s\right)\right) = g\left(E_t\right) g\left(\log\left(s\right)\right) = A\exp\left\{-B_{E_t}\left(E_{tm} - E_t\right)^2\right\} \exp\left\{-B_s\left(\log\left(s_m\right) - \log\left(s\right)\right)^2\right\}. \tag{10.17}$$

---

[8] In Section 10.3.3.1 it was demonstrated that at least one of the peaks may in fact be described by tunneling (Equations 10.15a, 10.15b), rather than a Gaussian but the error involved by assuming a Gaussian instead of Equations 10.15a, 10.15b may not be large.

$E_{t1}$ and $E_{t2}$, and $s_1$ and $s_2$ are the ranges over which $E_t$ and $s$ were varied and the resolutions for $E_t$ and $s$ were $n$ $E_t$-values and $m$ $s$-values, respectively, such that there were $k = n$ x $m$ glow peaks used to describe the net glow curve. McKeever and Sholom used $n = 89$ and $m = 87$ for a total of $k = 7,743$. $E_t$ and $s$ were then varied independently until a best fit was obtained to the experimental glow curve. The whole exercise was repeated for several heating rates, and for a "pseudo" glow curve using the sum of glow curves obtained at different heating rates to ensure self-consistency. An example best fit for the "pseudo" TL curve (FOM = 0.70%) from McKeever and Sholom (2021) is shown in Figure 10.26 with final results listed in Table 10.1.

Peak 4 is responsible for the main TL peak near 540 K. Examination of Table 10.1 reveals that the $E_{tm}$ and $s_m$ values for this peak (peak 4) are very similar to those estimated using the expression for tunneling for the main TL signal in Figure 10.23.

**Table 10.1**  Fitted $E_{tm}$, $s_m$ and other trapping parameter values extracted from the best fit to the "pseudo" TL glow curve shown in Figure 10.26a, with the best fit $E_{tm}$ and $s_m$ distributions shown in Figure 10.26b; a.u. = arbitrary units. (Data from McKeever and Sholom, 2021.)

|  | Peak 1 | Peak 2 | Peak 3 | Peak 4 |
|---|---|---|---|---|
| $E_{tm}$ (eV) | 0.72 | 1.13 | 1.72 | 1.38 |
| $s_m$ (s$^{-1}$) | 7.64x10$^9$ | 1.63x10$^{13}$ | 4.92x10$^{15}$ | 1.70x10$^{12}$ |
| $B_{E_t}$ (a.u.) | 544.89 | 29.99 | 12.05 | 1166.92 |
| $A$ (a.u.) | 7.08x10$^2$ | 2.66x10$^3$ | 1.80x10$^3$ | 5.58x10$^3$ |
| $B_s$ (a.u.) | 1.87 | 40.58 | 3.55 | 0.90 |

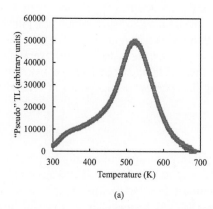
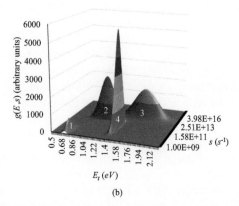

(a)   (b)

**Figure 10.26**  (a) A "pseudo" TL glow curve (data points) formed by adding the glow curves obtained from six different heating rates (0.1 K.s$^{-1}$, 0.25 K.s$^{-1}$, 0.5 K.s$^{-1}$, 1.0 K.s$^{-1}$, 2.0 K.s$^{-1}$, and 5.0 K.s$^{-1}$, see Chapter 5 for discussion of "pseudo" TL glow curves), and its best fit using Equations 10.16 and 10.17. The best fit results are given in Table 10.1. (b) $E_t$ and log($s$) distributions determined from the best fit results of part (a). (Adapted from McKeever and Sholom (2021).)

### 10.3.3.4  TL Fading

Since tunneling appears to be a strong feature of both OSL and TL production, and contributes to OSL fading, one might expect it to also contribute to TL fading. Experimental data seem to confirm this. Figure 10.27a shows the fading of the 473 K to 513 K region of the glow curves from several aluminosilicate glasses from a variety of smart phones. There appears to be more variation in the fading rates making the definition of a "universal" fading curve more difficult

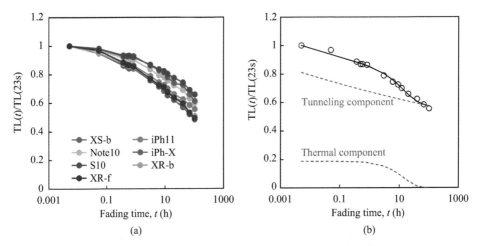

**Figure 10.27** (a) Fading of the TL from the 470 K to 513 K region of the glow curve from several aluminosilicate glasses obtained from different smartphones. (b) Fading curve for the iPhone 11 glass from part (a), fitted to Equation 10.13, and separated into thermal fading and tunneling components. (Original data kindly provided by S. Sholom.)

than for OSL. Figure 10.27b shows the fading data for iPh11 fitted to Equation 10.13 illustrating that tunneling appears to dominate the fading with a minor thermal component. It should be clear, however, that attempting to describe the thermal component with a single exponential when the analysis of the glow curve indicates a wide distribution of trapping states (Figure 10.26) is an overly simplistic approximation and serves no more than to indicate that thermal processes are required in addition to tunneling processes to describe the overall fading. However, attempting a more robust description of thermal fading by including a distribution of thermal components is not warranted considering the sparsity of data points in Figure 10.27 and the weakness of the thermal component.

### 10.3.3.5   Potential Uses in Radiation Dosimetry

Several research groups have demonstrated the feasibility of using aluminosilicate glasses from smartphones as potential TL dosimetry materials (Discher et al. 2013; Bassinet et al. 2014; Chandler et al. 2019; Kim et al. 2019). Each has demonstrated the linearity of the dose-response characteristics, the need for fading correction, the sensitivity to optical excitation, and the need to remove confounding, non-radiation-induced background signals. Nevertheless, despite the difficulties, the potential for use in retrospective dosimetry has been clearly demonstrated.

## 10.4   Final Remarks

The aluminosilicates, both natural, crystalline materials (feldspars) and engineered glasses, provide excellent examples of potential retrospective dosimetry materials (used in dating and emergency dosimetry, respectively) in which the mechanisms of TL, OSL, and RPL production are complex, interactive, and difficult to unravel. Excellent research reports are available in the published literature, only a fraction of which have been referenced here, that have helped to peel back the layers of complexity that hide the physics behind the luminescence processes in these materials. Undoubtedly, TL and OSL research on aluminosilicate glasses have much to learn from the much longer history of TL and OSL research on feldspars. Fortunately, there is much overlap between the two research communities and cross-fertilization is possible.

Studying these two material types is a formidable challenge, and much yet needs to be done, but at the same time it is a wonderful educational exercise. Examination of the properties of aluminosilicate materials not only allows demonstrations of the fundamental principles of TL, OSL and RPL, but also when to apply those principles and, crucially, when greater complexity is needed.

It is perhaps at this point that it is worth recalling the words of Alfred North Whitehead with which this chapter began.

# 11

# Concluding Remarks

## *The Possibilities for Imperfection Engineering*

*As a matter of fact, the construction of defects in solid materials is an extremely complicated process, in which different types of defects always co-exist in [the] material lattice.*

N. Zhang, C. Gao, Y. Xiong 2019

## 11.1 The Importance of Defects

### 11.1.1 The Ideal Luminescence Dosimeter

We have come full circle. As we have plowed our way through the phenomena, the materials, the processes, the kinetics, the equations, and the analyses, the most critical aspect to emerge from a study of TL, OSL, and RPL phenomena is that everything depends on the defects in the materials. Prof. Peter Townsend's quote at the beginning of Chapter 2 rings truer and truer as each layer of understanding is peeled back and we catch a clearer glimpse at the inner workings of the best luminescence dosimeters available. In this vein, and armed with the knowledge of how luminescence dosimeters work, it is appropriate to probe what it is that enables some materials to be efficient luminescence dosimeters and others not. If one were designing a TLD, OSLD, or RPLD material, what features would one try to build in? Certainly the following properties would be desirable:

- High sensitivity (high luminescence output for a small dose input).
- A wavelength of luminescence that matched the quantum efficiency of modern detection equipment.
- A linear dose-response function over a wide dose range (wide dynamic range).
- A stable signal (no fading of the signal either during or after irradiation).

The above would be the minimum requirements. Additionally, the following would be desirable:

*A Course in Luminescence Measurements and Analyses for Radiation Dosimetry,* First Edition.
Stephen W.S. McKeever.
© 2022 John Wiley & Sons Ltd. Published 2022 by John Wiley & Sons Ltd.
Companion Website: www.wiley.com/go/mckeever/luminescence-measurements

- The same radiation photon-energy dependence as human tissue (tissue equivalence).
- LET independence (a good sensitivity to charged particles independent of the particle's energy, charge, and mass).

In many ways, such a luminescence dosimeter would be ideal, but how might one approach the problem of engineering a material to achieve, or approximate, these properties?

Firstly, it is instructive to examine what is meant by a sensitive luminescence dosimeter. Just how efficient could a luminescence dosimeter be? Bos (2001) examined this question for TLDs and it is useful to outline his arguments in order to estimate the "ideal" target efficiency.

Considering a TLD (or, equivalently, an OSLD) as an energy converter, in which the energy from an incident radiation field is converted to the energy of visible luminescence photons, the intrinsic efficiency $\eta_i$ may be defined as the ratio of the energy emitted as TL photons (or OSL photons) to the energy absorbed from the initial radiation (i.e. the dose). The efficiency can be written:

$$\eta_i = n_{eh}\eta_{tr}\eta_L E_p \eta_{esc},\tag{11.1}$$

where $n_{eh}$ is the number of electron-hole pairs generated by the absorbed dose, per unit mass per unit dose, $\eta_{tr}$ is the fraction of these that have been captured and that contribute to TL or OSL, $\eta_L$ is the efficiency with which released electrons from traps recombine and produce a luminescence photon, and $\eta_{esc}$ is the probability that the photons produced escape from the sample without being absorbed by another defect. $E_p$ is the energy per luminescence photon emitted.

The term $\eta_L$ is important and may be defined as:

$$\eta_L = pSQ\tag{11.2}$$

where $p$ is the probability per second of release from the trap ($p = s\exp\{-E_t/kT\}$ for TL; $p = \sigma\Phi$ for OSL), $S$ is the efficiency with which the released electron is transported to the radiative recombination center, and $Q$ is the quantum efficiency of the recombination center.

Bos (2001) gives a theoretical upper limit ($p = 1$, $\eta_{tr} = 1$, $Q = 1$, $S = 1$, $\eta_{esc} = 1$) for $\eta_i$ as:

$$\eta_i = E_p \beta E_g,\tag{11.3}$$

where the constant $\beta$ relates the number of electron-hole pairs per absorbed unit of energy to the material's energy band gap, $E_g$. The value of $\beta$ depends on the material. For example, for ionic compounds (typical of many TLDs or OSLDs), $\beta = 1.4$ to $2.0$. Using these considerations, Bos calculates theoretical upper limits for the efficiency of a variety of materials. For example, for LiF:Mg,Ti the limit is 13%, a number which equates approximately to the average theoretical efficiency for several of the major TLD materials examined by Bos.

Actual measured efficiencies, however, are quite different. For LiF:Mg,Ti (TLD-100) it is a mere 0.032%. LiF:MCP is much better at 0.91%, while for $Al_2O_3$:C (as a TL dosimeter, for which $Q$ is certainly <1) it is 0.84%. For $CaF_2$:Dy (TLD-200) the value is an extremely high 4.7%, or only a factor of 2.7 below the theoretical upper limit. Two additional materials discussed by Bos also show extremely high efficiencies, namely $CaF_2$:Cu,Ho and $KMgF_3$:Ce, for which $\eta_i$ has measured values of 4.5% and 4.1%, respectively. Why do these latter materials display such high sensitivities compared to the other TLD materials?

The key parameter appears to be the factor $S$ in Equation 11.2. If the TL (or OSL) is produced via a delocalized recombination process, one can imagine that the freed electron has many other recombination and/or retrapping opportunities available to it other than finding a recombination site to produce a TL or OSL photon. This is the competition process referred to in earlier chapters. Competition is the cause of supralinearity in the dose-response curve where supralinearity should be correctly thought of not as being an "over response" at high doses but rather as being an "under response" at low doses. It is in the low-dose range where the competition (non-radiative) processes are strong and where $S < 1$ (or even $<<1$).

In contrast, the value of the efficiency term $S$ for localized recombination is greater than, or very much greater than, the value of $S$ for delocalized recombination since, for the former, there are no competing processes. In the $CaF_2$:Dy, $CaF_2$:Cu,Ho, and $KMgF_3$:Ce examples the trapping and recombination sites are spatially located near each other in defect complexes such that transport of the freed charge from the trap to the recombination center does not involve the delocalized band. For example, in $KMgF_3$:Ce, $Ce^{3+}$ ions substitute for $K^+$ with two $K^+$ vacancies nearby. The $Ce^{3+}$-divacancy complex emits at 360 nm and irradiation is assumed to form $Ce^{2+}$ via electron trapping with a trapped hole at one of the cation-vacancy sites. TL results when a hole is released during heating, recombining directly with $Ce^{2+}$ to create an excited $Ce^{3+}$ ion, which in turn relaxes with 360 nm emission (Bos 2001). Similar processes of localized recombination are believed to occur in all high-sensitivity TLD and OSLD materials. The key point here is to produce an engineered material in which the traps and recombination centers are closely associated so that the need for delocalized recombination is removed and replaced by highly efficient localized recombination.

Along with a high efficiency (and high sensitivity) when the trap and recombination center are closely associated will come a linear dose-response curve. This is because there will be no competition effects, as noted above. Furthermore, if the concentration of trap/recombination-center complexes can be made sufficiently high without disturbing the nature of the complex (e.g. by forming more-complicated clusters or even precipitates) then the linear dose-response curve may be made to extend to high doses, giving the dosimeter a wide dynamic range.

Under these circumstances the dosimeter will also have an excellent response to charged particles since a changing efficiency to charged particles, as a function of LET, is due to both supralinear and saturation effects in the photon dose-response curve. Of course, the doses within a charged-particle track are so high that saturation effects cannot be avoided; a drop in efficiency with increase in LET will be inevitable and will always occur ultimately, but the effect may be pushed to higher LET values with optimal defect, or imperfection, engineering.

However, a problem that may be faced with cluster formation between the trap and the recombination center is the possibility of tunneling. If the trap and the recombination site are close enough such that tunneling through the potential barrier between them becomes a viable recombination transition route at room temperature, even when the dosimeter is held in the dark, then fading of the TL and/or OSL signal will result. The materials engineering challenge then becomes one of ensuring that the traps and recombination centers are close enough for localized recombination, but not close enough for tunneling.

Interestingly, for RPLDs the opposite considerations may apply. RPL results from a defect that does not require excitation of an electron from the trap but instead only requires excitation of the electron to an excited state within the trapping center, followed by relaxation to the ground state and the emission of a photon (RPL). Any mechanism that causes the excited electron to leave the trap before relaxation to the ground state will result in the loss of luminescence emission and result in decreased efficiency. Therefore, close association between the trap and any other center should be avoided wherever and whenever possible such that localized transport and/or tunneling from the trap does not occur. Clustering may be good for TL and OSL, but not necessarily so for RPL.

### 11.1.2 How to Detect Defect Clustering and Tunneling

Townsend et al. (2021) summarized the evidence from the literature to indicate that clustering of defects is much more prevalent than is perhaps normally considered. To quote these authors: "In reality, no part of a crystal is 'perfect' and the concept of a truly 'point' defect should be well and truly consigned to history." There are several experimental clues that can indicate that defect clustering and/or tunneling processes are affecting the TL and/or OSL properties of the dosimetry material being studied. Some of the most prominent are the following:

#### 11.1.2.1  $E_t$ and s Analysis

As considered by Piters and Bos (1993b) and discussed in Chapter 7, and as discussed by Townsend et al. (2021), consider a reversible interaction between a defect $A^*$ (defect $A$ with a trapped electron) and defect $B$ thus, $A^* + B \leftrightarrow C^*$, where TL results from the thermal release of the electron from $C^*$, that is, $C^* \rightarrow C + e^-$. Piters and Bos (1993b) showed that a first-order, TL peak can still be formed and very-well fitted by the Randall-Wilkins equation to yield an effective activation energy $E_{teff}$ and effective frequency factor $s_{eff}$. They further showed that these values will change with heating rate in a manner dependent upon the relative values of the trap depth and association/dissociation energies involved in the reaction between $A^*$, $B$, and $C^*$. These reactions also take place during isothermal annealing and are dependent upon the storage times, temperatures, and cooling rates. The overall result is that a slower fading rate for the TL peak can be predicted if the fading rate is calculated from $E_{teff}$ and $s_{eff}$ than if it is measured during actual isothermal decay.

Correlated processes such as these (e.g. disassociation of a cluster followed by charge release) can results in abnormal apparent $E_t$ and $s$ values. The $E_{teff}$ and $s_{eff}$ values obtained from conventional glow curve analysis methods can be very high in these circumstances (~2.0 eV to 2.2 eV and $10^{22}$ s$^{-1}$ to $10^{24}$ s$^{-1}$, respectively). An alternative example of correlated processes is thermally activated tunneling when thermal activation of the trapped electron from the ground state to an excited state is followed by tunneling from the excited state to the recombination center. Here, entirely different $E_t$ and $s$ values can be expected. In this case a small energy value can be found along with a frequency factor which is less than, or even very much less than, the values expected from consideration of the lattice vibrational frequencies. In general, abnormal $E_t$ and $s$ values (high or low) are an indication of trap/recombination-center clustering effects in the dosimetry material.

#### 11.1.2.2  TL and OSL Curve Shapes

Since clustering between the trap and the recombination center can yield high $E_t$ and $s$ values correspondingly narrow TL peaks can be produced. LiF:Mg,Ti TL peak 5 is perhaps the archetypical material demonstrating this. Tunneling, on the other hand, may yield smaller $E_t$ and $s$ values and broad TL peaks may be produced. In addition, other differences in the TL peak shape can be expected. Kitis and Pagonis (2013) simulated the TL glow curve shapes for thermally assisted, excited-state tunneling for TL production and showed that the characteristic shape of such TL peaks is that they are broader than expected for first-order, Randall-Wilkins-shaped peaks and are characterized by a long, high-temperature tail. With peak overlap, however, such distinctions may be difficult to discern. Nevertheless, apparent broad TL peaks are a characteristic of tunneling and something to be looked for.

With OSL, optically assisted, excited-state tunneling gives the OSL decay curves a characteristic almost-straight-line shape when plotted as log($I_{CW\text{-}OSL}$)-versus-log($t$), indicative of an approximately $t^{-k}$ decay. The shape of the decay curve can be well-described by the analytical expression derived by Kitis and Pagonis (2013) for optically assisted, excited-state tunneling, but is difficult to describe by first-order kinetics and conventional delocalized recombination kinetics.

### 11.1.2.3 Fading

If excited-state tunneling is occurring during TL and/or OSL measurement, the possibility for ground-state tunneling during storage in the dark at low temperature (room temperature) must be suspected. If present, this will manifest itself by fading of the signal at a rate faster than that predicted from measurement of the $E_t$ and $s$ values for the signal. Such athermal, or anomalous, fading is a strong indicator of tunneling between the trap and the recombination center. The loss of signal during storage will also follow an approximately $t^{-k}$ law, although some degree of thermal fading may also be present.

### 11.1.2.4 Spectral Measurements

Close association between the trap and the recombination center can also be detected by measuring the emission spectra, both in OSL and, especially, in TL. The best way to proceed is to measure TL intensity-versus-wavelength-versus-temperature while heating the sample and to display the data as an isometric plot or a contour plot. An example for LiF was discussed in the early work by Fairchild et al. (1978a, 1978b) and more specific and wider examples have been discussed by Townsend et al. (2021). For broad band emissions, clustering can be detected by monitoring small shifts in the maximum of the emission as a function of the glow peak temperature. LiF:Mg,Ti again provides the classic example where emission from all main peaks is believed to be due to Ti(OH)$_n$ complexes closely associated with $\left[\mathrm{Mg}^{2+}/\mathrm{Li}_{\mathrm{vac}}^{+}\right]$ complexes. The emission from peak 2, associated with $\left[\mathrm{Mg}^{2+}/\mathrm{Li}_{\mathrm{vac}}^{+}\right]$ dipoles, is seen to be slightly different from that of peak 5, associated with $\left[\mathrm{Mg}^{2+}/\mathrm{Li}_{\mathrm{vac}}^{+}\right]$ trimers. The shift in the emission maximum is interpreted as a change in the crystal field due to distortion of the lattice caused by the association of the different $\left[\mathrm{Mg}^{2+}/\mathrm{Li}_{\mathrm{vac}}^{+}\right]$ complexes with the Ti(OH)$_n$ defects.

In rare-earth-doped systems, where sharp emission lines are evident, the spacing between the lines can be seen to alter with the temperature of the emission peak, indicating crystal field differences, and the temperature at which given emission lines appear can be seen to be different in some systems. Furthermore, changing the rare-earth ion not only changes the emission wavelength but can also change the TL peak position. These observations indicate that not only are the emitters associated with the recombination sites, but they are also associated with the trapping sites, suggesting close association between the two. Other subtle changes in the emission bands can sometimes be observed, such as asymmetric bulges in broad band emissions as a function of temperature, again indicating trap/recombination-center associations.

---

### Exercise 11.1   TL from LiF:Mg,Ti

In Chapter 7 the possibilities for localized and delocalized transitions in the production of TL in TLD-100 were discussed. Some characteristics of the main TL peak ("peak 5") from TLD-100 are:

- Significant supralinearity.
- Significant sublinearity.
- Intrinsic efficiency (0.032%) much less than the theoretical limit (13%).

> (a) What do these characteristics suggest to you about delocalized versus localized transitions in the production of TL peak 5 in this material?
> (b) Are these characteristics consistent with other known properties regarding peak 5 in this material ($E_t/s$ values, emission spectra, etc.)?

## 11.2    The Prospects for "Designer" TLDs, OSLDs, and RPLDs

The above discussion is all very well in that it helps with the detection of defect clustering and/or tunneling, but it is "after the fact," i.e. after the dosimeter has been produced. Controlling these properties such that they can be engineered into the material in the first place is a different proposition. This question was addressed in an introductory essay to their book *Thermoluminescence Dosimetry Materials: Properties and Uses* by McKeever et al. (1995) who looked at the issue from the point of view of thermodynamics and noted that TL, to which we may add OSL and RPL, phenomena are described by large numbers of coupled, non-linear, differential equations, including not only descriptions of electronic charge transfer between the various quantum states in the system, but also of interactions between the defects themselves that can react to create new defects of different types, which can in turn affect the luminescence properties of the material. As noted by these authors, the solutions to such complex systems are deterministic but unpredictable, and if one adds uncertainty (noise) to various key parameters (impurity concentrations, temperature fluctuations, annealing time variations, heating and cooling rate uncertainties, optical stimulation rate uncertainties, etc.) the outcomes can be very difficult to predict and difficult to reproduce. The prospect of designing a given dosimeter to behave in a specific way appears daunting.

However, since 1995 much has been learned about how luminescence dosimetry systems work. The intentional addition of specific impurities (e.g. the lanthanide series) into suitable hosts is known to give rise to specific localized states that can be predicted beforehand (Dorenbos 2003a, 2003b). Aggregation of defects can be intentionally initiated by thermal processing (e.g. LiF:Mg,Ti) and the distance between defects can be controlled by varying the dopant concentrations (Dobrowolska et al. 2014). Emission wavelengths can be manipulated by judicious selection of impurity types (Yukihara et al. 2013). In other words, all the tools are available to move forward with more optimism than was available at the time of McKeever et al.'s 1995 essay. Perhaps some readers of this book will be willing to take on the challenge.

# References

Afouxenidis, D., Polymeris, G.S., Tsirlingas, N.C., and Kitis, G. (2012). Computerized curve deconvolution of TL/OSL curves using a popular spreadsheet program. *Radiat. Prot. Dosim.* 149: 363–370.

Agersnap Larsen, N., Bøtter-Jensen, L., and McKeever, S.W.S. (1999). Thermally stimulated conductivity and thermoluminescence from $Al_2O_3$:C. *Radiat. Prot. Dosim.* 84: 87–90.

Aitken, M.J. (1985). *Thermoluminescence Dating*. ISBN 9780120463800. London: Academic Press.

Aitken, M.J. (1998). *An Introduction to Optical Dating*. ISBN 9780198540922. Oxford: Oxford University Press.

Akselrod, A.E. and Akselrod, M.S. (2002). Correlation between OSL and the distribution of TL traps in $Al_2O_3$:C. *Radiat. Prot. Dosim.* 100: 217–220.

Akselrod, G.M., Akselrod, M.S., Benton, E.R., and Yasuda, N. (2006a). A novel $Al_2O_3$ fluorescent nuclear track detector for heavy charged particles and neutrons. *Nucl. Instrum. Meth. Phys. Res. B* 247: 295–306.

Akselrod, M.S., Agersnap Larsen, N., Whitley, V., and McKeever, S.W.S. (1998). Thermal quenching of *F*-center luminescence in $Al_2O_3$:C. *J. Appl. Phys.* 84: 3364–3373.

Akselrod, M.S. and Akselrod, A.E. (2006). New $Al_2O_3$:C,Mg crystals for radiophotoluminescent dosimetry and optical imaging. *Radiat. Prot. Dosim.* 119: 218–221.

Akselrod, M.S., Akselrod, A.E., Orlov, S.S., Sanyal, S., and Underwood, T.H. (2003). Fluorescent aluminum oxide crystals for volumetric optical data storage and imaging applications. *J. Fluor.* 13: 503–511.

Akselrod, M.S., Fomenko, V.V., Bartz, J.A., and Ding, F. (2014b). FNTD radiation dosimetry system enhanced with dual-color wide-field imaging. *Radiat. Meas.* 71: 166–173.

Akselrod, M.S., Fomenko, V.V., Bartz, J.A., and Haslett, T.L. (2014a). Automatic neutron dosimetry system based on fluorescent nuclear track detector technology. *Radiat. Prot. Dosim.* 161: 86–91.

Akselrod, M.S., Fomenko, V.V., and Harrison, J. (2020). Latest advances in FNTD technology and instrumentation. *Radiat. Meas.* 133: 106302.

Akselrod, M.S. and Kortov, V. (1990). Thermoluminescent and exoemission properties of new, high-sensitivity TLD $\alpha$-$Al_2O_3$:C crystals. *Radiat. Prot. Dosim.* 33: 123–126.

Akselrod, M.S., Kortov, V.S., and Gorelova, E.A. (1993). Preparation and properties of alpha-$Al_2O_3$:C. *Radiat. Prot. Dosim.* 47: 159–164.

Akselrod, M.S. and Kouwenberg, J. (2018). Fluorescent nuclear track detectors – review of past, present and future of the technology. *Radiat. Meas.* 117: 35–51.

Akselrod, M.S. and McKeever, S.W.S. (1999). A radiation dosimetry method using pulsed optically stimulated luminescence. *Radiat. Prot. Dosim.* 81: 167–176.

*A Course in Luminescence Measurements and Analyses for Radiation Dosimetry,* First Edition.
Stephen W.S. McKeever.
© 2022 John Wiley & Sons Ltd. Published 2022 by John Wiley & Sons Ltd.
Companion Website: www.wiley.com/go/mckeever/luminescence-measurements

Akselrod, M.S. and Sykora, G.J. (2011). Fluorescent nuclear track detector technology – a new way to do passive solid state dosimetry. *Radiat. Meas.* 46: 1671–1679.

Akselrod, M.S., Yoder, R.C., and Akselrod, G.M. (2006b). Confocal fluorescent imaging of tracks from heavy charged particles utilising new $Al_2O_3$:C,Mg crystals. *Radiat. Prot. Dosim.* 119: 357–362.

Ando, M.F., Benzine, O., Pan, Z., Garden, J.-L., Wondraczek, K., Grimm, S., Schuster, K., Wondraczek, L. (2018). Boson peak, heterogeneity and intermediate-range order in binary $SiO_2$-$Al_2O_3$ glasses. *Sci. Rep.* 8: 5394.

Andreeva, N.Z., Dmitryuk, A.V., Perminov, A.S., Ptrovskii, G.T., Savvina, G.Ch., and Vil'chinskaya, N.N. (1985). Influence of activator concentration on the dosimetric properties of radiophotoluminescent glasses. *Translation from Atomnaya Énergiya.* 58, 132–135.

Antonov-Romanovsky, V.V., Keirum-Markus, I.F., Poroshina, M.S., and Trapeznikova, Z.A. (1955). Dosimetry of ionizing radiation with the aid of infra-red sensitive phosphors. *Conference of the Academy of Sciences of the U.S.S.R. on the Peaceful Use of Atomic Energy, Moscow, USAEC Report AEC-tr-2435 (pt.1)*, 239–250.

Aramu, F., Maxia, V., Spano, G., and Cortese, C. (1975). Kinetics of the carrier storage in thermoluminescent phosphors. *J. Lumin.* 11: 197–206.

Ayu, N.I.P., Kartini, E., Prayogi, L.D., and Faisal, M. (2016). Crystal structure analysis of $Li_3PO_4$ powder prepared by wet chemical reaction and solid-state reaction by using X-ray diffraction (XRD). *Ionics* 22: 1051–1057.

Babbage, C. (1864). *Passages from the Life of a Philosopher.* London: Longman, Green, Longman, Roberts and Green.

Bacon, F. (1605). *The Advancement of Learning*, in *The Works of Francis Bacon*, Spedding, J., Ellis, R., and Heath, D. (eds.) (1887–1901) 3: 293. Quoted in *Oxford Dictionary of Scientific Quotations*. Bynum, E.F., and Porter, R. (eds.). ISBN 0198584091. Oxford, Oxford University Press, 2005.

Bailey, R.M. (2001). Towards a general kinetic model for optically and thermally stimulated luminescence of quartz. *Radiat. Meas.* 33: 17–45.

Baldacchini, G., Bollanti, S., Bonfigli, F., Flora, F., Di Lazzaro, P., Lai, A., Marolo, T., Montereali, R.M., and Murra, D. (2005). Soft x-ray submicron imaging detector based on point defects in LiF. *Rev. Sci. Instrum.* 76: 113104.

Bartz, J.A., Kodaira, S., Kurano, M., Yasuda, N., and Akselrod, M.S. (2014). High resolution charge spectroscopy of heavy ions with FNTD technology. *Nucl. Instrum. Meth. Phys. Res. B* 335: 24–30.

Bassinet, C., Pirault, N., Baumann, M., and Clairand, I. (2014). Radiation accident dosimetry: TL properties of mobile phone screen glass. *Radiat. Meas.* 71: 461–465.

Becker, K. (1968). Recent progress in radiophotoluminescence dosimetry. *Health Phys.* 14: 17–32.

Becquerel, E. (1843). Des effets produits sur les corps par les rayons solaires. *Ann de Chim. et de Phys.* 9: 257–322.

Becquerel, H. (1883). Maxima et minima d'extinction de la phosphorescence sous l'influence des radiations infra-rouges. *Comptes. Rendus.* 96: 1853–1856.

Bellec, M., Royon, A., Bourhis, K., Choi, J., Bousquet, B., Treguer, M., Cardinal, T., Videau, J.-J., Richardson, M., and Canioni, L. (2010). 3D patterning at the nanoscale of fluorescent emitters in glass. *J. Phys. Chem. C* 114: 15584–15588.

Bellec, M., Royon, A., Bousquet, B., Bourhis, K., Treguer, M., Cardinal, T., Richardson, M., and Canioni, L. (2009). Beat the diffraction limit in 3D direct laser writing in photosensitive glass. *Optics Exp.* 17: 10304–10318.

Benavente, J.F., Gómez-Ros, J.M., and Romero, A.M. (2019). Thermoluminescence glow curve deconvolution for discrete and continuous trap distributions. *Appl. Radiat. Isot.* 153: 108843.

Berger, T. and Hajek, M. (2008). TL-efficiency – overview and experimental results over the years. *Radiat. Meas.* 43: 146–156.

Bernard, C. (1872). *Pathologie Expérimentale*, 72. Quoted in *Oxford Dictionary of Scientific Quotations*. Bynum, E.F., and Porter, R. (eds.). ISBN 0198584091. Oxford, Oxford University Press, 2005.

Betts, D.S., Couturier, L., Khayrat, A.H., Luff, B.J., and Townsend, P.D. (1993a). Temperature distribution in thermoluminescence experiments. I: experimental results. *J. Phys. D: Appl. Phys.* 26: 843–848.

Betts, D.S. and Townsend, P.D. (1993b). Temperature distribution in thermoluminescence experiments. II: some calculational models. *J. Phys. D: Appl. Phys.* 26: 849–857.

Biderman, S., Horowitz, Y., and Oster, L. (2002). Investigation of the emission spectra of LiF:Mg,Ti (TLD-100) during thermoluminescence. *Radiat. Prot. Dosim.* 100: 369–373.

Bilski, P. (2002). Lithium Fluoride: from LiF:Mg,Ti to LiF:Mg,Cu,P. *Radiat. Prot. Dosim.* 100: 199–206.

Bilski, P., Budzanowski, M., and Olko, P. (1996). A systematic evaluation of the dependence of glow curve structure on the concentration of dopants in LiF:Mg,Cu,P. *Radiat. Prot. Dosim.* 65: 195–198.

Bilski, P. and Marczewska, B. (2017). Fluorescent detection of single tracks of alpha particles using lithium fluoride crystals. *Nucl. Instrum. Meth. Phys. Res. B* 392: 41–45.

Bilski, P., Marczewska, B., Giezczyk, W., Kłosowski, M., Zhydachevskyy, Y., and Kodaira, S. (2019). Luminescent properties of LiF crystals for fluorescent imaging of nuclear particles tracks. *Opt. Mater.* 90: 1–6.

Bilski, P., Marczewska, B., Kłosowski, M., Giezczyk, W., and Naruszewicz, M. (2018). Detection of neutrons with LiF fluorescent nuclear track detectors. *Radiat. Meas.* 116: 35–39.

Bilski, P., Marczewska, B., and Zhydachevskii, Y. (2017). Radiophotoluminescence spectra of lithium fluoride TLDs after exposures to different radiation modalities. *Radiat. Meas.* 97: 14–19.

Bilski, P., Obryk, B., Olko, P., Mandowska, E., Mandowski, A., and Kim, J.L. (2008). Characteristics of LiF:Mg,Cu,P thermoluminescence at ultra-high dose range. *Radiat. Meas.* 43: 315–318.

Bilski, P., Olko, P., Puchalska, M., Obryk, B., Waligorski, M.P.R., and Kim, J.L. (2007). High-dose characterization of different LiF phosphors. *Radiat. Meas.* 42: 582–585.

Biswas, R.H., Bhatt, B.C., and Singhvi, A.K. (2013). Effect of optical bleaching on the dosimetric TL peak in $Al_2O_3$:C for blue and UV emissions. *Radiat. Meas.* 59: 37–43.

Biswas, R.H., Murani, M.K., and Singhvi, A.K. (2009). Dose-dependent change in the optically stimulated luminescence decay of $Al_2O_3$:C. *Radiat. Meas.* 44: 543–547.

Blakemore, J.S. and Rahimi, S. (1984). Models for mid-gap states in GaAs. In: *Semi-insulating GaAs: Semiconductors and Semimetals*, Vol. 20 (ed. R.K. Willardson and A.C. Beer), 233–361. ISBN 9780127521206. Orlando: Academic Press.

Böer, K.W. (1990). *Survey of Semiconductor Physics: Electrons and Other Particles in Bulk Semiconductors*. ISBN 2001026953. New York: Van Nostrand Reinhold.

Bos, A.J.J. (2001). On the energy conversion in thermoluminescence dosimetry materials. *Radiat. Meas.* 33: 737–744.

Bos, A.J.J. and De Haas, J.T.M. (1998). Temperature dependent absorption spectrometry on LiF:Mg,Ti. *Radiat. Meas.* 29: 349–353.

Bos, A.J.J. and Dielhof, J.B. (1991). The analysis of thermoluminescent glow peaks in $CaF_2$: Tm (TLD-300). *Radiat. Prot. Dosim.* 37: 231–239.

Bos, A.J.J. and Piters, T.M. (1993). Success and failure of the Randall-Wilkins model for thermoluminescence in LiF (TLD). *Radiat. Prot. Dosim.* 47: 41–47.

Bos, A.J.J., Piters, T.M., Gómez-Ros, J.M., and Delgado, A. (1993). An intercomparison of glow curve analysis computer programs: I. Synthetic glow curves. *Radiat. Prot. Dosim.* 47: 473–477.

Bos, A.J.J., Piters, T.M., Gómez-Ros, J.M., and Delgado, A. (1994). An intercomparison of glow curve analysis computer programs: II. Measured glow curves. *Radiat. Prot. Dosim.* 51: 257–264.

Bos, A.J.J., Vijverberg, R.N.M., Piters, T.M., and McKeever, S.W.S. (1992). Effects of cooling and heating rate on the trapping parameters in LiF:Mg,Ti crystals. *J. Phys. D: Appl. Phys.* 25: 1249–1257.

Bøtter-Jensen, L. and McKeever, S.W.S. (1996). Optically stimulated luminescence dosimetry using natural and synthetic materials. *Radiat. Prot. Dosim.* 65: 273–280.

Bøtter-Jensen, L., McKeever, S.W.S., and Wintle, A.G. (2003). *Optically Stimulated Luminescence Dosimetry*. ISBN 0444506845. Amsterdam: Elsevier.

Bowman, S.G.E. and Chen, R. (1979). Superlinear filling of traps in crystals due to competition during irradiation. *J. Lumin.* 18/19: 345–348.

Boyle, R. (1664). *Experiments and Considerations Touching Colours*. London: Transactions of the Royal Society, 413–421.

Boyle, R. (1669). *Physiological Essays*, 20. Quoted in *Oxford Dictionary of Scientific Quotations*. Bynum, E.F., and Porter, R. (eds.). ISBN 0198584091. Oxford, Oxford University Press, 2005.

Boyle, R. (1680). *The Aerial Noctiluca: Or, Some New Phenomena, and a Process of a Factitious Self-Shining Substance*. London: N. Ranew. Early English Books Online, Text Creation Partnership, 2011, http://name.umdl.umich.edu/a28938.0001.001 (accessed October 2019).

Bradbury, M.H. and Lilley, E. (1977). Precipitation reactions in thermoluminescent dosimetry crystals (TLD-100) (and ionic conduction). *J. Phys. D: Appl. Phys.* 10: 1261–1266.

Bräunlich, P. (ed.) (1979). *Thermally Stimulated Relaxation in Solids*. ISBN 9783662312605. Berlin: Springer-Verlag.

Bräunlich, P. and Scharmann, A. (1964). Ein einfaches modell für die deutung der thermolumineszenz und der thermisch stimulierten leitfähigkeit von alkalihalogeniden. *Z. Physik* 177: 320–336.

Brow, R.K. (1993). Nature of alumina in phosphate glass: I, structure of sodium aluminophosphate glass. *J. Am. Ceram. Soc.* 76: 913–918.

Brow, R.K. (2000). Review: the structure of simple phosphate glasses. *J. Non-Cryst. Sol.* 263/264: 1–28.

Brow, R.K., Kirkpatrick, R.J., and Turner, G.L. (1993). Nature of alumina in phosphate glass: II, structure of sodium aluminophosphate glass. *J. Am. Ceram. Soc.* 76: 919–928.

Bull, R.K., McKeever, S.W.S., Chen, R., Mathur, V.K., Rhodes, J.F., and Brown, M.D. (1986). Thermoluminescence kinetics for multipeak glow curves produced by the release of electrons and holes. *J. Phys. D: Appl. Phys.* 19: 1321–1334.

Bulur, E. (1996). An alternative technique for optically stimulated luminescence (OSL) experiment. *Radiat. Meas.* 26: 701–709.

Bulur, E. (2000). A simple transformation for converting CW-OSL curves to LM-OSL curves. *Radiat. Meas.* 32: 141–145.

Butts, J.J. and Katz, R. (1967). Theory of RBE for heavy ion bombardment of dry enzymes and viruses. *Radiat. Res.* 30: 855–871.

Carrejo, D.J. and Marshall, J. (2007). What is mathematical modelling? Exploring prospective teacher's use of experiments to connect mathematics to the study of motion. *Math. Educ. Res. J.* 19: 45–76.

Chandler, J., Sholom, S., McKeever, S.W.S., and Hall, H. (2019). Thermoluminescence and phototransferred thermoluminescence dosimetry on mobile phone protective touchscreen glass. *J. Appl. Phys.* 126: 074901.

Chang, W., Koba, Y., Katayose, T., Yasui, K., Omachi, C., Hariu, M., and Saitoh, H. (2017). Correction of stopping power and LET quenching for radiophotoluminescent glass dosimetry in a therapeutic proton beam. *Phys. Med. Biol.* 62: 8869–8881.

Chatterjee, A. and Schaefer, H.J. (1976). Microdosimetric structure of heavy ion tracks in tissue. *Radiat. Environ. Biophys.* 13: 215–227.

Chen, R. (1969a). On the calculation of activation energies and frequency factors from glow curves. *J. Appl. Phys.* 40: 570–585.

Chen, R. (1969b). Glow curves with general order kinetics. *J. Electrochem. Soc.* 116: 1254–1257.

Chen, R. and Kirsh, Y. (1981). *Analysis of Thermally Stimulated Processes*. ISBN 9781483285511. Oxford: Pergamon Press.

Chen, R. and McKeever, S.W.S. (1994). Characterization of nonlinearities in the dose dependence of thermoluminescence. *Radiat. Meas.* 23: 667–673.

Chen, R. and McKeever, S.W.S. (1997). *Theory of Thermoluminescence and Related Phenomena*. ISBN 9789810222956. Singapore: World Scientific.

Chen, R. and Pagonis, V. (2011). *Thermally and Optically Stimulated Luminescence: A Simulation Approach*. ISBN 9780470749272. Chichester: Wiley and Sons.

Chen, R. and Pagonis, V. (2013). On the expected order of kinetics in a series of thermoluminescence (TL) and thermally stimulated conductivity (TSC) peaks. *Nucl. Instrum. Meth. Phys. Res. B* 312: 60–69.

Chen, R. and Pagonis, V. (2014). The role of simulations in the study of thermoluminescence (TL). *Radiat. Meas.* 71: 88. 8–14.

Chen, R. and Pagonis, V. (eds.) (2019). *Advances in Physics and Applications of Optically and Thermally Stimulated Luminescence*. ISBN 97816786345783. New Jersey: World Scientific.

Chernov, V., Mironenko, S., Rogalev, B., Bos, A.J.J., de Haas, J.Th.M., and Delgado, A. (1998). Optical and thermoluminescence properties of LiF:Cu, LiF:Mg,Cu and LiF:Mg,Cu,P single crystals. *Radiat. Meas.* 29: 365–372.

Chernov, V., Piters, T.M., Okuno, E., and Yoshimura, E.M. (2001). Photoluminescence and thermal stability of 5.5 eV and Ti centres in gamma irradiated LiF:Mg,Ti crystals. *Radiat. Meas.* 33: 793–796.

Chruścińska, A. (2010). On some fundamental features of optically stimulated luminescence measurements. *Radiat. Meas.* 45: 991–999.

Chruścińska, A. (2016). Optical trap depths in $Al_2O_3$:C determined by the variable energy of stimulation OSL (VES-OSL) method. *Radiat. Meas.* 90: 94–98.

Chruścińska, A. (2019). Recent experiments and theory of OSL. In: *Advances in Physics and Applications of Optically and Thermally Stimulated Luminescence* (ed. R. Chen and V. Pagonis), Chapter 6, 205–241. ISBN 9781786345783. London and Singapore: World Scientific.

Coleman, A.C. and Yukihara, E.G. (2018). On the validity and accuracy of the initial rise method investigated using realistically simulated thermoluminescence curves. *Radiat. Meas.* 117: 70–79.

Curie, M. (1904). *Radioactive Substances*. English translation of Doctoral Thesis presented to the Faculty of Science, Paris. Greenwood Press, Westpoint, 1961.

Dallas, G.I., Polymeris, G.S., Stefanaki, E.C., Afouxenidis, D., Tsirliganis, N.C., and Kitis, G. (2008). Sample dependent correlation between TL and LM-OSL in $Al_2O_3$:C. *Radiat. Meas.* 43: 335–340.

Daniels, F., Boyd, C.A., and Saunders, D.F. (1953). Thermoluminescence as a research tool. *Science* 117: 343–349.

Delgado, L. and Delgado, A. (1984). Photoluminescence and thermoluminescence of Ti centers in LiF (TLD-100). *J. Appl. Phys.* 55: 515–518.

Di Bartolo, B. (1968). *Optical Interactions in Solids*. ISBN 9789814295741. New York: Wiley.

Discher, M., Woda, C., and Fiedler, I. (2013). Improvement of dose determination using glass display of mobile phones for accident dosimetry. *Radiat. Meas.* 56: 240–243.

Dmitryuk, A.V., Paramzina, S.E., Perminov, A.S., Solov'eva, N.D., and Timofeev, N.T. (1996). Influence of glass composition on the properties of silver-doped radiophotoluminescent phosphate glasses. *J. Non-Cryst. Sol.* 202: 173–177.

Dmitryuk, A.V., Paramzina, S.E., Savvina, O.Ch., and Yashchurzhinskaya, O.A. (1989). Nature of radiophotoluminescence centers in silver-activated phosphate glasses. *Opt. Spectrosc. (USSR)* 66: 626–629.

Dmitryuk, A.V., Perminov, A.S., and Savvina, O.Ch. (1986). Spectroscopic consequences of thermal and photochemical conversion of $Ag^0$ centers in phosphate glasses. *Opt. Spectrosc. (USSR)* 60: 63–65.

Dobrowolska, A., Bos, A.J.J., and Dorenbos, P. (2014). Electron tunnelling phenomena in $YPO_4$:Ce,Ln(Ln=Er,Ho,Nd,Dy). *J. Phys. D: Appl. Phys.* 47: 335301.

Dorenbos, P. (2003a). f→d transition energies of divalent lanthanides in inorganic compounds. *J. Phys.: Condens. Matter* 15: 575–594.

Dorenbos, P. (2003b). Systematic behaviour in trivalent lanthanide charge transfer energies. *J. Phys.: Condens. Matter* 15: 8417–8434.

El-Kinawy, M., El-Nashar, H., and El-Faramawy, N. (2019). New handling of thermoluminescence glow curve deconvolution expressions for different kinetic orders based on OTOR model. *J. Phys.: Conf. Series* 1253: 012012.

Eller, S.A., Ahmed, M.F., Bartz, J.A., Akselrod, M.S., Denis, G., and Yukihara, E.G. (2013). Radiophotoluminescence properties of $Al_2O_3$:C,Mg crystals. *Radiat. Meas.* 56: 179–182.

Engels, F. (1878). *Herr Eugen Dühring's Revolution in Science (Anti-Dühring)*. First Publication. Trans. Burns, E. and ed. Dutt, C.P., 27–28. Quoted in *Oxford Dictionary of Scientific Quotations*. Bynum, E.F., and Porter, R. (eds.). ISBN 0198584091. Oxford, Oxford University Press, 2005.

Evans, B.D. (1993). A review of the optical properties of anion lattice vacancies, and electrical conduction in α-$Al_2O_3$: their relation to radiation-induced electrical degradation. *J. Nucl. Mats.* 219: 202–223.

Evans, B.D., Pogatshnik, G.J., and Chen, Y. (1994). Optical properties of lattice defects in α-$Al_2O_3$. *Nucl. Inst. Meth. Phys. Res. B* 91: 258–262.

Fairchild, R.G., Mattern, P.L., Lengweiler, K., and Levy, P. (1978a). Thermoluminescence of LiF TLD-100: emission spectra measurements. *J. Appl. Phys.* 49: 4512–4522.

Fairchild, R.G., Mattern, P.L., Lengweiler, K., and Levy, P. (1978b). Thermoluminescence of LiF TLD-100: glow-curve kinetics. *J. Appl. Phys.* 49: 4523–4533.

Fan, S., Yu, C., He, D., Li, K., and Hu, L. (2011). Gamma rays induced defect centers in phosphate glass for radio-photoluminescence dosimeter. *Radiat. Meas.* 46: 46–50.

Firszt, F., Męczyńska, H., Łęgowski, S., Zakrzewski, J., Strzałkowski, K., and Wróbel, M. (2004). Persistent photoconductivity and optically stimulated luminescence in $Zn_{0.8}Mg_{0.2}Se$ mixed crystals. *Phys. Stat. Sol. (C)* 1: 916–920.

Fowler, J.F. (1963). Solid state dosimetry. *Phys. Med. Biol.* 8: 1–32.

Fuerstner, M., Brunner, I., Forkel-Wirth, D., Mayer, S., Menzel, H., and Vinke, H. (2005). First calibration of alanine and radio-photo-luminescence dosemeters to a hadronic radiation environment. *Proc. 2005 Particle Accelerator Conf., Knoxville, Tennessee, USA. IEEE*, 3097–3099.

Furetta, C. and Weng, P.-S. (1998). *Operational Thermoluminescence Dosimetry*. ISBN 9810234686. Singapore: World Scientific.

Garlick, G.F.J. and Gibson, A.F. (1948). The electron trap mechanism of luminescence in sulphide and silicate phosphors. *Proc. Phys. Soc.* 60: 574–590.

Gavartin, J.L., Shidlovskaya, E.K., Shluger, A.L., and Varaskin, A.N. (1991). Structure and interaction of impurity-vacancy ($Mg^{2+}$-$V_c^-$) dipoles in crystalline LiF. *J. Phys. Condens. Matter* 3: 2237–2245.

Gaza, R., Yukihara, E.G., McKeever, S.W.S., Avila, O., Buenfil, A.-E., Gamboa-deBuen, I., Rodríguez-Villafuerte, M., Ruiz-Trejo, C., and Brandan, M.-E. (2006). Ionization density dependence of the optically stimulated luminescence dose-response of $Al_2O_3$:C to low-energy charged particles. *Radiat. Prot. Dosim.* 119: 375–379.

Geiß, O.B., Krämer, M., and Kraft, G. (1998). Efficiency of thermoluminescent detectors to heavy charged particles. *Nucl. Instrum. Meth. Res. B* 142: 592–598.

Gieszczyk, W., Bilksi, P., Obryk, B., Olko, P., and Bos, A.J.J. (2013). Measurements of high-temperature emission spectra of highly irradiated LiF:Mg,Cu,P (MCP-N) TL detectors. *Radiat. Meas.* 56: 183–186.

Gieszczyk, W., Bilksi, P., Olko, P., and Obryk, B. (2014). Radial distribution of dose within heavy charged particle tracks – models and experimental verification using LiF:Mg,Cu,P TL detectors. *Radiat. Meas.* 71: 242–246.

Gobrecht, H. and Hofmann, D. (1966). Spectroscopy of traps by fractional glow technique. *J. Phys. Chem. Solid.* 27: 509–522.

Gong, Z., Li, B., and Li, S.-H. (2012). Study of the relationship between infrared stimulated luminescence and blue light stimulated luminescence for potassium-feldspar from sediments. *Radiat. Meas.* 47: 841–845.

Granville, D.A., Sahoo, N., and Sawakuchi, G.O. (2014). Calibration of the $Al_2O_3$:C optically stimulation luminescence (OSL) signal for linear energy transfer (LET) measurements in therapeutic proton beams. *Phys. Med. Biol.* 59: 4295–4310.

Grimmeis, H.G. and Ledebo, L.-Å. (1975a). Photo-ionization of deep impurity levels in semiconductors with non-parabolic bands. *J. Phys. C: Sol. Stat. Phys.* 8: 2615–2626.

Grimmeis, H.G. and Ledebo, L.-Å. (1975b). Spectral distribution of photoionization cross sections by photoconductivity measurements. *J. Appl. Phys.* 46: 2155–2162.

Grossberg, M., Krustok, J., Jagomägi, A., Leon, M., Arushanov, E., Nateprov, A., and Bodnar, I. (2007). Investigation of potential and compositional fluctuations in $CuGa_3Se_5$ crystals using photoluminescence spectroscopy. *Thin Solid Films* 515: 6204–6207.

Grosshans, D.R., Duman, J.G., Gaber, M.W., and Sawakuchi, G. (2018). Particle radiation induced neurotoxicity in the central nervous system. *Int. J. Part. Ther.* 5: 74–83. Epub 2018, PMC6871599; PMID: 31773021.

Halperin, A., Braner, A.A., Ben-Zvi, A., and Kristianpoller, N. (1960). Thermal activation energies in NaCl and KCl crystals. *Phys. Rev.* 117: 416–422.

Harvey, E.N. (1957). *A History of Luminescence from Earliest Times until 1900*. ISBN 9780486442587. Philadelphia: The American Philosophical Society.

Hayes, W. and Stoneham, A.M. (1985). *Defects and Defect Processes in Nonmetallic Solids*. ISBN 9780471897910. New York: Wiley-Interscience.

Hoogenstraaten, W. (1958). Electron traps in zinc sulphide phosphors. *Philips Res. Rep.* 13: 515–693.

Hornyak, W.F. and Chen, R. (1989). Thermoluminescence and phosphorescence with a continuous distribution of activation energies. *J. Lumin.* 44: 73–81.

Horowitz, A. and Horowitz, Y.S. (1990). Optimisation of LiF:Mg,Cu,P for radiation protection dosimetry. *Radiat. Prot. Dosim.* 33: 267–270.

Horowitz, Y.S. (ed.) (1984). *Thermoluminescent and Thermoluminescent Dosimetry, Vols, I, II and III*. ISBNs 9780849356643, 9780429292248 and 9780849356667. Boca Raton: CRC Press.

Horowitz, Y.S. (2001). Theory of thermoluminescence gamma dose response: the unified interaction model. *Nucl. Instrum. Meth. Phys. Res. B* 184: 68–84.

Horowitz, Y.S., Fuks, E., Datz, H., Oster, L., Livingstone, J., and Rosenfeld, A. (2011). Mysteries of LiF TLD response following high ionization density irradiation: nanodosimetry and track structure theory, dose response and glow curve shapes. *Radiat. Prot. Dosim.* 145: 356–372.

Horowitz, Y.S., Mahajna, S., Rosenkrantz, M., and Yossian, D. (1996a). Unified theory of gamma and heavy charged particle TL supralinearity: the track/defect interaction model. *Radiat. Prot. Dosim.* 65: 7–12.

Horowitz, Y.S., Moscovitch, M., and Wilt, M. (1986). Computerized glow curve deconvolution applied to ultralow dose LiF thermoluminescence dosimetry. *Nucl. Instrum. Meth. Phys. Res. A* 244: 556–564.

Horowitz, Y.S., Oster, L., Biderman, S., and Einav, Y. (2002b). The composite structure of peak 5 in the glow curve of LiF:Mg,Ti (TLD-100): confirmation of peak 5a arising from a locally trapped electron-hole configuration. *Radiat. Prot. Dosim.* 100: 123–126.

Horowitz, Y.S., Oster, L., and Eliyahu, I. (2019). Modeling the effects of ionization density in thermoluminescence mechanisms and dosimetry. In: *Advances in Physics and Applications of Optically and Thermally Stimulated Luminescence* (ed. R. Chen and V. Pagonis), Chapter 3, 83–129. ISBN 9781786345783. London and Singapore: World Scientific.

Horowitz, Y.S. and Rosenkrantz, M. (1990). Track interaction theory for heavy charged particle TL supralinearity. *Radiat. Prot. Dosim.* 31: 71–76.

Horowitz, Y.S., Rosenkrantz, M., Mahajna, S., and Yossian, D. (1996b). The track interaction model for alpha particle induced thermoluminescence supralinearity: dependence of the supralinearity on the vector properties of the alpha particle radiation field. *J. Phys. D: Appl. Phys.* 29: 205–214.

Horowitz, Y.S., Satinger, D., Brandon, M.-E., Avila, O., and Rodriguez-Villafuerte, M. (2002a). Supralinearity of peaks 5a, 5 and 5b in TLD-100 following 6.8 MeV and 2.6 MeV He ion irradiation: the Extended Track Interaction Model. *Radiat. Prot. Dosim.* 100: 95–98.

Horowitz, Y.S., Satinger, D., Yossian, D., Brandon, M.-E., Buenfil, A.E., and Gamboa-deBuen, I. (1999). Ionization density effects in the thermoluminescence of TLD-100: computerized $T_m$-$T_{stop}$ glow curve analysis. *Radiat. Prot. Dosim.* 84: 239–242.

Horowitz, Y.S., Siboni, D., Oster, L., Livingstone, J., Guatelli, S., Rosenfeld, A., Emfietzoglou, D., Bilski, P., and Obryk, B. (2012). Alpha particle and proton relative thermoluminescence efficiencies in LiF:Mg,Cu,P: is track structure theory up to the task? *Radiat. Protect, Dosim.* 150: 359–374.

Hsu, S.M., Yung, S.W., Brow, R.K., Hsu, W.L., Lu, C.C., Wu, F.B., and Ching, S.H. (2010). Effect of silver concentration on the silver-activated phosphate glass. *Mat. Chem. Phys.* 123: 172–176.

Huntley, D.J. (2006). An explanation of the power-law decay of luminescence. *J. Phys.: Condens. Matter* 18: 1359–1365.

Huntley, D.J., Godfrey-Smith, D.I., and Thewalt, M.L.W. (1985). Optical dating of sediments. *Nature* 313: 105–107.

Huntley, D.J., Short, M.A., and Dunphy, K. (1996). Deep traps in quartz and their use for optical dating. *Can. J. Phys.* 74: 81091.

Hütt, G., Jack, I., and Tchonka, J. (1988). Optical dating: K-feldspars optical response stimulation spectra. *Quat. Sci. Rev.* 7: 381–385.

ICRU. (2019). International Commission on Radiation Units and Measurements, Report 94: methods for initial-phase assessment of individual doses following acute exposure to ionizing radiation. *J. ICRU* 19: 1–162.

Iwao, M., Takase, H., Shiratori, D., Nakauchi, D., Kato, T., Kawaguchi, N., and Yanagida, T. (2021). Ag-doped phosphate glass with high weathering resistance for RPL dosimeter. *Radiat, Meas.* 140: 106492.

Jain, M., Bøtter-Jensen, L., and Thomsen, K.J. (2007). High local ionization density effects in x-ray excitations deduced from optical stimulation of trapped charge in $Al_2O_3$:C. *J. Phys.: Condens. Matter* 19: 116201.

Jain, M., Guralink, B., and Anderson, M.T. (2012). Stimulated luminescence emission from localized recombination in randomly distributed defects. *J. Phys.: Condens. Matter* 24: 385402.

Jain, M., Sohbati, R., Guralnik, B., Murray, A.S., Kook, M., Lapp, T., Prasad, A.K., Thomsen, K.J., and Buylaert, J.P. (2015). Kinetics of infrared stimulated luminescence from feldspars. *Radiat. Meas.* 81: 242–250.

Jaros, M. (1977). Wave functions and optical cross sections associated with deep centers in semiconductors. *Phys. Rev. B* 16: 3694–3706.

Jiang, H.X. and Lin, J.Y. (1990). Percolation transition of persistent photoconductivity in II-VI mixed crystals. *Phys. Rev. Letts.* 64: 2547–2550.

Jones, W. (1781). *Physiological Disquisitions*, 148. Quoted in *Oxford Dictionary of Scientific Quotations*. Bynum, E.F., and Porter, R. (eds.). ISBN 0198584091. Oxford, Oxford University Press, 2005.

Kars, R.H., Poolton, N.R.J., Jain, M., Ankjærgaard, C., Dorenbos, P., and Wallinga, J. (2013). On the trap depth of the IR-sensitive trap in Na- and K-feldspar. *Radiat. Meas.* 59: 103–113.

Kelly, P., Laubitz, M.J., and Bräunlich, P. (1971). Exact solutions of the kinetics equations governing thermally stimulated luminescence and conductivity. *Phys. Rev. B* 4: 1960–1968, and Errata (1972). *Phys. Rev. B*, 5, 3370.

Kiisk, V. (2013). Deconvolution and simulation of thermoluminescence glow curves with Mathcad. *Radiat. Prot. Dosim.* 156: 261–267.

Kim, H., Kim, M.C., Lee, J., Chang, I., Lee, S.K., and Kim, J.-L. (2019). Thermoluminescence of AMOLED substrate glasses in recent mobile phones for retrospective dosimetry. *Radiat. Meas.* 122: 53–56.

Kitis, G., Gomez-Ros, J.M., and Tuyn, J.W.N. (1998). Thermoluminescence glow-curve deconvolution functions for first, second and general orders of kinetics. *J. Phys. D: Appl. Phys.* 31: 2636–2641.

Kitis, G. and Pagonis, V. (2008). Computerized curve deconvolution analysis for LM-OSL. *Radiat. Meas.* 43: 737–741.

Kitis, G. and Pagonis, V. (2013). Analytical solutions for stimulated luminescence emission from tunneling recombination in random distributions of defects. *Radiat. Meas.* 137: 109–115.

Kitis, G. and Pagonis, V. (2019). On the resolution of overlapping peaks in complex thermoluminescence glow curves. *Nucl. Instrum. Meth. Phys. Res.* A913: 78–84.

Klasens, H.A. (1946). Transfer of energy between centers in zinc sulphide phosphors. *Nature* 158: 306–307.

Kodaira, S., Kusumoto, T., Kitamura, H., Yanagida, Y., and Koguchi, Y. (2020). Characteristics of fluorescent nuclear track detection with $Ag^+$-activated phosphate glass. *Radiat. Meas.* 132: 106252.

Kook, M., Kumar, R., Murray, A.S., Thomsen, K.J., and Jain, M. (2018). Instrumentation for the non-destructive optical measurement of trapped electrons in feldspar. *Radiat. Meas.* 120: 247–252.

Kortov, V.S., Milman, I.I., Moiseykin, E.V., Nikiforov, S.V., and Ovchinnikov, M.M. (2006). Deep-trap competition model for TL in $\alpha$-$Al_2O_3$:C heating stage. *Radiat. Prot. Dosim.* 119: 41–44.

Kortov, V.S., Milman, I.I., and Nikiforov, S.V. (1999). The effect of deep traps on the main features of thermoluminescence in dosimetric $\alpha$-$Al_2O_3$:C crystals. *Radiat. Prot. Dosim.* 84: 35–38.

Kortov, V.S., Milman, I.I., and Nikiforov, S.V. (2002). Thermoluminescent and dosimetric properties of anion-defective $\alpha$-$Al_2O_3$:C single crystals with filled deep traps. *Radiat. Prot. Dosim.* 100: 75–78.

Kouwenberg, J.J.M., Kremers, G.J., Slotman, J.A., Wolterbeek, H.T., Houtsmuller, A.B., Denkova, A.G., and Bos, A.J.J. (2018). Alpha particle spectroscopy using FNTD and SIM super-resolution microscopy. *J. Microsc.* 270: 326–334.

Kouwenberg, J.J.M., Ulrich, L., Jäkel, O., and Greilich, S. (2016). A 3D feature point tracking method for ion radiation. *Phys. Med. Biol.* 61: 4088–4104.

Kovacs, A., Baranyai, M., Wojnarovits, L., McLaughlin, W.L., Miller, S.D., Miller, A., Fuochi, P.G., Lavalle, M., and Slezsak, I. (2000). Application of the Sunna dosimeter film in gamma and electron beam radiation processing. *Radiat. Phys. Chem.* 57: 691–695.

Krbetschek, M.R., Götze, J., Dietrich, A., and Trautmann, T. (1997). Spectral information from minerals relevant for luminescence dating. *Radiat. Meas.* 27: 695–748.

Kristianpoller, N., Chen, R., and Israeli, M. (1974). Dose dependence of thermoluminescence peaks. *J. Phys. D: Appl. Phys.* 7: 1063–1072.

Kumar, R., Kook, M., and Jain, M. (2020). Understanding the metastable states in K-Na aluminosilicates using novel site-selective excitation-emission spectroscopy. *J. Phys. D: Appl. Phys.* 53: 465301.

Kumar, R., Kook, M., Murray, A.S., and Jain, M. (2018). Towards direct measurement of electrons in metastable states in K-feldspar: do infrared-photoluminescence and radioluminescence probe the same trap? *Radiat. Meas.* 120: 7–13.

Kurobori, T., Yanagida, Y., Kodaira, S., and Shirao, T. (2017). Fluorescent nuclear track images of Ag-activated phosphate glass irradiated with photons and heavy charged particles. *Nucl. Instrum. Meth. Phys. Res. A* 855: 25–31.

Kurobori, T., Zheng, W., Miyamoto, Y., Nanto, H., and Yamamoto, T. (2010). The role of silver in the radiophotoluminescent properties in silver-activated phosphate glass and sodium chloride crystal. *Opt. Mat.* 32: 1231–1236.

Landreth, J.L. and McKeever, S.W.S. (1985). Some observations on the optical absorption bands in LiF:Mg,Ti. *J. Phys. D: Appl. Phys.* 18: 1919–1933.

Landsberg, P.T. (2003). *Recombination in Semiconductors*. ISBN 0521543436. Cambridge: Cambridge University Press.

Lavon, A., Eliyahu, I., Oster, L., and Horowitz, Y.S. (2015). The modified unified interaction model: incorporation of dose-dependent localized recombination. *Radiat. Prot. Dosim.* 163: 362–372.

Lawless, J.L., Chen, R., and Pagonis, V. (2009). Sublinear dose dependence of thermoluminescence and optically stimulated luminescence prior to the approach to saturation level. *Radiat. Meas.* 44: 606–610.

Leverenz, H.W. (1949). Luminescent solids (phosphors). *Science* 109: 183–195.

Levita, M., Schlesinger, T., and Friedland, S.S. (1976). LiF dosimetry based on radiophotoluminescence (RPL). *IEEE Trans. Nucl. Sci.* 23: 667–674.

Levy, P.W. (1984). Thermoluminescence systems with two or more glow peaks described by anomalous parameters. *Nucl. Instrum. Meth. Phys. Res. B* 1: 436–444.

Lewandowski, A.C., Markey, B.G., and McKeever, S.W.S. (1994). Analytical description of thermally stimulated luminescence and conductivity without the quasiequilibrium approximation. *Phys. Rev. B* 49: 8029–8045.

Lewandowski, A.C. and McKeever, S.W.S. (1991). Generalized description of thermally stimulated processes without the quasiequilibrium approximation. *Phys. Rev. B* 43: 8163–8178.

Li, B. and Li, S.-H. (2013). The effect of band-tail states on the thermal stability of the infrared stimulated luminescence from K-feldspar. *J. Lumin.* 136: 5–10.

Li, Y.P. and Ching, W.Y. (1985). Band structures of polycrystalline forms of silicon dioxide. *Phys. Rev. B* 31: 2172–2179.

Lommler, B., Pitt, E., and Scharmann, A. (1992). Detection of neutron-induced heavy charged particle tracks in RPL glasses. *Radiat. Prot. Dosim.* 44: 375–377.

Lucovsky, G. (1964). On the photoionization of deep impurity centers in semiconductors. *Sol. State Commun.* 3: 299–302.

Mahajni, S. and Horowitz, Y.S. (1997). The unified interaction model applied to the gamma ray induced supralinearity and sensitization of peak 5 in LiF:Mg,Ti (TLD-100). *J. Phys. D: Appl. Phys.* 30: 2603–2619.

Mandowska, E., Bilski, P., Obryk, B., Mandowski, A., Olko, P., and Kim, J. (2010). Spectrally resolved thermoluminescence of highly irradiated LiF:Mg,Cu,P detectors. *Radiat. Meas.* 45: 579–582.

Mandowska, E., Bilski, P., Ochab, E., Świątek, J., and Mandowski, A. (2002). TL emission spectra of differently doped LiF:Mg detectors. *Radiat. Prot. Dosim.* 100: 451–454.

Mandowski, A. (2004). Semi-localized transitions model for thermoluminescence. *J. Phys. D: Appl. Phys.* 38: 17–21.

Mandowski, A. (2006). Topology-dependent thermoluminescence kinetics. *Radiat. Prot. Dosim.* 119: 23–28.

Mandowski, A. (2008). Semi-localized transitions model – general formulation and classical limits. *Radiat. Meas.* 43: 199–202.

Markey, B.G., Colyott, L.E., and McKeever, S.W.S. (1995). Time-resolved optically stimulated luminescence from $\alpha$-$Al_2O_3$:C. *Radiat. Meas.* 24: 457–463.

Markey, B.G., McKeever, S.W.S., Akselrod, M.S., Bøtter-Jensen, L., Agersnap Larsen, N., and Colyott, L.E. (1996). The temperature dependence of optically stimulated luminescence from $\alpha$-$Al_2O_3$:C. *Radiat. Prot. Dosim.* 65: 185–190.

Masillon, -J.L.G., Gamboa-deBuen, I., and Brandan, M.E. (2006). Onset of supralinear response in TLD-100 exposed to $^{60}$Co gamma rays. *J. Phys. D: Appl. Phys.* 39: 262–268.

Masillon, -J.L.G., Johnston, C.S.N., and Kohanoff, J. (2019). On the role of magnesium in a LiF:Mg,Ti thermoluminescent dosimeter. *J. Phys.: Condens. Matter* 31: 025502.

May, C.E. and Partridge, J.A. (1964). Thermoluminescent kinetics of alpha-irradiated alkali halides. *J. Chem. Phys.* 40: 1401–1408.

McKeever, J., Macintyre, D.J., Taylor, S.R., McKeever, S.W.S., Horowitz, A., and Horowitz, Y.S. (1993a). Diffuse reflectance and transmission measurements on LiF:Mg,Cu,P powders and single crystals. *Radiat. Prot. Dosim.* 47: 123–127.

McKeever, J., Walker, F.D., and McKeever, S.W.S. (1993b). Properties of the thermoluminescence emission from LiF(Mg,Cu,P). *Nucl. Tracks. Radiat. Meas.* 21: 179–183.

McKeever, S.W.S. (1980). On the analysis of complex thermoluminescence glow-curves: resolution into individual peaks. *Phys. Status Solidi. (A)* 62: 331–340.

McKeever, S.W.S. (1984). Optical absorption and luminescence in lithium fluoride TLD-100. *J. Appl. Phys.* 56: 2883–2889.

McKeever, S.W.S. (1985). *Thermoluminescence of Solids*. ISBN 0521368111. Cambridge: Cambridge University Press.

McKeever, S.W.S. (1990). 5.5 eV optical absorption, supralinearity and sensitization of thermoluminescence in LiF TLD-100. *J. Appl. Phys.* 68: 724–731.

McKeever, S.W.S. (1991). Measurements of emission spectra during thermoluminescence (TL) from LiF:Mg,Cu,P TL dosimeters. *J. Phys.D:Appl. Phys.* 24: 988–996.

McKeever, S.W.S. and Akselrod, M.S. (1999). Radiation dosimetry using pulsed optically stimulated luminescence of $Al_2O_3$:C. *Radiat. Prot. Dosim.* 84: 317–320.

McKeever, S.W.S., Akselrod, M.S., Colyott, L.E., Agersnap, Larsen, N., Polf, J.C., and Whitley, V. (1999). Characterization of $Al_2O_3$: C for use in thermally and optically stimulated luminescence dosimetry. *Radiat. Prot. Dosim.* 84: 163–168.

McKeever, S.W.S., Akselrod, M.S., and Markey, B.G. (1996). Pulsed optically stimulated dosimetry using alpha-$Al_2O_3$:C. *Radiat. Prot. Dosim.* 65: 267–272.

McKeever, S.W.S., Moscovitch, M., and Townsend, P.D. (1995). *Thermoluminescence Dosimetry Materials: Properties and Uses.* ISBN 1870965191. Ashford, Kent: Nuclear Technology Publishing.

McKeever, S.W.S. and Sholom, S. (2019). Luminescence measurements for retrospective dosimetry. In: *Advances in Physics and Applications of Optically and Thermally Stimulated Luminescence* (ed. R. Chen and V. Pagonis), Chapter 9, 319–362. ISBN 97816786345783. World Scientific Publishing Europe Ltd.

McKeever, S.W.S. and Sholom, S. (2021). Trap level spectroscopy of disordered materials using thermoluminescence: an application to aluminosilicate glass. *J. Lumin.* 234: 117950.

McKeever, S.W.S., Sholom, S., and Chandler, J.R. (2020b). Developments in the use of thermoluminescence and optically stimulated luminescence from mobile phones in emergency dosimetry. *Radiat. Prot. Dosim.* 192: 205–235.

McKeever, S.W.S., Sholom, S., and Shrestha, N. (2019). Observations regarding the build-up effect in radiophotoluminescence of silver-doped phosphate glasses. *Radiat. Meas.* 123: 13–20.

McKeever, S.W.S., Sholom, S., Shrestha, N., and Klein, D.M. (2020a). Build-up of radiophotoluminescence (RPL) in Ag-doped phosphate glass in real-time both during and after exposure to ionizing radiation: a proposed model. *Radiat. Meas.* 132: 106246.

McKinlay, A.F. (1981). *Thermoluminescence Dosimetry.* ISBN 9780852745205. Bristol: Adam Hilger.

McLaughlin, W.L., Lucas, A.S., Kapsar, B.M., and Miller, A. (1979). Electron and gamma-ray dosimetry using radiation-induced color centers in LiF. *Radiat. Phys. Chem.* 14: 467–480.

Meijvogel, K., Bos, A.J.J., Bilski, P., and Olko, P. (1995). Thermoluminescence emission characteristics of LiF:Mg,Cu,P with different dopant concentrations. *Radiat. Meas.* 24: 411–416.

Miller, S.D. and Endres, G.W.R. (1990). Laser-induced optically stimulated M centre luminescence in LiF. *Radiat. Prot. Dosim.* 33: 59–62.

Mirov, S.B. and Dergachev, A.Y. (1997). Powerful, room-temperature stable LiF:$F_2^{+**}$ tunable laser. *SPIE* 2986: 162–173.

Mirov, S.B. and Basiev, T. (1995). Progress in color center lasers. *IEEE J. Selected Topics in Quantum Electronics* 1: 22–30.

Mische, E.F. and McKeever, S.W.S. (1989). Mechanisms of supralinearity in lithium fluoride thermoluminescence dosimeters. *Radiat. Prot. Dosim.* 29: 159–175.

Mishra, D.R., Soni, A., Rawat, N.S., Kulkarni, M.S., Bhatt, B.C., and Sharma, D.N. (2011). Method of measuring thermal assistance energy associated with OSL traps in $\alpha$-$Al_2O_3$:C phosphor. *Radiat. Meas.* 46: 635–642.

Miyamoto, Y., Kinoshita, K., Koyama, S., Takei, Y., Nanto, H., Yamamoto, T., Sakakura, M., Shimotsuma, Y., Miura, K., and Hirao, K. (2010a). Emission and excitation mechanism of radiophotoluminescence in $Ag^+$-activated phosphate glass. *Nucl. Instrum. Meth. Phys. Res. A.* 619: 71–74.

Miyamoto, Y., Nanto, H., Kurobori, T., Fujimoto, Y., Yanagida, T., Ueda, J., Tanabe, S., and Yamamoto, T. (2014). RPL in alpha particle irradiated $Ag^+$-doped phosphate glass. *Radiat. Meas.* 71: 529–532.

Miyamoto, Y., Takei, Y., Nanto, H., Kurobori, T., Konnai, A., Yanagida, T., Yohikawa, A., Shimotsuma, Y., Sakakura, M., Miura, K., Hirao, K., Nagashima, Y., and Yamamoto, T. (2011). Radiophotoluminescence from silver-doped phosphate glass. *Radiat. Meas.* 46: 1480–1483.

Miyamoto, Y., Yamamoto, T., Kinoshita, K., Koyama, S., Takei, Y., Nanto, H., Shimotsuma, Y., Sakakura, M., Miura, K., and Hirao, K. (2010b). Emission mechanism of radiophotoluminescence in Ag-doped phosphate glass. *Radiat. Meas.* 45: 546–549.

Montereali, R.M., Almaviva, S., Bonfigli, F., Cricenti, A., Faenov, A., Flora, F., Gaudio, P., Lai, A., Martellucci, S., Nichelatti, E., Pikuz, T., Reale, L., Richetta, M., and Vincenti, M.A. (2010). Lithium fluoride thin-film detectors for soft X-ray imaging at high spatial resolution. *Nucl. Instrum. Meth. Phys. Res., A* 623: 758–762.

Morse, H.W. (1905). The spectra of weak luminescences. *Astro. Phys. J.* XXI: 83–100.

Morthekai, P., Thomas, J., Pandian, M.S., Balaram, V., and Singhvi, A.K. (2012). Variable range hopping mechanism in band-tail states of feldspars: a time-resolved IRSL study. *Radiat. Meas.* 47: 857–863.

Mott, N.F. (1978). Recombination: a survey. *Solid State Electronics* 21: 1275–1280.

Mott, N.F. and Davis, E.A. (2012). *Electronic Processes in Non-Crystalline Materials*, 2e. ISBN 97801909645336. Oxford: Oxford University Press.

Mott, N.F. and Gurney, R.W. (1948). *Electronic Processes in Ionic Crystals*. ISBN 9780486611839. Oxford: Clarendon Press.

Mysovsky, S., Rogalev, B., and Chenov, V. (1995). Stabilization of *H*-centres in irradiated LiF:Mg crystals. *Radiat. Eff. and Def. in Solids* 134: 493–497.

Nail, I., Horowitz, Y.S., Oster, L., and Biderman, S. (2002). The Unified Interaction Model applied to LiF:Mg,Ti (TLD-100): properties of the luminescent and competitive centers during sensitization. *Radiat. Prot. Dosim.* 102: 295–304.

Nakajima, T., Murayama, Y., and Matsuzawa, T. (1979). Preparation and dosimetric properties of a highly sensitive LiF thermoluminescent dosimeter. *Health Phys.* 36: 79–82.

Nakajima, T., Murayama, Y., Matsuzawa, T., and Koyano, A. (1978). Development of a new highly sensitive LiF thermoluminescence dosimeter and its applications. *Nucl. Instrum. Meth.* 157: 155–162.

Nichols, E.L. and Merritt, E. (1912). *Studies in Luminescence*. ISBN 9781347330654. Carnegie Institute of Washington.

Nikiforov, S.V., Milman, I.I., and Kortov, V.S. (2001). Thermal and optical ionization of *F*-centers in the luminescence mechanism of anion-defective corundum crystals. *Radiat. Meas.* 33: 547–551.

Noras, J.M. (1980). Photoionization and phonon coupling. *J. Phys. C: Solid St. Phys.* 13: 4779–4789.

Oberhofer, M. and Scharmann, A. (eds.) (1981). *Applied Thermoluminescence Dosimetry*. ISBN 0852745443. Bristol: Adam Hilger.

Obryk, B., Bilski, P., Glaser, M., Fuerstner, M., Budzanowski, M., Olko, P., and Pajor, A. (2010). The response of TL lithium fluoride detectors to 24 GeV/c protons and for doses ranging up to 1 MGy. *Radiat. Meas.* 45: 643–645.

Olko, P., Bilski, P., Budzanowski, M., Waligórski, M.P.R., and Reitz, G. (2002). Modeling the response of thermoluminescence detectors exposed to low- and high-LET radiation fields. *J. Radiat. Res.* 43: S59–S62.

Pagonis, V. (2019). Recent advances in the theory of quantum tunneling for luminescence phenomena. In: *Advances in Physics and Applications of Optically and Thermally Stimulated Luminescence* (ed. R. Chen and V. Pagonis), Chapter 2, 37–81. ISBN 9781786345783. London and Singapore: World Scientific.

Pagonis, V., Morthekai, P., and Kitis, G. (2014). Kinetic analysis of thermoluminescence glow curves in feldspar: evidence for a continuous distribution of energies. *Geochronometria* 41: 168–177.

Pagonis, V., Morthekai, P., Singhvi, A.K., Thomas, J., Balaram, V., Kitis, G., and Chen, R. (2012). Time-resolved infrared stimulated luminescence signals in feldspars: analysis based on exponential and stretched exponential functions. *J. Lumin.* 132: 2330–2340.

Pagonis, V., Phan, H., Ruth, D., and Kitis, G. (2013). Further investigations of tunneling recombination processes in random distributions of defects. *Radiat. Meas.* 58: 66–74.

Parisi, A. (2020). Further clarification on the microdosimetric $d(z)$ model in response to 'The recent success of microdosimetry' by Y.S. Horowitz. *Radiat. Prot. Dosim.* 189: 534–538.

Parisi, A., Dabin, J., Schoonjans, W., Van Hoey, O., Mégret, P., and Vanhavere, F. (2019b). Photon energy response of LiF:Mg,Ti (MTS) and LiF:Mg,Cu,P (MCP) thermoluminescent detectors: experimental measurements and microdosimetric modeling. *Radiat. Phys. Chem.* 163: 67–73.

Parisi, A., Olko, P., Swakón, J., Horwacik, T., Jabłonski, H., Malinowski, L., Nowak, T., Struelens, L., and Vanhavere, F. (2020). Mitigation of the proton-induced low temperature anomaly of LiF:Mg,Cu,P detectors using a post-irradiation readout thermal protocol. *Radiat. Meas.* 132: 106233.

Parisi, A., Van Hoey, O., Mégret, P., and Vanhavere, F. (2019a). Microdosimetric modeling of the relative luminescence efficiency of LiF:Mg,Cu,P (MCP) detectors exposed to charged particles. *Radiat. Prot. Dosim.* 183: 172–176.

Parisi, A., Van Hoey, O., and Vanhavere, F. (2018). Microdosimetric modeling of the relative luminescence efficiency of LiF:Mg,Ti (MTS) detectors exposed to charged particles. *Radiat. Prot. Dosim.* 180: 192–195.

Perry, J.A. (1987). *RPL Dosimetry: Radiophotoluminescence in Health Physics*. ISBN 085274272X. Bristol and Philadelphia: Adam Hilger.

Piesch, E., Burgkhardt, B., Fischer, M., Röber, H.G., and Ugi, S. (1986). Properties of radiophotoluminescent glass dosemeter systems using pulsed laser UV excitation. *Radiat. Prot. Dosim.* 17: 293–297.

Piesch, E., Burgkhardt, B., and Vilgis, M. (1990). Photoluminescence dosimetry: progress and present state of art. *Radiat. Prot. Dosim.* 33: 215–226.

Piters, T.M. and Bos, A.J.J. (1993a). Thermoluminescence emission spectra of LiF (TLD-100) after different thermal treatments. *Radiat. Prot. Dosim.* 47: 91–94.

Piters, T.M. and Bos, A.J.J. (1993b). A model for the influence of defect interactions during heating on thermoluminescence in LiF:Mg,Ti (TLD-100). *J. Phys. D: Appl. Phys.* 26: 2255–2265.

Piters, T.M., Meulemans, W.H., and Bos, A.J.J. (1993). An automated research facility for measuring thermoluminescence emissions spectra using an optical multichannel analyzer. *Rev. Sci. Instrum.* 64: 109–117.

Polf, J.C. (2000). *The role of oxygen vacancies in thermoluminescence processes in $Al_2O_3$:C*. M.S. Thesis, Oklahoma State University.

Polymeris, G.S. and Kitis, G. (2019). Thermally assisted optically stimulated luminescence (TA-OSL). In: *Advances in Physics and Applications of Optically and Thermally Stimulated Luminescence* (ed. R. Chen and V. Pagonis), Chapter 4, 131–171. ISBN 9781786345783. London and Singapore: World Scientific.

Polymeris, G.S., Pagonis, V., and Kitis, G. (2017). Thermoluminescence glow curves in preheated feldspar samples: an interpretation based on random defect distributions. *Radiat. Meas.* 97: 20–27.

Poolton, N.R.J., Bøtter-Jensen, L., and Johnsen, O. (1996). On the relationship between luminescence and excitation spectra and feldspar mineralogy. *Radiat. Meas.* 26: 93–101.

Poolton, N.R.J., Kars, R.H., Wallinga, J., and Bos, A.J.J. (2009). Direct evidence for the participation of band-tails and excited-state tunnelling in the luminescence of irradiated feldspars. *J. Phys.: Condens. Matter* 21: 485505.

Poolton, N.R.J., Ozanyan, K.B., Wallinga, J., Murray, A.S., and Bøtter-Jensen, L. (2002b). Electrons in feldspars II: a consideration of the influence of conduction band-tails states on luminescence processes. *Phys. Chem. Minerals* 29: 217–225.

Poolton, N.R.J., Wallinga, J., Murray, A.S., Bulur, E., and Bøtter-Jensen, L. (2002a). Electrons in feldspars I: on the wavefunction of electrons trapped at simple lattice defects. *Phys. Chem. Minerals* 29: 210–216.

Portal, G. (1981). Preparation and properties of principal TL products. In: *Applied Thermoluminescence Dosimetry* (ed. M. Oberhofer and A. Scharmann), Chapter 6, 97–122. ISBN 0852745443. Bristol: Adam Hilger Ltd.

Prasad, A.M., Lapp, T., Kook, M., and Jain, M. (2016). Probing luminescence centers in Na rich feldspar. *Radiat. Meas.* 90: 292–297.

Prasad, A.M., Poolton, N.R.J., Kook, M., and Jain, M. (2017). Optical dating in a new light: a direct, non-destructive probe of trapped electrons. *Sci. Rep.* 7: 12097.

Przibram, K. (1923). Verfärbung und Lumineszenz durch Becquerelstrahlen. *Zeit. Physik* 20: 196–208.

Przibram, K. (1927). Über die Verfärbung des gepreßten Steinsalzes. In: *Mitteilungen aus dem Institut für Radium forschung, Nr. 196*, 43–56. Hölder-Pichler-Tempsky AG.

Przibram, K. and Kara-Michailowa, E. (1922). Über Radiolumineszenz und Radiophoto lumineszenz. *Wiener Ber.* 131: 511.

Puchalska, M. and Bilski, P. (2006). GlowFit – a new tool for thermoluminescence glow-curve deconvolution. *Radiat. Meas.* 41: 659–664.

Radzhabov, E.A. and Nepomnyachikh, A.I. (1981). Neutral magnesium atoms on anion sites in LiF. *Phys. Stat. Sol. (B)* 108: k75–k78.

Ramírez, R., Tardío, M., González, R., Chen, Y., and Kokta, M.R. (2005). Photochromism of vacancy-related defects in thermochemically reduced $\alpha$-$Al_2O_3$:Mg single crystals. *Appl. Phys. Lett.* 86: 081914.

Randall, J.T. and Wilkins, M.H.F. (1945a). Phosphorescence and electron traps – I: the study of trap distributions. *Proc. Roy. Soc. A* 184: 366–389.

Randall, J.T. and Wilkins, M.H.F. (1945b). Phosphorescence and electron traps – II: the interpretation of long-period phosphorescence. *Proc. Roy. Soc. A* 184: 390–407.

Rasheedy, M.S. (1993). On the general-order kinetics of the thermoluminescence glow peak. *J. Phys.: Condens. Matter* 5: 633–636.

Redfield, D. (1963). Effect of defect fields on the optical absorption edge. *Phys. Rev.* 130: 916–918.

Remy, K., Sholom, S., Obryk, B., and McKeever, S.W.S. (2017). Optical absorption in LiF, LiF:Mg, LiF:Mg,Cu,P irradiated with high gamma and beta doses. *Radiat. Meas.* 106: 113–117.

Ridley, R.K. (1988). *Quantum Processes in Semiconductors*. ISBN 0198511701. Oxford: Clarendon Press.

Riedesel, S., King, G.E., Prasad, A.K., Kumar, R., Finch, A.A., and Jain, M. (2019). Optical determination of the width of the band-tail states, and the excited and ground state energies of the principal dosimetric trap in feldspar. *Radiat. Meas.* 125: 40–51.

Rogalev, B.I., Mysovsky, S., Nepomnyaschikh, A.I., and Chernov, V.G. (1990). Mechanism of storage of ionizing radiation energy in LiF:Mg,Ti crystals. *Radiat. Prot. Dosim.* 33: 15–18.

Rose, A. (1963). *Concepts in Photoconductivity and Allied Problems*. ISBN 9780882755687. New York: Interscience.

Rossi, H.H. and Zaider, M. (1996). *Microdosimetry and Its Applications*. ISBN 9783642851865. Berlin, Heidelberg: Springer-Verlag.

Rudlof, G., Becherer, J., and Glaefeke, H. (1978). Behaviour of the fractional glow technique with first-order detrapping processes, traps distributed in energy or frequency factor. *Phys. Stat. Sol. (A)* 49: K121–K124.

Rutherford, E. (1913). *Radioactive Substances and Their Radiations*. Cambridge: Cambridge University Press.

Sadek, A.M., Eissa, H.M., Basha, A.M., Carinou, E., Askounis, P., and Kitis, G. (2015). The deconvolution of thermoluminescence glow-curves using general expressions derived from one trap-one recombination (OTOR) level model. *Appl. Radiat. Isotop.* 95: 214–221.

Sagastibelza, F. and Alvarez Rivas, J.L. (1981). Thermoluminescence in LiF (TLD-100) and LiF crystals irradiated at room temperature. *J. Phys. C: Sol. St. Phys.* 14: 1873–1889.

Sanyal, S. and Akselrod, M.S. (2005). Anisotropy of optical absorption and fluorescence in $Al_2O_3$:C,Mg crystals. *J. Appl. Phys.* 98: 033518.

Sato, T., Niita, K., Matsuda, N., Hashimoto, S., Iwamoto, Y., Furuta, T., Noda, S., Ogawa, T., Iwase, H., Nakashima, H., Fukahori, T., Okumura, K., Kai, T., Chiba, S., and Sihver, L. (2015). Overview of particle and heavy ion transport code system PHITS. *Ann. Nucl. Energy* 82: 110–115.

Sature, K.R., Patil, B.J., Dahiwale, S.S., Bhoraskar, V.N., and Dhole, S.D. (2017). Development of computer code for deconvolution of thermoluminescence glow curve and DFT simulation. *J. Lumin.* 192: 486–495.

Sawakuchi, G.O. (2007). *Characterization and modelling of relative luminescence efficiency of optically stimulated luminescence detectors exposed to heavy charged particles*. Ph.D. Theses, Oklahoma State University.

Sawakuchi, G.O., Ferreira, F.A., McFadden, C.H., Hallacy, T.M., Granville, D.A., Sahoo, N., and Akselrod, M.S. (2016). Nanoscale measurements of proton tracks using fluorescent nuclear track detectors. *Med. Phys.* 43: 2485–2490.

Sawakuchi, G.O., Sahoo, N., Gasparian, P.B.R., Rodriguez, M.G., Archambault, L., Titt, U., and Yukihara, E.G. (2010). Determination of average LET of therapeutic proton beams using $Al_2O_3$:C optically stimulated luminescence (OSL) detectors. *Phys. Med. Biol.* 55: 4963–4976.

Sawakuchi, G.O., Sahoo, N., Gasparian, P.B.R., Rodriguez, M.G., Archambault, L., Titt, U., and Yukihara, E.G. (2014). Corrigendum: determination of average LET of therapeutic proton beams using $Al_2O_3$:C optically stimulated luminescence (OSL) detectors. *Phys. Med. Biol.* 59: 3239–3240.

Sawakuchi, G.O., Yukihara, E.G., McKeever, S.W.S., and Benton, E.R. (2008a). Overlap of heavy charged particle tracks and the change in shape of optically stimulated luminescence curves of $Al_2O_3$:C dosimeters. *Radiat. Meas.* 43: 194–198.

Sawakuchi, G.O., Yukihara, E.G., McKeever, S.W.S., and Benton, E.R. (2008b). Optically stimulated luminescence fluence response of $Al_2O_3$:C dosimeters exposed to different types of radiation. *Radiat. Meas.* 43: 450–454.

Schneckenburger, H., Regulla, D.F., and Unsöld, E. (1981). Time-resolved investigations of radiophotoluminescence in metaphosphate glass dosimeters. *Appl. Phys A* 26: 23–26.

Schön, M. (1951). Über die strahlungslosen übergänge in sulfidphosphoren. *Z. Naturforsch.* 6a: 251–255.

Schulman, J.H., Ginther, R.J., Klick, C.C., Alger, R.S., and Levy, R.A. (1951). Dosimetry of x-rays and gamma-rays by radiophotoluminescence. *J. Appl. Phys.* 22: 1479–1483.

Sfampa, I.K., Polymeris, G.S., Pagonis, V., Theodosoglou, E., Tsirliganus, N.C., and Kitis, G. (2015). Correlation of basic, TL, OSL and IRSL properties of ten K-feldspar samples of various origins. *Nucl. Instrum. Meth. Phys. Phys. Res. B* 359: 89–98.

Shenker, D. and Chen, R. (1972). Numerical solution of the glow curve differential equations. *J. Comput. Phys.* 10: 272–283.

Sholom, S. and McKeever, S.W.S. (2020). High-dose dosimetry with Ag-doped phosphate glass: applicability test with different techniques. *Radiat. Meas.* 132: 106263.

Short, M.A. (2005). The simultaneous measurement of luminescence and photocurrent of an irradiated K-feldspar excited with 1.45 or 2.41 eV photons. *Radiat. Meas.* 39: 197–201.

Shoushan, W. (1988). Dependence of thermoluminescence response and glow curve structure of LiF(Mg,Cu,P) TL materials on Mg, Cu, P dopants concentration. *Radiat. Prot. Dosim.* 25: 133–136.

Soni, A. and Mishra, D.R. (2016). Mathematical formulation of $T_{max}$-$T_{stop}$ method for LM-OSL and its experimental validation on $\alpha$-Al$_2$O$_3$:C. *Nucl. Instrum. Meth. Phys. Res. B* 375: 87–92.

Soni, A., Mishra, D.R., Bhatt, R.C., Gupta, S.K., Rawat, N.S., Kulkarni, M.S., and Sharma, D.N. (2012). Characterization of deep energy level defects in $\alpha$-Al$_2$O$_3$:C using thermally assisted OSL. *Radiat. Meas.* 47: 111–120.

Spooner, N.A. (1994). On the optical dating signal from quartz. *Radiat. Meas.* 23: 593–600.

Stoebe, T.G. and DeWerd, L.A. (1985). Role of hydroxide impurities in the thermoluminescent behavior of lithium fluoride. *J. Appl. Phys.* 57: 2217–2220.

Sun, F., Jiao, L., Wu, J., Yang, Z., Yuan, S., and Dai, G. (1994). X ray diffraction and ESR studies on LiF:Mg,Cu,P phosphor. *Radiat. Prot. Dosim.* 51: 183–189.

Sunta, C.M., Ayta, W.E.F., Chubaci, J.F.D., and Watanabe, S. (2001). A critical look at kinetic models of thermoluminescence: I. First-order kinetics. *J. Phys. D: Appl. Phys.* 34: 2690–2698.

Sunta, C.M., Ayta, W.E.F., Chubaci, J.F.D., and Watanabe, S. (2005). A critical look at the kinetic models of thermoluminescence - II. Non-first order kinetics. *J. Phys. D: Appl. Phys.* 38: 95–102.

Sweet, M.A.S. and Urquhart, D. (1981). A new procedure for analyzing complex thermoluminescence. *J. Phys. C: Solid State Phys.* 14: 773–781.

Sykora, G.J. and Akselrod, M.S. (2010a). Spatial frequency analysis of fluorescent nuclear track detectors irradiated in mixed neutron-photon fields. *Radiat. Meas.* 45: 1197–1200.

Sykora, G.J. and Akselrod, M.S. (2010b). Novel fluorescent nuclear track detector technology for mixed neutron-gamma fields. *Radiat. Meas.* 45: 594–598.

Sykora, G.J. and Akselrod, M.S. (2010c). Photoluminescence study of photochromically and radiochromically transformed Al$_2$O$_3$:C,Mg crystals used for fluorescent nuclear track detectors. *Radiat. Meas.* 45: 631–634.

Sykora, G.J., Akselrod, M.S., Benton, E.R., and Yasuda, N. (2008a). Spectroscopic properties of novel fluorescent nuclear track detectors for high and low LET charged particles. *Radiat. Meas.* 43: 422–426.

Sykora, G.J., Akselrod, M.S., Salasky, M., and Marino, S.A. (2007). Novel Al$_2$O$_3$:C,Mg fluorescent nuclear track detectors for passive neutron dosimetry. *Radiat. Prot. Dosim.* 126: 278–283.

Sykora, G.J., Akselrod, M.S., and Vanhavere, F. (2009). Performance of fluorescence nuclear track detectors in mono-energetic and broad spectrum neutron fields. *Radiat. Meas.* 44: 988–991.

Sykora, G.J., Salasky, M., and Akselrod, M.S. (2008b). Properties of novel fluorescent nuclear track detectors for use in passive neutron dosimetry. *Radiat. Meas.* 43: 1017–1023.

Tanaka, H., Fujimoto, Y., Koshimizu, M., Yanagida, T., Yahaba, T., Saeki, K., and Asai, K. (2016). Radiophotoluminescence properties of Ag-doped phosphate glasses. *Radiat. Meas.* 94: 73–77.

Taylor, G.C. and Lilley, E. (1978). The analysis of thermoluminescent glow peaks in LiF (TLD100). *J. Phys. D: Appl. Phys.* 11: 567–582.

Taylor, G.C. and Lilley, E. (1982a). Clustering and precipitation in LiF (TLD100). I. *J. Phys. D: Appl. Phys.* 15: 1243–1252.

Taylor, G.C. and Lilley, E. (1982b). Effect of clustering and precipitation on thermoluminescence in LiF (TLD-100) crystals. II. *J. Phys. D: Appl. Phys.* 15: 1253–1263.

Taylor, G.C. and Lilley, E. (1982c). Rapid readout rate studies of thermoluminescence in LiF (TLD-100) crystals. III. *J. Phys. D: Appl. Phys.* 15: 2053–2065.

Templer, R. (1986). *Thermoluminescence techniques for dating zircon inclusions.* D. Phil. Thesis, Univ. Oxford.

Townsend, P.D. (1992). *The Need for Imperfections.* Professorial Lecture. Brighton: University of Sussex.

Townsend, P.D., Ahmed, K., Chandler, P.J., McKeever, S.W.S., and Whitlow, H.J. (1983). Measurements of the emission spectra of LiF during thermoluminescence. *Radiat. Eff.* 72: 245–257.

Townsend, P.D. and Kelly, J.C. (1973). *Colour Centres and Imperfections in Insulators and Semiconductors.* ISBN 9780856210105. London: Chatto and Windus, Sussex University Press.

Townsend, P.D., Wang, Y., and McKeever, S.W.S. (2021). Spectral evidence for defect clustering: relevance to radiation dosimetry materials. *Radiat. Meas.* 147: 106634.

Trowbridge, J. and Burbank, J.E. (1898). Phosphorescence produced by electrification. *Am. J. Sci.* 5: 25–26.

Urbach, F. (1930). Zur lumineszenz der alkalihalogenide. I-III. *Wiener Ber.* 139: 353–495.

Van den Eeckhout, K., Bos, A.J.J., Poelman, D., and Smet, P.F. (2013). Revealing trap depth distributions in persistent phosphors. *Phys. Rev. B* 87: 045126.

Van Dijk, J.W.E. (2006). Thermoluminescence glow curve deconvolution and its statistical analysis using the flexibility of spreadsheet programs. *Radiat. Prot. Dosim.* 119: 332–338.

Visocekas, R. (2002). Tunnelling in afterglow: its coexistence and interweaving with thermally stimulated luminescence. *Radiat. Prot. Dosim.* 100: 45–54.

Von Liebeg, J. (1834). *Liebeg to Berzelius, July 22, 1834,* in *Berzelius und Liebeg: ihre Briefe,* Carrière, J. (ed.) (1898), 94. Trans. Brock, W.H. Quoted in *Oxford Dictionary of Scientific Quotations.* Bynum, E.F., and Porter, R. (eds.). ISBN 0198584091. Oxford, Oxford University Press, 2005.

Waligórski, M.P.R., Hamm, R.N., and Katz, R. (1986). The radial distribution of dose around the path of a heavy ion in liquid water. *Nucl. Tracks Radiat. Meas.* 11: 309–319.

Walker, F.D., Colyott, L.E., Agersnap Larsen, N., and McKeever, S.W.S. (1996). The wavelength dependence of light-induced fading of thermoluminescence from $\alpha$-Al$_2$O$_3$:C. *Radiat. Meas.* 26: 711–718.

Wang, Y. and Townsend, P.D. (2013). Potential problems in collection and data processing of luminescence signals. *J. Lumin.* 142: 202–211.

Whitehead, A.N. (1920). *The Concept of Nature*, 163. Quoted in *Oxford Dictionary of Scientific Quotations*. Bynum, E.F., and Porter, R. (eds.). ISBN 0198584091. Oxford, Oxford University Press, 2005.

Whitley, V., Agersnap Larsen, N., and McKeever, S.W.S. (2002). Determination of ionization energies and attempt-to-escape factors using thermally stimulated conductivity. *Radiat. Prot. Dosim.* 100: 147–151.

Whitley, V. and McKeever, S.W.S. (2000). Photoionization of deep centers in $Al_2O_3$. *J. Appl. Phys.* 87: 249–256.

Whitley, V. and McKeever, S.W.S. (2001). Linear modulated photoconductivity and linear modulated optically stimulated luminescence measurements on $Al_2O_3$. *J. Appl. Phys.* 90: 6073–6083.

Wiedemann, E. (1889). Zur Mechanik des leuchtens. *Annal. Phys.-Berlin* 37: 177–248.

Wiedemann, E. and Schmidt, G.C. (1895). Über Lumineszenz. *Annal. Physik Chemie, Neue Folge* 54: 604–625.

Xu, Y.N. and Ching, W.Y. (1991). Electronic and optical properties of all polymorphic forms of silicon dioxide. *Phys. Rev. B* 44: 11048–11059.

Yamamoto, T. (2011). RPL dosimetry: principles and applications. In: *Concepts and Trends in Medical Radiation Dosimetry* (ed. A.B. Rosenfeld, T. Kron, F. d'Errico and M. Moscovitch). *American Institute of Physics Conference Proceedings*, 1345, 217–230.

Yamamoto, T., Yanagida-Miyamoto, Y., Iida, T., and Nanto, H. (2020). Current status and future prospect of RPL glass dosimeter. *Radiat. Meas.* 136: 106363.

Yang, X.-B., Li, H.-J., Bi, Q.-Y., Cheng, Y., Tang, Q., and Xu, J. (2008). Influence of carbon on the thermoluminescence and optically stimulated luminescence of $\alpha$-$Al_2O_3$:C crystals. *J. Appl. Phys.* 104: 123112.

Yasuda, H. and Kobayashi, I. (2001). Optically stimulated luminescence from $Al_2O_3$:C irradiated with relativistic heavy ions. *Radiat. Prot. Dosim.* 95: 339–343.

Yasuda, H., Kobayashi, I., and Morishima, H. (2002). Decaying patterns of optically stimulated luminescence from $Al_2O_3$:C for different quality radiations. *J. Nucl. Sci. Technol.* 39: 211–213.

Yin, L., Townsend, P.D., Wang, Y., Khanlary, M.R., and Yang, M. (2020). Comparisons of thermoluminescence signals between crystal and powder samples. *Radiat. Meas.* 135: 106380.

Yokota, R. and Imagawa, H. (1967). Radiophotoluminescent centers in silver-activated phosphate glass. *J. Phys. Soc. Japan.* 23: 1038–1048.

Yokota, R. and Nakajima, S. (1965). Improved fluoroglass dosimeter as personnel monitoring dosimeter and microdosimeter. *Health Phys.* 11: 241–253.

Yokota, R., Nakajima, S., and Sakai, E. (1961). High-sensitivity silver-activated phosphate glass for the simultaneous measurement of thermal neutrons, gamma-, and/or beta-rays. *Health Phys.* 5: 219–224.

Yuan, X.L. and McKeever, S.W.S. (1988). Impurity clustering effects in magnesium-doped lithium fluoride. *Phys. Stat. Sol. (A)* 108: 545–551.

Yukihara, E.G. (2020). A review on the OSL of BeO in light of recent discoveries: the missing piece of the puzzle? *Radiat. Meas.* 134: 106291.

Yukihara, E.G. and McKeever, S.W.S. (2006). Spectroscopy and optically stimulated luminescence of $Al_2O_3$:C using time-resolved measurements. *J. Appl. Phys.* 100: 083512.

Yukihara, E.G. and McKeever, S.W.S. (2011). *Optically Stimulated Luminescence; Fundamentals and Applications*. ISBN 9780470697252. Chichester: Wiley.

Yukihara, E.G., Milliken, E.D., Oliveira, L.C., Orante-Barrón, V.R., Jacobsohn, L.G., and Blair, M.W. (2013). Systematic development of new thermoluminescence and optically stimulated luminescence materials. *J. Lumin.* 133: 203–210.

Yukihara, E.G., Sawakuchi, G.O., Guduru, S., McKeever, S.W.S., Gaza, R., Benton, E.R., Yasuda, N., Uchihori, Y., and Kitamura, H. (2006). Application of the optically stimulated luminescence (OSL) technique in space dosimetry. *Radiat. Meas.* 41: 1126–1135.

Yukihara, E.G., Whitley, V.H., McKeever, S.W.S., Akselrod, A.E., and Akselrod, M.S. (2004). Effect of high-dose irradiation on the optically stimulated luminescence of $Al_2O_3$:C. *Radiat. Meas.* 38: 317–330.

Zha, Z., Wang, S., Shen, W., Zhu, J., and Cai, G. (1993). Preparation and characteristics of LiF:Mg,Cu,P thermoluminescent material. *Radiat. Prot. Dosim.* 47: 111–118.

Zhang, N., Gao, C., and Xiong, Y. (2019). Defect engineering: a versatile tool for tuning the activation of key molecules in photocatalytic reactions. *J. Energy Chem.* 37: 43–57.

Zhu, J., Muthe, K.P., and Pandey, R. (2014). Stability and electronic properties of carbon in $\alpha$-$Al_2O_3$. *J. Phys. Chem. Sol.* 75: 379–383.

Zimmerman, J. (1971). The radiation-induced increase of thermoluminescence sensitivity of the dosimetry phosphor LiF (TLD-100). *J. Phys. C: Sol. Stat. Phys.* 4: 3277–3291.

# Index

Page numbers in **bold** refer to **Tables**
Page numbers in *italics* refer to *Figures*
Page numbers in ***bold-italics*** refer to ***Exercises***

---